Society of Economic Geologists Annual Conference (SEG 100)

Celebrating a Century of Discovery

Whistler, Canada and Online
14 – 17 September 2021

ISBN: 978-1-7138-6509-4

Printed from e-media with permission by:

Curran Associates, Inc.
57 Morehouse Lane
Red Hook, NY 12571

Some format issues inherent in the e-media version may also appear in this print version.

Printed with permission by Curran Associates, Inc. (2023)

Society of Economic Geologists (SEG)
7811 Shaffer Parkway
Littleton, CO 80127-3732
USA

Phone: +1(720) 981-7882
Fax: +1(720) 981-7874

www.segweb.org

Additional copies of this publication are available from:

Curran Associates, Inc.
57 Morehouse Lane
Red Hook, NY 12571 USA
Phone: 845-758-0400
Fax: 845-758-2633
Email: curran@proceedings.com
Web: www.proceedings.com

TABLE OF CONTENTS

BASINS THROUGH TIME - LINKING PROCESS AND ORE SYSTEMS

BEYOND 2021 - THE NEXT HUNDERED YEARS

DIVERSITY IN ORE DEPOSITS

GAME CHANGERS - THE FIRST HUNDRED YEARS

GOLD 2020S - GOLDEN PAST, PRECIOUS FUTURE

LINDGREN'S LEGACY - ORE DEPOSITS IN DEPTH

ORE DEPOSIT STRUCTURE - PROCESSES, PATTERNS, AND INNOVATIONS

TECTONOMAGMATISM AND PORPHYRY-EPITHERMAL METALLOGENY

B1
Global distribution of sediment-hosted metals controlled by steps in lithospheric thickness

Fred Richards[1], Mark Hoggard[2], Karol Czarnota[2, 3], David Huston[2], Lynton Jaques[3], Sia Ghelichkhan[2]
1. Department of Earth Science & Engineering, Imperial College London, London, United Kingdom, 2. Research School of Earth Sciences, Australian National University, Canberra, ACT, Australia, 3. Geoscience Australia, Canberra, ACT, Australia

The ongoing transition to a greener economy is driving unprecedented global demand for base metals. With the pace of deposit discovery currently declining, there is a pressing need to open up new frontiers for mineral exploration and improve success rates. A consensus is emerging that many of the largest remaining continental discoveries are likely to lie beneath post-mineralisation cover in ancient sedimentary basins. However, these basins cover ~75% of the continental surface and the key mineral system ingredients of evaporites associated with brine formation, felsic and mafic volcanic rocks for sourcing metals, and organic-rich shale precipitation sites, are widespread. Accordingly, an additional geological control is required to explain the spatial distribution of existing deposits and narrow the search-space for discoveries. Here, we exploit recent laboratory experiments that constrain the relationship between seismic velocity and temperature at upper mantle conditions to derive global maps of lithospheric thickness from seismic tomographic models. We find that 85% of sediment-hosted base metals, including all giant deposits (>10 Mt of metal), occur within 200 km of the transition between thick and thin lithosphere at the edges of cratons. The probability of obtaining this result if sediment-hosted deposits were randomly distributed is less than 1 in 1012, making the correlation exceptionally robust. This relationship can be explained by considering the thermomechanical structure of thick lithosphere and its effect on basin dynamics. We find that melt depletion-induced reduction in the density of cratonic roots leads to greater subsidence during rifting, while the relative thickness of the lithospheric template results in more modest geothermal gradients and lower heat flow into the base of the sediment pile. These effects combine to enlarge the depth extent over which hydrothermal circulation within the basin is favourable to metal transport and precipitation, promoting the formation of giant deposits. Significantly, mineralisation ages span the last 2 billion years, which implies the cratonic edges that result from this rifting process are generally stable and suggests a genetic link between deep Earth processes and near-surface hydrothermal mineral systems. This discovery provides a global framework for identifying fertile regions for targeted mineral exploration, reducing the search-space for new deposits by two thirds on this lithospheric thickness criterion alone. We suggest that integration of this new constraint with existing knowledge of other mineral system components could dramatically decrease risk of exploration in frontier areas such as the undercover southern portion of the Mount Isa Province and previously underexplored basins along the margin of the Kimberley and Pilbara cratons in Australia.

B2

Sequence stratigraphy in stratiform sediment-hosted base metals exploration: An example from the ca. 1640 Ma Barney Creek Formation, McArthur Basin, Australia

Marcus Kunzmann[1], Vincent Crombez[2], Teagan N. Blaikie[1], Octavian Catuneanu[3], Rodney King[4], Galen P Halverson[5], Susanne Schmid[1], Samuel C. Spinks[1]

1. CSIRO Mineral Resources, Australian Resources Research Centre, Kensington, WA, Australia, 2. CSIRO Energy, Deep Earth Imaging Future Science Platform, Australian Resources Research Centre, Kensington, WA, Australia, 3. Department of Earth and Atmospheric Sciences, University of Alberta, Edmonton, AB, Canada, 4. Baldivis, WA, Australia, 5. Department of Earth and Planetary Sciences/Geotop, McGill University, Montreal, QC, Canada

The McArthur Basin is part of a Proterozoic basin system on the North Australia Craton that represents one of the most prospective provinces for clastic-dominated Zn-Pb deposits in the world. The ca. 1640 Ma Barney Creek Formation is the most important target unit in the McArthur Basin, hosting the McArthur River and Teena Zn-Pb deposits. The Barney Creek Formation was deposited in a highly compartmentalized basin, characterized by sub-basins and paleohighs on the scale of mostly <10 km. The mineralization in the Barney Creek Formation is typically stratiform and hosted by pyritic, organic matter-rich, dolomitic siltstone deposited in sub-basin depocenters. The timing of mineralization is debated, with traditional SEDEX models favoring syngenetic mineralization on the seafloor, and emerging diagenetic-epigenetic models focusing on sub-seafloor replacement of carbonate laminae. The Barney Creek Formation is 10—900 m thick. Facies analysis indicates it was mostly deposited by turbidity currents below storm wave base, although shoaling to shallow subtidal carbonate environments locally occurred on paleohighs. The formation comprises one complete (B1) and one incomplete (B2) depositional sequence. The transgressive systems tract of depositional sequence B1 records deepening from peritidal carbonate environments of the underlying Teena Dolostone to offshore environments in the lower Barney Creek Formation. It culminates in a maximum flooding surface (B1 MFS), which is developed as organic-rich, weakly sulfidic dolomitic siltstone on paleohighs and in shallow parts of sub-basins. In contrast, it can be developed as highly organic-matter rich and strongly sulfidic black shale in sub-basin depocenters. Where the B1 MFS is a highly organic-rich and sulfidic black shale, it is expected to act as redox-chemical trap in traditional syngenetic models for mineralization. In contrast, following compaction due to burial, it would have acted as a seal to ascending metalliferous brines in recent diagenetic-epigenetic models for the McArthur River and Teena deposits. This scenario implies that the mineralization should occur in the transgressive strata below the maximum flooding surface. Regardless of the preferred model for mineralization, the relationship between the transgressive strata, maximum flooding surface, and mineralization, demonstrate that it is critical to I) map this maximum flooding surface across the basin, and II) create facies maps that show where in the basin this surface is developed as mudrock (i.e., chemical trap or seal). Integration of these data with the modern-day structural architecture of the basin, and with information on the distribution of the metal source, suitable aquifer facies, and potential fluid flow conduits, can significantly improve targeting (Fig. 1). The principles observed from sequence stratigraphic analysis and the mineralization in the Barney Creek Formation can be generalized for other sediment-hosted ore deposit types. Fig. 1: Three-dimensional cross-section across the southern McArthur Basin connecting all facies and gamma-ray logged drill cores of this study. Shown is also the B1 MFS and whether it is developed as black shale (i.e., ideal trap or seal facies).

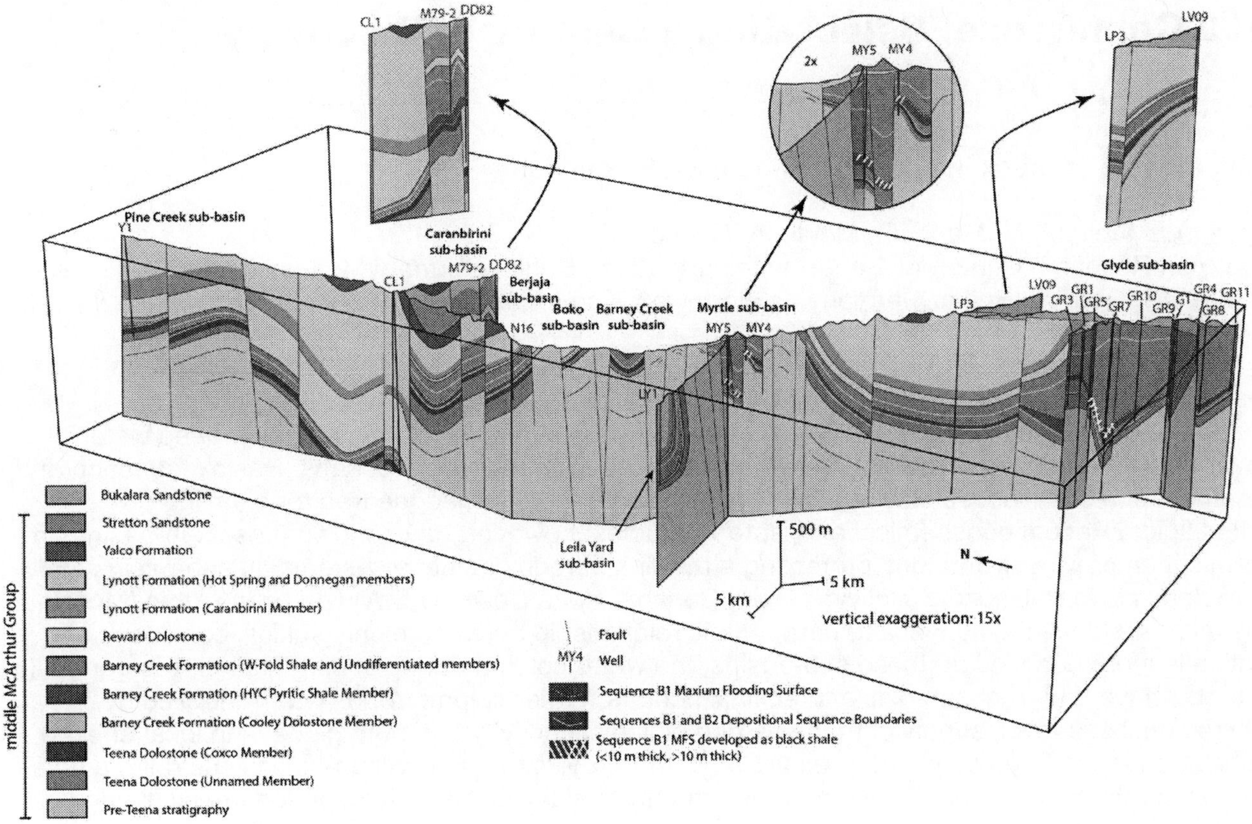

CL1
M79-2
DD82
2x
MY5
MY4
LP3
LV09
Pine Creek sub-basin
Y1
Caranbirini sub-basin
CL1
M79-2
DD82
Berjaja sub-basin
N16
Boko sub-basin
Barney Creek sub-basin
Myrtle sub-basin
MY5
MY4
LY1
LP3
LV09
Glyde sub-basin
GR3
GR1
GR5
GR7
GR10
GR9
G1
GR4
GR8
GR11
Leila Yard sub-basin
500 m
5 km
5 km
N
vertical exaggeration: 15x
Bukalara Sandstone
Stretton Sandstone
Yalco Formation
Lynott Formation (Hot Spring and Donnegan members)
Lynott Formation (Caranbirini Member)
Reward Dolostone
Barney Creek Formation (W-Fold Shale and Undifferentiated members)
Barney Creek Formation (HYC Pyritic Shale Member)
Barney Creek Formation (Cooley Dolostone Member)
Teena Dolostone (Coxco Member)
Teena Dolostone (Unnamed Member)
Pre-Teena stratigraphy
middle McArthur Group
Fault
MY4
Well
Sequence B1 Maxium Flooding Surface
Sequences B1 and B2 Depositional Sequence Boundaries
Sequence B1 MFS developed as black shale
(<10 m thick, >10 m thick)

B3

Re-thinking the role of redox in clastic-dominant Zn-Pb deposits

Sarah Gleeson[1, 2], Joseph M. Magnall[1], Merilie A. Reynolds[3]
2. GFZ, German Research Center for Geosciences, Potsdam, Berlin, Germany, 2. Institute for Geological Sciences, Freie Universität, Berlin, Germany, 3. Northwest Territories Geology Survey, Yellowknife, NT, Canada

The largest and highest value Zn reserves are hosted in carbonaceous mudstones. For many years it was considered that these mudstones ("shales") formed by pelagic settling of organic matter in deep water settings over long time periods, in basins that were restricted from the global oceans. Basin stagnation and the development of a stratified, sulfidic (euxinic) water column then formed the trap for exhaling hydrothermal fluids. Recent geochemical and isotopic studies, however, have shown these sub-basins are more dynamic than previously thought, prompting a re-think of redox in these systems. In sub-basins that host the ore deposits in Palaeozoic (Selwyn Basin, Canada; Kuna Basin, U.S.A) and Proterozoic (McArthur Basin, Australia) systems, the spatial and paragenetic relationship between highly sulfidic conditions and mineralization is more nuanced. In these sub-basins, the water column during sedimentation was generally anoxic, but may have had oxic, sub-oxic and euxinic periods. Water column redox was controlled by changes in regional sea-level, supply of nutrients and biological activity (i.e. both global and local effects), which will have changed through time. For example, in the Proterozoic Teena sub-basin (Australia) a unit with high concentrations of diagenetic pyrite formed during a relatively short-lived period of euxinia and is found stratigraphically above the hydrothermal system; this unit can be recognised as a regional feature across the McArthur basin. Major redox boundaries can also develop during diagenesis in the pre-ore environment of major CD-type deposits. For example, recent work at the Anarraaq (Red Dog) and Macmillan Pass deposits has shown that the sulphide ore zones overprinted a diagenetic assemblage of barite and pyrite ± calcite. These phases formed when methane, generated in the sediment pile, mixed with downward diffusing seawater sulphate, resulting in the anaerobic oxidation of methane (AOM). This process is biologically mediated, and can be associated with the genesis of H2S, which can ultimately result in stratiform pyrite formation. AOM has been suggested to be an important process in the formation of thick sequences of diagenetic, hanging-wall pyrite at Macmillan Pass, Howards Pass and in the Anarraaq deposit. At Macmillan Pass and in the Anarraaq deposit, diagenetic barite dissolution released SO4, which was then reduced and the S was incorporated into the ore minerals. In summary, these deposits formed in sub-basins that are characterised by 1) fluctuating sea-level, 2) high biological productivity and fluxes of organic material to the seafloor and 3) high, but variable, rates of sedimentation. Such conditions are found near continental margins or in epicontinental basins with relatively shallow water. These are optimal settings for processes that generate abundant reduced S as a "trap" for the hydrothermal system, but which can also result in the formation of stratiform, barren, diagenetic pyrite; the latter poses a challenge for exploration programs. There are no known modern analogues for these mineral systems but areas like the Guaymas Basin, Baltic Sea and the South China Sea may be appropriate modern analogues for the ore deposit "trap" environment.

B4
Cryptic Alteration Haloes in Sediment-Hosted Ore Deposits: New Tools, New Exploration Vectors, New Genetic Models

Shaun L. Barker[1], Benjamin S. Andrew[2], Rocky D. Barker[3]
1. University of British Columbia, Vancouver, BC, Canada, 2. Mount Isa Mines, Mount Isa, QLD, Australia, 3. Imdex Limited, Hamilton, New Zealand

Sediment-hosted ore deposits of various types often have "cryptic" alteration haloes—that is, visible alteration is difficult to recognise and use to vector towards mineralization. Over the last 15 years, a series of new analytical tools and approaches have made recognising and quantifying alteration in sediment-hosted deposits possible. This includes the routine use of 4-acid ICPMS geochemistry (e.g., Halley et al., 2016), automated mineralogy using scanning electron microscopy, the use of shortwave infrared (SWIR) and longwave infrared (LWIR) core imaging (Barker et al., 2021), and routine, inexpensive stable isotope analyses (Barker et al., 2011; Beinlich et al., 2017). Collectively, these methods can be utilised to recognise and quantify alteration footprints at varying distances from mineral deposits. In the the case of stable isotopes, the isotopic footprint can extend 5-10 kilometres around mineralization. For whole-rock lithogeochemistry, haloes can extend hundreds of metres to kilometres around mineralization, while mineralogical alteration can be tracked tens of metres to more than a kilometre from ore bodies. In this presentation, we will overview the development and use of these methods. Then we will show two case studies, from the Four Mile Carlin-type gold deposit in Nevada and the Mount Isa Copper deposit in Australia, to demonstrate how these tools can be integrated to define mineralogical, geochemical, and isotopic alteration haloes which provide vectors towards mineralization and ultimately inform genetic models.

At Mount Isa, newly acquired petrographic and geochemical data collected across a range of scales identified an extensive zoned alteration system that manifests as a series of interpreted reaction fronts, which extends at least 1500 m beyond mineralisation. Copper mineralisation is contained within a zone of visible mineral alteration, hydrothermal brecciation, and veining, locally known as the "silica-dolomite." The silica-dolomite is characterised by silicification and brecciation of shales, recrystallisation of dolomite, and intense ^{18}O-depletion, from $\delta^{18}O \approx 22‰$ VSMOW in the least altered rocks to $\delta^{18}O \approx 10‰$ in the most altered zone. A cryptic halo of K- and Ca-depletion extends from the inferred fluid input zones to include the region outboard of the visible mineral alteration envelope. Beyond the region of Ca-depletion and K-depletion, a large halo of cryptic potassic alteration is identified by whole-rock geochemical analysis. Potassium responsible for this alteration is interpreted to have been remobilised from zones of silicification and K-depletion at the core of the hydrothermal system. The change from dolomite- to calcite-dominated vein cement within individual veins is interpreted to be driven by increases in the relative activity of calcium during the evolution of the hydrothermal fluid. Consequently, ore-stage veins with mixed dolomite-calcite vein cement potentially represent a distal expression of silicification and decalcification within the core of the mineralising system. The alteration haloes described above, both visible and cryptic, are all contained with the broad zone of ^{18}O-depletion, representing the most spatially extensive alteration halo to copper mineralisation at Mount Isa (Andrew, 2021).

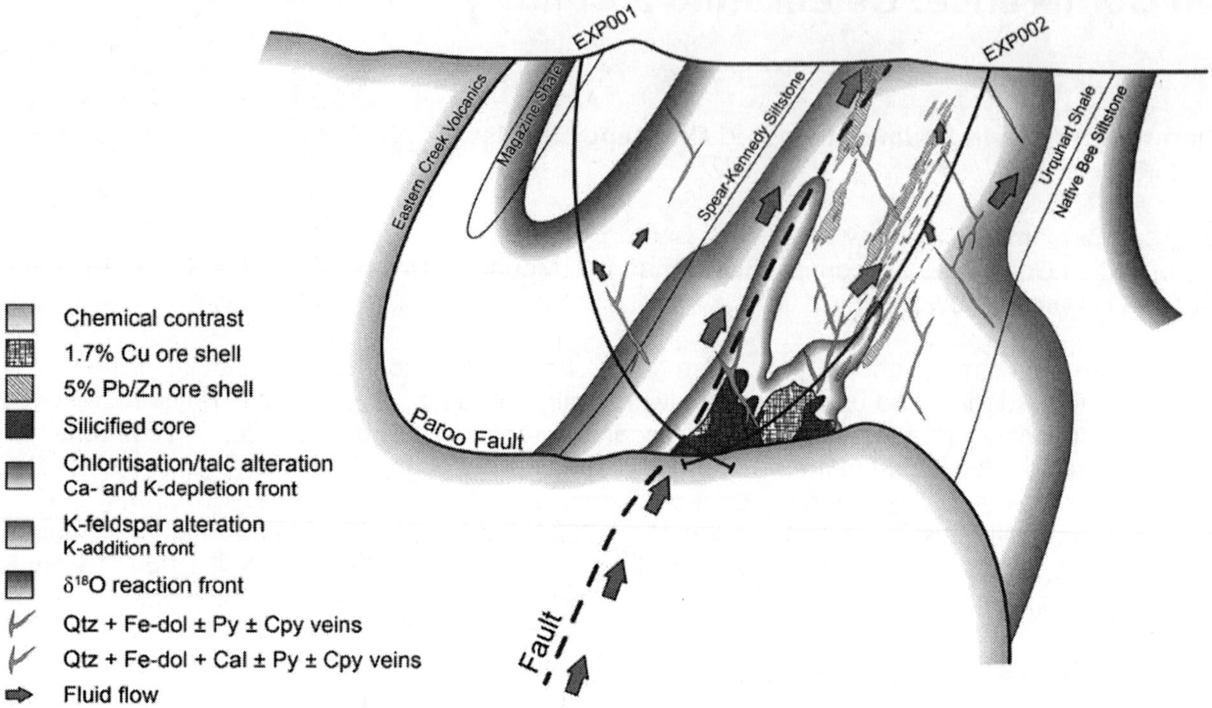

EXP001
EXP002
Eastern Creek Volcanics
Magazine Shale
Spear-Kennedy Siltstone
Urquhart Shale
Native Bee Siltstone
Paroo Fault
Fault
Chemical contrast
1.7% Cu ore shell
5% Pb/Zn ore shell
Silicified core
Chloritisation/talc alteration
Ca- and K-depletion front
K-feldspar alteration
K-addition front
δ18O reaction front
Qtz + Fe-dol ± Py ± Cpy veins
Qtz + Fe-dol + Cal ± Py ± Cpy veins
Fluid flow

B5

Structural controls on geometry, continuity and mineralisation for the Kamoa-Kakula deposits of the Western Foreland shelf domain of the Central African Copperbelt

George Gilchrist[1], Shane Nielsen[1], Timothy Brooks[1], Innocent Mushobekwa [2], Franck Twite[2], David Edwards[1]
1. Ivanhoe Mines SA, South Africa, 2. Kamoa Copper, Democratic Republic of the Congo (Kinshasa)

The Central African Copperbelt (CACB) is contained in the Katangan basin, an intracratonic rift straddling Zambia and the Democratic Republic of Congo (DRC). Rifting commenced at ~840 Ma, characterised by episodic pulses of extension and rapid sediment accumulation, followed by periods of gradual sediment accumulation and near basin-wide evaporite sheet accumulation.

Mineralisation occurred when a copper-bearing fluid passed up through an oxidized aquifer to interact with a reducing horizon. In the DRC sector, laterally extensive salt sheets prevented this process occurring until halokinetic modification allowed deeper-level brines and possibly downward percolating brines access to the intrasalt stratigraphic levels to interact with reducing horizons, principally the carbonate and siltstones of the Mines Subgroup deposited during the latter stage of the initial rift phase. The second phase of rifting marked an abrupt transition to sub wave-base sedimentation during the Mwashya Subgroup, before deposition of a distinctive glaciogenic diamictites of the Nguba Group equated with an age of ~740Ma.

Occurring outside the traditional limit of the CACB, Kamoa-Kakula occurs in a much higher stratigraphic position at the base of the Nguba Group. It occupies a salt-marginal position within the Western Foreland shelf domain, is relatively undeformed, and thus much easier to reconcile the fundamental control exerted by the rifting geometry.

Aquifer quality exerts a fundamental control on fluid flow. At Kamoa-Kakula, the Mwashya Subgroup (R4.2) is a medium to coarse grained subarkose, with variable proportions of siltstone and conglomerate. Textural changes within the R4.2 can be abrupt suggesting they are controlled by local rift architecture. These facies can locally be linked to facies changes in the overlying Nguba sediments, suggesting that fundamental sub-basin-bounding elements were inherited and multiply reactivated through these depositional periods.

The Nguba marks a transition to deposition under largely reducing conditions, with a combination of diamictite units and laterally discontinuous sub-basin-controlled siltstone and sandstone units. Rift climax events are marked by strongly sulfidic intervals within two laterally continuous, siltstones (the Ng1.1.2 and Ng1.1.4), the Ng1.1.4 marking the uppermost pyrite-rich siltstone. Overlying diamictites can be pyritic due to ingestion of the upper portion of the siltstone during emplacement, redistributing primary sulfide.

There is strong evidence for structural controls on mineralisation. The distribution of copper remains fundamentally controlled by the interplay of hydrologic and redox architecture; principal sulfidic hosts formed in isolated sub-basins would have required very particular, but very subtle fault configurations, ensuring sediment starvation and hydrological proximity to the aquifer, whilst also amplifying permeability and connectivity within the aquifer.

Faults can be discontinuous with variable offset and reflect the complex interplay of the WNW Monwezi, ENE Kansuki and NNW Manika regional structural influences. These fundamental elements were multiply reactivated throughout lower Nguba sedimentation, influencing thickness and facies variation, giving each deposit unique characteristics.

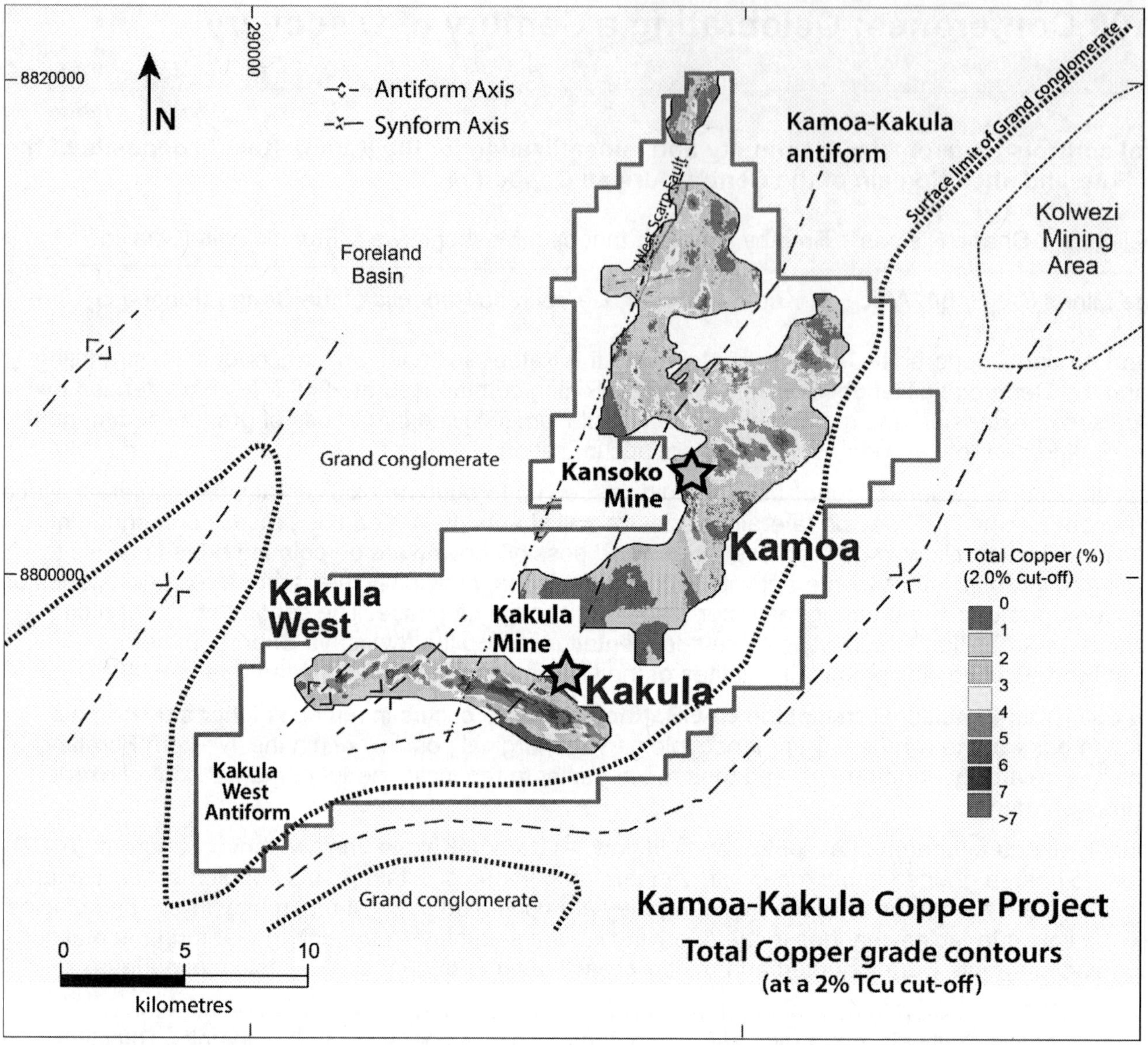
290000
8820000
N
Antiform Axis
Synform Axis
Kamoa-Kakula antiform
Surface limit of Grand conglomerate
West Scarp Fault
Kolwezi Mining Area
Foreland Basin
Grand conglomerate
Kansoko Mine
Kamoa
8800000
Kakula West
Kakula Mine
Kakula
Kakula West Antiform
Total Copper (%) (2.0% cut-off)
0
1
2
3
4
5
6
7
>7
Grand conglomerate
0
5
10
kilometres
Kamoa-Kakula Copper Project
Total Copper grade contours
(at a 2% TCu cut-off)

B6
Metal Sources in the Central Africa Copperbelt: Assessing the Mafic Contribution

Daryl E. Blanks, David A. Holwell
University of Leicester, Leicester, United Kingdom

Mafic intrusions, primed with base metals including Ni, Cu, and Co, are often an abundant component of sedimentary rift basins, particularly those along craton margins. The Central African Copperbelt represents a Neoproterozoic rift basin along the southern margin of the Congo Craton within present day Zambia and the Democratic Republic of the Congo, and it is one of the richest resources of sediment-hosted Cu, Co, and other base metals. The estimated contained Cu resource continues to increase with new discoveries, yet these resources cannot be fully accounted for solely by models that propose redbeds or basement rocks as the source. Thus, given the elevated inherent base metal content of mafic rocks, they provide an intriguing potential answer to this mass balance problem. Mafic rocks are known to be an important or even major source of metals in some sediment-hosted Cu systems, such as the Yeneena and McArthur Basins, Australia. Although the deposits and their host stratigraphy have been extensively explored and well documented, the volume of mafic bodies within the Copperbelt remains relatively poorly constrained.

The Domes region in the Zambian Copperbelt is one area where mafic intrusions have been recognised and mapped more extensively and, notably, in close proximity to the largest deposit, Kansanshi. Modelling of the well-constrained occurrences at Kansanshi has highlighted the morphology, abundance, and volume of mafic bodies within the area, which have preferentially exploited evaporitic horizons and occur mostly as flattened pods rather than continuous sills. The intrusions occur at multiple lithostratigraphic depths, are of differing sizes from <10 m^3 up to 10^7 m^3, and have been extensively and variably altered. The intense alteration of mafic rocks highlights the likelihood that significant volumes of metals may have been leached from them and transferred into the circulating, basinal ore-forming fluids.

The association of the mafic rocks with evaporitic strata may be an important factor in their contribution as a metal source. It is well documented that the ore-forming fluids in the Copperbelt were highly saline and likely sourced, at least in part, from evaporites that have subsequently been destroyed by later metamorphism. Alteration of the mafics is characterised by scapolite-rich assemblages, indicating likely interaction with evaporate-derived fluids. Chemically, the mafic rocks show signatures consistent with extensive alkali metasomatism and associated net losses of Cu and Co. We evaluate this data in the context of the basin as a whole and the relative contribution that the mafic rocks have had in sourcing base metals in the Copperbelt. Furthermore, we assess the role that highly aggressive fluids in some parts of the region may have also liberated Ni from mafic sources to form deposits such as Enterprise.

B7
The Role of Cu- and Co-rich Potassic Brines in Zambian Copperbelt Mineralisation

James Davey[1, 2, 3], Stephen Roberts[2], Jamie Wilkinson[3]
1. SRK Consulting UK, Cardiff, United Kingdom, 2. University of Southampton, Southampton, United Kingdom, 3. Natural History Museum, London, United Kingdom

The Central African Copperbelt is the largest repository of sediment-hosted copper and cobalt on Earth. The criticality of these elements in battery technology and electricity transmission establishes them as fundamental components of the carbon-free energy revolution. As such, understanding the processes through which they are sourced, transported, and precipitated in sedimentary basin environments is fundamental to securing future supply. In the Central African Copperbelt, the relative timing of mineralization, and the nature and origins of fluids involved, is controversial. As essential components of the mineral system, constraining these characteristics is central to developing an understanding of the unique metal endowment of the region.

Microthermometry, scanning electron microscopy, and Laser Ablation ICP-MS analysis of an extensive suite of barren and mineralized vein-hosted fluid inclusions indicate that base metal concentrations vary by 1 to 2 orders-of-magnitude between "barren" and "ore" fluids, with concomitant distinctions in the major salt chemistry of fluids.

Across three deposits in the Zambian Copperbelt (Nkana-Mindola, Nchanga, and Lumwana), anomalously high temperature-salinity brines characterised by $K/Na \geq 0.8$ host elevated metal concentrations (10^2 to 10^3 ppm Cu and Co) within mineralised pre- to syn-kinematic veins. Conversely, lower temperature and salinity fluids, characterised by $K/Na \ll 0.8$ are typically barren, and occur as primary fluids within post-kinematic veins as well as secondary inclusions in all vein types.

Crush-leach analysis of fluids from several Central African Copperbelt deposits indicates that early potassic ore fluids were bittern brines with an advanced evaporitic origin, whereas many later sodic fluids represent halite-dissolution brines which interacted extensively with evaporites hosted within the Katangan stratigraphy.

A model invoking high-temperature, potentially sulfate-deficient, metalliferous and ultra-potassic bittern brines residing at sub-salt levels correlates with pervasive potassic metasomatism, where K-feldspar and phlogopite are intimately associated with Cu-Co-sulphides. From the onset of halokinesis through basin closure, lower temperature, halite-undersaturated fluids interacted with evaporites, forming structurally-controlled sodic metasomatism.

These new data suggest that the extended residence of potentially sulfate-deficient, ultra-potassic bittern brines within a high temperature depocenter was a critical factor in mobilizing anomalous concentrations of base metals in the Central African Copperbelt.

Figure 1: Scanning electron microscope images of fluid inclusion cavities. Net intensity element maps display the relative masses of various solid daughter phases using backscattered electron imagery. A-C) Intergrown halite and sylvite forming anhedral salt masses in primary fluid inclusions hosted by two mineralized, syn-kinematic veins from the Nchanga Upper Orebody. D) Primary inclusion in a mineralized, pre- to syn-kinematic vein from the ore shale at Nkana. Sylvite and halite form the primary solid phases, with a minor Ba(CO3?) phase (left).

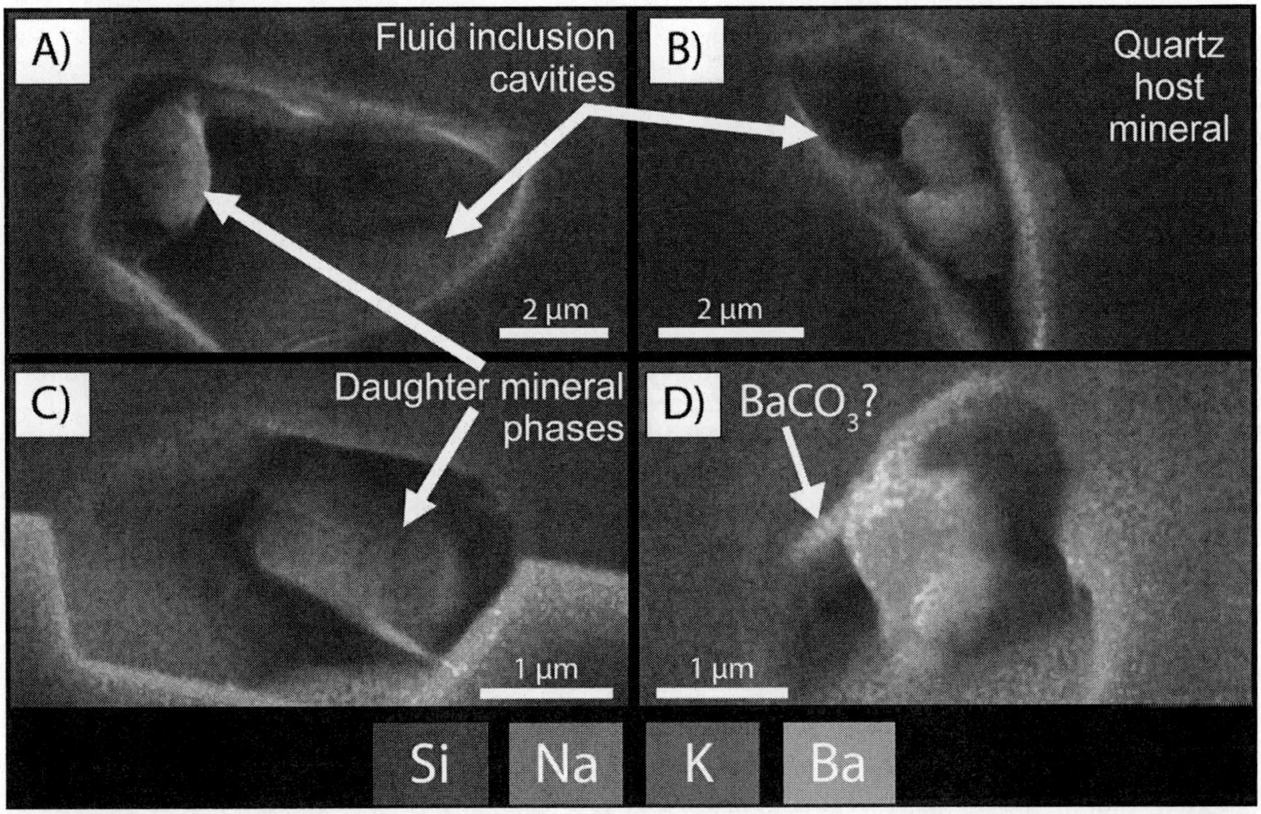

B8
Connecting Proterozoic Carbonate-Hosted Zn-Pb-Cu-Ag Mineral Systems Between Two Continents Using Whole-Rock Pb Isotope Geochemistry

Neil A. Fernandes[1, 2], Dan Layton-Matthews[2], Alexandre Voinot[2], Matthew Leybourne[2]
1. Teck Resources, Vancouver, BC, Canada, 2. Queens University, Kingston, ON, Canada

In carbonate-hosted mineral deposit systems, a link can be made between the Pb isotope compositions of ore minerals (galena, sphalerite, pyrite), host rocks (limestones, dolostones), and underlying continental crust that sedimentary basins were deposited upon (e.g., Tosdal et al., 1999). Previous workers have suggested that ore sulfide minerals in the Proterozoic carbonate-hosted Vazante district, on the southwestern margin of the São Francisco Craton (SFC) in Brazil, and in Proterozoic basins on the margins of the Congo Craton in Africa (Kipushi, DRC; Kabwe, Zambia; Tsumeb, Namibia) could have formed during a common metallogenetic event or events, related to the Neoproterozoic Brasiliano/Pan-African orogeny (800–550 Ma). Although a physical connection between the SFC and the Congo Craton is relatively well-established, little work has been done to directly compare the continental crust between these geographically-disparate regions of the world. In this work, we integrate recent studies of the provenance, geochemistry, and metallogenetic evolution of the sedimentary rocks in the Vazante Zn-Pb-Cu-Ag mineral system (Fernandes et al., 2019a, b) with whole-rock Pb isotope geochemistry of the meta-siliciclastic sequence underlying the Vazante ore-hosting carbonate rocks (Fernandes et al., unpub. data) to further test the Afro-Brazilian intercontinental connection.

The meta-siliciclastic sequence underlying the Vazante carbonate rocks was mostly derived from Paleoproterozoic material and lithogeochemistry of these meta-siliciclastic rocks is supportive of a juvenile, intermediate to felsic continental arc origin for sedimentary detritus (Fernandes et al., 2019b). This meta-siliciclastic sequence has also locally undergone a hydrothermal metal enrichment event that predates the orogenic events associated with carbonate-hosted mineralization (Fernandes et al., 2019a). Present-day whole-rock Pb isotope ratios ($^{206}Pb/^{204}Pb$, $^{207}Pb/^{204}Pb$, $^{208}Pb/^{204}Pb$) of the meta-siliciclastic sequence (Fernandes et al., unpub. data) show a remarkable overlap with present-day Pb isotope ratios of African Paleoproterozoic volcanic arc rocks rimming the Congo Craton, such as the Orange River Group (ORG; Figure 1; Reid et al., 1987). The ORG was one of the major sources of siliciclastic material and metals in Pan-African carbonate-hosted mineral systems (Frimmel et al., 2004).

The already-recognized, similar deposit-scale characteristics between the mineral deposits of the Vazante District in Brazil and various carbonate-hosted Pan-African districts, along with our new insights into the origin, composition, and whole-rock Pb isotope ratios of the continental crust underneath both of these systems, lead us to suggest that these 1st-order Proterozoic metal-endowed sedimentary basins were indeed floored and connected by continental arc material of similar origin and Paleoproterozoic continental arcs dominated the margins of a composite São Francisco-Congo continent during its evolution. This study highlights the power of whole-rock Pb isotope geochemistry to be used in mineral exploration and in connecting mineral systems across the cratonic to intercratonic scale.

Figure 1: Comparison of Serra do Garrote Formation whole-rock Pb isotope compositions with those of galena from carbonate-hosted zinc deposits in the Vazante District, Brazil, galena from Pan-African carbonate-hosted deposits, and whole-rock Pb isotope ratios of Paleoproterozoic Orange River volcanic Arc, Africa (a) $^{206}Pb/^{204}Pb$ vs. $^{207}Pb/^{204}Pb$ and (b) $^{206}Pb/^{204}Pb$ vs. $^{208}Pb/^{204}Pb$.

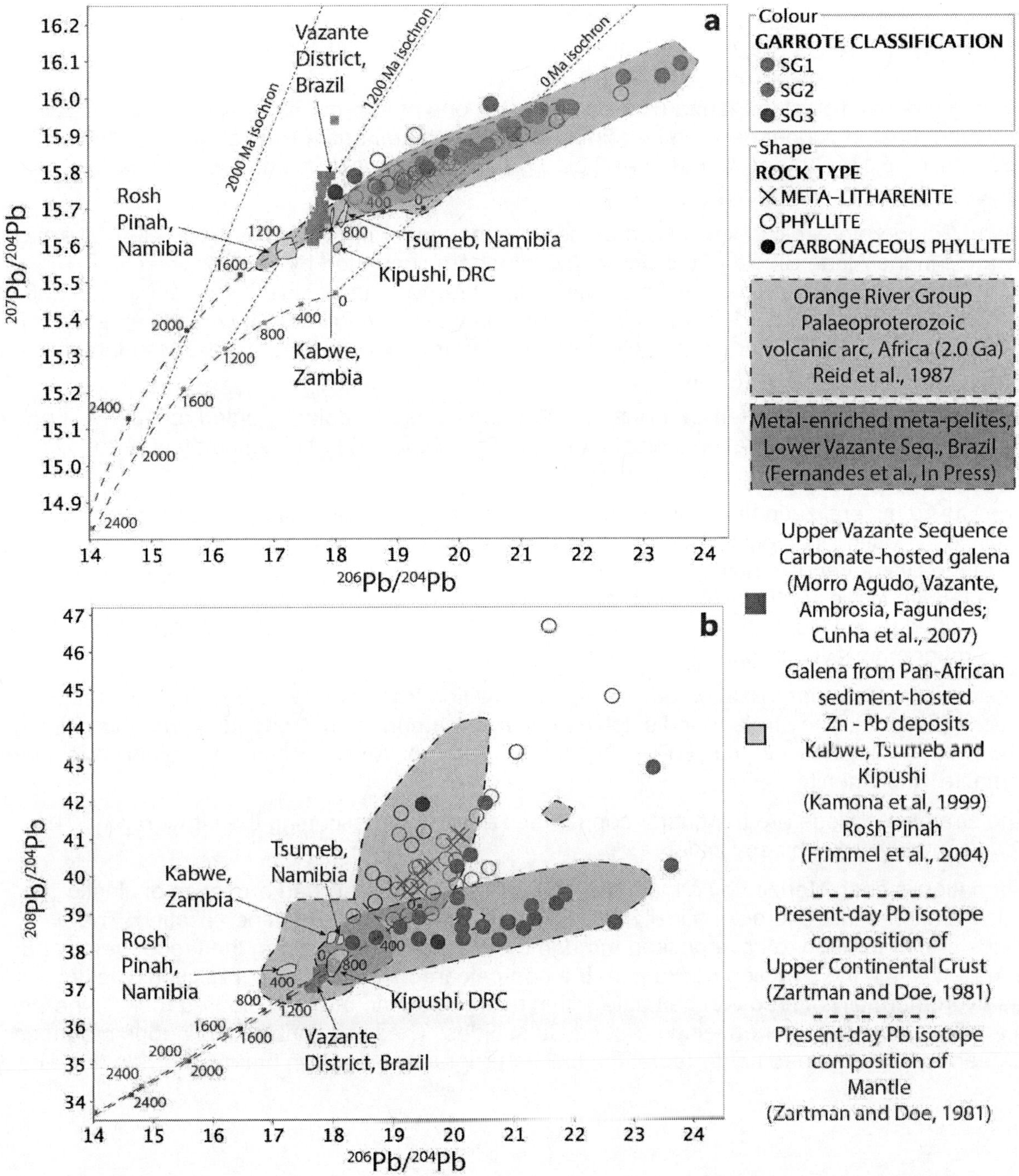

B9
Kodar-Udokan Basin, Siberia (Russia): New Advances in the World-Class Copper District

Alexander Yakubchuk, Michael Hopley
Azarga Metals Corp, White Rock, BC, Canada

The Kodar-Udokan mineral district in Siberia (Russia; Fig. 1) is one of four globally important basins with world-class sediment-hosted copper deposits. A mining complex at the largest Udokan deposit, with resources of 25 Mt Cu (@1% Cu) and 805 Moz Ag (@10 g/t), is being constructed. Several nearby prospects are currently being re-explored.

The 200- x 60-km Paleoproterozoic basin is situated near the boundary between the 3.4-2.9 Ga Olekma and Tynda terranes in the Aldan Shield of the Siberian Craton, amalgamated by 2.6 Ga. The Cu mineralization occurs at several stratigraphic levels within the 12,000-m-thick clastic sequence (2.2-2.06 Ga Udokan Supergroup), deformed at ca 1.9 Ga. This deformation was soon followed by emplacement of 1876 Ma A-type granite, 1867 Ma Fe-Ti-V and Cu-Ni-PGE-bearing layered gabbro, and a 1863 Ma diabase dike swarm, as well as Neoproterozoic mafic dikes.

U-Pb dating of titanite from disseminated Cu ore and Cu-quartz veins at Udokan yielded an 1896 Ma age of mineralization, indicating a syndeformational emplacement of Cu-Ag-bearing fluids shortly prior to emplacement of A-type granites and layered gabbro.

Structurally, the Cu-Ag mineralization is preferentially controlled by the limbs and closures of synclines. The relationships between hosting lithologies, mineralization, and alteration were recently documented by Azarga Metals at the Unkur deposit, hosting >50 Mt of mineralization grading 0.59% Cu and 40 g/t Ag over a strike length of 6.6 km in the southwestern limb of the 10- x 5-km Unkur syncline. Copper oxide minerals were recently identified over 1- to 3-km length on the opposite (northeast) limb of the syncline (Fig. 1A-B) at more than one stratigraphic level, requiring drill testing.

In the southwestern limb, the mineralization occurs at the transition from non-calcareous to calcareous sedimentary rocks (Fig. 1C). The stratabound sulphide minerals comprise chalcopyrite, pyrite, bornite, chalcocite, and covellite oxidized into malachite, chrysocolla, and brochantite. Accessory minerals include magnetite, hematite, and ilmenite.

Underneath the stratabound lode are the quartz-copper feeder veins, crosscutting the stratigraphy. They host simpler chalcopyrite-pyrite mineralization.

The hosting formation is characterized by a magnetic high, reflecting widespread presence of clastic magnetite, which is altered into characteristically coloured hematite near Cu-Ag mineralization. The mineralized fluids are interpreted to have passed through the clastic package along the feeder veins and then were deposited within the stratified package at the non-calcareous/calcareous transition. Clastic magnetite was hydrothermally altered into hematite within a large volume of pink-coloured rocks, whereas pyrite formed an IP-detectable wider envelope around Cu sulfides. These regionally continuous magnetic lows and chargeability highs help to better target the drilling programme through the overburden moraine.

Figure 1: (A) Satellite image, (B) reduced-to-pole magnetic map, and (C) lithological-structural setting of two types of Cu-Ag mineralization at Unkur prospect in the Kodar-Udokan basin, Russia.

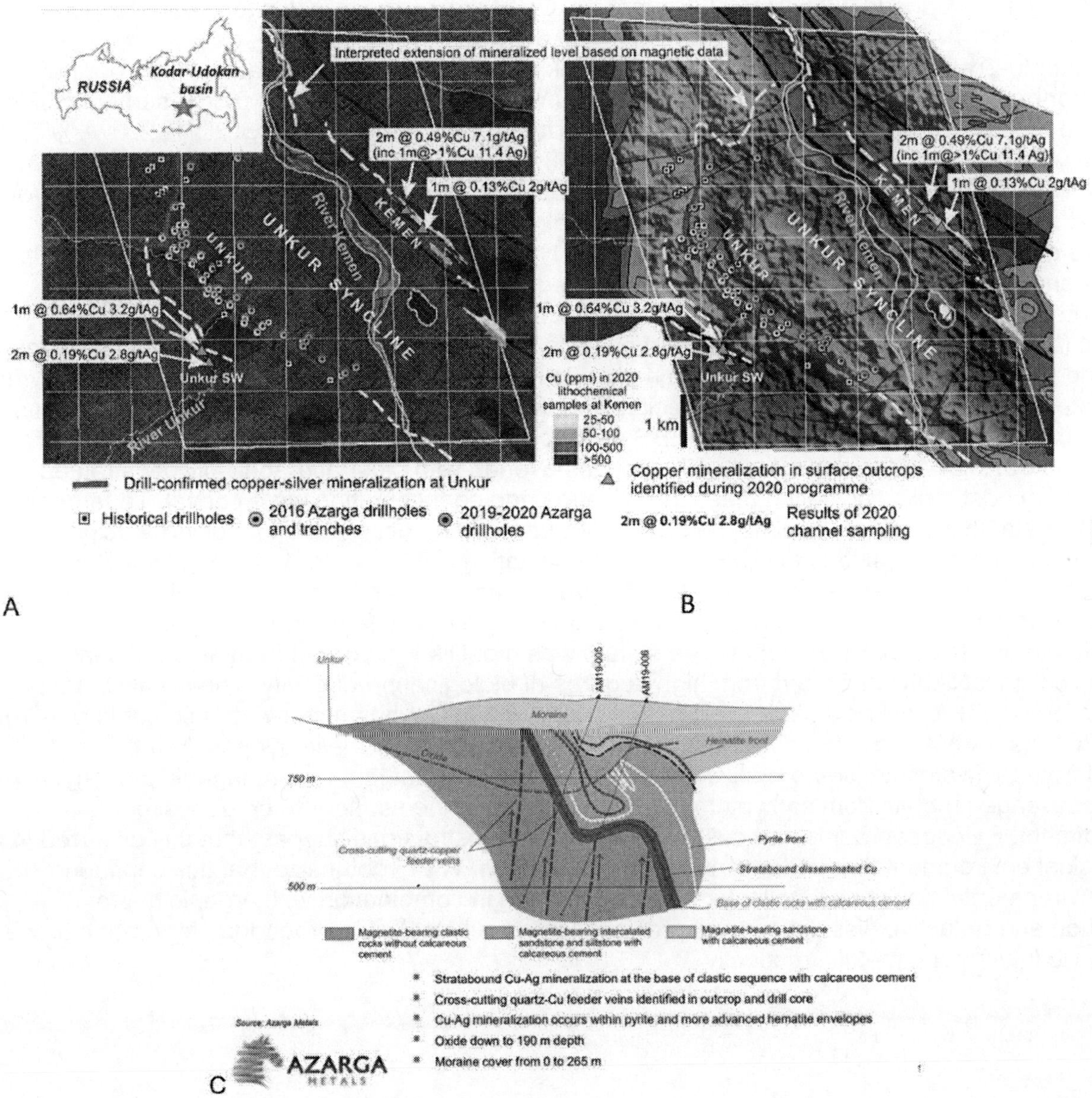

B10
Genesis of polymetallic hyper-enriched black shale mineralization in the northern Canadian Cordillera

Michael G. Gadd[1], Jan M. Peter[1], Dan Layton-Matthews[2]
1. Geological Survey of Canada, Ottawa, ON, Canada, 2. Queens University, Kingston, ON, Canada

Polymetallic (Ni-Mo-Zn-Pt-Pd-Au-Re) hyper-enriched black shale (HEBS) mineralization in the northern Cordillera of northern Yukon and northeastern British Columbia consists of thin (3–10 cm), semi-massive sulfides interbedded with black shale. High-grade mineralization contains up to 7.4 wt.% Ni, 2.7 wt.% Zn, 0.38 wt.% Mo, 400 ppb Pt, 250 ppb Pd, 160 ppb Au and 58.5 ppm Re. The HEBS in northern Yukon was studied at the Nick, Peel River, Monster River and Moss localities. Mineralization discontinuously outcrops at the regional stratigraphic contact between the black shales of the Road River Group and Canol Formation in a geographically widespread area within the Richardson and Blackstone troughs. The sulfide mineral paragenesis, similar for all localities, was protracted, initiated during syngenesis to earliest diagenesis and persisted into later diagenesis. Laser ablation-inductively coupled plasma-mass spectrometric (LA-ICP-MS) analyses indicates that pyrite is the principal host of platinum group elements (PGE), Au and Re. These (and other) metals are hosted in framboidal, nodular and cryptocrystalline pyrite, and these textural varieties reflect diagenetic precipitation. Ni sulfide minerals host subordinate amounts of these metals. Rhenium-Os geochronology of HEBS at Nick and Peel River indicate the age of mineralization is 390.7±5.1 and 387.3±4.4 Ma, respectively. These ages overlap with biostratigraphically constrained ages of the sedimentary host rocks, indicating that mineralization and sedimentation were coeval. Together with the LA-ICP-MS data, this suggests minimal remobilization or resetting occurred after sulfide precipitation. The bulk sulfur isotope composition of HEBS is uniformly negative (?34S = -29 to -10‰), indicating that bacterial sulfate reduction generated reduced sulfur to precipitate sulfide minerals. The high S contents and low negative sulfur isotope compositions of the HEBS imply long-lived sulfate reduction in an environment with ample sulfate and labile organic matter. The sulfate was most likely sourced from ambient seawater, and the organic C most likely originated from high degrees of biological productivity. The initial Os ratios at Peel River (0.25 ± 0.07) and Nick (0.32 ± 0.20) overlap with Middle Devonian seawater and continent-crust normalized PGE patterns are similar to modern seawater, and collectively these suggest that the metals were derived from seawater. Shale-normalized U and Mo ratios suggest that ferromanganese oxyhydroxide particulates scavenged metals from seawater and shuttled them to the seafloor to be deposited as a chemical sediment. Paleogeographic and eustatic sea level reconstructions suggest that this occurred in a favorable basinal environment that underwent periodic restriction. We hypothesize that this confluence of ideal paleoenvironmental conditions facilitated HEBS formation in combination with organic matter remineralization and reductive dissolution (and precipitation as sulfides) of hydrogenous ferromanganese particles supplied sulfur and metals, respectiv

B11

Relevance of Hydrocarbon-Water Trace Metal Partitioning and Re-Os Geochronology for Sedimentary-Hosted Ore Deposits and Petroleum Systems

Nicole C. Hurtig[1, 2], Holly J. Stein[2, 3], Judith L. Hannah[2, 3]
1. Dept. of Earth and Environmental Science, New Mexico Tech, Socorro, NM, USA, 2. AIRIE Program, Colorado State University, Fort Collins, CO, USA, 3. Institute of Geosciences, University of Oslo, Oslo, Norway

Oil-producing basins commonly host sedimentary ore deposits and the same permeability networks are often utilized by formation waters, ore-forming fluids, and hydrocarbons. This results in a complex framework of fluid-rock-hydrocarbon interaction along fluid migration paths. The occurrence of hydrocarbons has been documented in a variety of ore-forming systems including sedimentary-hosted Pb-Zn(-Ag), U, Au, and platinum group elements deposits. The role of hydrocarbons for ore formation can vary widely from 1) hydrocarbons acting as reductants to facilitate sulfide precipitation from metalliferous basinal brines; 2) hydrocarbons indirectly contributing to ore formation by producing H_2S (sour gas) that can either mix with metalliferous fluids resulting in sulfide precipitation or oxidize on mixing with formation waters resulting in the formation of highly acidic brines; and 3) hydrocarbons acting as ore fluids by transporting metals. Hydrocarbons within ore deposits have elevated ore metal contents and are characterized by specific trace metal signatures reflecting brine-hydrocarbon or hydrocarbon-rock/mineral interaction. The Re-Os geochronometer has been used to date bitumen in sulfide deposits providing an accurate age for ore formation. It follows that water-hydrocarbon interaction may be an integral part of sedimentary-hosted ore formation and that hydrocarbons may provide important information on ore-forming processes.

In this study, we experimentally determined fundamental controls on trace metal fractionation during water-oil interaction for Cd, Co, U, As, Ag, Ni, Mo, Ba, Re, Os, V, and other trace metals (Figure 1). The experimental approach is based on the basic chemical principle that trace metals with different chemical properties fractionate across the water-oil interface and become variably enriched or depleted in hydrocarbons. Therefore, quantifying trace metal partitioning between water and hydrocarbons provides new insights for understanding what controls metal signatures in hydrocarbons. Experiments interacting waters and oils with varying initial compositions were conducted at 70°, 150°, and 220°C. In general, the behavior of different trace metals is highly variable with preferential partitioning of some elements (i.e., Ba, Fe) into the aqueous solution and retention and/or enrichment of other elements (i.e., Mo, U, Os, Re, Sb) in the oil. Increasing the salinity and pH of aqueous solutions results in stabilization of aqueous complexes and leaching of metals from the oil and limiting uptake of metals from the spiked aqueous solutions (Figure 1). The behavior of Re and Os during water-oil interaction is of special interest because the Re-Os geochronometer is invaluable for directly determining the timing of water-oil interaction. Results show that Re and Os fractionate across the water-oil interface with Os preferentially partitioning into the oils over Re. We further demonstrate that the Re-Os geochronometer can only be reset due to water-oil interaction for starting waters with sufficient Re contents and low pH at temperatures <220°C, underpinning the application of Re-Os dating of hydrocarbons associated with ore deposits in sedimentary basins.

Figure 1: Trace metal systematics of three different crude oils (black squares) and patterns for reacted crude oils that were equilibrated with trace metal spiked waters at different temperatures (colored circles). The y-axis represents metal concentration in the crude oil.

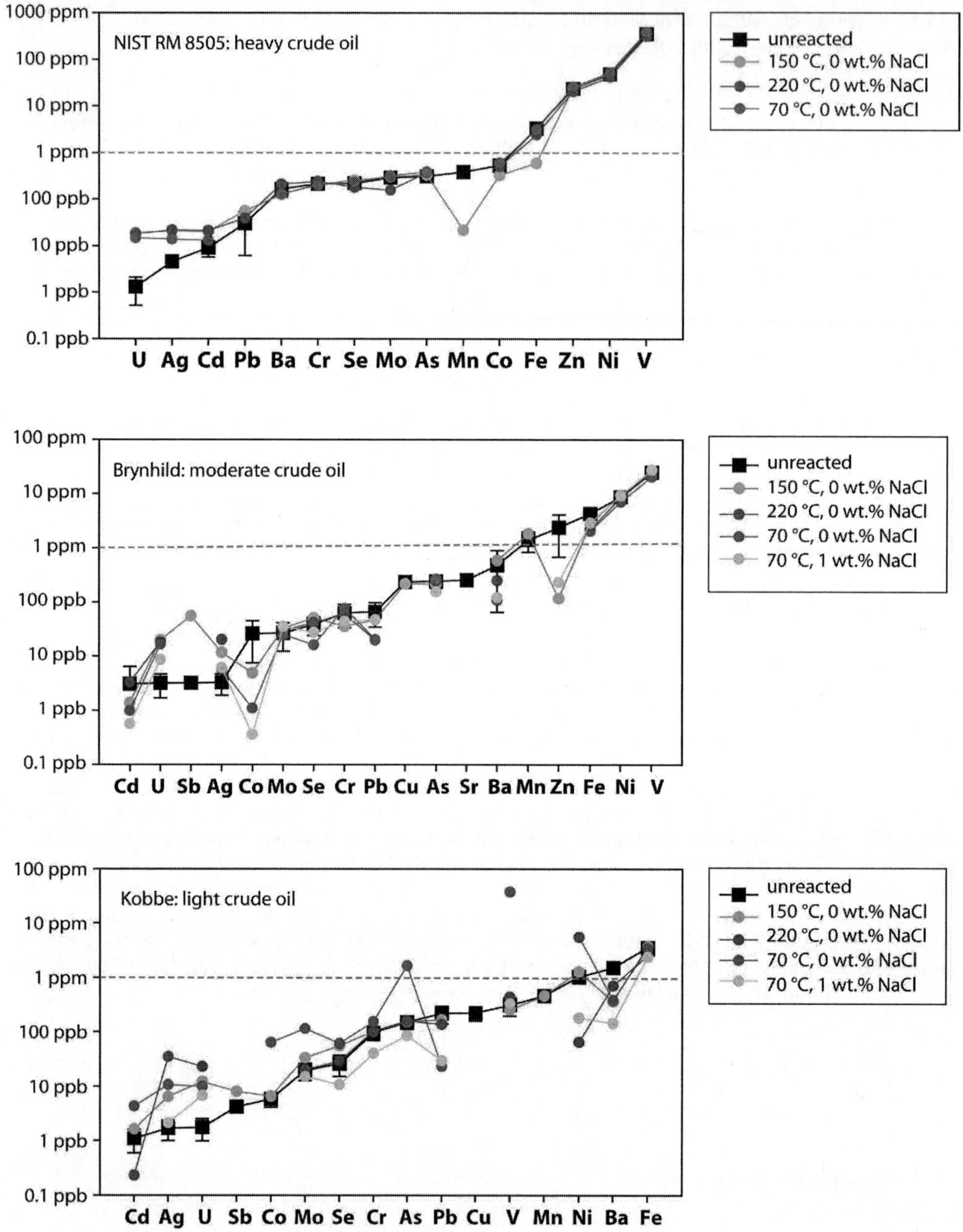

B12
Mapping the Evolution of Diagenetic Fluids using LA-ICPMS of Carbonate Cements and Implications for Redox-Sensitive Metals

Christopher Reed[1], Malcolm Wallace[2], Ashleigh Hood[2]
1. Teck Resources, West Perth, WA, Australia, 2. The University of Melbourne, Melbourne, VIC, Australia

Carbonate cements can be used to constrain the history of pore fluids from the seafloor to later burial. Reed and Wallace (2001 and 2004) documented a consistent cathodoluminescence paragenesis for the Waulsortian Limestone in Ireland: 1) early, non-luminescent, Mn-Fe-poor cement, 2) a bright luminescent, Mn-rich cement, and, finally, 3) dull-luminescent, Mn- and Fe-rich cement. This paragenesis represents a progression from near surface oxidized conditions to reduced burial conditions (Barnaby and Rimstidt, 1989). With LA-ICPMS, it is now possible to determine the trace element composition that occurs with this transition and, therefore, the evolution of diagenetic fluids. Here, we present new LA-ICPMS data from two localities in Ireland.

At Ballinalack, MVT-style Zn-Pb mineralization occurs synchronous with bright luminescent cements. A second locality 15 km west is barren with no known mineralization but preserves the same calcite cement paragenesis.

Both localities record similar Mn and Fe profiles that correlate with the non-bright-dull cement generations consistent with a transition to reduced burial conditions. A difference is that the Fe profile is more depressed relative to Mn at Ballinalack.

The Mg profile shows a decrease at the beginning of the bright luminescent cement. This is consistent with the relative timing of the regional dolomite as constrained by previous paragenetic studies (Reed and Wallace, 2001, 2004). This trend is most pronounced at the unmineralized locality.

The concentration of Zn also decreases to near zero coincident with the bright luminescent cement. This is consistent with the initiation of Zn-Pb mineralization. While the Zn profiles are similar for both localities, the dull-luminescent cement shows more metal variability after the main mineralization event at Ballinalack.

Zinc and Fe have a high affinity for reduced sulfur and will form sulfides in the presence of even small concentrations (Jacobs et al., 1987). The reduction in Zn and Fe during mineralization suggests that sulfur has been introduced into the system, causing sulfide precipitation.

We suggest the period of mineralization records the transition to reduced conditions and the availability of reduced sulfur, triggering the formation of Zn-Pb-Fe sulfides. This event is coeval with dolomitization, which is fueled by the drawdown of Mg from pore fluids. The lack of any significant increase in metals (or Mg) prior to this event and the relatively early timing of the sulfide precipitation suggest that regional subsurface fluids are involved in supplying metals. There is evidence for continued influx of metalliferous fluids during the dull-luminescent cement stage, however, this postdates the mineralization event.

This work also demonstrates the application of LA-ICPMS for constraining the nature of burial fluids within carbonate cements.

Figure

(A) Barren locality and (B) Ballinalack Deposit. For each locality, the cathodoluminescence traverse (top), Mn-Fe profile (middle), and Mg-Zn profile (bottom) are shown. The x-axis is presented as time along the traverse and the y-axis as element concentration.

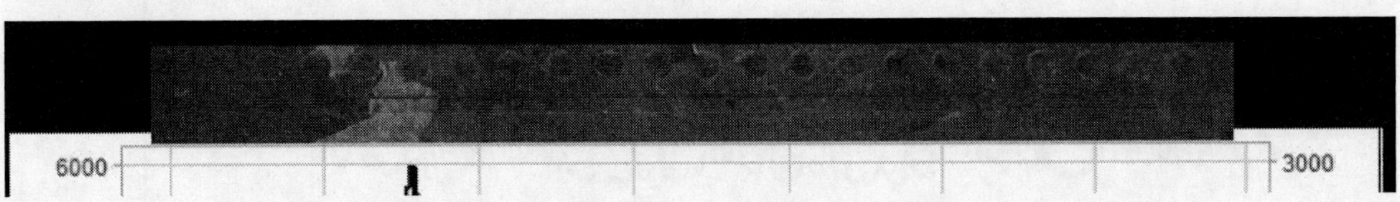

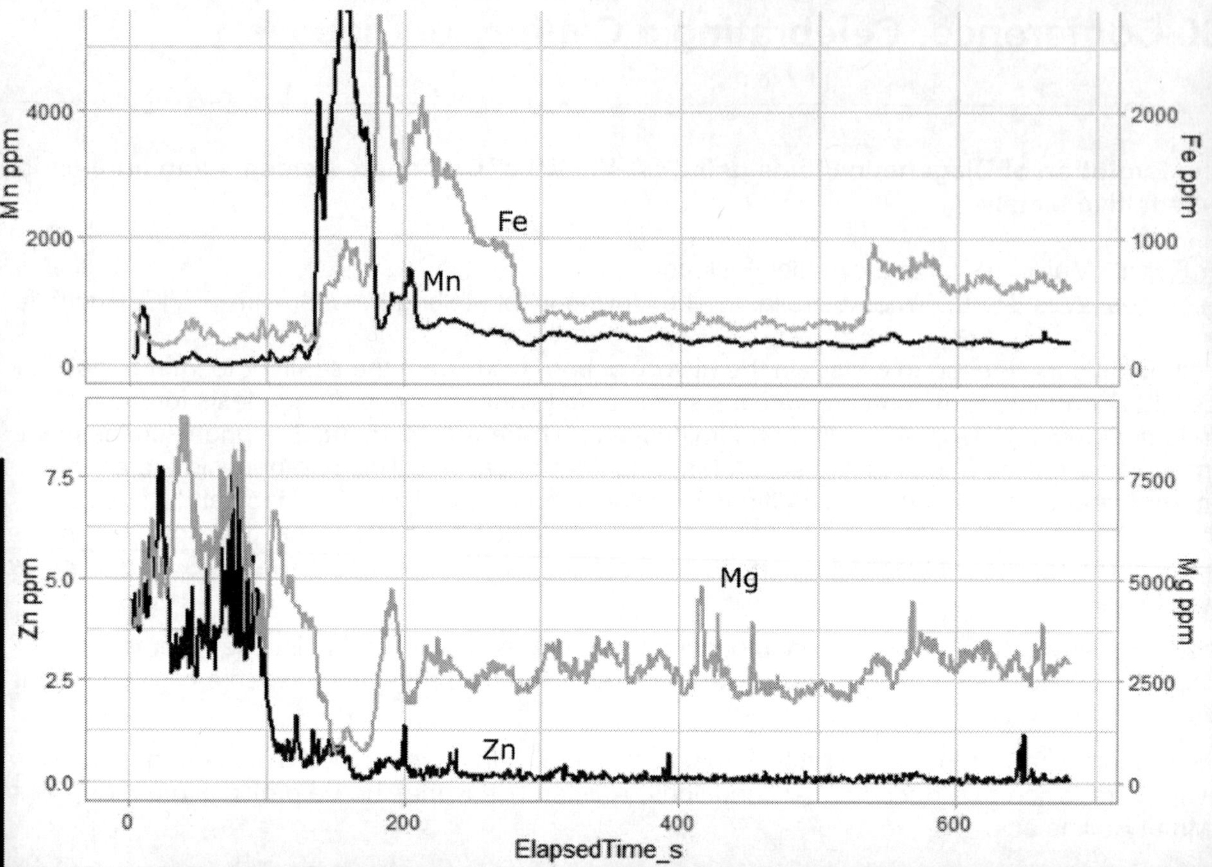
Mn ppm
Fe ppm
Fe
Mn
Zn ppm
Mg ppm
Mg
Zn
ElapsedTime_s

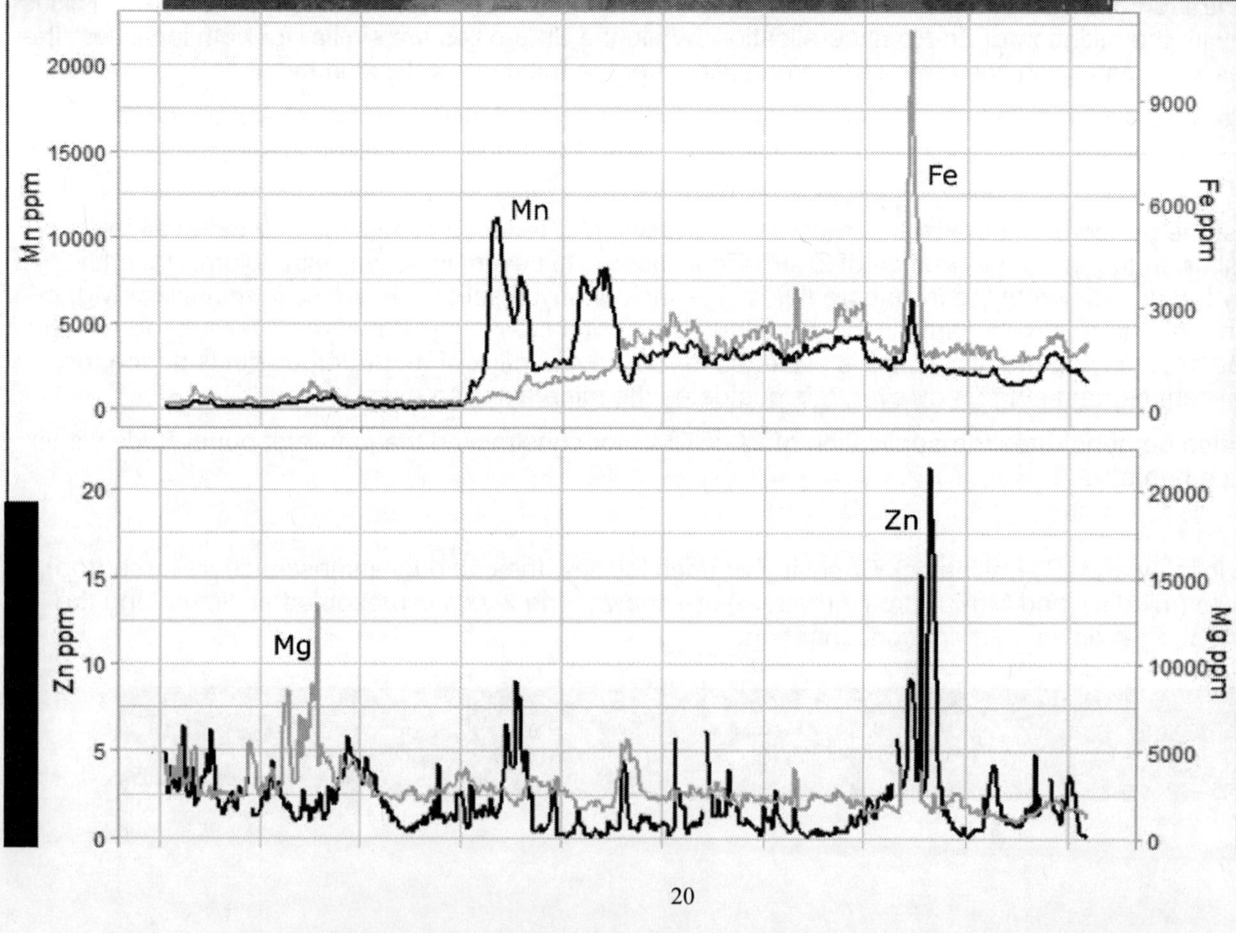
Mn ppm
Fe ppm
Fe
Mn
Zn ppm
Mg ppm
Mg
Zn

0
100
200
300
400
ElapsedTime_s

B13
Banded Iron Formations: Geochemical and Stratigraphic Markers of Ore-Forming Processes?

David Diekrup, Mark D. Hannington
University of Ottawa, Ottawa, ON, Canada

Algoma-type Banded Iron Formation (BIF) is a common lithology in Archean greenstone belts. They are products of slow chemical sedimentation of iron-oxyhydroxide-silica gel during breaks in volcanic activity when contemporaneous volcanogenic massive sulfide deposits (VMS) could form. However, the origin of Algoma-type BIF, its relationship to ore-forming hydrothermal systems, and controls on the timing and stratigraphic position in the depositional basin and the host volcanic sequence are commonly not well constrained. One model favours a biogenic origin and rhythmic changes in physicochemical conditions to account for the characteristic Fe- and Si-rich banding. Another considers the Fe and Si to be direct precipitates from hydrothermal vents (so-called "exhalite"), often along strike from VMS deposits.

A detailed mineralogical and geochemical study of one of the type-localities of Algoma-type BIF, in the 2.73 Ga Temagami Greenstone Belt, provides direct evidence of primary deposition as a water-saturated gel, followed by diagenetic sorting, recrystallization, and replacement to produce the distinctive hematitic chert (jasper) and magnetite bands. The internal layering is a product of an initially homogenous Fe-Si gel that underwent compaction and dewatering, mobilizing Si to the uppermost layers of the sedimentary package and leaving Fe behind. Early-formed nanophase hematite was progressively recrystallized to magnetite, with some layers of primary hematite preserved in quartz and individual crystals of magnetite showing original Si-rich cores. Trace element signatures of the BIF layers were previously interpreted in terms of changing hydrothermal input (e.g., elevated Ge/Si ratios in magnetite layers and smaller La, Eu, and Y anomalies and more abundant REE in magnetite compared to chert layers). We interpret these signals to reflect preferential incorporation of Ge in magnetite during diagenetic growth and the inheritance of REE from early authigenic apatite contained in the original jasper layers, rather than signals of direct hydrothermal input. Metal concentrations in the hematitic and magnetite layers are orders of magnitude lower than in similar BIF associated with VMS mineralization. There are few indicators of undiscovered VMS deposits proximal to the Temagami BIF. Instead, the trace element geochemistry suggests Temagami BIF deposition in an open ocean environment influenced by distal hydrothermal sources and providing a large reservoir of REE and other components with long residence times, as opposed to localized, direct hydrothermal input. We suggest that Algoma-type BIFs represent a continuum of deposit types from VMS-related “exhalites” to “open ocean” deposition, and the distinction (including the different behaviour of Ge, REE, and transition metals) provides a useful tool in exploration.

B14
The lithogeochemical and mineralogical alteration footprint of the George Fisher CD-type Zn-Pb massive sulfide deposit

Philip Rieger[1, 2], Joseph M. Magnall[1], Sarah Gleeson[1, 2]
1. GFZ German Research Centre for Geosciences, Potsdam, Germany, 2. Freie Universität Berlin, Institute for Geological Sciences, Berlin, Germany

Clastic dominated (CD-type) massive sulfide Zn-Pb deposits (formerly known as SEDEX deposits) comprise a large proportion of global Zn-Pb reserves. Many of the world's highest value CD-type deposits (>100 Mt at >10 % Zn+Pb) are located in the Carpentaria province in northern Australia. These deposits are typically stratabound and hosted in fine-grained sedimentary rocks. In the 100 years since the discovery of the first giant Carpentaria CD-type Zn-Pb deposit at Mount Isa, debate over the genesis of these systems was generally focussed on whether mineralization formed via syn-sedimentary exhalative (SEDEX) processes or due to epigenetic host rock replacement. In contrast, there have been relatively few systematic studies on the mineralogy and lithogeochemistry of the mineralization and stratigraphically equivalent un-mineralized host rocks. In this study, we present mineralogical, lithogeochemical, and petrographic data of drill core samples from drill holes that intersected the main ore bodies at the George Fisher deposit (169 Mt at 8.9% Zn, 3.4% Pb) and from a drill hole that intersected correlative un-mineralized host rock lithologies of the Urquhart Shale Formation. Relative to the barren Urquhart Shale, hydrothermal alteration during the first stage of sulfide mineralization (stratabound, Zn-dominated) has led to albite, chlorite, and calcite depletion and dolomite and K-phyllosilicate enrichment in the host rocks. These mineralogical changes are consistent with minor and trace element enrichment (Tl and Mn) and depletion (Na and Sr) at George Fisher. A geochemical alteration index based on these elements (George Fisher index = 10((400Tl+Mn)/(10Sr+Na))) is highly effective in differentiating mineralized from barren host rocks and may be a useful tool for future geochemical exploration programs in the Carpentaria province. This study highlights the benefit of establishing the baseline composition of barren host rocks by combining geochemical, mineralogical, and petrographic data in order to investigate the alteration footprint of basin-hosted mineral systems.

B15
Why Is the Proterozoic Athabasca Basin Endowed with Rich and Large Unconformity-Related Uranium Deposits?

Guoxiang Chi[1], Eric G. Potter[2], Morteza Rabiei[1], Yumeng Wang[1]
1. University of Regina, Regina, SK, Canada, 2. Geological Survey of Canada, Ottawa, ON, Canada

The Proterozoic Athabasca Basin in northern Canada hosts a number of high-grade (mostly >1%, some >10% U) and large-tonnage uranium deposits, making Canada the second largest producer of uranium in the world. These deposits are mostly located near the basal unconformity between the basin and crystalline basement and are termed unconformity-related uranium (URU) deposits. The deposits are spatially associated with reactivated basement faults and hydrothermal alteration halos characterized by chlorite, tourmaline, kaolinite, and illite. They are generally interpreted to have resulted from reaction between oxidizing, uranium-bearing, basin-derived brines and reducing lithologies in, or reducing fluids derived from, the basement. Although this "diagenetic-hydrothermal" model adequately describes formation processes and conditions of the URU deposits, it does not explain why the Athabasca Basin is exceptionally endowed in uranium mineralization; indeed, although URU deposits occur elsewhere in the world, few of them (e.g. Kombolgie Basin in Australia) are comparable to those in the Athabasca Basin.

Chemical analyses of fluid inclusions in quartz overgrowths in sandstones of the Athabasca Basin, and drusy quartz in both mineralized and barren areas along structures hosting URU deposits, suggest basin-wide development of uranium-rich brines. Such unusual, large-scale formation of uranium-rich brines, which is possibly the critical reason for the rich endowment of uranium deposits in this basin, may be attributed to several geological factors. First, the basin is flanked by two orogens, the Taltson Magmatic Zone-Thelon Orogen to the west and the Trans-Hudson Orogen to the east, both of which contain large amounts of granitic rocks that may have provided abundant uranium-rich detritus to the basin. Second, the low mud content of the sandstones, related in part to the lack of land vegetation in the Proterozoic, enhanced fluid circulation. Third, increased atmospheric oxygen levels in the late Paleoproterozoic, as recorded by the red beds in the basin, enhanced extraction of uranium from the sediments. Fourth, basinal brines derived from seawater evaporation, as indicated by evaporitic carbonate rocks above the sandstones and high-salinity fluid inclusions in quartz overgrowths, increased uranium extraction. Finally, elevated temperatures, as recorded by fluid inclusions, clay geothermometry, and zircon thermochronology, facilitated fluid convection and uranium extraction. Fluid convection is reflected by basin-scale quartz precipitation-dissolution patterns and demonstrated by reactive mass transport modeling. The elevated temperatures may be related to high-radiogenic heat-producing felsic intrusions at depth or upwelling of the asthenosphere in relation to the breakup of the Nuna supercontinent. Based on regional geological and geochronological data, we propose that that these conditions came together in the Athabasca Basin at ca. 1550 Ma, which, together with reactivation of basement-rooted structural zones channelling deep-sourced reducing fluids toward the unconformity, resulted in large-scale URU mineralization. Although individual factors may be satisfied in many other basins, the combination of them all is rare. Therefore, the unusually rich uranium deposits in the Athabasca Basin are the result of coupling of multiple favorable geological factors, including both shallow (basinal) and deep (thermo-tectonic) processes, at a specific time and location in the Earth's crust.

ST.001
Sulfide Mineral Chemistry Integrated with Principal Component Analysis to Constrain the Ore Genesis of the Morro Agudo Pb-Zn District, Brazil

Colin Aldis[1], Gema R. Olivo[1], Ilkay Cevik[1], Brian Joy[1], Jessica Arruda[2]
1. Queen's University, Kingston, ON, Canada, 2. Nexa Resources, Paracatu, MG, Brazil

The Vazante-Paracatu Pb-Zn mineral belt in central Brazil accounts for the majority of Brazilian base metal production. The Mesoproterozoic (1.3-1.1 Ga) Upper Sequence of the Vazante Group hosts distinct styles of mineralization, including the world-class silicate zinc deposits (Vazante and North Extension) and sulfide deposits (Morro Agudo, Fagundes, Ambrósia, and Bonsucesso) and occurrences (Bento Carmelo, Sucuri, and Morro do Capão). The mineralization is hosted in dolomitic rocks in the northern (sulfide zinc-lead) and southern (silicate-zinc) parts of the belt. This study focuses on the Morro Agudo District where the interpretation of the processes and timing related to the formation of the sulfide deposit is controversial (coeval with sedimentation or with the Neoproterozoic Brasiliano Orogeny) and little is known about the other occurrences. The main ore minerals throughout the district are sphalerite and galena associated with micritic and sparitic dolomite, hydrothermal dolomite, pyrite, and quartz. The compositions of sphalerite and pyrite were determined by electron microprobe analyses (EMPA) with the objective to understand the processes related to ore formation during the evolution of the district. Sphalerite grains were selected from each of the occurrences and from the various orebodies at Morro Agudo Mine (Basal, JKL, M, and N) in various sections of the deposit (483, 580, 350) to identify spatial variability. To identify element associations and populations within the district, principal component (PCA) and cluster analyses were applied. Three end-member populations of sphalerite were identified: Fe-rich, Cd-rich, and Cd-Fe-poor. The variation of the sphalerite composition was not controlled by host rock composition, indicating that high fluid-rock ratio and additional factors such as physical-chemical conditions of the metalliferous brine or mixing of fluids influenced the composition. The lower orebodies of Morro Agudo, the JKL and the Basal, display great variability in composition but primarily fall between the Fe-rich and Cd-rich end-members, suggesting that fluid physical-chemical conditions changed during deposition. This could be related to multiple episodes/events of fluid-rock interactions or fluid mixing. However, the Basal orebody in section 580 and the JKL orebody in section 350 are poor in Fe and Cd. This composition is similar to samples from Morro do Capão and likely formed from a distinct fluid composition compared to other orebodies. The upper orebodies, M and N, display lesser variability between Fe-rich and Cd-rich end-members. The sphalerite composition of Sucuri is similar to those of Morro Agudo and likely formed from similar fluids. Bento Carmelo has a unique mineral chemistry with intermediate contents of Zn and Cd and elevated contents of Hg, Cu, and Ni, suggesting the fluid-rock interactions and processes related to ore formation were unique at this location. This is consistent with whole-rock lithogeochemistry and ore-related mineralogy. The results show that the history of metalliferous fluids is complex and likely represents either episodic/multiple fluid events or fluid mixing.

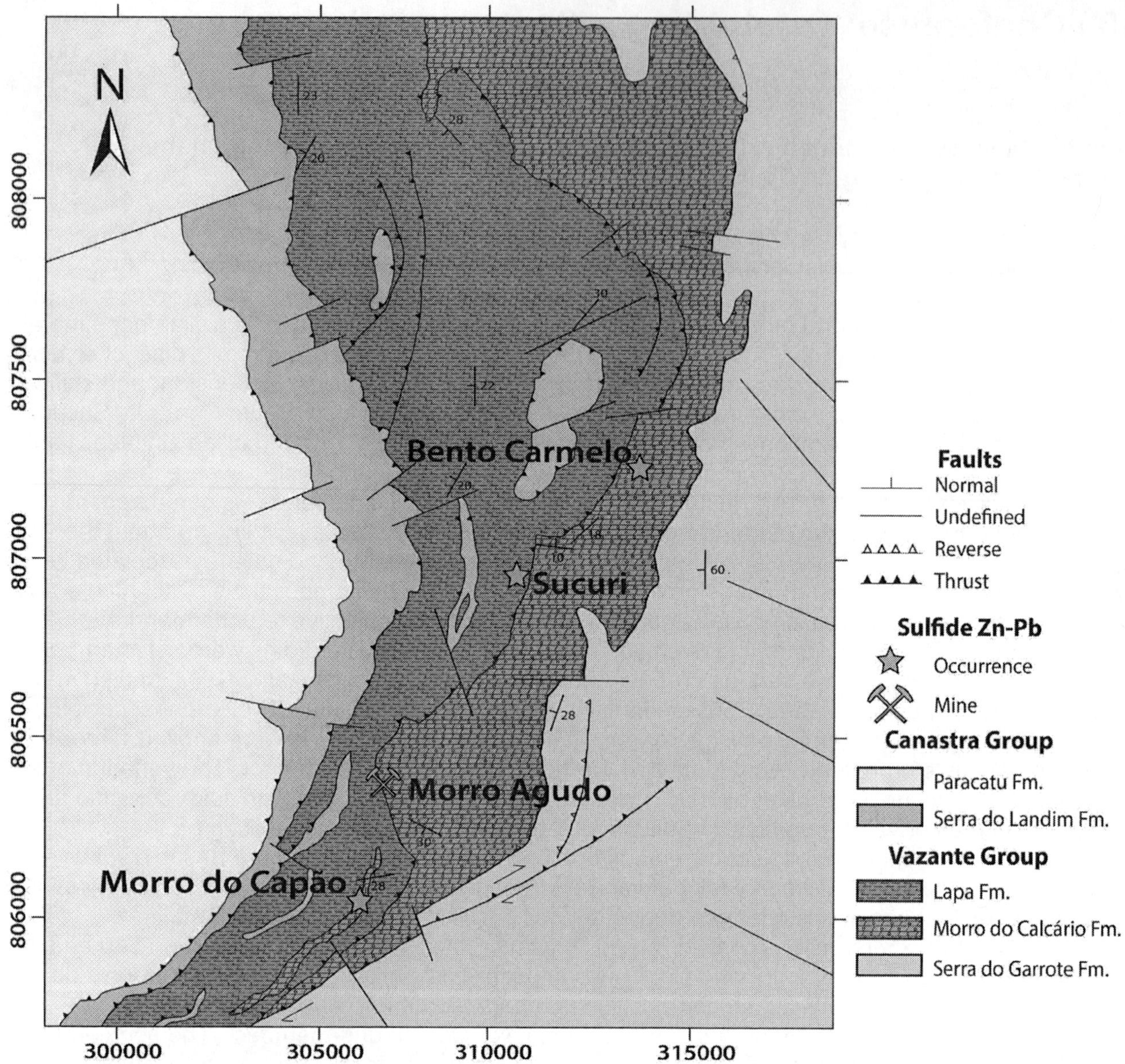
N
Bento Carmelo
Sucuri
Morro Agudo
Morro do Capão
808000
807500
807000
806500
806000
300000
305000
310000
315000
Faults
Normal
Undefined
Reverse
Thrust
Sulfide Zn-Pb
Occurrence
Mine
Canastra Group
Paracatu Fm.
Serra do Landim Fm.
Vazante Group
Lapa Fm.
Morro do Calcário Fm.
Serra do Garrote Fm.

ST.003
Deducing Different Mineralization Styles of Multi-Stage Sediment-Hosted Cu-Co Deposits: The Dolostone Ore Formation Deposit, Northwestern Namibia

Viktor Bertrandsson Erlandsson[1], Frank Melcher[1], Daniela Wallner[1], Rainer Ellmies[2], Johann G. Raith[1]
1. Montanuniversität Leoben, Leoben, Austria, 2. Gecko Namibia, Swakopmund, Namibia

The Dolostone Ore Formation (DOF) is a sediment-hosted Cu-Co mineralization hosted in Neoproterozoic carbonates and shales of the Ombombo Subgroup, Damara Supergroup (Allen, 2016; Hoffmann et al., 2002). Sediment-hosted Cu(-Co) deposits (such as the central European Kupferschiefer and the Central African Copperbelt) commonly comprise multiple styles of mineralization, most predominantly disseminations and veins. Constraining the paragenetic relationship between different mineralization styles in one mineralization commonly requires a multi-method approach, based largely on microscopic observations and fluid inclusion data.

In the pursuit of better understanding the sulfide geochemistry of the DOF, in situ LA-ICP-MS trace element analysis was done on sphalerite, chalcopyrite, pyrite, and pyrrhotite. Six mineralization styles were identified and analyzed for this study: disseminated, clustered, nodular, "sigma," veins, and "events". Disseminations are typically monomineralic sulfide crystals in the host rock. Clusters comprise anhedral iron-sulfide with intergrown sphalerite and chalcopyrite. Nodules are predominantly iron-sulfide mineralized, with interstitially grown chalcopyrite and sphalerite. The "sigma" shows distinct pressure shadows making it texturally different enough from the typical nodule style. The "sigma" has a core clast composed of pyrite, whilst chalcopyrite, sphalerite, and gangue minerals are exclusively limited to the pressure shadows around the pyrite core. Veins can be either sulfide or gangue mineral dominated. Veins are not exclusive to the DOF horizon(s) and can be found unmineralized throughout the stratigraphy. The so-called "events" are still enigmatic and portray ductile and brittle textures. "Events" were interpreted as soft-sediment deformation structures, but also show great likeness to crack-seal veins. "Events" are the only style so far where pyrite and pyrrhotite coexist. Linnaeite is the dominant Co-phase of the DOF and is only absent in the "sigma" and vein styles. Accessory galena, cobaltite, and pentlandite has been identified in some of the styles.

Elements such as Fe, Co, Ni, Ga, Se, and Cd in sphalerite and Co, Ni, Se, and Bi in chalcopyrite show a distinct grouping of concentrations between (1) dissemination, cluster, nodule, and "event" styles and (2) "sigma" and vein styles. The Ge-Ga-In-Mn-Fe-in-sphalerite thermometer (Frenzel et al., 2016) was utilized on sphalerites from each mineralization style, and each style resulted in sphalerite formation temperatures above the system closing temperature of 310 ± 50°C. Trace element compositions of coexisting euhedral pyrite and anhedral pyrrhotite (of the "event" style) suggest that the pyrrhotite may be the metamorphic product from the earlier pyrites. Elements such as Cu, As, Sb, Tl, and Pb are enriched in the "event" pyrite compared to the pyrrhotite, whilst Co, Ni, and Se show similar concentrations in both iron sulfides. This is a trend commonly seen in metamorphically recrystallized iron sulfides.

This work by no means infers that trace element data should, or can, replace the more conventional methods used for paragenetic studies. Our data demonstrates that trends in sulfide trace elements can be utilized to identify groups of mineralization styles, thus suggested to have formed contemporaneously, and also to infer the association of metamorphism to the formation of these sulfides.

ST.004

The Occurrence of Vein and Matrix Kyanite Hosted in Carbonates of the Menda Prospect of the Congolese Copperbelt

Anna Bidgood[1], Murray Hitzman[1], Helen Twigg[1], David Chew[3], Sean Johnson[2]
1. iCRAG, University College Dublin, Dublin, Ireland, 2. Boliden, Skellefteå, Sweden, 3. Trinity College Dublin, Dublin, Ireland

Kyanite, the high-pressure Al_2SiO_5 polymorph, is traditionally considered a high-pressure, high-temperature mineral. Common kyanite-forming reactions such as pyrophyllite or chlorite dehydration take place at ~400° and >600°C, respectively. In this study, kyanite is observed within low-temperature dolomite-magnesite carbonates and carbonaceous siliciclastic lithologies, hosted in postevaporitic breccia, and associated with abundant chalcopyrite. Kyanite occurs as matrix porphyroblasts as well as coarse-grained, vein-hosted crystals intergrown with quartz, dolomite, chalcopyrite, and monazite. Corroded crystal shapes indicate breakdown of kyanite to Mg-chlorite. The occurrence of kyanite in low-pressure-temperature rocks indicates unusual conditions of kyanite growth. Cathode luminescence and element mapping by laser ablation mass spectrometry are used in this study to unravel the complex vein paragenesis and to characterise the multiple textural generations of kyanite.

The Central African Copperbelt (CACB) is the world's primary source of copper and cobalt, producing about 70% of the metal that is critical for the production of batteries needed to help decarbonise our societies. Highly saline fluids are associated with extensive mineralisation and alteration in the CACB. Intimate association of kyanite-bearing veins and Cu-sulphide mineralisation indicates that the unusual thermodynamic conditions encountered play a role in the understanding of the Copperbelt Cu-Co system. The occurrence of kyanite in Al-poor carbonates and in veins suggests that Al, thought to be relatively immobile, has been transported. The preservation of sedimentary features, indicative of an original grainstone, also suggests that there was a compositional control on kyanite growth. Understanding the paragenesis of kyanite in relation to Cu, Ni, and U mineralisation at Menda, as well as the unusual thermodynamic conditions required for low-temperature kyanite growth and Al-mobilisation, has implications for the interpretation of the kyanite-bearing assemblages in metamorphic assemblages in the Domes region of the Zambian Copperbelt, as well as for metal mobility in highly saline systems.

Figure caption: There are many different textural generations of kyanite which are summarised in the schematic. Generation (i) is distributed along compositional layers which also act as conduits for fluid flow. Generations (ii), (v), and (vi) are related to brittle fractures and vein margins. Generations (iii) and (iv) are found within dolomite, quartz, monazite veins.

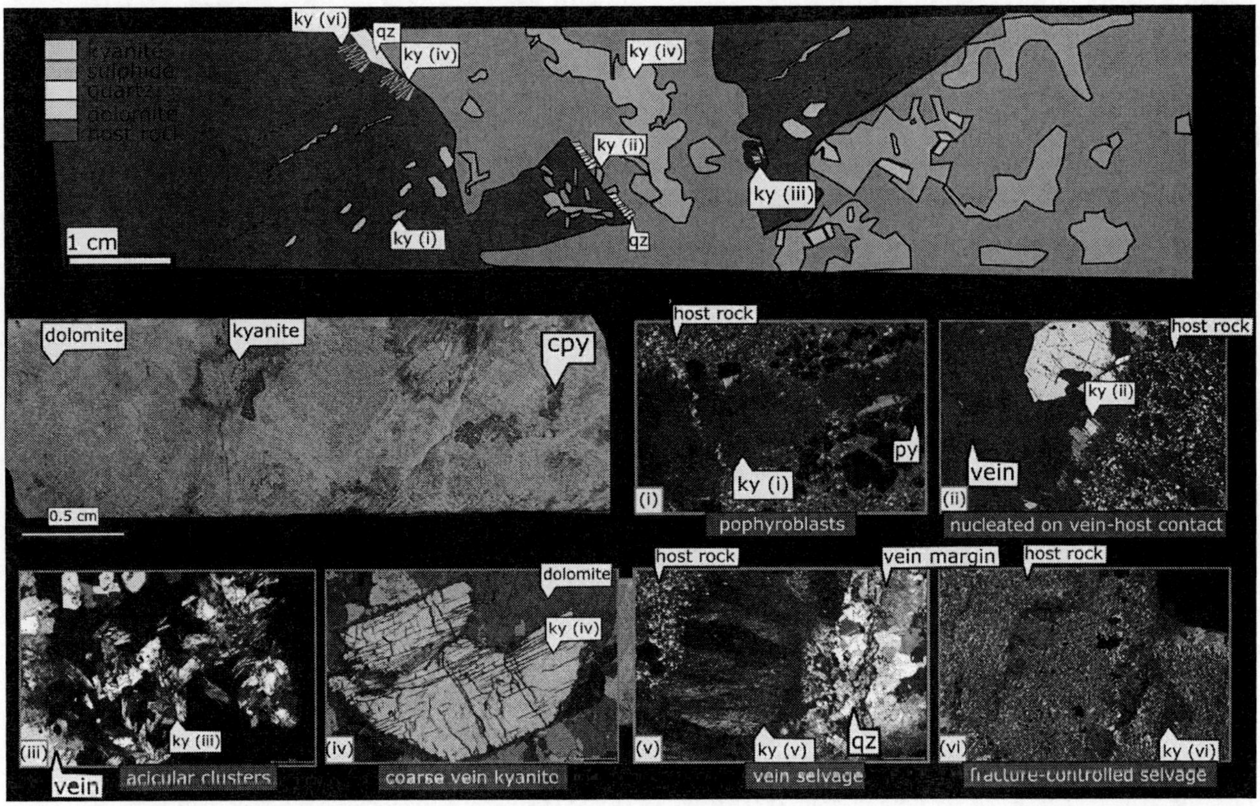

ST.005

Potential for Volcanogenic Massive Sulfide Mineralization in the Upper Hazelton Group, North-Central British Columbia

Stefanie M. Brueckner[1], Gregory Johnson[1], Harold Gibson[2], Stephanie Wafforn[3], Ross Sherlock[2], Ken McNaughton[4]
1. Dept of Geological Sciences, University of Manitoba, Winnipeg, MB, Canada, 2. Harquail School of Earth Sciences, Laurentian University, Sudbury, ON, Canada, 3. Pretivm Resources Inc, Vancouver, BC, Canada, 4. P2Gold, Vancouver, BC, Canada

The A6 Anomaly, located circa 15 km northeast of the epithermal Brucejack Au mine and circa 30 km southeast of the Au-rich volcanogenic massive sulfide (VMS) Eskay Creek deposit, is a greenfields exploration project which was drilled in 2019-2020 to investigate its VMS potential. Observations from surficial mapping and drill core puts the A6 Anomaly into the Lower to Middle Jurassic Upper Hazelton Group. The stratigraphy shows similarities to Eskay Creek with an andesitic tuff unit overlain by quartz-sericite ± chlorite ± pyrite-altered massive and flow-banded coherent felsic rocks and minor volcaniclastic intercalations. The felsic units are overlain by a volcano-sedimentary package of mudstone to polymictic conglomerate intercalated with basaltic flows, sills, and hyaloclastite.

Lithogeochemical results on altered felsic flows indicate a rhyodacite to trachyte composition and calc-alkaline affinity similar to FII rhyolites. Immobile elements have constant ratios, indicating a common precursor of these rocks, but the ratios are different to the tholeiitic, strongly altered Eskay Creek rhyolite. Basaltic flows and sills in the hanging wall of the altered felsic flows have calc-alkaline to tholeiitic affinities.

Alteration of the felsic flows is dominated by quartz and illite with minor Fe-chlorite (chamosite) and trace pyrite that occurs as either disseminated crystals or pyrite stringers. Illite composition was determined by X-ray diffraction and electron probe microanalyses yielding an average illite composition of $(K_{0.67}, Na_{0.03}, Ca_{0.01})_{\Sigma 0.70}(Al_{1.38}, Fe_{0.10}, Mg_{0.15})_{\Sigma 1.63}(Al_{0.65}, Si_{2.94})_{\Sigma 3.59}(OH)_{1.48}$ that is deficient in Si, Al and interlayer cations, presumably due to intergrowth with another clay component of unknown composition. However, Tschermak substitution (i.e., changing from phengite to celadonite composition) is not present. Illite alteration is moderate and a product of feldspar destruction by mildly acidic, relatively low-temperature hydrothermal fluids.

The fluid temperature was determined from oxygen isotope compositions of illite measured by secondary ion mass spectrometry. Illite has an average $d^{18}O = 10.7 \pm 3.0$ ‰ VSMOW; geothermometry was applied using an assumed $d^{18}O = 0.5\text{-}2.3$‰ VSMOW of the hydrothermal fluid. An average fluid temperature of 150°-180°C was determined to have caused the pervasive quartz-sericite (illite) ± chlorite ± pyrite alteration in calc-alkaline rhyodacites to trachytes. This is similar to fluid temperatures responsible for alteration and Au-rich sulfosalt mineralization at the Eskay Creek deposit and lower than hydrothermal fluid temperatures reported for the majority of VMS deposits.

The A6 Anomaly was most likely formed within an immature (back-arc) rift in which felsic flows and calc-alkaline basalts assimilated some crustal components, followed by a widening of the rift resulting in relatively shallow, tholeiitic mafic volcanism. Moreover, the rift was subject to several turbiditic mass movements into which calc-alkaline and tholeiitic basaltic flows and sills intruded prior to consolidation of sedimentary sequences. Although mineralization is not prominent and limited to pyrite in the drilled holes, stratigraphy, lithogeochemistry, and alteration mineralogy and chemistry indicate potential for a bimodal-felsic VMS-style deposit formed in a (back-arc) rift similar to other calc-alkaline hosted VMS deposits of which the A6 Anomaly may represent the distally altered, low-temperature footwall.

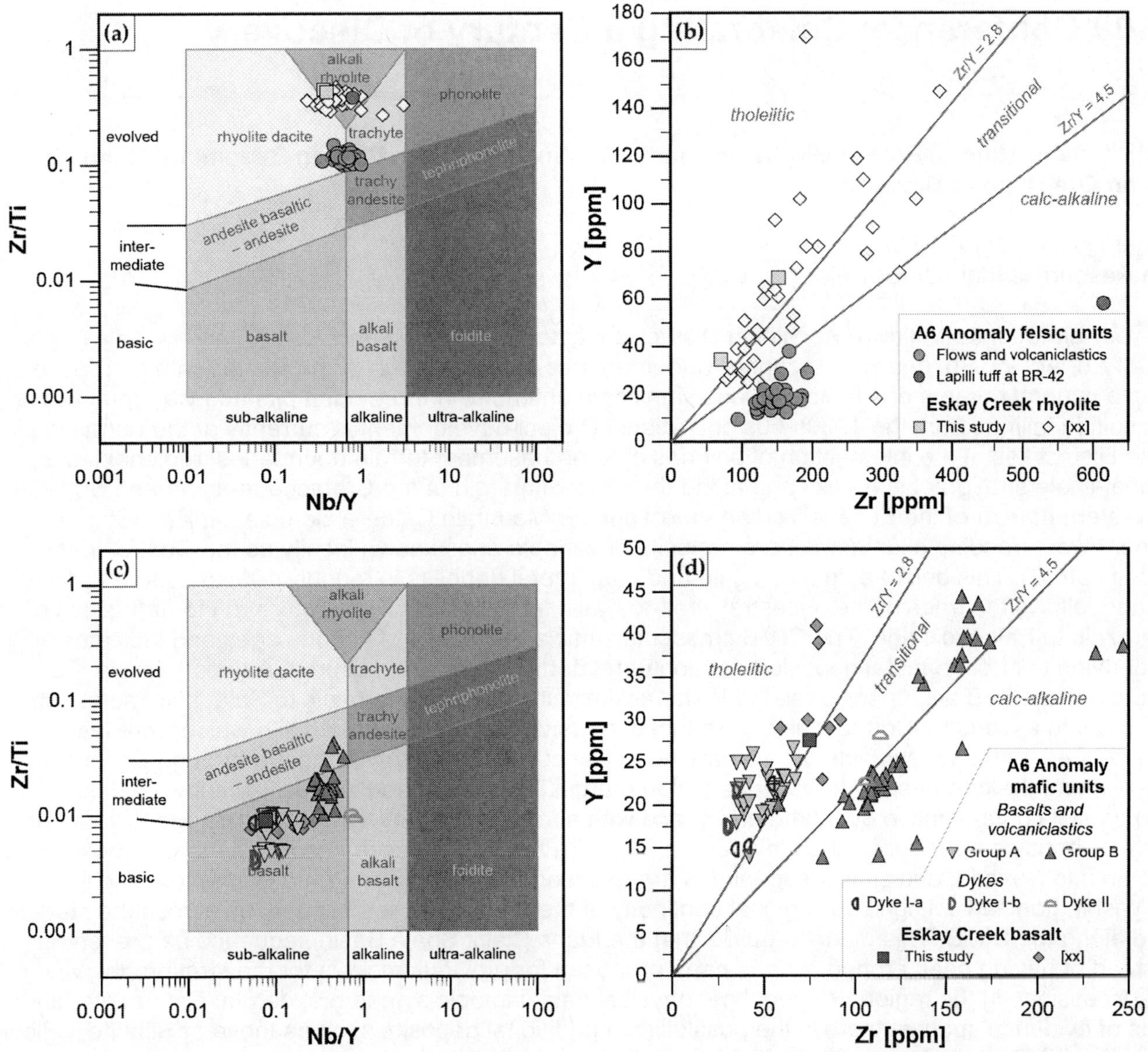

(a)
Zr/Ti
Nb/Y
alkali rhyolite
phonolite
evolved
rhyolite dacite
trachyte
tephriphonolite
trachy andesite
andesite basaltic – andesite
inter-mediate
basalt
alkali basalt
foidite
basic
sub-alkaline
alkaline
ultra-alkaline
(b)
Y [ppm]
Zr [ppm]
Zr/Y = 2.8
Zr/Y = 4.5
tholeiitic
transitional
calc-alkaline
A6 Anomaly felsic units
Flows and volcaniclastics
Lapilli tuff at BR-42
Eskay Creek rhyolite
This study
[xx]
(c)
Zr/Ti
Nb/Y
(d)
Y [ppm]
Zr [ppm]
A6 Anomaly mafic units
Basalts and volcaniclastics
Group A
Group B
Dykes
Dyke I-a
Dyke I-b
Dyke II
Eskay Creek basalt
This study
[xx]

ST.006

The Central Idaho Mineral Belt- A Big Basin, Basin Busting Batholith, Broken Basement, Bustling Brines, and Ore Deposit Brewing

Christopher Dail, Austin Zinsser
Perpetua Resources Idaho, Inc., Boise, ID, USA

The 90+-year-old Stibnite-Yellow Pine Mining District is located in central Idaho, U.S.A., and is a major past producer of Au, Ag, W, Sb, and Hg. Recent development has led to a rebirth of the District with a proposal to develop a mineral reserve of 4.8 Moz Au and significant antimony in three open pits and via reprocessing of former sulfide tailings from the 1930s-50s operations. Project development is currently at the permitting stage. The District lies at the intersection of a series of large basement terrain boundaries, regional faults, and regional-scale stratigraphic breaks and along the eastern margin of the Cretaceous-age Idaho Batholith and the western margin of the Eocene Tertiary-age Thunder Mountain Caldera Complex. It lies within a discrete northwest-trending alignment of ore deposits of variable ages known locally as the Central Idaho Mineral Belt (CIMB). This belt of epigenetic gold and base metal deposits is roughly 100 km long and 10-20 km wide and follows the trace of the ancestral shelf-to-basin transition of "Windermere" rift to drift facies of Neoproterozoic to Paleozoic age. The CIMB crosscuts surficial rock units of various ages and indicates some underlying control on metal deposit formation unrelated to surficial exposures. A hypothesis is presented that the CIMB alignment is related to the assimilation of Neoproterozoic to Paleozoic metal-rich basinal rocks into ascending Cretaceous age reduced intrusives of the Idaho Batholith which were later uplifted, extended, and then subjected to Tertiary heat flow associated with Paleogene through Eocene magmatism along the former shelf-to-basin transition margin. The Cretaceous intrusives show evidence for increased crustal contamination over time and space with a transition from early metaluminous to younger, more peraluminous compositions. The Idaho Black Shale Belt is well exposed in south-central Idaho and farther south into Nevada, and similar sequences are exposed in the northern Idaho Panhandle and in eastern Washington, indicating likely regional continuity of the reduced facies sequences across the state of Idaho and along the CIMB. This model requires that the Idaho Black Shale Basin sequence be present at depth instead of having been eroded, as has been proposed for previous models for the stratigraphic and tectonic successions in the region. Regional geophysical data, isotopic signatures, detrital zircon data, and other lines of evidence suggest there is the possibility of additional deposits such as those at Stibnite-Yellow Pine along the CIMB, and potentially into B.C. as well, that resulted from similar basement controls for ascending magmas, stratigraphic sources for metals, and structural plumbing systems for hydrothermal systems of variable age.

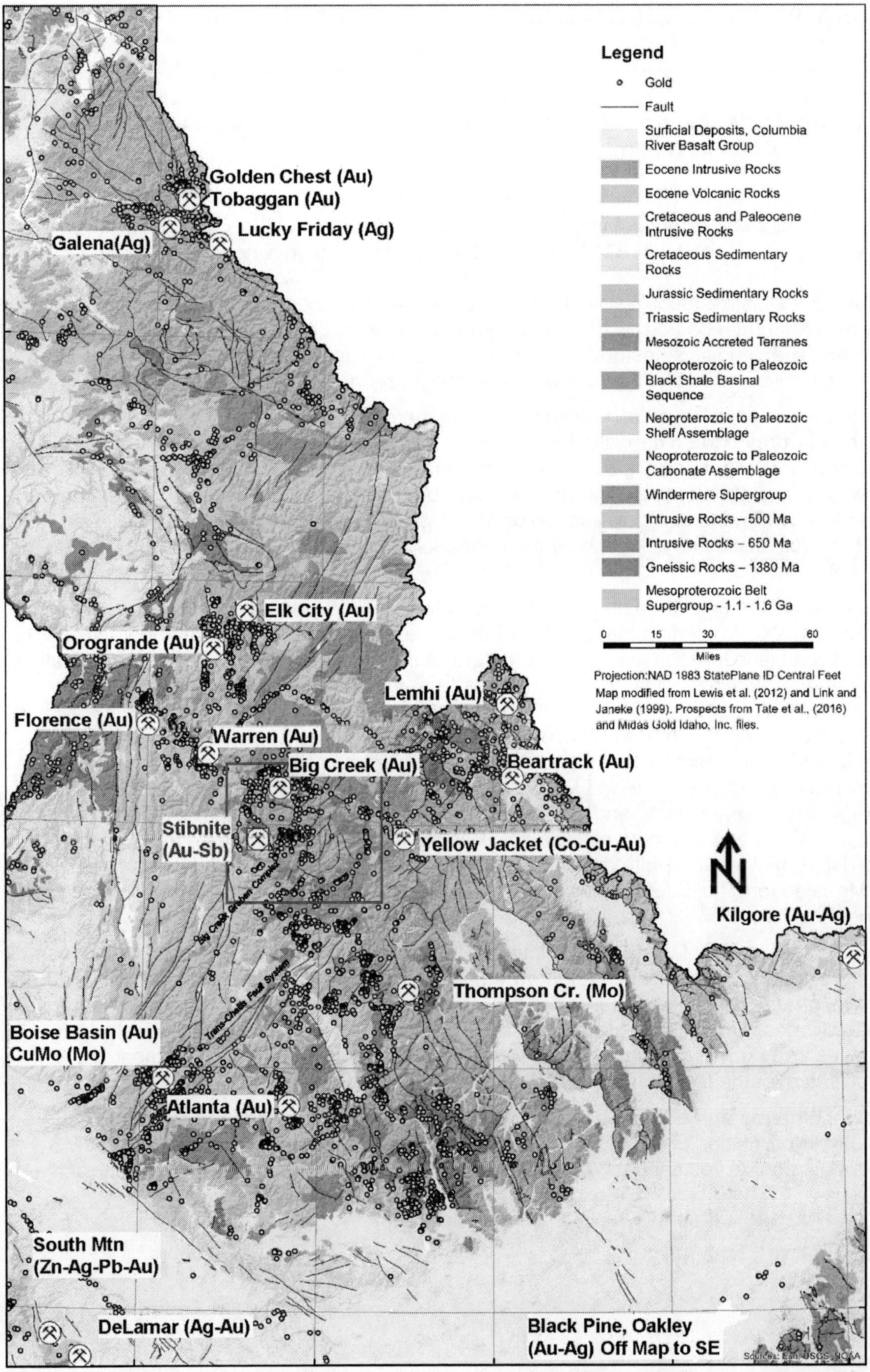

Legend
Gold
Fault
Surficial Deposits, Columbia River Basalt Group
Eocene Intrusive Rocks
Eocene Volcanic Rocks
Cretaceous and Paleocene Intrusive Rocks
Cretaceous Sedimentary Rocks
Jurassic Sedimentary Rocks
Triassic Sedimentary Rocks
Mesozoic Accreted Terranes
Neoproterozoic to Paleozoic Black Shale Basinal Sequence
Neoproterozoic to Paleozoic Shelf Assemblage
Neoproterozoic to Paleozoic Carbonate Assemblage
Windermere Supergroup
Intrusive Rocks – 500 Ma
Intrusive Rocks – 650 Ma
Gneissic Rocks – 1380 Ma
Mesoproterozoic Belt Supergroup - 1.1 - 1.6 Ga
0 15 30 60
Miles
Projection:NAD 1983 StatePlane ID Central Feet
Map modified from Lewis et al. (2012) and Link and Janeke (1999). Prospects from Tate et al., (2016) and Midas Gold Idaho, Inc. files.
Golden Chest (Au)
Tobaggan (Au)
Galena(Ag)
Lucky Friday (Ag)
Elk City (Au)
Orogrande (Au)
Lemhi (Au)
Florence (Au)
Warren (Au)
Big Creek (Au)
Beartrack (Au)
Stibnite (Au-Sb)
Yellow Jacket (Co-Cu-Au)
Kilgore (Au-Ag)
Big Creek Graben Complex
Trans-Challis Fault System
Thompson Cr. (Mo)
Boise Basin (Au)
CuMo (Mo)
Atlanta (Au)
South Mtn (Zn-Ag-Pb-Au)
DeLamar (Ag-Au)
Black Pine, Oakley (Au-Ag) Off Map to SE
Sources: Esri, USGS, NOAA
48°0'N
47°0'N
46°0'N
45°0'N
44°0'N
43°0'N
116°0'W
115°0'W
114°0'W
113°0'W
112°0'W

ST.007

Decoding the Extensive Hydrothermal Alteration at the Kudz Ze Kayah Replacement-Style VMS Deposit, Finlayson Lake District, Yukon

Nikola Denisová[1], Stephen J. Piercey[1], Robin Black[2], Robert Burke[2], Neil Martin[3]
1. Department of Earth Sciences, Memorial University of Newfoundland, St. John's, NL, Canada, 2. BMC Minerals Ltd, Vancouver, BC, Canada, 3. BMC (UK) Limited, London, United Kingdom

The Kudz Ze Kayah deposit is a bimodal-felsic, replacement-style volcanogenic massive sulfide (VMS) deposit that contains a mineral resource of 19.1 Mt grading @ 6.6 wt % Zn, 0.9 wt % Cu, 2.0 wt % Pb, 1.4 g/t Au, and 156 g/t Ag. The deposit is hosted by rocks of the Yukon-Tanana terrane in the Finlayson Lake District in south-central Yukon, a district that contains >40 Mt of polymetallic VMS mineralization.

The deposit is hosted within Late Devonian volcanosedimentary rocks of the upper Kudz Ze Kayah formation, which are interpreted to be deposited in a continental back-arc environment. Volcaniclastic rocks (tuffs, crystal tuffs, and lapilli tuffs) and argillites are interbedded, and coherent volcanic rocks (mafic sills and felsic sills, flows and cryptodomes) are interbedded with or emplaced within the volcanosedimentary pile. Replacement-style VMS mineralization and associated hydrothermal alteration were facilitated by rapid deposition rates of the volcaniclastic rocks and by the contrasts in porosity and permeability between the volcaniclastic and coherent rocks.

Drill core and outcrop observations in the deposit area indicate that VMS-related hydrothermal alteration extends several kilometres beyond the mineralization. Pervasive white mica and chlorite alteration occur proximal to the massive sulfide lenses (within 10s of meters) both in the hanging wall and the footwall, and zones of strong and moderate white mica alteration ± chlorite-carbonate extend laterally and into the hanging wall and footwall stratigraphy (100s - 1,000s of meters). Alteration indices in lithogeochemical data (e.g., AI, CCPI, Al_2O_3/Na_2O, Ba/Sr), mass balance calculations, and petrographic observations indicate that widespread feldspar destruction (low Na and Ca, mass losses of Na) and formation of white mica (mass gains of K) and chlorite (mass gains of Fe and Mg) were the main alteration processes. Barite is a common gangue mineral within the mineralization, and elevated Ba values occur proximal to the mineralized lenses. Electron microprobe (EMPA) and scanning electron microscopy (SEM) results indicate that Ba is a minor constituent in K-feldspar, biotite, and white micas in addition to barite. Volatile metals in whole rock samples (e.g., Sb, Tl, and Mo) also form halos with elevated values up to 200 - 300 m into the hanging wall and footwall to the mineralization.

Petrographic and EMPA analyses of white mica, biotite, chlorite, and carbonate show that Mg-rich varieties are more common proximal to mineralization and formed earlier in the deposit paragenesis than the Fe-rich varieties. Short wave infrared data for mica and chlorite, however, shows no clear spatial trends across the deposit area. Phyllosilicates with distinct Mg- or Fe-rich compositions are juxtaposed proximal to the massive sulfide lenses and likely formed at different temperatures; this is interpreted to indicate multiple overprinting fluid pulses over the lifetime of the hydrothermal system.

Integration of fieldwork, petrography, lithogeochemistry, and EMPA-SEM methods offers unique opportunities to study the evolution of an ancient hydrothermal system and a replacement-style VMS deposit; the results will improve vectoring towards VMS mineralization in similar, extensively altered environments.

ST.008
Stages in the History of Brines in the Creede, Colorado Hydrothermal System From Laser Microprobe Noble Gas Mass Spectrometry of K, Cl, Br, I, Ar, Kr, and Xe in Fluid Inclusions.

James J. Irwin
Sattva Global Partners, Vancouver, BC, Canada

Previously published Laser Microprobe Noble Gas Mass Spectrometry (LMNGMS) analyses of fluid inclusions (FI) at Creede are evaluated in the context of the extensive data base for halogen and noble gas abundances in rocks, fluids, FI, and the geology of the Creede system published after the FI data. Representative FI from Creede stages B, C, D, the Bondholder district, and the North Amethyst vein were studied by standard micro-thermometric techniques prior to pile irradiation, laser targeting, and mass spectrometry. Creede FI are large, aqueous, primary or pseudosecondary assemblages, nearly ideal for the technique employed. The range and mean Br/Cl in all the FI studied is 1.5- 2.08 x 10^{-3} and 1.84 x 10^{-3} molar, respectively, similar to FI in porphyries and volcanic calderas in the western U.S., but I/Cl in B, C, and D stage FI is an order of magnitude less than in these systems (1 - 4.4 x 10^{-6}). This level of Iodine depletion is caused by removal of I from brine by organic matter, which must have occurred in the Creede moat. B-stage FI contain ^{36}Ar and ^{84}Kr concentrations equal to fresh air saturated water (ASW; 6.7 x 10^{-8} molal ^{36}Ar, ^{84}Kr/^{36}Ar = 0.035). FI in C-stage fluorite from Bulldog Mountain and D-stage sphalerite from the OH vein contain substantially less ^{36}Ar and ^{84}Kr than fresh ASW and B-stage FI (1.5 and 0.74 x 10^{-8} molal ^{36}Ar, ^{84}Kr/^{36}Ar of 0.036 and 0.027 respectively), equal to some liquids in Norris Geyser Basin (NGB) at Yellowstone, a system studied specifically because of its history of boiling. The liquids at NGB that contain the same gas concentrations as the Creede FI are a product of early boiling and subsequent mixing with ASW, processes also known to be associated with mineralization in these stages. Sphalerite from the Bondholder district contains FI with Br/Cl within the range of magmas (1.92 x 10^{-3}) but I/Cl greater than other stages, (1.39 x 10^{-5}), evidence of a different fluid history. ^{36}Ar and ^{84}Kr abundances are much less than in fresh ASW and a factor of ~3 less than in C and D-stage FI (~0.2 x 10^{-8} molal), similar to the lowest gas concentrations measured in the NGB system. The North Amethyst vein contains FI that are mostly vapor. ^{36}Ar and ^{84}Kr in these FI are much higher than any others in this system (^{36}Ar > 50 x 10^{-8} molal), similar to the vapor phase in some modern systems. ^{84}Kr/^{36}Ar is less than ASW (0.03), as expected for the vapor phase of a fluid that boiled. Stages in the history of fluids in this system include the following: 1) some FI originated or had a stage as fresh ASW, 2) halogens in the Creede moat were acquired from a magmatic source, 3) Iodine was removed from these fluids by reaction with organic matter in the moat, 4) boiling of fluids occurred prior to trapping, and 5) some fluids that boiled were subsequently mixed with ASW. A sequence of stages is implied.

ST.009

Zinc Sulfide Mineralization Hosted in Marbles of the Proterozoic Grenville Supergroup, Canada: Implications for Base Metal Exploration in High-Grade Terranes

Hannah Lang[1], Gema Olivo[1], André Tessier[2]
1. Queen's University, Kingston, ON, Canada, 2. Delta Resources Limited, Kingston, ON, Canada

The demand for zinc is expected to exceed supply this decade, unless new deposits are discovered. Over the course of the past century, the success of discovered mineral deposits has decreased, making it more important to expand our knowledge of exploration to cover complex terranes such as those of high metamorphic grade. This study examines marble-hosted zinc in the high-grade metamorphosed Proterozoic Grenville supergroup at the Salerno Lake and Calumet-Sud occurrences of Ontario and Quebec. The purpose of this study is to determine the stratigraphic controls and characteristics of the multiple styles of mineralization present at these two locations in comparison to other local sediment-hosted zinc deposits. The carbonate sequences of the southeastern Central Metasedimentary Belt (CMB) in the Grenville Province host several zinc deposits and occurrences (e.g., Balmat-Edwards, Franklin and Sterling Hill, Renprior/Cadieux, Calumet-Sud, Salerno Lake, Long Lake, Maniwaki). Most of the zinc mineralization is hosted in high-grade metamorphosed dolomitic (and locally calcitic) marbles, regionally controlled by the basin-margin faults that border the CMB, containing calc-silicate minerals such as diopside, tremolite, serpentine, phlogopite, and talc. The sulfide zinc deposits have been interpreted to have been formed from exhalative brines during sedimentation of the carbonate sequence (SEDEX deposits). However, the results from detailed mapping, paragenesis, mineral chemistry, and lithogeochemistry at the Salerno Lake (Ontario) and Calumet-Sud (Quebec) occurrences revealed that syenite, carbonatite, and felsic intrusions occur proximal to and may have contributed to the zinc mineralization.

At the Salerno Lake occurrence, yellow (Fe- and Mn-poor), intrusion-proximal sphalerite is associated with apatite displaying elevated concentrations of F, Na_2O, and SO_3, similar to apatite in the intrusive rocks. Moreover, both the proximal hydrothermally altered marble and meta-syenite intrusion contain elevated K_2O, Al_2O, TiO_2, Ba, Sr, Rb, Ce, and Nd. Red sphalerite (Fe- and Cd-richer) occurs as overgrowths on yellow sphalerite in the proximal zones, extends along bands in the distal zones, and is associated with apatite and graphite. Calumet-Sud shares similar mineralogical relationships. It is proposed that the marble-hosted zinc mineralization is epigenetic and had input of fluids that exsolved or interacted with the alkaline magmatic rocks, particularly in the proximal zones. The stratabound controls on sphalerite mineralization are interpreted to be due to the interaction of magmatic and possible basinal mineralizing fluids with favorable reduced-carbonate layers rich in organic matter and phosphate. Amphibolite-facies metamorphism and deformation have caused changes in the textures of the ore, but the compositional signature of the mineralized zones is interpreted to represent that of the original hydrothermal system.

This study brings new insights to understand processes related to zinc mineralization in the Grenville Supergroup, revealing the relevance of the association with intrusive rocks, that should be taken into consideration when exploring in similar tectonic/geological setting. Furthermore, it provides a robust approach integrating mineralogy, lithogeochemistry, and mineral chemistry to document processes that occurred prior to high-grade metamorphic conditions, which could be applied as vectors to exploration in high-grade metamorphosed terranes globally.

ST.012
δ18O Isotope Review of the Irish Midlands and Implications for Models of Dolomitization

Christopher Reed[1], Malcolm Wallace[2], John Ashton[3]
1. Teck Resources, West Perth, WA, Australia, 2. The University of Melbourne, Melbourne, VIC, Australia, 3. Consultant, Navan, Ireland

A regional compilation of available dolomite ^{13}C and ^{18}O isotope data for the Irish Midlands has been undertaken to assess potential regional fluid pathways and histories. The database contains over 700 georeferenced points from public sources and includes samples from drillcore, outcrop, and underground samples from Zn-Pb deposits. The rocks included are the Waulsortian and Pale Beds; here, we present results from just the Waulsortian Limestone. The large spatial distribution afforded by this database allows regional trends to be evaluated over a significant area of the Irish Carboniferous.

A consistent paragenesis has been assigned based on the sample descriptions. This is summarized as an early Mn-Fe-poor planar dolomite, followed by a coarser-grained, grey to white, Mn- and Fe-rich saddle dolomite. The Zn-Pb mineralization typically occurs synchronously with the early planar dolomite and may overlap with the saddle dolomite, which typically occludes the remaining porosity. Based on this, the dolomites were categorized as pre-, syn-, and postmineralization (Reed and Wallace, 2001, 2004).

Previous authors have suggested that the dolomitizing event was driven by a south-to-north directed fluid flow event triggered by the Hercynian Orogeny (Hitzman and Beaty, 1998). This conclusion was supported by a progressive shift in ^{18}O values from lighter in the south to heavier in the north. The transition from light to heavy oxygen isotope compositions is attributed to progressive host-rock buffering as the fluids migrated through the carbonate sequence. Conversely, Gregg et al (2001) argued against a single flow event and documented a more complex history involving multiple fluid generations.

As can be seen in this more extensive dataset (figure 1), a south-to-north trend is not evident. Significantly, the data do not display any significant deviations in values in samples close to or associated with Zn-Pb mineralization. The transition to more depleted ^{18}O values in the later diagenetic phase (saddle dolomite) is consistent with increasing burial temperature.

We suggest that dolomitization is a diagenetic event from a pore fluid that was uniformly buffered by the host limestone as burial temperatures increased, and there is no evidence for a north-directed fluid flow event. Instead, it's possible to envisage more localized fluid sources that are not dependent on a single distal source or hydrothermal fluid. This has implications for models for the Zn-Pb mineralization as it does not require a single fluid flow event.

FIGURE 1

$\delta^{18}O$ values (n = 405) for the syn-mineral dolomites within the Waulsortian Limestone (y-axis). X-axis is northing in Irish National Grid (metres). Data has been summarized as box plots with 30-km bin size. Black dots are the outliers. Jitter applied to points for visualization.

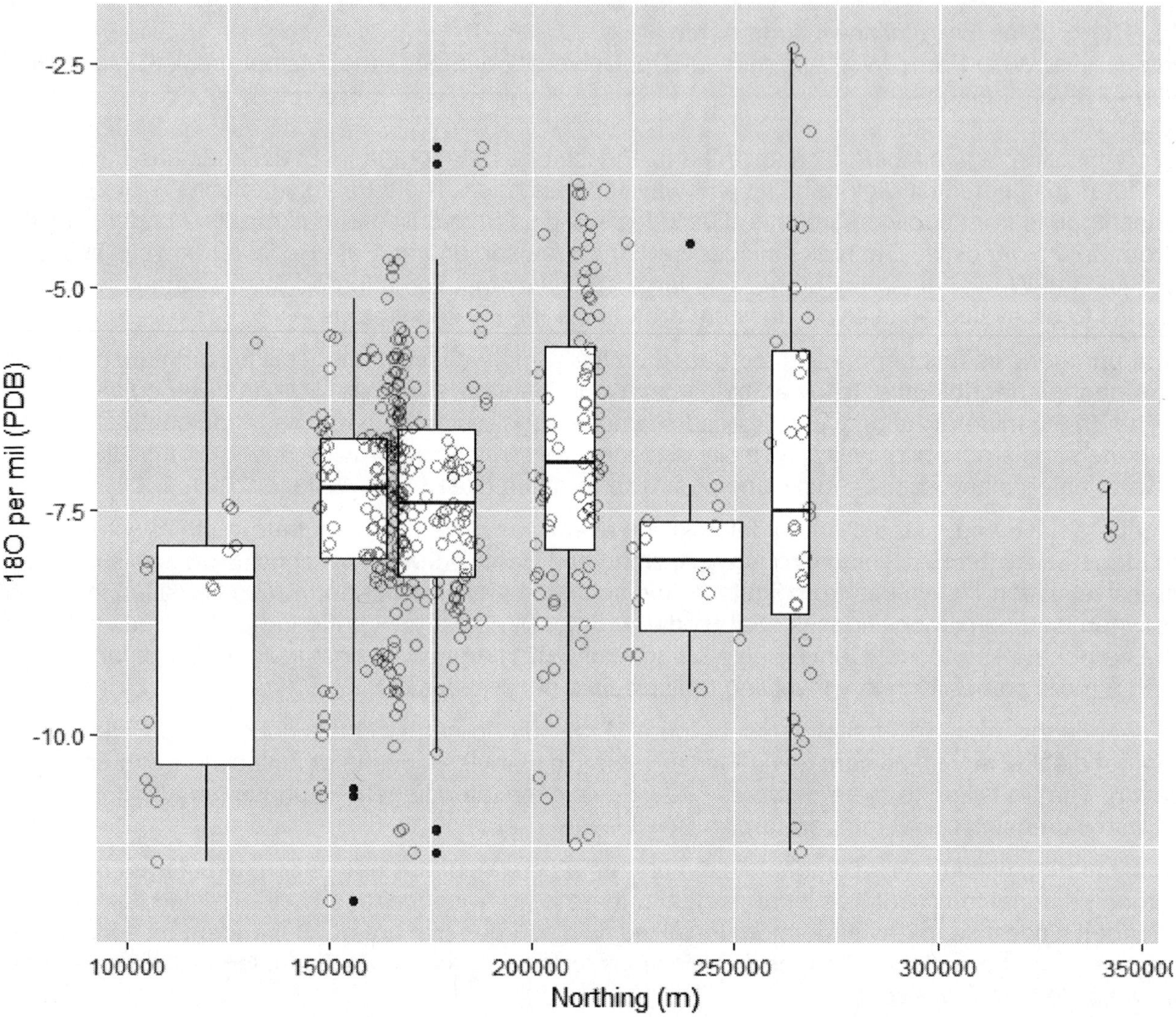

ST.015
Pb-Pb Sulfide Age Constraints on the VHMS Zn-Pb Deposit of the Mundo Novo Greenstone Belt, NE Brazil, and the Metallogenic Implications Along the Contendas-Jacobina Lineament

Ricardo R. Spreafico
Mining Company of Bahia State, Salvador, BA, Brazil

Thermal ionization mass spectrometry (TIMS) Pb-Pb geochronologic data allow the determination of the timing and genesis of the volcanic-hosted massive sulfide (VHMS) Zn-Pb deposit located at Coqueiro Farm and hosted in the Neoarchean Mundo Novo greenstone belt (MNGB), eastern São Francisco Craton, NE Brazil. The mineral deposit is inserted in the Rhyacian-Orosirian Contendas-Jacobina Lineament, which extends approximately 300 km at a north-south trend between Paleoarchean tectonic blocks, causing a broad metallogenic interest along this regional structural feature, in which is also located other important Neoarchean terrain named the Contendas-Mirante metavolcano-sedimentary belt (CMB). The basement of the MNGB comprises tonalite-trondhjemite-granodiorite orthogneisses, migmatites, and subordinate Paleoarchean metarhyolites and metagranites tectonically emplaced between supracrustal rocks. The metavolcanosedimentary rocks of the VHMS deposit were initially formed in an ocean floor setting and comprise the western metabasalts, calc-silicate rocks, aluminous schists, metacherts, banded iron formations and tremolitites of the middle sequence, and the metasedimentary siliciclastic rocks of the upper sequence of the MNGB. The western metabasalts host two hydrothermal alteration zones: one carbonate (calc-silicate rock), proximal, hosting massive sulfides (8 m thick) composed mainly of sphalerite and galena with minor chalcopyrite; and another sericite-chlorite (aluminous schists), distal, hosting mainly disseminated chalcopyrite and traces of galena and sphalerite. Pb-Pb galena, chalcopyrite, and sphalerite data on the massive and disseminated zones in the Coqueiro Farm deposit yield model ages of 2804 ± 11.15, 2794 ± 11.2, and 2767 ± 11.1 Ma, respectively, and the Pb contained in these minerals was sourced from the upper crust based on uranogenic and thorogenic diagrams. The Pb-Pb isochron crystallization age of 2747 ± 16 Ma obtained from the chalcopyrite and sphalerite samples from the massive and disseminated zones suggests that the sulfides are coeval and were not strongly affected by later metamorphic-hydrothermal events. The crystallization age of the sulfides is also in agreement with the previous Neoarchean crystallization age obtained from MNGB volcanic rocks. Therefore, the VHMS deposit in the MNGB would have formed from Neoarchean ocean floor volcanic exhalative processes, and the Rhyacian-Orosirian tectonic event compressed the deposit between Paleoarchean blocks along the Contendas-Jacobina Lineament, preserving the sulfides from remobilization processes. Evidence of Neoarchean hydrothermal alteration zones coeval with ocean floor exhalative processes, all amalgamated along the Rhyacian-Orosirian Contendas-Jacobina Lineament, are important rules for the discovery of new settings that host VHMS-type deposits similar to those described in the MNGB. In this sense, the new metallogenic perspectives opened along the Contendas-Jacobina Lineament extension include the Neoarchean volcanic lithological associations formerly described in the CMB in addition to other similar and cogenetic settings that eventually may be discovered along the lineament based on the new exploratory parameters resulting from MNGB studies.

Fig. 1. A) São Francisco Craton, NE Brazil. B) MNGB and CMB in the Contendas-Jacobina Lineament and the structural lineaments of the eastern São Francisco Craton. C) Geologic setting in which the MNGB and VHMS deposit are inserted and the Paleoarchean blocks of the eastern São Francisco Craton.

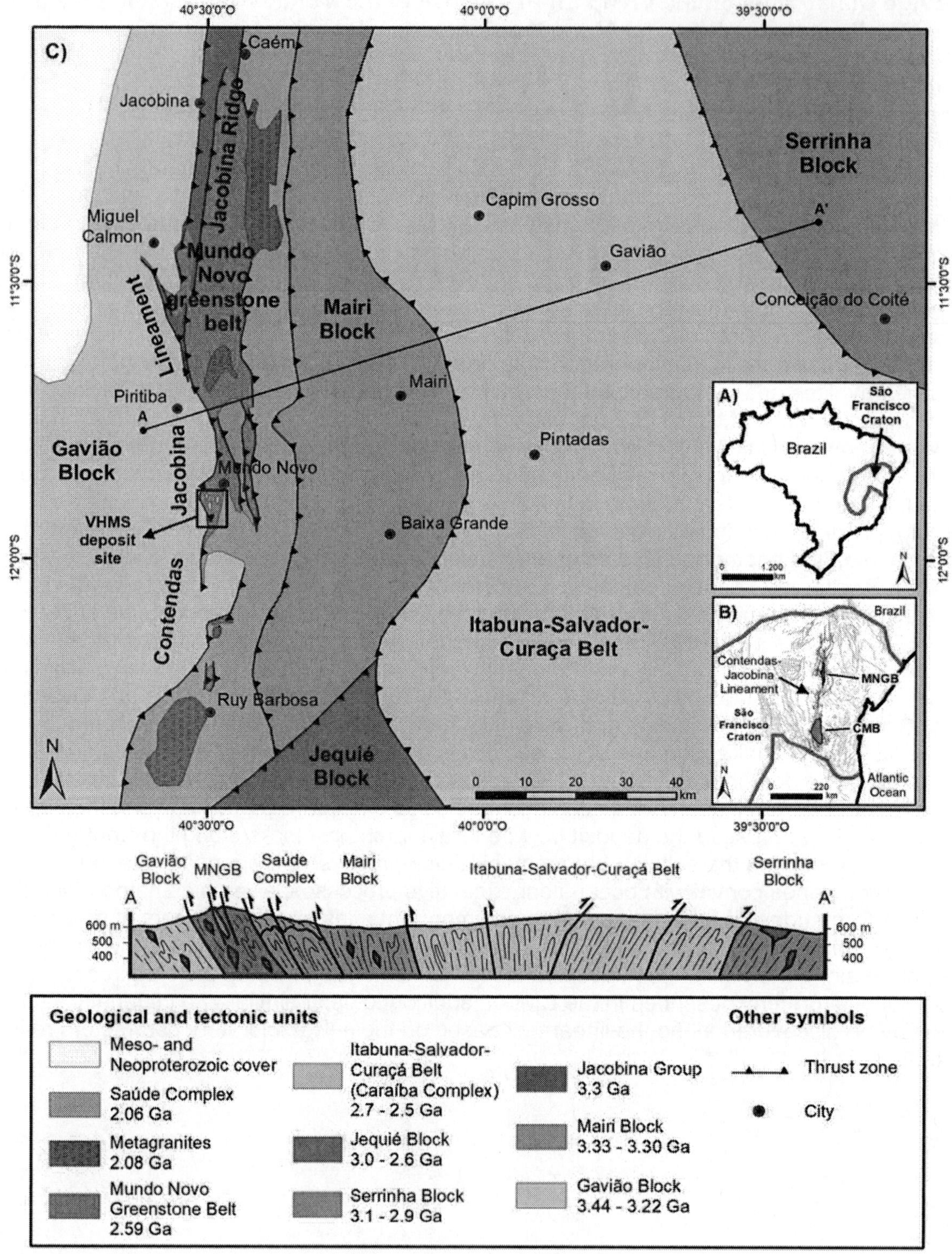

ST.016
Fault-Controlled Stratigraphic Architecture of the Fishtie Cu-Co Deposit in Zambia

Subaru Tsuruoka, Murray Hitzman, Koen Torremans
Irish Centre for Research in Applied Geosciences (iCRAG), University College Dublin, Dublin, Ireland

Cu-Co deposits of the Central African Copper Belt are hosted in rocks of the Neoproterozoic Katangan Supergroup, which comprises a rift-phase clastic sequence (Lower Roan Group), a platformal carbonate-rich sequence (Upper Roan Group), and a shallow marine-origin clastic sequence (Mwasha Subgroup, Nguba, and Kundelungu groups). In the Fishtie area, to the southeast of the Zambian Copperbelt, the Lower Roan and Upper Roan sequences are, at least locally, absent, and Cu-Co sulfides are hosted in a ~300-m-thick succession of Nguba Group rocks. The stratigraphy consists of Grand Conglomérat, which is interpreted as a Sturtian-age glacial diamictite, overlying Kaponda Formation siltstone, and Kakontwe Formation dolostone. The area has undergone significant deep weathering.

The Grand Conglomérat (Mwale Formation) at Fishtie is comprised of five diamictite lithofacies that show distinct clast sizes and two siltstone lithofacies with distinctive sedimentary textures. These lithofacies are in turn classified into five facies associations (FAs) based on stratigraphic location, occurrence of iron formation, and Ti-Zr geochemical facies. These five FAs are interpreted as five discrete flow units of diamictite, each of which is capped by a siltstone layer. The two lowermost flow units (GC1 and GC2) are present only in the northeastern portion of the current drilling area. The stratigraphically higher units (GC3, GC4, and GC5) occur across the drilled Fishtie area. The Grand Conglomérat commonly grades into overlying calcareous Kaponda Formation siltstone. The Kaponda Formation siltstone is typically <5 m thick and carbonaceous. The Kaponda Formation siltstone grades upwards into the bedded silty dolostone of the lowermost Kakontwe Formation dolostone. The Kakontwe Formation at Fishtie is comprised of four lithofacies including silty dolostone, massive dolostone, carbonaceous siltstone, and laminated dolostone.

Individual units of the Mwale and Kakontwe formations display significant thickness variations across the Fishtie area. Thickness variations suggest that deposition of the Mwale and Kakontwe formations occurred within accommodation space formed locally in the hanging walls of three E-W-trending, segmented, high-angle, syn-sedimentary normal fault arrays and a NE-SW-trending normal fault.

Sulfides occur within the basement rocks, the Mwale Formation, and the overlying Nguba Group rocks. Chalcopyrite and bornite are the dominant sulfides and occur as disseminations and in dolomite-quartz veins. Minor chalcocite, pyrite, and sphalerite are also present. Higher Cu grades occur in proximity to the interpreted faults and in carbonaceous units of the Kaponda and Kakontwe formations. Authigenic pyrite in carbonaceous units was replaced and overgrown by the Cu sulfides. Biotite, chlorite, and iron carbonate are spatially associated with the Cu sulfides. Cobalt occurs in cobaltite as disseminations and in characteristic mosaic-textured dolomite-quartz veins. The fact that the highest grade zones of Cu and Co do not occur together suggests Cu and Co mineralization were not coincidental. Although no systematic change in alteration mineral assemblage is currently recognized with respect to distance from the faults, the observations suggest that the syn-sedimentary faults served as conduits for the Cu-Co mineralizing fluids and the carbonaceous units served as reductants.

ST.017
Interpreting Basin Processes at Local and District Scales Using Aeromagnetic Surveys Combined with Soil and Drill Hole Geochemistry

Helen Twigg, Murray Hitzman
iCRAG, University College Dublin, Dublin, Ireland

The Neoproterozoic Central African Copperbelt located in southern region of the Democratic Republic of Congo (DRC) and the northwestern Zambia contains 48% of the world's cobalt reserves and significant resources of copper, zinc, nickel, and gold. The energy-critical elements are hosted in the Katanga Supergroup carbonate and siliciclastic metasedimentary rocks which were deposited in an epicontinental basin formed during the breakup of Rhodinia and subsequently inverted. Multiphase hypogene mineralization in the Katangan basin occurred over a time frame of >300 million years linked to basin rift and inversion events and salt evacuation and dissolution.

This study uses early-stage exploration data (soil geochemistry and airborne geophysical survey) to understand sedimentary basin processes and target first pass drilling more effectively. Integrated multielement geochemistry and high-resolution airborne magnetic surveys are used as a mapping tool. The syn- and postsalt sediments are divided into prehalokinematic, synhalokinematic, and posthalokinematic strata by tracking unconformities observed in the aeromagnetic data. Chemo marker horizons delineated from soil geochemistry surveys further refine stratigraphic mapping. In addition to mapping stratigraphic units and lithology, it is possible to detect facies changes. This data can be used to interpret the structural development of local sub-basins created by rifting and/or halokinematics.

The mapping technique was developed in the southern Congolese Copperbelt terrane, which covers the Mwashya Subgroup and the Nguba and Kundelungu groups. The study area has poor surface exposure, but is covered by high-resolution geophysical surveys, stratigraphically extensive drill hole coverage, and robust multielement sampling of drill holes and soil surveys. Geochemical and lithofacies logging of drill holes provides "ground truthing" for the soil geochemical data sets and allows documentation of facies and sediment provenance changes during sedimentation of the original sedimentary sequence. This method has been applied to a second location in the Central Copperbelt terrane, covered by lower-resolution geophysical surveys, a widely spaced soil geochemical survey, and extensive drill hole coverage of the Mines Subgroup but limited drill hole intersections outside the Roan Group.

Camp-scale, district-scale, and potential basin-wide processes are interpreted from these data. In the Southern Congolese Copperbelt, local, small-scale (<5-km-long) unconformities can be seen nucleating from discordant megabreccia bodies after evaporites which are identified as highly contrasting and chaotic zones of extreme low and high magnetic response and interpreted as salt diapirs. District scale unconformities at the base of the Katete Formation and the Kundelungu Group are coincident with changes in sediment provenance signals identified in soil geochemistry. Stratigraphic geochemical signatures are comparable between study locations in the central Congolese and the southern Congolese Copperbelt. Comparing the stratigraphic location of unconformities at different locations in the Katangan basin will aid in unraveling the relative influence of eustatic sea level changes associated with the Marinoan and Sturtian glaciations versus basin halokinematic, rift, or inversion processes, which can reroute sediment supply.

ST.018

The Geodynamics and Ore-Forming Process of the Makeng-Type Deposits in Southwestern Fujian Depression Belt, South China

Absai Vatuva[1, 2], Da Zhang[2], Hai-Bin Feng[2], Yuan Yuan[2], Sen Wang[2]
1. University of Namibia, Keetmanshop, Namibia, 2. China University of Geosciences, Beijing, China

The Southwestern Fujian Depression Belt (SFDB) is located in the southeast margin of South China Block. Approximately 98% of the Fe deposits in Fujian Province are hosted in this belt, such as the Makeng, Dapai, Pantian, Luoyang, Yangshan, Tangquan, and Zhongjia deposits. An association of iron, lead, zinc, copper, and molybdenum, hosted by the Late Paleozoic carbonate and clastic rocks intruded by Mesozoic granites, characterizes the deposits. They are collectively named Makeng-type deposits, after Makeng, the largest Fe polymetallic deposits in the SFDB. There are divergent views among scholars about the genesis and the role of granitoids on mineralization. We combine data from previous scholars on the Makeng type deposits with our Rb-Sr and U-Pb isotope geochronology and S-O-Pb isotopes studies of the Dapai Fe-Pb-Zn-Cu-Mo deposit to understand the geodynamics and ore-forming processes in the SFDB. The Dapai Fe-Pb-Zn-Cu-Mo deposit orebodies are stratabound; it has a visible metal zonation that comprises Pb-Zn ores above Fe ores. Cu ores are present between the Pb-Zn and Fe ores but are often associated with Pb-Zn mineralization. The minor Mo ores are sporadically disseminated throughout the deposit.

Pyrite and sphalerite grains from the Pb-Zn-Cu ore bodies yield a well-constrained ^{87}Rb-^{86}Sr isochron age of 175.5 ± 3.3 Ma. A porphyritic granodiorite in the Dapai ore district yielded U-Pb zircon ages of 131.72 ± 0.41 Ma while granodiorite yielded 146.6 ± 2.3 Ma. Five porphyritic granodiorite zircon grains yielded a ^{206}Pb / ^{238}U model age cluster of about 150 Ma. Furthermore, six granodiorite zircon grains yielded a ^{206}Pb / ^{238}U model age clustering around 180 Ma. The 175.5 ± 3.3 and 180 Ma ages represent a Pb-Zn mineralization period. The 131.72 ± 0.41 and 146.6 ± 2.3 Ma ages are related to the Mo and Fe periods of mineralization, respectively; similar ages were reported in Pantian, Luoyang, Yangshan, and Makeng. The $\delta^{34}S_{CDT}$ values of galena, pyrite, and sphalerite (from -2.6 to 1.5‰) and their $^{206}Pb/^{204}Pb$, $^{207}Pb/^{204}Pb$, and $^{208}Pb/^{204}Pb$ ratios (18.486-18.537, 15.665-15.712, and 38.823-38.979‰, respectively) indicate that the source of sulfur and metals for Pb-Zn-Cu ore is mainly crust-derived magmas slightly mixed with mantle magmas. The $\delta^{18}O$ values of magnetite grains (average value of 3.53‰) and garnet grains (average value of 4.8‰), intimately associated with magnetite, suggest that the Fe metals and fluid are related to magma mixing. The Re contents of the Mo orebody coupled with the geochemical signatures of the ~130 Ma old granite suggest that the significant source of fluids and metals of the Molybdenite body is mainly crustal magmas.

Three periods of mineralization are recognized based on the deposit characteristics, microscope observation, and robust isotopic ages (Pb-Zn-Cu: 180-160, Fe: 150-135, and Mo: 135-125 Ma). We propose a multi-stage hydrothermal metallogenetic model for the Makeng-type deposits. Different metal associations may have been formed by multiple stages of magmatism that were stimulated by compression, transformation, and extension tectonics widely accepted in the area. The system is driven by the subduction and rollback of the paleo-Pacific plate under South China.

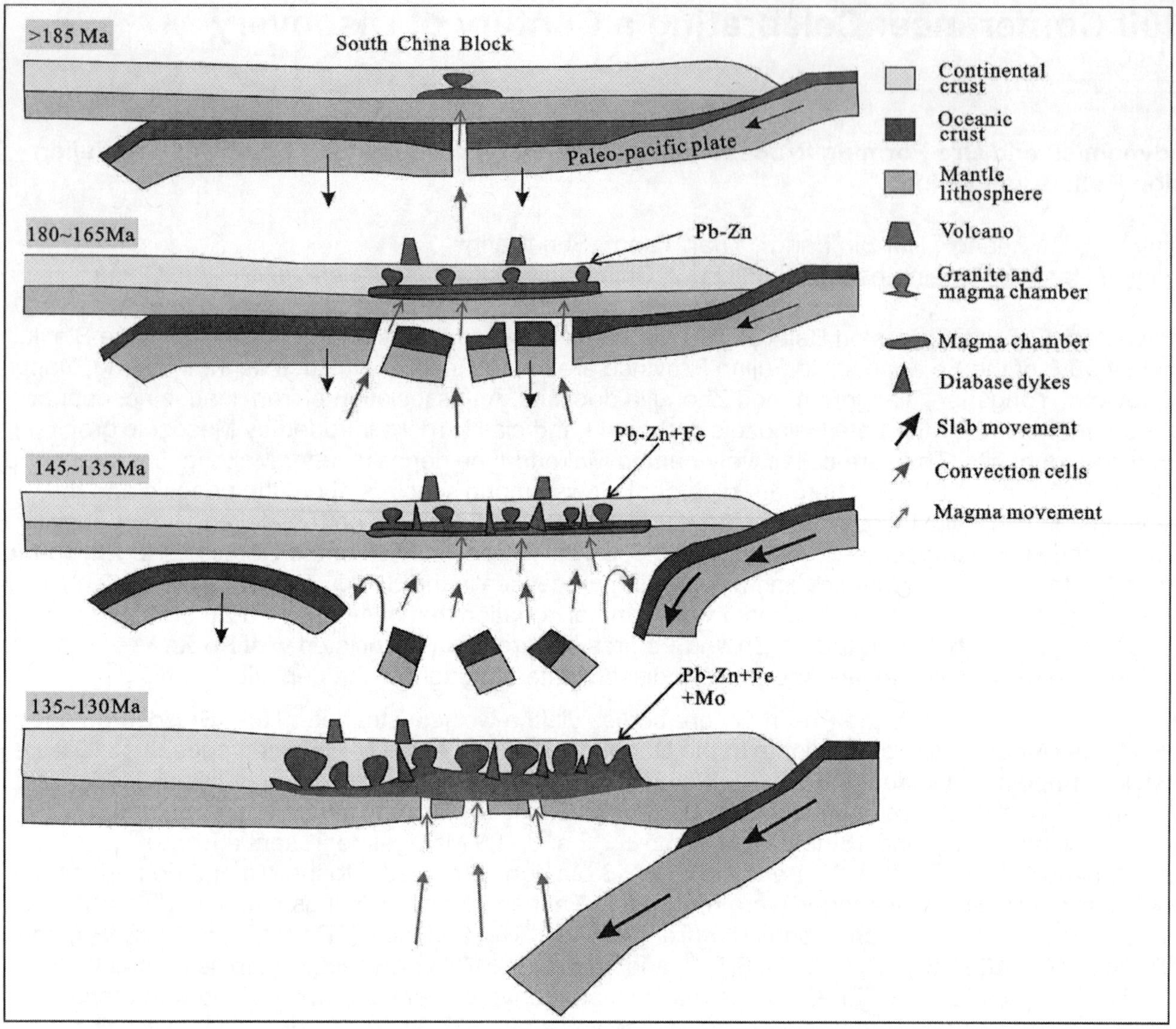

>185 Ma
South China Block
Paleo-pacific plate
180~165Ma
Pb-Zn
145~135Ma
Pb-Zn+Fe
135~130Ma
Pb-Zn+Fe
+Mo
Continental crust
Oceanic crust
Mantle lithosphere
Volcano
Granite and magma chamber
Magma chamber
Diabase dykes
Slab movement
Convection cells
Magma movement

ST.019
Structural and Geometric Evaluation of the Tom Deposit, Selwyn Basin, Yukon

Quinton Willms[1], Kenneth Hickey[1], Jack Milton[2]
1. The University of British Columbia, Department of Earth, Ocean & Atmospheric Sciences (EOAS), Vancouver, BC, Canada, 2. Fireweed Zinc Ltd., Vancouver, BC, Canada

Stratiform Pb-Zn orebodies that make up shale-hosted massive sulphide (SHMS) deposits account for 40% of Zn and 60% of Pb reserves globally. SHMS deposits form at shallow depths and low temperatures in reduced sedimentary basins from deep convection of hydrothermal fluid within the host sedimentary basin. Fluid circulation is dependent on active faults to provide vertical permeability across sedimentary layering. Defining the fault architecture of a basin is a critical step in reconstructing basin geometry and identifying prospective areas for SHMS mineralization along these fluid pathways and their respective trap sites.

The Selwyn Basin in Yukon, Canada, comprises Neoproterozoic to earliest Mississippian passive margin clastic sedimentary strata deposited along the continental margin of Laurentia. From the Late Devonian, the Laurentian passive margin transitioned into a convergent margin with eastward-directed subduction. Subduction was accompanied by back-arc extension and the formation of back-arc basins into which Devonian to Mississippian sediments were deposited. The Macmillan Pass Tom and Jason SHMS deposits are hosted in one such Mid-Devonian to Mississippian extensional sub-basins where mineralization occurred along syngenetic fault structures. Continued subduction and plate convergence through the Jurassic and Cretaceous resulted in the Cordilleran orogeny that deformed the earlier SHMS deposits. Within the Selwyn basin, the Cordilleran orogeny is manifest as a foreland fold and thrust belt that is up to 130 km wide. The Macmillan Pass SHMS deposits occur at the eastern edge of this belt, west of a transition into a fold and thrust belt in Mackenzie platform strata. These deposits are hosted in Devonian strata that have been extensively folded by mid-Cretaceous crustal shortening with deformation manifest as thrust duplexes in the northern portion of the Macmillan Pass area and up to 3 phases of upright folds in its central portion. This study's principal objective is to decipher the structural evolution of the Macmillan Pass area and evaluate the fault geometry and structure of the Tom deposit within the central portion of Macmillan Pass to identify faults that may have controlled hydrothermal fluid flow and SHMS mineralization. This project intends to improve upon previous local mapping in the Macmillan Pass area and advance the structural understanding of the Cordilleran deformation's effect on SHMS mineralization in the Selwyn Basin. Evaluating the Macmillan Pass area's structural controls and identifying fault geometry and kinematic histories further define the structural geometry, spatial variation in strain accommodation, overprinting relationships, and current state of SHMS mineralization seen at the Tom Deposit. This would provide an important constraint on the regional distribution of SHMS deposits within the Selwyn Basin. Analysis of the Cordilleran overprint will assess the extent to which deformation has affected the geometry and grade distribution of SHMS deposits at Macmillan Pass and will bring an understanding to the degree to which the sulphides, their siliciclastic host rocks, and ore-controlling structures have been overprinted and/or reactivated by Cordilleran deformation.

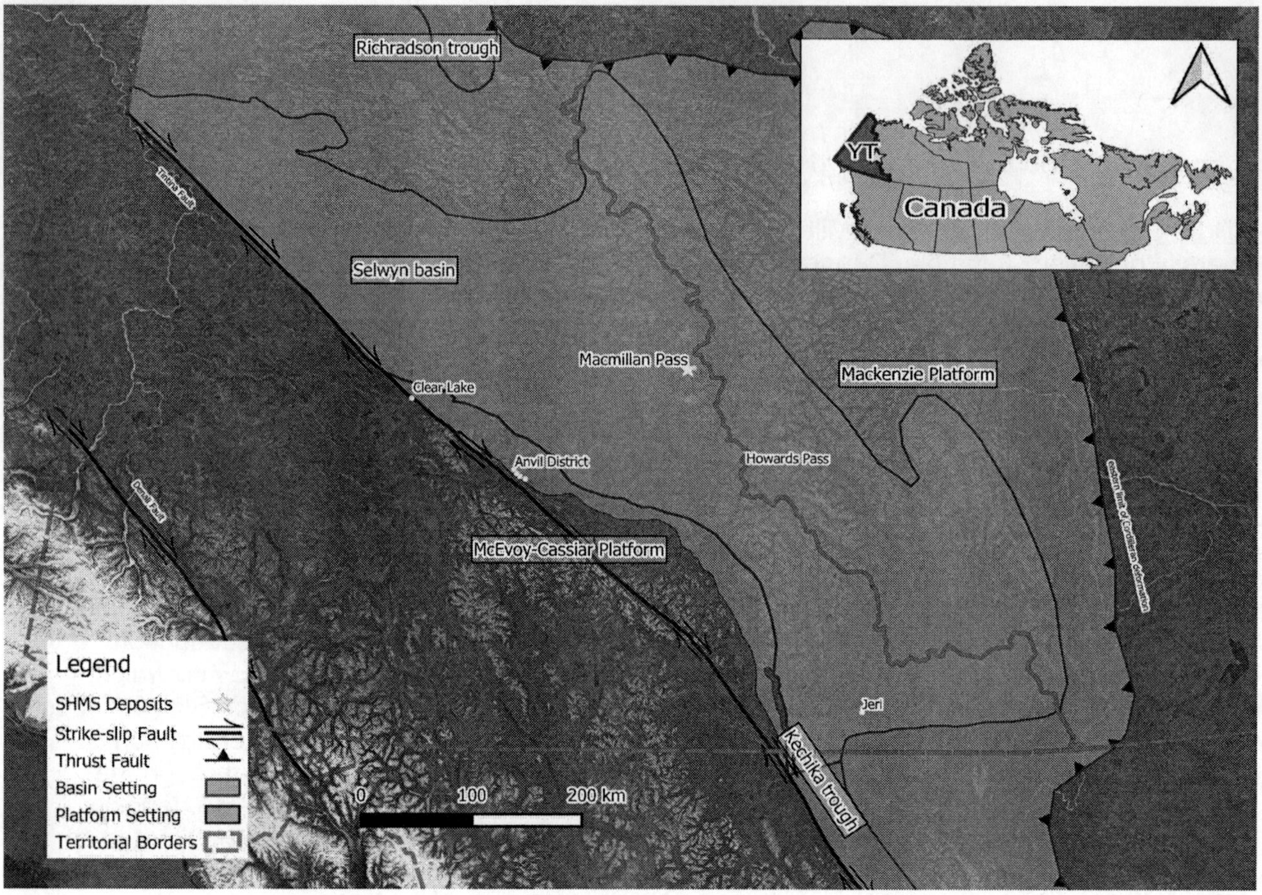
Richradson trough
Tintina Fault
Denali Fault
Selwyn basin
Macmillan Pass
Clear Lake
Mackenzie Platform
Anvil District
Howards Pass
eastern limit of Cordilleran deformation
McEvoy-Cassiar Platform
Jeri
Keckika trough
YT
Canada
Legend
SHMS Deposits
Strike-slip Fault
Thrust Fault
Basin Setting
Platform Setting
Territorial Borders
0
100
200 km

ST.037
The Copper Kitchen: Observing the Generation of a Cu-Bearing Brine in the Midcontinent Rift System, USA

Simon Jones[1], Tony Prave[2], Tim Raub[2], Jonathan Cloutier[3]
1. First Quantum Minerals, West Perth, WA, Australia, 2. School of Earth and Environmental Sciences, University of St Andrews, St Andrews, United Kingdom, 3. CODES, University of Tasmania, Hobart, TAS, Australia

Redbed sandstones are widely recognized as the principal source of metals in low-temperature sediment-hosted stratiform copper deposits. They form in a variety of depositional settings and commonly contain a large component of detrital feldspar or volcanic lithic fragments. Commencing upon surface exposure and continuing during burial, oxidized pore fluids interact with sediments causing dissolution of labile minerals, and the development of characteristic iron-oxide grain coatings. This oxidation process liberates Cu, Ag, Co and other metals into formation waters. Migration and focused flow of these 'pregnant' fluids to a reductant (e.g. shale or hydrocarbon trap) induces copper sulfide precipitation. A less well understood aspect of redbed leaching is the conditions, pathway and rate at which minerals decompose and release their metals. This creates an obvious impediment to evaluating the likelihood that a volume of redbeds has released metals into pore fluids, and thus achieved the first step in a chain of events required to form sediment-hosted stratiform copper mineralization. Here, we present a case study from the 1.1 Ga Midcontinent Rift System, Michigan, USA, of the metal-leaching process responsible for the formation of the White Pine deposit (~250 Mt @ 1.1 % Cu). The Copper Harbor Formation redbed is a sequence of conglomerates and sandstones containing ~50–95% basaltic detrital grains derived from the underlying Portage Lake Volcanics. The depositional setting was a marine-influenced braided fluvial-sabkha environment. Differential exhumation of the Copper Harbor Formation along a 120 km segment the Keweenaw Fault provides a window into diagenesis at depths ranging from ~3-7 km. We demonstrate the onset of metal leaching began in the surface environment with the development of leached weathering rinds and continued during eodiagenesis (~50-100°C, 1-4 km depth) with the partial replacement of rhyolitic and basaltic lithic fragments by beidellite, montmorillonite and interlayered illite-smectite. Dissolution of volcanic grains accelerated rapidly in the telodiagenetic zone (>180°C, 6-7 km depth) with the formation of non-expandable illite-smectite and chlorite-smectite, incipient feldspar dissolution and laumontite precipitation. In addition to liberating ore metals, mineral dissolution and precipitation resulted in a continual exchange of components between mineral phases and pore fluids. Dissolution of evaporites (anhydrite and isolated halite) increased pore fluid salinity, progressive de-watering of interlayered clays contributed water, illitization of smectite consumed Al + K and released Si producing quartz overgrowths, and the combined effects of alkali feldspar dissolution and laumontite precipitation probably created pore fluids with an excess of Mg, Na, and K relative to Ca. We conclude that whilst metals in the White Pine deposit were derived from the Copper Harbor Formation, they were predominantly sourced not from the immediate local, but from a "copper kitchen" some 10s of kms to the north and northeast, at greater depths (>7 km), in a process analogous to petroleum maturation in an "oil kitchen". Characterizing diagenetic reactions such as those outlined here may provide mineral explorers with a means of evaluating the efficiency of metal leaching in a sedimentary basin; thus helping to mitigate a major risk factor in sediment-hosted stratiform copper exploration.

F1
Five Myths of Data Science for Exploration

Kurt House
KoBold Metals, San Francisco, CA, USA

The application of statistical and scientific computing to all industrial sectors and areas of study has accelerated appreciably over the last decade. Mining companies have been heavily investing in data science to improve upstream and processing operations. Yet, to date, mineral exploration activities have received proportionally less data science investment. KoBold Metals is a full-stack battery metals exploration company developing and deploying machine learning and other scientific computing techniques to improve the efficacy and efficiency of exploration programs. In this talk, we'll address five common misconceptions about the application of data science to mineral exploration: (1) there isn't sufficient data, (2) prospectivity mapping is the principal output, (3) machine learning is a black box, (4) data science should be mostly applied for drill targeting, and (5) there's a tension between physical models and statistical models.

F2

What Can Machine Learning Do for Me? Improved Targeting, 3D Modelling, and Resource Estimation: A Case Study from the Castelo de Sonhos Paleoplacer Gold Deposit, Brazil

Britt Bluemel[1], Vivien Janvier[1], Shawn Hood[1], Mo Srivastava[2]
1. GoldSpot Discoveries Corp., Montreal, QC, Canada, 2. Tristar Gold Inc., Toronto, ON, Canada

As we rapidly increase the speed and volume of data collection in mineral exploration campaigns, mines, and near-mine environments, our data processing methods must become more sophisticated to efficiently deal with this data deluge. Machine learning workflows and machine-assisted modeling will become routine parts of the mining value chain; integrating traditional geoscience knowledge with data science expertise will be critical to successful algorithm deployment.

The most important components of any successful mineral exploration campaign are a robust, reliable geological map and a serviceable geological model. In this case study, the chemostratigraphy of the Esperança South paleoplacer gold deposit of the Castelo De Sonhos (CDS) plateau, Brazil, has been successfully delineated using machine-assisted modeling of geochemical data.

The dominant lithologies at CDS were previously subdivided into four informal units based on granulometry: the lower coarsening-upwards sandstone and pebble conglomerate unit, the middle cobble conglomerate unit, the upper fining-upwards sandstone and pebble conglomerate unit and the top sandstone-dominated unit. The challenge faced when attempting to accurately discriminate stratigraphy at CDS is the textural similarity of the metasedimentary units present across the plateau. Unfortunately, continuity between the four logged lithological units could not be confidently interpreted between drill holes plotted on sections or otherwise mapped in a way to better interpret and predict geology in the area.

This issue was addressed by TriStar Gold and GoldSpot Discoveries Corp. by introducing a four-acid digestion geochemical sampling strategy, which enabled previously unattainable chemostratigraphic definition. The new chemostratigraphy includes seven main lithological units, and four subunits. These were validated spatially (using continuity on cross sections and observations from drilling). Results have been used to create a robust surficial geological map, as well as a 3D geological model to inform exploration and resource domain modeling.

As new data are obtained during resource definition drilling, machine learning classification algorithms assign rock types to new data, and the 3D geological model is quickly and efficiently updated. Resource estimation is enhanced both by the inclusion of more accurate geologically based grade domains and dynamic anisotropy models based on the resulting interpreted erosional surfaces (which are geologically relevant to the placer-style gold mineralisation).

Finally, and perhaps most importantly, incorporation of the chemostratigraphic model almost doubled the indicated resources at CDS because the ranges of the variograms could be extended due to better correlation along the stratigraphy. The project-wide indicated resource estimate (Table 1) has increased from 17.7 Mt in September 2018 to 40.1 Mt in April 2021.

References: Srivastava, M. (2020), “Mineral Resources Update for the Castelo de Sonhos Project, Pará State - Brazil”, TriStar Gold National Instrument 43-101 Technical Report.

Table 1. Comparison of current and previous resources estimates for the Castelo de Sonhos Project

	Classification	Tonnage (Mt)	Grade (g/t Au)	Metal Content (Moz)
Current Project Total	Indicated	40.1	1.2	1.5
	Inferred	22.2	1.0	0.7
Previous Project Total	Indicated	17.7	1.2	0.7
	Inferred	39.8	1.0	1.3

F3

Machine Learning for Geochemical Exploration: Classifying Magma Fertility in Arcs and Insights into Porphyry Copper Deposit Formation

Chetan Nathwani[1, 2], Jamie Wilkinson[1, 2], Robin Armstrong[1], Daniel Smith[3], Christian Ihlenfeld[4]
1. Natural History Museum, London, United Kingdom, 2. Imperial College London, London, United Kingdom, 3. University of Leicester, Leicester, United Kingdom, 4. Anglo American, London, United Kingdom

Machine learning algorithms provide a powerful new approach to help in the discovery of ore deposits that are not exposed at the Earth's surface. A current exploration focus is the development of tools, based on large geochemical data sets, that will enable the identification of magmatic districts predisposed to host porphyry copper deposits. In this study, we train and test four common supervised machine learning techniques: logistic regression, support vector machines, artificial neural networks, and Random Forest™ to classify magma fertility in arc rocks using a global data compilation. Several properties of geochemical data sets, such as class imbalance, high multicollinearity, sparsity, missing values, and compositional data effects, preclude the application of traditional statistics and many machine learning techniques. Application of machine learning techniques on geochemical data therefore requires critical preprocessing steps. We evaluate the classification accuracy of each supervised machine learning technique using cross-validation, showing high accuracy for each model. A test data set, comprising three deposits of varying size, metal endowment, and tectonic setting, shows similar classification performance in all cases. We further illustrate the efficacy of our approach by demonstrating that our models outperform those derived from bivariate fertility indicators currently used by the exploration industry. This highlights the value in utilising the high dimension space available in geochemical data for classification problems. Feature analysis of our models reveals that the most important parameters for classification reflect early and abundant fractionation of amphibole and/or garnet and suppression of plagioclase during lower crustal, hydrous magma evolution. We cannot exclude that some signatures may additionally reflect relatively subtle hydrothermal alteration effects. Empirically, our models underscore the potential of using supervised machine learning approaches in mineral exploration, and that the "black box" nature of these techniques can be overcome.

F4

Performance of Predictive Supervised Classification Models of Trace Elements in Magnetite for Mineral Exploration

Émilie Bédard[1, 2], Victor de Bronac de Vazelhes[1, 2], Georges Beaudoin[1, 2]
1. Université Laval, Québec, QC, Canada, 2. Centre E4m - Centre de recherche sur la géologie et l'ingénierie des ressources minérales, Québec, QC, Canada

Magnetite is a reliable indicator mineral for exploration because it records petrogenetic processes and discriminates deposit types. Binary discriminant diagrams have limits to accurately predict deposit types (prediction accuracy ~40%). This study aims to determine the best predictive supervised multivariate classification method using the geochemical composition of magnetite in order to provide a model usable by industry for mineral exploration and by government for resource assessments. After screening a worldwide database of ~30k magnetite analyses, ~17k observations are selected for study. These data are from 303 different deposits that belong to nine major deposit types (BIF, Fe-Ti, IOCG, IOA, Ni-Cu-PGE, porphyry, VMS, skarn, V) and a variety of unmineralized rocks aggregated in the class "country rocks." We tested three most supervised machine learning algorithms (Naive Bayes, K-Nearest Neighbor, Random Forest) with two open-source statistical computation platforms: Orange and R. The Random Forest (RF) algorithm yield the best predictive outcome on untransformed data, with prediction accuracies of 80% with Orange and 81% with R. In R, classification precision is always higher than 70%, reaching 94% for Fe-Ti deposits. We tested our RF model on three case studies: 1) IOCG-like deposits, 2) China porphyry Cu-Au and Cu skarn, and 3) Scandinavian Ni-Cu-PGE deposits. Our model was able to effectively predict the deposit type for the first two case studies to the large families of porphyry/skarn and IOCG/IOA. For the third case study, almost 61% of the observations were correctly identified as belonging to Ni-Cu-PGE deposits. To obtain the best prediction statistically, it is essential to analyze the largest number of magnetite samples possible. The RF model developed in this study is accurate enough to be used with confidence for mineral exploration for these nine deposit types. Future work includes a continuous update of the magnetite chemistry database as well as the development of new models based on other indicator minerals to strengthen model predictions and cover a wider number of deposit types.

Figure: A) Spatial distribution of magnetite data according to the deposit type. B) Prediction accuracies for magnetite composition for three studied algorithms using R.

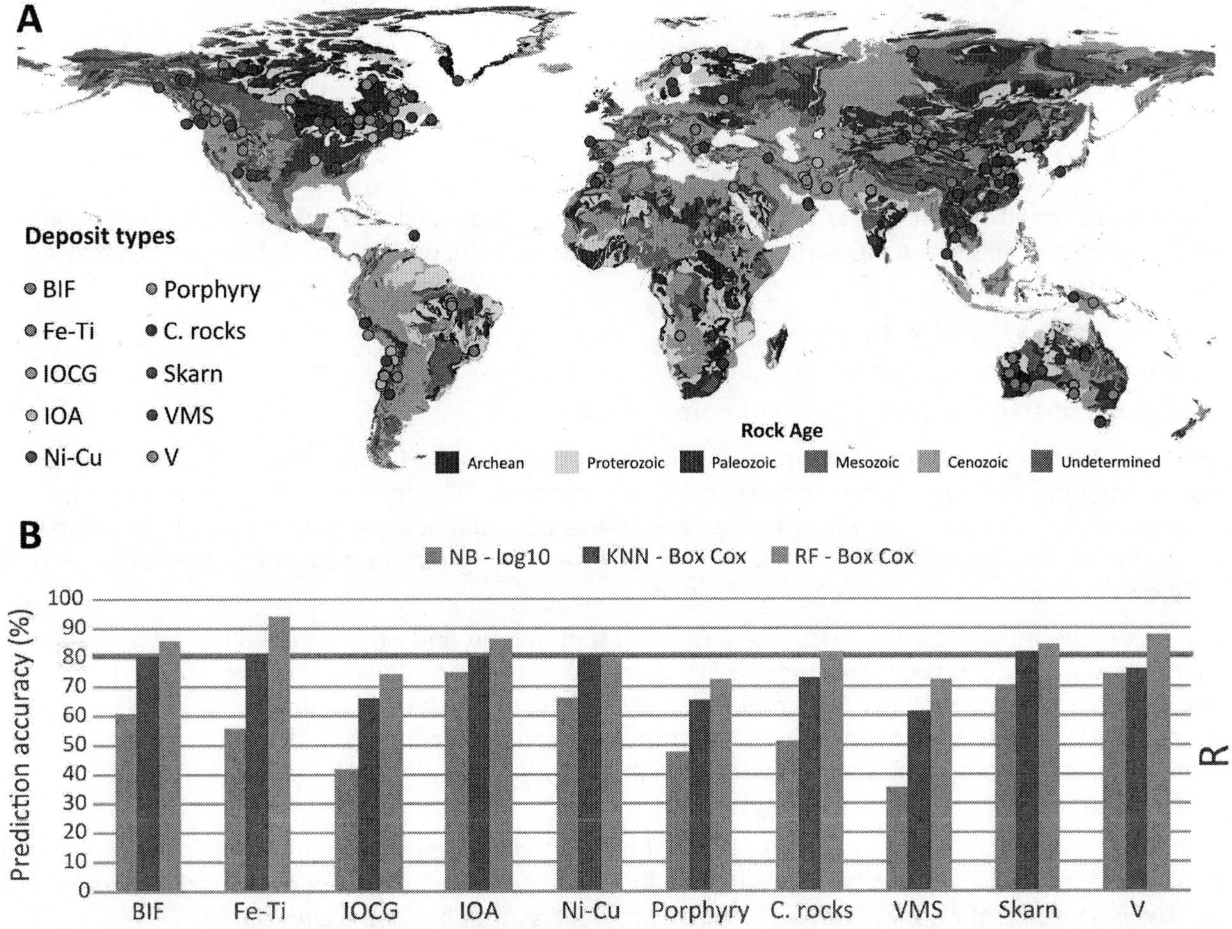

F5
Machine Learning in Earth Science Is Hard; Now What?

Matt Hall
Agile Scientific, Mahone Bay, NS, Canada

Practitioners are discovering that applying machine learning to complex systems like the earth is difficult, but what does that imply about technical communities and earth science education? I think we have three tasks:

- Raise our community's digital literacy level.
- Build a toolkit around open code and open data.
- Grow a culture of open collaboration and entrepreneurship.

We must urgently raise the level of digital literacy in subsurface science and engineering, both among practitioners and students and among the people who train, manage, QC, and regulate. Our community needs formal educational efforts, both at universities and in the workplace, but we also need accessible short courses, tutorials, and hackathons. Technical societies have a big role to play in this retooling effort, if they want it. One or two are already making a difference.

Alongside education, we must grow the ecosystem of open technology and open data with which people are building, training, and evaluating new tools, especially machine learning tools. Awareness and understanding of open-source software, experimental replicability, open data licensing, and almost every aspect of this domain are quite low, especially in the corporate domain (as the continual modification of Equinor's Volve data set licence illustrates). This growth must happen among academics, corporate practitioners, entrepreneurs, and government agencies.

Digital literacy and an open toolkit are necessary but insufficient conditions for a high-functioning community of digital innovators. They also need to be connected with others with similar and complementary goals. We must seek out ways to connect people at every scale, from teams to global communities.

Building on the foundation afforded by the previous three points, we will be able to more effectively deepen our scientific ideas, the developing theoretical framework, and our applied practice in industry. These ideas include

- Novel neural network architectures;
- Ways to develop data-driven models alongside physics-driven models;
- The relative importance of natural and synthetic data in training sets;
- The design of cost functions that reflect what we know about the earth.

Participants at a subsurface hackathon, Copenhagen, 2018.

F6
Metals Fuel the Circular Economy: Process Metallurgy a Key Enabler

Markus Reuter
SMS Group GmbH, Düsseldorf, Germany

Metals lie at the heart of the circular economy as the Metal Wheel explains[1]. Process metallurgy is therefore a key enabler of a circular economy (CE); its digitalization is the metallurgical Internet of Things (m-IoT)[2]. Metallurgy and its reactors are therefore at the heart of an economically functioning CE, particularly also as metals have strong intrinsic recycling potentials and have key enabling functional roles, as these are the fabric of the renewable energy and sustainable societal infrastructures.

The wheels of a sustainable society are based on a multitude of elements as the figure above shows[3], all segments from the processing system, i.e., the carrier metallurgy of steel, aluminum, copper, zinc, lead, tin, etc., are required for a smoothing cycling through the circular economy, not only globally but especially also locally or within a larger economic region. Embracing policy that ensures no segment breaks form the wheel is of utmost importance. The metallurgical process reactors recover the green elements in the segments of each metal processing infrastructure and require a processing segment, called the carrier metal processing infrastructure. The yellow elements are recovered into alloys or intermediate materials/compounds if the economics make sense, and red if the elements are lost in intermediate materials, alloys, or residues (including dissipative losses) and inevitable losses from the system.

This paper will discuss within the context of circular economy, as shown in the figure below, the simulation of metallurgical process reactors that produce and recycle different metals. To maximize the recovery of the elements into appropriate phases in the metallurgical reactors, a deep understanding of the energy and mass transfer between very complex multielement/compound molten (can reach above 2,000°C) and gas phases is required. CFD simulation and dynamic reactor models will show the state of the art based on copper bath smelting (concentrate, scrap, eWaste, batteries, slags, etc.) reactors[4]. Various fundamental parameters effecting mass and energy transfer in the reactors is among others discussed. This is embedded in system simulation models for the CE that quantifies the environmental footprint and exergy dissipation of large CE systems[5]. Of special importance is understanding the downward spiral of the system and, therefore, understanding how to mitigate the exergy dissipation. Large-scale system simulation models provide this detail understanding and will be discussed also in relationship of minerals be these geologically created minerals or the "minerals" called consumer products, energy infrastructure, etc.

References:

1. MA Reuter et al. (2019): Challenges of the Circular Economy - A material, metallurgical and product design perspective. Annual Review of Materials Research, 49, 253-274.
2. MA Reuter (2016): Digitalizing the Circular Economy - Circular Economy Engineering defined by the metallurgical Internet of Things-, 2016 TMS EPD Distinguished Lecture, USA, Metallurgical Transactions B, 47(6), 3194-3220.
3. https://kuleuven.sim2.be/wp-content/uploads/2019/02/SOCRATES-Policy-Brief-2019-Lead.pdf & https://new-mine.eu/wp-content/uploads/2021/02/SOCRATES-Policy-Brief-2020-final.pdf
4. SMS Blue Technologies: https://www.sms-group.com/br/press-media/media/downloads/#item:987
5. NJ Bartie, YL Cobos-Becerra, M Fröhling, R Schlatmann, MA Reuter (2021): The resources, exergetic and environmental footprint of the silicon photovoltaic circular economy: Assessment and opportunities, Environmental Impacts, Resources, Conservation and Recycling, 49 169, 105516

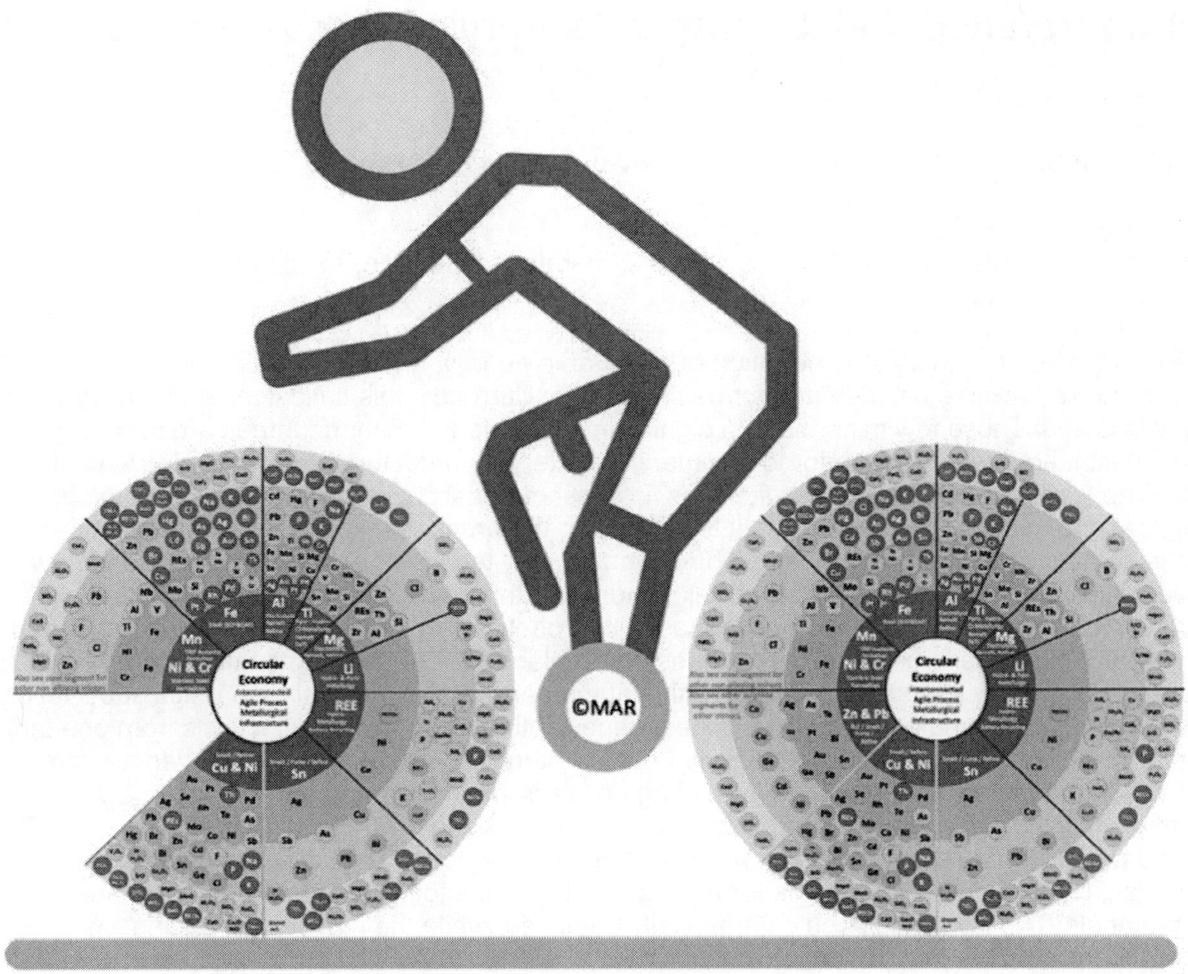
©MAR
Circular Economy
Fe
Al
Ti
Mg
Li
REE
Sn
Cu & Ni
Ni & Cr
Mn
Zn & Pb

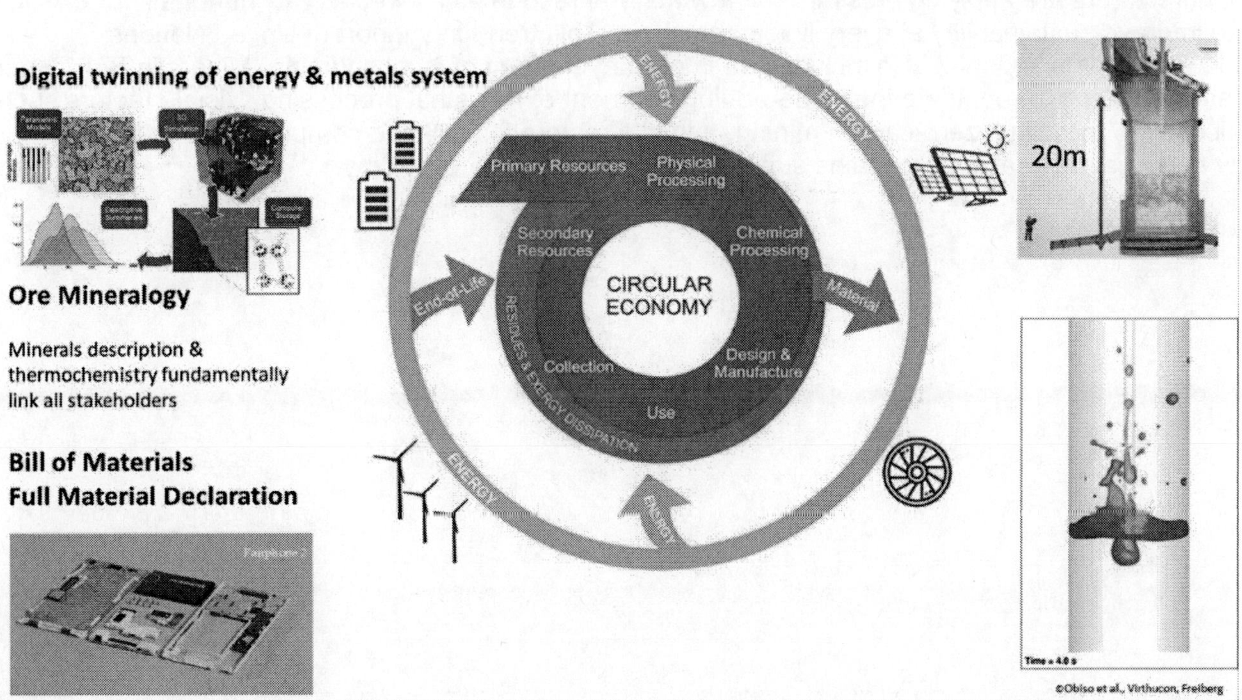
Digital twinning of energy & metals system
Ore Mineralogy
Minerals description & thermochemistry fundamentally link all stakeholders
Bill of Materials
Full Material Declaration
ENERGY
Primary Resources
Physical Processing
Chemical Processing
Material
Design & Manufacture
Use
Collection
Secondary Resources
End-of-Life
CIRCULAR ECONOMY
RESIDUES & EXERGY DISSIPATION
20m
©Obiso et al., Virthucon, Freiberg

F7

The Critical Mineral Supply Chain: Evolution and Revolution as New Battery Technologies Emerge

Edith Newton Wilson[1], Jesse Edmondson[2], Paige Johnson[3]
1. Rock Whisperer LLC, Tulsa, OK, USA, 2. US Critical Minerals, Fayetteville, AR, USA, 3. Ten-Nine Technologies, Tulsa, OK, USA

Energy storage is the key to widespread adoption of renewable energy, and the resulting exponential growth in battery production creates demand for raw materials. Currently, this critical mineral supply chain resides mainly in China. Those few minerals with significant extraction in other countries are largely processed to industrial grade in China prior to shipment to battery manufacturers, of which 75 percent are in China as well. The lead time on design specifications for most materials is 3-5 years, so the use of the lithium ion battery and the supply chain behind it is baked into the production of batteries for electric vehicles for the coming decade. But what about the advent of new technology, including the use of new metals? Current research is in place to substitute aluminum for lithium, copper foam for liquid electrolyte, and iron fluoride plus polymer for cobalt- or nickel-based cathodes. Companies who mine platinum group sulfides are researching ways to use them in batteries and replace the declining autocatalyst market. Many reformulations rely on chemical and physical properties at the nano scale. In addition to metal substitutions, redox flow batteries (RFBs) create charge from excess renewable energy to draw electrons from one tank to another. Return to steady state generates electricity from the stored solutions. These batteries are large and not very portable, but have the advantage of separating power from energy, and are thus safer and easily scalable. Vanadium, iron, chromium, and zinc are common metals used in RFBs, while nonaqueous RFBs use polycyclic aromatic hydrocarbons to improve energy density. Concentration of solar-voltaic energy in either silica or salt, to be released later in the form of light or heat, is another storage mechanism reliant on mineral raw materials. The latter has been commercially deployed, while the former is in testing. As each of these battery technologies becomes cheaper and scales, the supply chain to support production will evolve. Battery manufacturers are applying pressure on new mineral resources as well as raw materials for existing batteries to improve sustainability at every link in the chain. This trend in support of United Nations Sustainable Development Goal 7 also includes an important element of local sourcing. Miners in Bolivia and Australia are beginning to repatriate the value-adding segment of industrial processing. Visible factors such as social licensing, recycling, zero-carbon mining, and site restoration will add complexity as well as profitability to critical mineral supply chains supporting the new energy economy.

Lithium primary batteries made with TENIX™ last twice as long

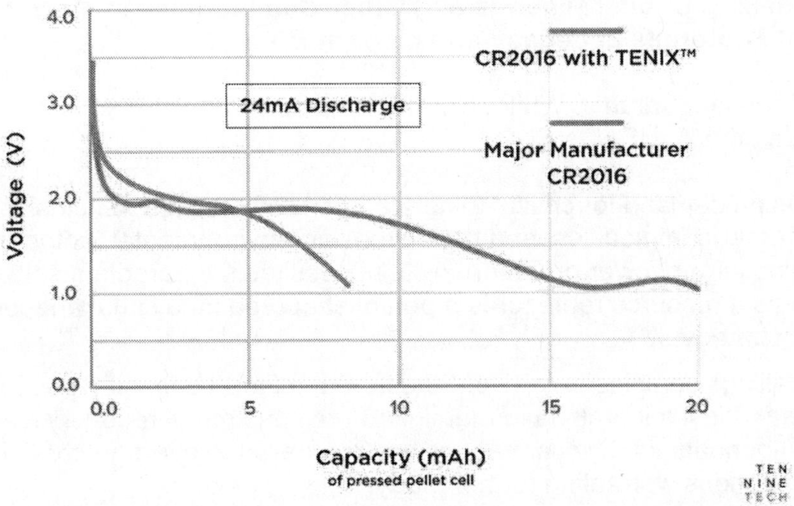

F8
Waste as a Resource: Seeking Critical Minerals and Other Commodities in Mine Waste and Reducing Environmental Footprints at Legacy and Modern Mines

Robert R. Seal, Nadine M. Piatak, Sarah J. White, Sarah M. Hayes
US Geological Survey, Reston, VA, USA

The combination of increasing demand for critical minerals, many of which are exclusively derived as by-products or coproducts of other commodities, and the increasing environmental footprint of mines due to growing emphasis on mining larger, lower-grade deposits necessitates a paradigm shift to improve sustainability. Mine waste as a resource represents a potential approach to simultaneously address these issues using a synergistic strategy.

A suite of nearly 100 mill tailings samples from over 20 different mineral deposit types has been assembled to evaluate the potential for critical mineral, base metal, and precious metal recovery. Deposit type clearly influences critical mineral abundances. Concentrations are compared to average crustal abundances to identify anomalous concentrations warranting further study.

A mass balance approach was applied more specifically to the ore-processing scheme at an active porphyry copper-molybdenum mine to better understand the deportment of critical minerals and other commodities in concentrates and tailings. This analysis indicates that roughly half of the gold, silver, zinc, platinum group metals, selenium, and antimony in the ore is lost to tailings. Approximately two-thirds of the tellurium and indium and most of the cobalt, germanium, gallium, and rare earth elements also reside in the tailings. The chalcophile nature of gold, silver, and many of the by-product critical mineral commodities suggests that they may be associated with pyrite—a hypothesis supported by ongoing mineralogical studies. Thus, the increasingly common practice of producing a pyrite concentrate at many porphyry copper mines to better manage acid drainage risks may also be an important step toward recovering valuable resources. This approach combines resource recovery with improved environmental management.

Any attempt to reprocess mine waste for recovery of critical minerals or other commodities, however, must account for the present-day hosts of those elements. Investigations at an abandoned lead-zinc Mississippi Valley-type mine that was an historically important by-product germanium producer highlight the influence of mine waste weathering on germanium deportment. Sphalerite was the original germanium host, but weathering of the waste piles produced hemimorphite as a secondary phase, which now hosts a majority of the germanium in wastes sampled.

Synergies between resource development and improved environmental protection also include consideration of the carbon footprint of mining. Mafic and ultramafic magmatic nickel-copper-platinum group metal deposits are predominantly hosted by high calcium and magnesium silicate minerals, such as olivine, pyroxene, and plagioclase. Extensive crushing and grinding create mafic silicate tailings with high surface area that are prime targets for mineral carbonation through the production of calcium and magnesium carbonate minerals. The voluminous amounts of mafic gangue minerals in tailings can sequester carbon dioxide and contribute to the development of carbon-neutral mines.

Waste streams at legacy and active mines should be reevaluated to identify additional resource potential. The most tenable scenarios will likely be situations where resource recovery can be combined with improved environmental management to enhance the remediation of abandoned mines and the sustainability of active mines. At abandoned mines, resource recovery can also help offset remediation costs.

F9
Economic Geology and Geomet for Ocean Resources: Sustainable Seafloor Metals

Nicholas R. Mitchell
Cornell University, Ithaca, NY, USA

The economic geology of deep-sea polymetallic nodules includes understanding ore deposit formation, characterization, and financial development in a regional geological context, just as it does for on-land deposits. Similarly, geomet models for these resources link geological understanding with process development for developing sustainable operations. It is generally appreciated that Ni, Cu, and Co deposits are of increasing importance given the projected demand curves for electric vehicles, battery energy storage, and other 21st-century technologies. While seafloor nodules have been known to be potential resources for these metals for decades, the economic geology and geomet of these deposits is less well understood because spatial metal deportment variability and processibility is not well constrained.

Seafloor polymetallic nodules from 13 sites across a 25,160-km^2 economic zone of the CCZ have been characterized for metal content, mineral deportment, and variability in terms of site-specific bathymetry, proximity to volcanic cones, fracture zones, abyssal plains, slope, relative crustal age, and nodule size classification. Micro-XRF analysis at the Cornell High Energy Synchrotron Source (CHESS) was conducted on nodules from each of the 13 sites in order to develop a better understanding of nodule mineral structure and element deportment variability. Data processing and cluster analysis reveal spatial and temporal distribution patterns within and among nodules, including hydrogenetic and diagenetic mineral species present across sequential layered growth structures (LGS). Fe-, Pb-, and potentially Co-rich hydrogenetic minerals (Fe-vernadite) follow finely laminated growth structures characteristic of turbostratically aligned phyllomanganate sheets around the exterior rims of multiple LGSs. Ni- and Cu-rich zones contain diagenetic minerals (7 À vernadite, 10 À vernadite, and todorokite) in repeated, delicate colloform accretionary growth textures radiating out from a central nucleus toward the interior of LGS zones. Todorokite is a topotactic transformation of 10 À vernadite when high levels of interlayer Mg+ cations are present. Characterizing todorokite within nodules is important because the microporous tunnels in todorokite are associated with edge-site Ni and Cu substitution for Mn3+ and Mn4+. The extent of topotactic transformation from 10 À vernadite to todorokite thus serves as a proxy for related metal distribution in a geomet framework for improved processing, since current flowsheets rely on bulk crush and metallurgy in the absence of understanding ore variability across the seafloor. Additional outcomes include parameters for environmental impact and potential remediation solutions.

These data assist in reducing some of the uncertainty surrounding deep-sea mining, which has seen a multidecadal regulatory holding period conducted by the International Seabed Authority (ISA) as designated in the UNCLOS. With finalization of the official ISA mining code set for 2024, such technical studies help to drive sustainable development of seafloor resources for the needs of society in the 21st century.

F10

Controls on Rare Metal Concentration in the Crust: Implications for Future Exploration

Anthony E. Williams-Jones, Olga V. Vasyukova
Department of Earth and Planetary Sciences, McGill University, Montreal, QC, Canada

Technological advances, many driven by the need to address the global warming issue, have created enormous demand for a variety of rare metals that will continue for the foreseeable future. These metals include the REEs, Co, Nb, and Zr. Here, we review the processes that concentrate rare metals to economic levels in the crust. Ultimately, they originate in the mantle, but their paths to the crust are quite different. Whereas Nd, Dy , Nb, and Zr are incompatible elements, Co is compatible and Sc partially compatible. Thus, Nd, Dy, Nb, and Zr partition strongly into magmas forming after small degrees of partial melting and concentrate in peralkaline igneous rocks and carbonatites as REE minerals, pyrochlore, and zirconosilicate minerals. In contrast, Co is released only after high degrees of partial melting and reaches its highest concentration in ultramafic magmas, commonly concentrating in immiscible sulphide liquids. The behaviour of scandium is more complex. Although it partitions strongly into magmas representing intermediate degrees of partial melting, it may be present in sufficient concentration in magmas from low degrees of partial melting to reach economic levels through other processes. Indeed, at Kovdor (Russia), Sc is interpreted to have reached economic concentrations in phoscorite derived by liquid immiscibility from carbonatitic magma. The same process may be applicable to Nb. Moreover, Nd and Dy are interpreted to have been concentrated by an immiscible fluoride liquid in the Strange Lake REE-Nb-Zr pegmatite deposit (Canada). Elsewhere, Sc has concentrated through fractional crystallisation of clinopyroxene into which it partitions preferentially. Despite the importance of magmatic processes, aqueous fluids play a pivotal role in the formation of economic rare metal deposits. Thus, both Co and Sc form exploitable deposits due to weathering of ultramafic rocks that releases them from olivine and clinopyroxene, respectively, to mildly acidic surface waters as simple ions. With increasing pH, they accumulate at depth in laterites. The REEs are commonly mobilised as chloride complexes by fluids exsolved from carbonatitic and alkaline silicate magmas and in some cases, e.g., Bayan Obo (China), this is the primary ore-forming process. Chloride complexation in hydrothermal fluids is also the major means by which Co is concentrated. Although Zr and Nb are much less mobile than other rare metals, they can be concentrated hydrothermally as hydroxyfluoride complexes. In summary, rare metals are mobilised by partial melting of the mantle to a degree consistent with their compatibility and transported by magmas into the crust. There, they are concentrated to mineable levels by these magmas, hydrothermal fluids, and weathering fluids. Understanding these processes is the key to their future successful exploration.

F11
Developing Capable Technical Professionals to Support the Future Resource Industry

Joanne Heyes
BHP, Melbourne, Australia

The resources industry is transforming due to digital disruption and the application of new technologies, discovery and extraction of increasingly deep and complex orebodies, a changing future-facing commodity and energy mix, scarcity of water resources, and environmental, heritage and social value commitments and constraints.

At the same time, an aging and experienced technical workforce is contemplating retirement; early-career resource scientists and engineers require upskilling; industry professionals require continuous development; universities face significant financial challenges; breadth of resources-industry career options isn't understood by students, educators, or careers advisors; and perception of the resources industry is at an all-time low. These factors are contributing to unprecedented low numbers of enrollments in resources-related degrees such as Geoscience, Geotechnical Engineering, Mining Engineering, and Metallurgical Engineering.

The future resources industry needs a consistent supply of diverse, quality geoscience and resources engineering graduates and access to workforce development programs to meet forecast technical, technology, and innovation requirements and societal/environmental demands.

This paper will discuss the challenges and collaboration opportunities for educators, governments, and industry to inform, influence, and support students to build careers in geoscience and resources engineering and to support career professionals for a lifetime of learning and development.

F12
Accelerating Innovations in Subsurface Modeling and Analysis: The Human Factor

Adam Pidlisecky
University of Calgary, Calgary, AB, Canada

When we look to the future of data and computation and how it will transform our understanding of the subsurface, it's easy to focus on the changes in technology and data we're experiencing today. For example, over the past decade we've been influenced by increasing volumes and types of geoscience data and have implemented artificial intelligence and machine learning.

Certain data and processing trends are clear—we will continue to see an uptake in sensor technologies and increased spatial and temporal sampling of data, and geoscience data will need to be shared across increasingly large groups of stakeholders. In data processing, technologies like cloud computing will enable complex levels of analysis, and we will be able to leverage technology development from other sectors that are tackling similar computational challenges.

However, the increase in data volumes and computing power represent only part of what is required to transform our understanding of the subsurface.

Unlike other applications of big data and high-performance computing (e.g., autonomous driving), geoscience data do not and will not—for the foreseeable future—contain enough information on its own to uniquely describe subsurface geology. As well, our understanding of geological theory alone is not sufficient to support deterministic modelling of geological systems. Geoscience is not exact. This is an interpretive science—a synthesis of theory and practical experience from our geologists, who provide a rich, tacit understanding of subsurface processes that cannot be described succinctly by deterministic equations.

This means as we move forward, we must turn our attention to technologies that

1. enable geoscientists to leverage maximum insights from technological innovations; and
2. codify the tacit understanding of geology that seasoned practitioners possess.

Enabling geoscientists in this way requires a shift in how we view innovation, to include transformative approaches in areas like the human-computer interface (HCI) and visual analytics. By investing in HCI, we ensure that geoscientists can experience multidimensional data in a way that allows them to draw insights and empowers them to communicate risks, uncertainty, and outcomes easily. Additionally, by focusing on visual analytics, we can move beyond simply providing a rich experience with data to learning how to understand and codify how a geoscientist sees and solves a problem.

This shift will allow the geoscience community to better share, scrutinize, and ultimately act on the insights provided by an individual geoscientist or learn from the differences in how geoscientists view problems.

In the next 100 years, technologies like machine learning and artificial intelligence will no doubt underpin an improved understanding of the subsurface. But for this technology to be truly transformative, we must ensure that technology enables geoscientists to get the most out of their data by focusing on the human side of the equation.

F13
New Developments: In-Field Measurements and Analyses

James Cleverley
Imdex Limited / REFLEX Instruments Europe Limited, Iford, United Kingdom

The process of ore discovery relies on the incremental building of rock knowledge, understanding four key parameters at multiple scales: geochemistry, mineralogy, texture, and spatial location. Traditionally this data was collected disparately and could take anywhere from weeks or months to collate in order to make decisions, often long after the drill rig had demobilised. Using geological sensing to collect in-field measurements, and the ability to deliver information where it is needed, reduces this time to decision and supports the geoscientist to make rapid informed, reliable, and auditable decisions (Cleverley et al., 2017). This change is happening at a time of rapid advancement in sensor hardware development for field deployment, automation, and the data infrastructure to deliver information to the geoscientist where and when it is needed.

The last 20 years has seen geosensing technology move from the laboratory into the field. In-field measurement technology is developing with innovation in 1) sensing technology, 2) hardware platforms to deliver the measurements closer to the sampling site, and 3) infrastructure ecosystems and analytics to capture, exchange, and turn data into information. Increasingly, products are a combination of all of these, implemented within or interfacing with other systems as an ecosystem. There has been a shift from single technologies to whole systems of sensors and open data exchange.

Geochemistry has benefited from improvement of direct-measurement sensing technology like portable XRF, but now new sensors such as laser-induced breakdown spectroscopy (LIBS) and Raman spectroscopy deliver elemental and molecular analysis that are physically not possible with XRF in the field. Mineralogy is mostly calculated by proxy through hyperspectral sensing or directly in X-ray diffraction using portable formats. Getting mineralogy is critical yet in both instances the processing of the raw signal and confidence in a result is where the innovation and development is centred. There are many new analytic developments that are speeding up the process of delivering the quantified result from spectral or X-ray analysis and the way the results are delivered to the field.

The methods by which the sensors are deployed to the field is centred on platform technologies, robotics, and automation. Core scanners have been developed as ways of collecting multiple sensing streams but also starting to automate the workflow for processing diamond drill core, delivering consistency compared to handheld point-and-shoot methods. In geochemical and mineralogical sensing, the quality and representivity of the sample is also critical to the result. Supporting technology to allow in situ spot or interval measurement, downhole, or sample preparation prior to analysis is important to getting a reliable, quality result supporting a quality decision and is as important as the sensing hardware itself.

For mineralogy, future research in the processing of the raw sensor data will lead to improvements, especially in the coprocessing or inversion of chemical and mineralogical data. Future advances in sensing technology will broaden the range and quality of variables that can be measures in-field to underpin quicker, more robust rock knowledge and lead to quicker, quality decisions in the discovery process.

F14
The Mining Map of the Future

Michel Jebrak
UQAM, Montreal, QC, Canada

Economic geologists are responsible for the discovery and development of new mines for satisfying the needs of humanity, with more and constraints to keep a habitable planet. The mining map is a representation of an equilibrium between the demand in minerals and the offer from production sites, including mines and mineral plants. In the next 50 years, could we define its main tendencies?

1. *Consumption* defines the nature, the quantity, and the location of the mining resources. In 2080, the electrification of transportation, the better management of energy, and a possible decrease in physical transportation for the most developed countries will increase the need for Cu, Al, and numerous strategic metals. Urban mining may contribute partly to the answer. As the largest mines have been always located in core economic countries whatever their geology, a shift of the production will occur toward the shore of the Indian Ocean, specialy East Africa, and along the new silk roads.
2. *Production* is moving toward lower grades and higher tonnages, implying newer types of mineral deposits. Following the Launay law (1913), we will exploit types of deposits closer and closer to source rocks. The integration of social and environmental externalities implies new classifications of minescapes, i.e., integrated mineral systems. The long-term tendency of hydrometallurgical processing will continue and could shift from factory to in situ processes, mobilizing catalysing biological processes. Costs and delays for production will increase because of the need to get social acceptability; they will be transmitted in fine to the consumers. Brownfield exploration will continue to be preeminent, as opening new mining districts by greenfield exploration will encounter more and more social and environmental opposition. The present frontier of development will be blocked by the conservation of the remnant wilderness. Present hydrocarbon-dependant countries will desperately shift toward new minerals resources. Artisanal-scale mining will remain an adjustment parameter.
3. *Transportation* of mineral resources will be governed by the geopolitical evolution of the world. Between 1945 and 2015, globalization had allowed construction of integrated value chains that led to maximization of the Ricardian specialization. Proximity to the world ocean was key for numerous metals. The ongoing deglobalization is developing continental value chains and increased state regulations. It will require the discovery of deposits near economic centers and along continental and oceanic roads. We could move from a Raleigh-Mahan oceanic geopolitical vision back to the McKinder-Radjell continental approach, increasing the growing world conflict potential.

These evolutions will impact economic geology. Our postmortem descriptive approach will need to be more explicative and consider the multiple steps of concentrations in a constructivist approach. The development of artificial intelligence already requires better organized sets of data for machine learning. New territories will offer new deposit types. Explorationists will used both empirical automatized approaches and process-based predictions and consider new geological environments, such as deep Proterozoic orogens.

F15
Inserting orebody knowledge into resource definition

Jelena Puzic
Teck Resources Limited, Vancouver, BC, Canada

Geoscience has an important role to play in maintaining and characterizing a global mineral supply for sustainable development. The last few years have brought significant technological advances in the area of rock and material characterization, thereby enabling related technical disciplines to add value and manage risk across a digitally supported mining value chain. As the industry races to modernize its existing mines and develop the next generation of operations, geoscientists will be enablers, leveraging their foundational knowledge while developing the new ideas and creative thinking necessary to bridge the old and the new.

Geoscientists are rapidly advancing and embracing new ways to collect, integrate, interpret, model, and communicate quantitative data streams across a range of scales and mining stages. Multidisciplinary scientific and technical collaboration with neighboring (and new!) disciplines is essential to ensure geoscience products are embedded into an evolving operational workflow. In particular, rapid advances in data science require collaboration with subject matter experts from the geosciences to ensure advanced analytics and artificial intelligence inputs are appropriate and that outcomes are geologically reasonable and consistent with the “on the rocks” data and knowledge. This talk will explore examples of the evolving geoscience workflow including new innovations and sensors fueling the change in how we interact, integrate, and interpret drivers of value and risk at our mineral resources—ultimately enabling us to make better decisions.

By leveraging data and learnings from the past and ensuring the right data at the right time in the present, we will build more reliable and sustainable orebody predictions for the future.

ST.020
Geophysical Data Interpretation in a 3D Environment Provides Insights into Porphyry Copper Systems near Silverton, Colorado

Eric D. Anderson, Douglas B. Yager, Maryla Desczc-Pan, Brian D. Rodriguez, Bruce D. Smith
US Geological Survey, Denver, CO, USA

The Silverton caldera in the San Juan Mountains of southwest Colorado hosts mineralization and alteration associated with porphyry copper systems. Most production came from polymetallic veins worked over a 121-year period that yielded a total of 16.4 million metric tonnes of gold, silver, copper, lead, and zinc ore. However, extensive postmineralization erosion has exposed significant vertical extents of the porphyry copper systems from shallow lithocaps to moderately deep, quartz-molybdenite stockwork veins and sericitic alteration, thereby indicating the potential for additional porphyry-style mineralization. Airborne electromagnetic and magnetic surveys provide relatively continuous observations that can geophysically characterize outcropped and concealed components of the porphyry copper systems. We interpret Versatile Time Domain Electromagnetic (VTEM) and magnetic data along with three-dimensional geologic modeling to better understand the ore systems. We model the porphyry copper systems along 1-km-spaced east-west profiles across a 9- x 10-km area.

The Silverton caldera is composed of Oligocene to Miocene volcanic and intrusive rocks that exhibit extensive propylitic alteration and more localized sericitic and advanced argillic alteration. Intrusions along the caldera margin host quartz-sericite-pyrite veins with lesser molybdenite exposed on Mt Moly and in surrounding valleys. To the north, mountain peaks in the Red Mountain district exhibit a lithocap around 3 km in diameter that overlies sericitic-altered rocks exposed in the surrounding incised valleys. Similar alteration patterns are present to the south along a ridge of vuggy quartz at Ohio Peak, and the alteration patterns continue south an additional 2.5 km to Anvil Mountain where the lithocap broadens to around 4 km.

Maps of VTEM calculated time constant (tau) indicate that conductive features underlie the lithocaps and sericitic-altered rocks, whereas propylitic-altered rocks are commonly resistive. Aeromagnetic data show magnetic highs over the intrusion-hosted molybdenite stockwork veins and sericitic-altered rocks at Mt Moly; however, low magnetic susceptibility measurements indicate that such rocks are not the source. Magnetic depth estimates suggest sources around 300- to 500-m depth near the mountain peak that shallow to 200 m in the valleys where the sericitic alteration is less intense, which may reflect the presence of magnetite-rich potassic or propylitic-altered rocks. VTEM 1D inversion results show changes in conductivity down to 250-m depth at Mt Moly where a trend of decreasing conductivity with decreasing intensity of sericitic alteration is observed from the mountain peak toward the valleys. Low conductivity occurs over a mapped zone of quartz-alunite-pyrophyllite alteration. The Red Mountain district lithocap exhibits magnetic lows, suggesting magnetite destructive alteration that is corroborated with very low magnetic susceptibility measurements. The northernmost east-west section occurs at the southern extent of the district. In this section, the VTEM inversion results show conductive sericitic alteration in the valleys beneath the lithocap where magnetic depth estimates indicate sources <200-m depth. The Ohio Peak-Anvil Mountain area shows moderate magnetic highs centered near Anvil Mountain with depth estimates generally <200 m. VTEM inversion results indicate conductive rocks beneath the lithocap and sericitic-altered rocks with the exception being the silicified rocks that exhibit very low conductivities.

ST.021
Carbonatites in India through Time and Space: A Review and Implications for REE Exploration

Malcolm Aranha[1], Alok Porwal[1, 3], Ignacio González-Álvarez[2, 3]
1. Indian Institute of Technology Bombay, Mumbai, MH, India, 2. Commonwealth Scientific and Industrial Research Organisation (CSIRO), Kensington, WA, Australia, 3. Centre for Exploration Targeting, University of Western Australia, Crawley, WA, Australia

Chronotectonic settings and petrogeochemistry of carbonatite occurrences of India are compiled and reviewed in relationship to global tectonic events associated with the amalgamation and breakup of supercontinents. Four chronotectonic domains, namely, southern domain, southeastern domain, northeastern domain and northwestern domain, are delineated based on the geographical distribution, tectonic settings, and temporal relationships among the carbonatite complexes. The southern domain comprises two subdomains—Paleoproterozoic and Neoproterozoic. The Paleoproterozoic subdomain is related to extension due to relaxation after the Southern Granulite Terrain-Dharwar accretion, while the Neoproterozoic subdomain is related to rifting related to the fragmentation of Rodinia. The southeastern domain is associated with the Mesoproterozoic fragmentation of Columbia. The northeastern domain is related to the Mid-Cretaceous breakup of Greater India from Australia-Antarctica driven by the Kerguelen mantle plume that also produced the Rajmahal-Sylhet Large Igneous Province (LIP). Finally, the northwestern domain is related to the Late Cretaceous Indo-Seychelles-Madagascar split and the passage of Greater India over the reunion hotspot, which also produced the Deccan LIP.

Petrogenetically, the Cretaceous carbonatites of the northeastern and northwestern domains are results of low-degree partial melting of a subcontinental lithospheric mantle (SCLM) that was metasomatised by a mantle plume. The older carbonatites of the southern and southeastern provinces show no geologic evidence of mantle-plume associations and are most likely results of low-degree partial melting of an SCLM, which was metasomatised by decompressional anatexis in an extensional regime. In each province, the carbonatites occur along major crust penetrating faults and are often polychronous. Geochemically, the Precambrian carbonatites display EMI or HIMU + EMI source features, whereas the post-Cambrian carbonatites are interpreted to be produced in a HIMU + EMII magmatic source—a trend that is observed in prominent carbonatites globally. In terms of the depth of the carbonatitic magma source, the southern domain likely had a shallow source, while the northeastern domain had a much deeper source.

In terms of REE exploration, presence of (1) metasomatised SCLM pockets, which form fertile source regions for REE-bearing fluids, (2) extensional geodynamic settings, mainly intracontinental rifts, with a possible association with mantle plumes, and (3) lithospheric-scale network of fluid plumbing architecture, which includes reactivated zones of structural weakness in the lithosphere and trans-lithospheric faults for tapping REE-rich fluids from SCLM and near-surface upper-crustal faults and joints for focusing the fluids to near-surface levels, are identified as prerequisites for the emplacement of REE-rich alkaline-carbonatite complexes. Further enrichments by metasomatic processes can be traced by geochemical sampling, radiometric surveys, and emplacement depth determination for direct detection of REE deposits.

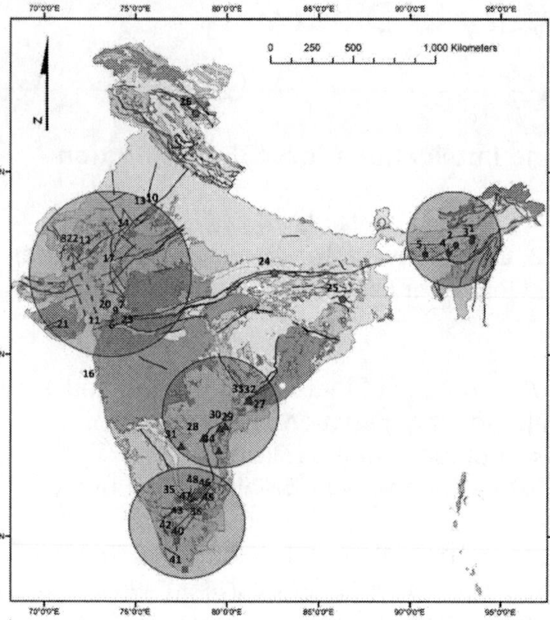
0 250 500 1,000 Kilometers
N
70°0'0"E
75°0'0"E
80°0'0"E
85°0'0"E
90°0'0"E
95°0'0"E
30°0'0"N
20°0'0"N
10°0'0"N

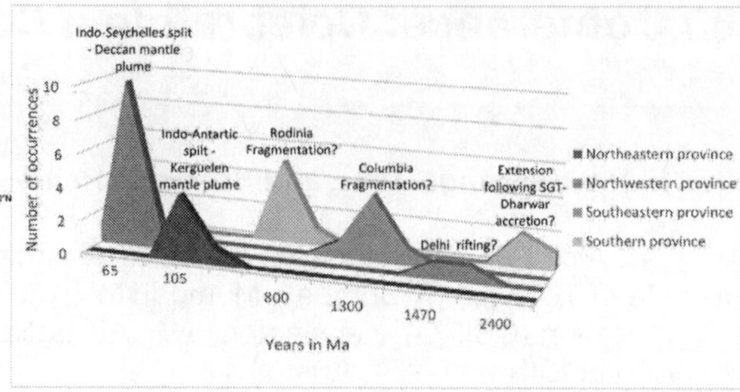
Indo-Seychelles split - Deccan mantle plume
Indo-Antartic spilt - Kerguelen mantle plume
Rodinia Fragmentation?
Columbia Fragmentation?
Delhi rifting?
Extension following SGT-Dharwar accretion?
Number of occurrences
Years in Ma
65
105
800
1300
1470
2400
Northeastern province
Northwestern province
Southeastern province
Southern province

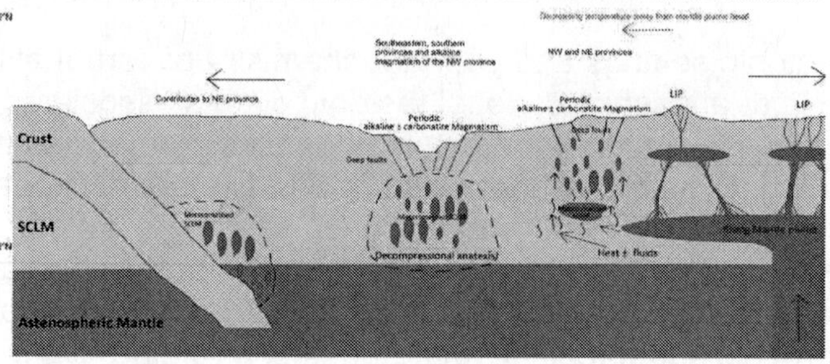
Crust
SCLM
Astenospheric Mantle
Periodic alkaline ± carbonatite Magmatism
Decompressional anatexis
LIP
Contributes to NE province
NW and NE provinces
Heat ± fluids

ST.022
Prospectivity Modeling of NE India for REE Deposits Associated with Carbonatites and Alkaline Complexes

Malcolm Aranha[1], Alok Porwal[1, 3], Ignacio González-Álvarez[2, 3]
1. Indian Institute of Technology Bombay, Mumbai, MH, India, 2. Commonwealth Scientific and Industrial Research Organisation (CSIRO), Kensington, WA, Australia, 3. Centre for Exploration Targeting, University of Western Australia, Crawley, WA, Australia

The rare earth elements (REEs) are a group of 17 metallic elements that are projected to be critical for future industrial development. Currently, China dominates the world production of REEs, contributing >55% of the global output. The entire production of REEs in India comes from the monazite-bearing beach sands along the eastern and western coasts. No primary economic-grade deposit of REEs has been discovered so far in India, despite the presence of a variety of enriched source rocks such as carbonatites, pegmatites, and alkaline complexes. In this contribution, we develop a GIS-based prospectivity model for exploration targeting of REEs associated with carbonatites and alkaline complexes in the Assam-Meghalaya plateau, NE India.

Drawing from McCuaig and Hronsky (2014) and Groves et al. (2020), a generalised mineral systems model for alkaline-carbonite-complexes-related REEs was developed to identify regional-scale targeting criteria for REEs in the study area. The main components of the mineral systems are (1) pockets of metasomatised SCLM, which form fertile source regions for REE-bearing fluids, (2) extensional geodynamic settings, mainly intracontinental rifts and large igneous provinces associated with mantle plumes, and (3) lithospheric-scale network of fluid plumbing architecture, which includes reactivated zones of structural weakness in the lithosphere and translithospheric faults for tapping REE-rich fluids from SCLM and near-surface upper-crustal faults and joints for focusing the fluids to near-surface levels. Finally, a postemplacement tectonic regime preserves the deposits. A series of predictor maps were generated from public domain geologic, geophysical, geochemical, and satellite data to represent the above components. A fuzzy inference system (FIS) was designed to model the interactions of the above REE mineral systems components. The outputs of the FIS were mapped to generate n REE prospectivity map of the study area. Furthermore, systemic and stochastic uncertainties in the model are estimated using the methods and procedure described by Lisitsin et al. (2014), Chudasama et al. (2017), and Porwal et al. (2003).

The following are the main findings of this exercise. Project-scale detailed ground exploration is recommended for the Sung Valley and Jasra Complexes, which show high prospectivity at all confidence levels. More data collection is recommended for the Mikir Hills in the areas surrounding Samchampi and Barpung Complexes. Similarly, more data should be collected around the swarms of dikes in the Garo Hills in the western part of the study area, where plugs or even a major intrusion of alkaline-carbonatite rocks may be present. The area northeast of Tura is geophysically similar to the Sung Valley and Jasra Complexes and is therefore a possible exploration target. Finally, to better understand the prospectivity of the Mawpyut ultramafic complex, detailed geochemical sampling and surface or airborne radiometric surveys should be carried out. This study is one of the first comprehensive GIS-based prospectivity models for exploring REEs associated with carbonatites and alkaline complexes that we are aware of in literature.

ST.023

Critical Minerals and the New Energy Enterprise for Oil and Gas and Mining Companies of the 21st Century: Why and How

Lindsay A. Ross[1], David Beckett[2]
1. Cherryvale Partners Pty Ltd, Wamuran, QLD, Australia, 2. Rockleigh Energy, Rockleigh, SA, Australia

Supplying the forecast demand for high-value minerals critical to emerging energy technology will require innovative methods of exploration and production, beyond the current reach and scale of traditional practice. To this end, the oil and gas industry and the mining industry may each, separately and competitively, leverage intrinsic technical competencies or alternatively, by way of strategic partnership become integrated future energy resource enterprises. Disruption and the progress we can make from it has a habit of allowing us to see some barriers rendered obsolete or at least blurred. To this end, importing, cross-skilling and developing new geoscientific capabilities under this concept requires engagement of some geoscience skill sets not usually considered at home in an oil company and likewise for a mining house. Cross-skilling and new team structures are required to enable this approach within the corporation. Additionally, we consider trajectories for corporations within the context of vertical and horizontal integration and modes for corporate evolution within the nonlinear space.

For the upstream oil and gas business the application of play based exploration (PBE) shifted critical focus from the prospect to the play—that is, the entire spatial, temporal, and geological system elements necessary to the conventional or unconventional hydrocarbon accumulation. PBE revolutionised oil and gas exploration, conceptually and commercially, and many features are directly applicable to the exploration for high-value critical minerals and complement the minerals systems analysis approach. In this presentation we discuss PBE for critical minerals including lithium and graphite.

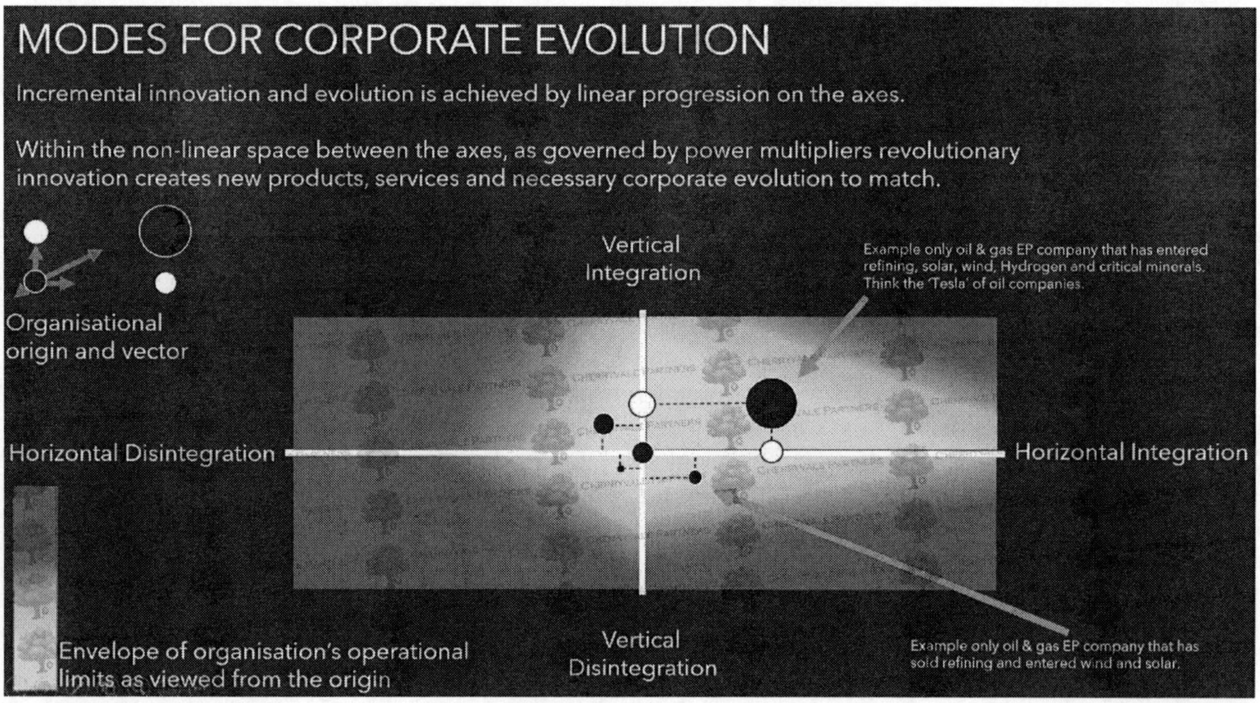

ST.024
Geochemical Neural Network Classification of Indicator Minerals Based on Associated Diamond Content

Diana Benz
Takom Exploration Ltd., Prince George, BC, Canada

The presence and abundances of diamond indicator minerals, as well as their major and minor element concentrations, are the standard in diamond exploration to determine the location of a diamond-bearing body and if the conditions were conducive to diamond formation. Despite compelling diamond indicator mineral geochemistry, some kimberlites, however, are found to contain few diamonds. This study examines the element concentrations of diamond indicator silicate minerals by creating a neural network classification model to differentiate between minerals associated with highly diamondiferous and poorly diamondiferous kimberlites.

Five types of diamond indicator silicate minerals were analysed using an electron microprobe: harzburgitic garnets, eclogitic garnets, clinopyroxenes, orthopyroxenes, and olivines. The suites of minerals were from kimberlites located in the Slave (NT, Canada) and Kaapvaal Cratons (South Africa): Beartooth (highly diamondiferous with 86.34% total octahedrals), Koala (highly diamondiferous with 43.75% total octahedrals), Panda (highly diamondiferous with 56.21% total octahedrals), Kit (noneconomic), New Elands (orangeite with highly resorbed diamond population), and Zero (conditions unsuitable for diamond formation). The minerals with the chemical compositions known only to diamond inclusion minerals and diamond-bearing kimberlitic xenocrysts, xenoliths, and phenocrysts were chosen for further analysis. Laser ablation-inductively coupled plasma-mass spectrometry (LA-ICP-MS) was used to determine major, minor, trace, and ultratrace abundances in the chosen minerals. A total of 494 diamond indicator silicate minerals were analysed with LA-ICP-MS resulting in maxima of 20 examples from each of the five minerals for the six different kimberlites pipes.

The LA-ICP-MS results were explored, summarized, treated as compositions using the log-ratio approach, and investigated for their element associations using RQ-mode principal component analysis. Based on the exploratory data analysis, neural network classification models for each type of indicator mineral were created using Keras, an open-source Python library. Each set of minerals were split into training, validation, and test subsets. Due to the small number of grains available, the stratified shuffle split was replicated an additional three times to cross-validate the models' effectiveness to separate minerals associated with highly diamondiferous and poorly diamondiferous kimberlites. As a result, five multielement neural network models were created to classify diamond indicator minerals based on their associated diamond content. Feature importance was calculated for each model, and percent probabilities were captured for each grain. An annotated coding companion was created in Jupyter Notebook. Future research would focus on larger data sets and examining the important features within the neural network as well as their relationships to diamondiferous kimberlites.

ST.025
Future Mineral Exploration

Herbert S. Jacobson[2], Diana M. Benz[1]
1. Takom Exploration, Prince George, BC, Canada, 2. Petrometal, Anthem, AZ, USA

Increased exploration and discovery are needed because shortages of some commodities are likely. Copper shortages are predicted beginning in 2025 due to the electric car boom. Recent exploration focused on exploring known areas by searching laterally or deeper in known districts. In Australia depth of discovery increased to 2,000 feet in 2010. The trend in recent years have been fewer discoveries at higher cost. There is a great potential for new discoveries under thin cover. Approximately 50% of the world's land area is under cover of recent unconsolidated sediments, jungles, deserts, glacial moraines, and lakes. Future exploration in these areas will increase using available tools. Diamond deposits were discovered in Canada under glacial moraines using indicator minerals in the till. Geochemical surveys can be used to explore for a variety of mineral deposits under glacial cover in the northern hemisphere. Side-looking airborne radar (SLAR) can detect target areas under jungle cover as was done in Brazil. Airborne radar surveys could be used to detect exploration targets in jungle areas in Africa and Southeast Asia. Airborne magnetic and electromagnetic geophysical surveys have detected target areas and in some localities have found mineral deposits. In Australia the Olympic Dam deposit was found by an aeromagnetic survey, and in Canada massive sulfide deposits were located by airborne electromagnetic surveys. Airborne surveys can detect areas of interest under deserts and under hundreds of feet of rock cover and could be used in many countries around the world. The use of drones has made airborne geophysics more efficient and less costly. Landsat multispectral data can be used to identify geological features related to subsurface mineralization as has been done in the Philippines. Mineral deposits created by the energy of large meteorite impacts as in the Sudbury, Canada, district can be detected by negative gravity anomalies similar to the Sudbury anomaly. We have examined Canadian regional gravity data and identified areas of possible mineralization created by meteorite impacts. Proposed future exploration will begin selecting areas based on known geological, geochemical, and geophysical data. Prime targets for selection are in Precambrian shields around the world. In the northern hemisphere large shield areas are covered by glacial moraines. Deserts in Australia, China, and Arabia and jungle cover in Africa and South America cover future mineral deposits. Future exploration will need to be done in joint ventures between private companies, governments, and research organizations as is in progress in Australia. The exploration will be promoted by mineral shortages and higher prices.

ST.028
Seismic Detection of Permeability and What It Could Mean for Exploration

Lawrence Cathles III[1], Charles Sicking[2], Peter C. Leary[3], Peter E. Malin[4]
1. Cornell University, Lansing, NY, USA, 2. Charles Sicking, Plano, TX, USA, 3. Peter C. Leary, Auckland, New Zealand, 4. Peter E Malin, Durham, NC, USA

In the spirit of 2021 and beyond, this talk addresses how new methods of seismic permeability mapping could aid ore deposit exploration. Ore deposits are related to fluid flow, and flow is controlled by permeability. The patterns of permeability that produced an ore deposit in a specific location likely remain long after the deposit formed. New methods for detecting the microslip events and very low amplitude harmonic resonances (humming) on fluid-filled fractures that appear to be related to present-day permeability could therefore be relevant to exploration. How seismic monitoring can map subsurface permeability, how and why it is thought that permeability is related to seismic "noise" that is detected in data today, and illustrative applications of seismic permeability mapping to exploration will be described.

ST.029

Road to the Future: What Ore Deposit Study Should Focus on Beyond 2020

Huayong Chen
Guangzhou Institute of Geochemistry, Chinese Academy of Sciences, Guangzhou, China

The future trends for ore deposit study should depend on the demands from our society and industry, which have clearly changed in the past decade and will continuously evolve in the future. In general, three major challenges on mineral exploration have been raised for all economic geologists, including (1) deep exploration, (2) high-efficiency exploration, and (3) environmentally friendly exploration, which we must take into serious consideration. The most crucial thing for the researchers is to determine what we should focus on in ore deposit study and what we can do to help when facing these challenges for mineral exploration.

Deep exploration is becoming increasingly important due to limited exposed metal sources on the surface and at shallow level. The combination of multiple newly developed methods will be critical for increased success during deep exploration. Except for improvement of geophysics such as DF-IP and WA-EM, approaches to geochemical vectoring using physical and chemical characteristics of alteration minerals are emerging and have good potential to be very useful in the future (e.g., chlorite and alunite in porphyry-epithermal systems). More accurate estimation of ore formation depth and erosion rates coupled with a well-defined preservation history should be paid much more attention as this will provide fundamental information for mineral exploration on both deposit and regional scales (e.g., new studies on the Paleozoic porphyry Cu deposits in NW China).

Tremendous knowledge from ore deposit studies has been accumulated in the past 100 years. However, applications of this valuable information were largely based on personal experience and limited highly technical approaches with low efficiency, compared with astonishing development in other fields such as environment and life sciences during the last decade. Obviously, a robust and more efficient AI-assisted exploration guiding system should be built in the future. Such a system will not work without careful establishment of an accurate database for each deposit type. This can only be achieved by the involvement of well-trained professional economic geologists. Among all the processes needed to build these data sets, before turning to AI assistance, establishing a comprehensive combination of valid identifying criteria (i.e., identification system) for each deposit type is crucial (e.g., breakthrough by GoldSpot on Au exploration in Canada and new approaches on porphyry Cu deposits in China).

Mineral exploration and mining can have negative impacts on the environment and are facing serious restrictive policies even in many developing countries. An evaluation system is required to more accurately justify if a newly discovered deposit is viable considering its environment impacts. Again, the involvement of economic geologists will be necessary in this work as each deposit type will have distinctive metal associations and mining approaches, which will be complicated by different landforms and weather conditions. Clearly, developing the subsystems for each deposit type in specific regions will be fundamental to establish an entire environment-protected mining evaluation system (e.g., new attempts on porphyry-epithermal deposits in strongly weathered mountain areas in SE China).

ST.032
Cobalt and MVT Zn-Pb-Ag in the Italian Alps

Marcello de Angelis
Energia Minerals (Italia) Srl, oltre il colle, Italy

The geology of Italy, a relatively small country of about 300,000 km^2, is characterised by a variety of geological features because of Italy's location in the middle of the Mediterranean Sea (a remnant of the Tethys Ocean), where lithologies were subjected to the complex geodynamic processes derived from the Triassic breakup of Pangea and the subsequent collision between the Eurasian and African plates that started in the Jurassic. These processes shaped the distinctive form of the Italian peninsula in orogenic and magmatic arcs flanked by basins of different structural significance, as well as contributing to a variety of minerogenetic environments.

In Italy, mining history dates to Neolithic age, but most of the mines that existed at the beginning of the century closed in the 1980s, and zinc, lead, nickel, copper, iron, antimony, gold, and cobalt minerals were left with resources still in the ground.

In this scenario, two projects of high interest have become the flagships of Alta Zinc Ltd (AZI), the Australian company holding mining and exploration permits in Italy:

The Gorno Zn-Pb-Ag Project

This project includes one mining permit and four exploration licences securing a number of Mississippi Valley-Type deposits of the Gorno Mineral District. The deposits are in the form of stratabound and columnar mineralisation hosted by Triassic limestone in the Central Alps.

The Gorno mines were known since Roman times for calamine and, in the 1970s, sphalerite/galena sulphides were extracted until 1980, when all activity was prematurely stopped.

Extensive diamond drilling in excess of 19 km done by AZI from 2015 to date resulted in the definition of JORC resources of 3.3 Mt at 6.1% Zn + Pb and 27 g/t Ag in the Colonna Zorzone. Current drilling carried out at Pian Bracca, the NE extension of Zorzone, has resulted in very promising stacked horizons at 12 to 19% Zn + Pb and 38 to 60 g/t Ag in intersected thickness of 10 to 12 metres for each level.

Following the definition of additional resources, a feasibility study is planned for mining development from 2022.

The Punta Corna Co Project

The project is located about 40 km north-northwest of Turin, in the Municipality of Usseglio, Piedmont Region, in one of numerous postorogenic hydrothermal systems found in the Italian Western Alps and exploited in the past.

The Co-Ni-As-rich veins are emplaced in the Piedmont Zone, an assemblage of metaophiolitic, ultramafic, and metasedimentary rocks representing fragments of oceanic lithosphere of Jurassic age.

The mineralisation is hosted by E-W-trending, postmetamorphic, subvertical veins of up to 6 m in width over a strike length of 6-7 kilometres. The wall rock always shows intense hydrothermal alteration.

AZI holds three exploration licences covering the known deposits and their lateral and downdip potential, which could extend for over 6,000 metres in strike and about 1,000 m downdip.

Work is continuing to define the lateral and downdip extent of the mineralisation.

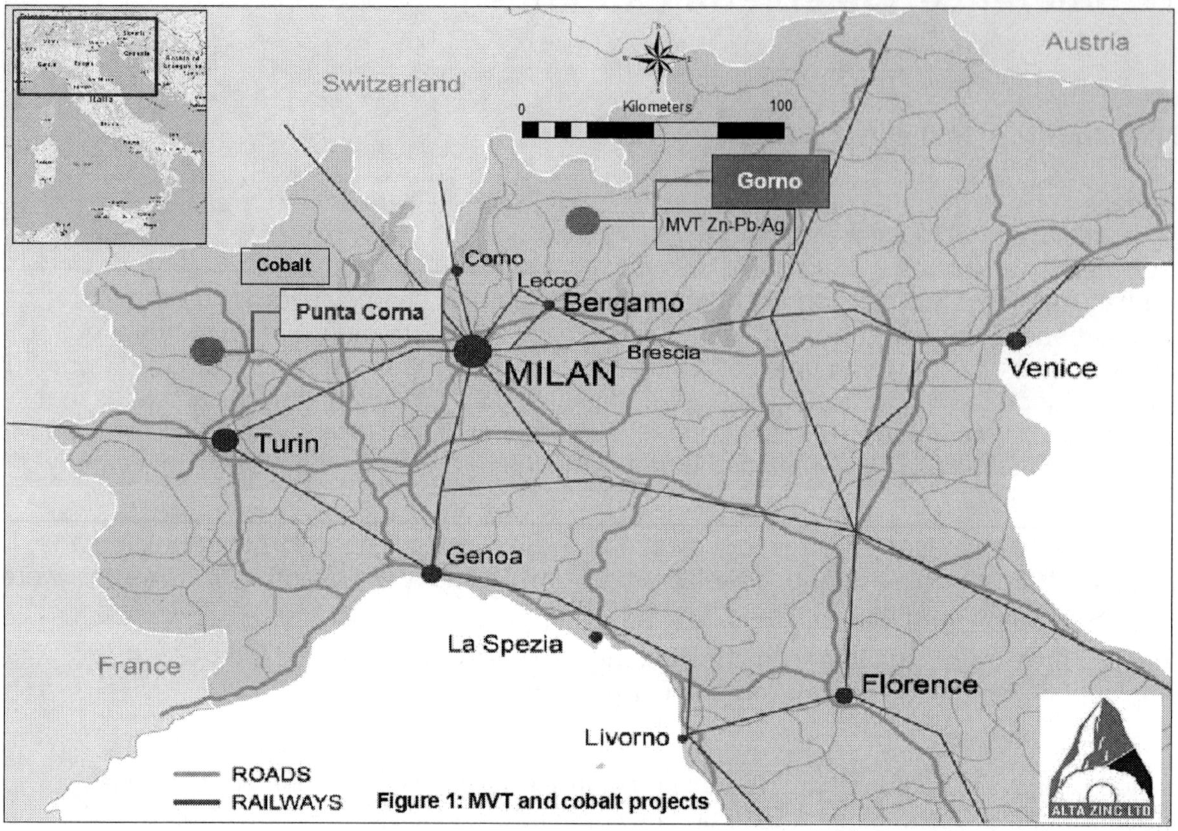

Figure 1: MVT and cobalt projects

ST.033
CO2 Storage Potential in Mine Tailing in Brazil: An Initial Approach

Saulo B. de Oliveira, Colombo C. Tassinari
Instituto de Energia e Ambiente, Universidade de São Paulo - IEE-USP, São Paulo, SP, Brazil

The geologic storage of CO_2 is encouraged by the IPCC (Intergovernmental Panel on Climate Change) to contain the increase in the average temperature of the planet compared to the preindustrial era. The research agenda published by the National Academies of Sciences Engineering Medicine indicate CO_2 underground storage in sedimentary formations and mineral carbonation (or carbon mineralization) as the most promising methods to achieve the goals proposed in the Paris Agreement, UNFCCC. Mineral carbonation consists of the precipitation of carbonate minerals, as calcite or magnesite, through the reaction of CO_2 with calcium- or magnesium-rich rocks. Mafic and ultramafic mine tailings have potential use for mineral carbonation as they consist of minerals rich in calcium and magnesium like olivine, wollastonite, brucite, and serpentine. Although the mine tailings have a higher surface reactivity than subsurface geologic formations, most minerals in mafic and ultramafic rocks have relatively slow reaction rates in surface conditions. Some enhanced methods that accelerate the carbonate precipitation, such as sparging air, air-saturated water, or CO_2-rich gas through the tailings or adding microorganisms, could be considered additionally. Here an initial approach of active and paralyzed mines of nickel, copper, chromium, and vanadium, whose rock hosts are gabbros, serpentinites, and ultramafic, in Brazil is presented. Tailings, most often in the form of a wet slurry, are conventionally stored above ground behind earthen dams. In Brazil, a comprehensive national registry of dams, including tailings dams, was recently generated by Agência Nacional de Mineração (ANM) (https://app.anm.gov.br/SIGBM/Publico). This initial study totaled 11 mines of different resources in distinct geologic provinces (Fig. 1) with a total sum of 84 million cubic meters of tailings with potential for mineral carbonation. In addition to tailings dams, waste rock piles should also be considered, as they would have a larger accessible area exposed for carbonate precipitation and because they are safer, given that the dams would need additional detailed monitoring to minimize the risk of leaks. Figure 1 also highlights the extensive occurrence of basalts in the southern and southeastern regions of Brazil referring to the Serra Geral Formation, where several small to medium gravel quarries occur. The known occurrence of the Serra Geral basalts in depth also presents the potential for in situ carbon mineralization through injection and circulation of CO_2-bearing fluids through this formation. Future detailed studies of mineralogical characterization in each case are necessary for a better evaluation of the best methods and techniques to be applied in each mineral deposit. Although mineral carbonation is a small storage opportunity in terms of CO_2 volume, the mines could still offset CO_2 emissions and provide local solutions for regions with concentrated CO_2 sources.

Fig. 1. Mines with potential for carbon mineralization for CO_2 storage in Brazil, with basalt, gabbro, and ultramafic formations.

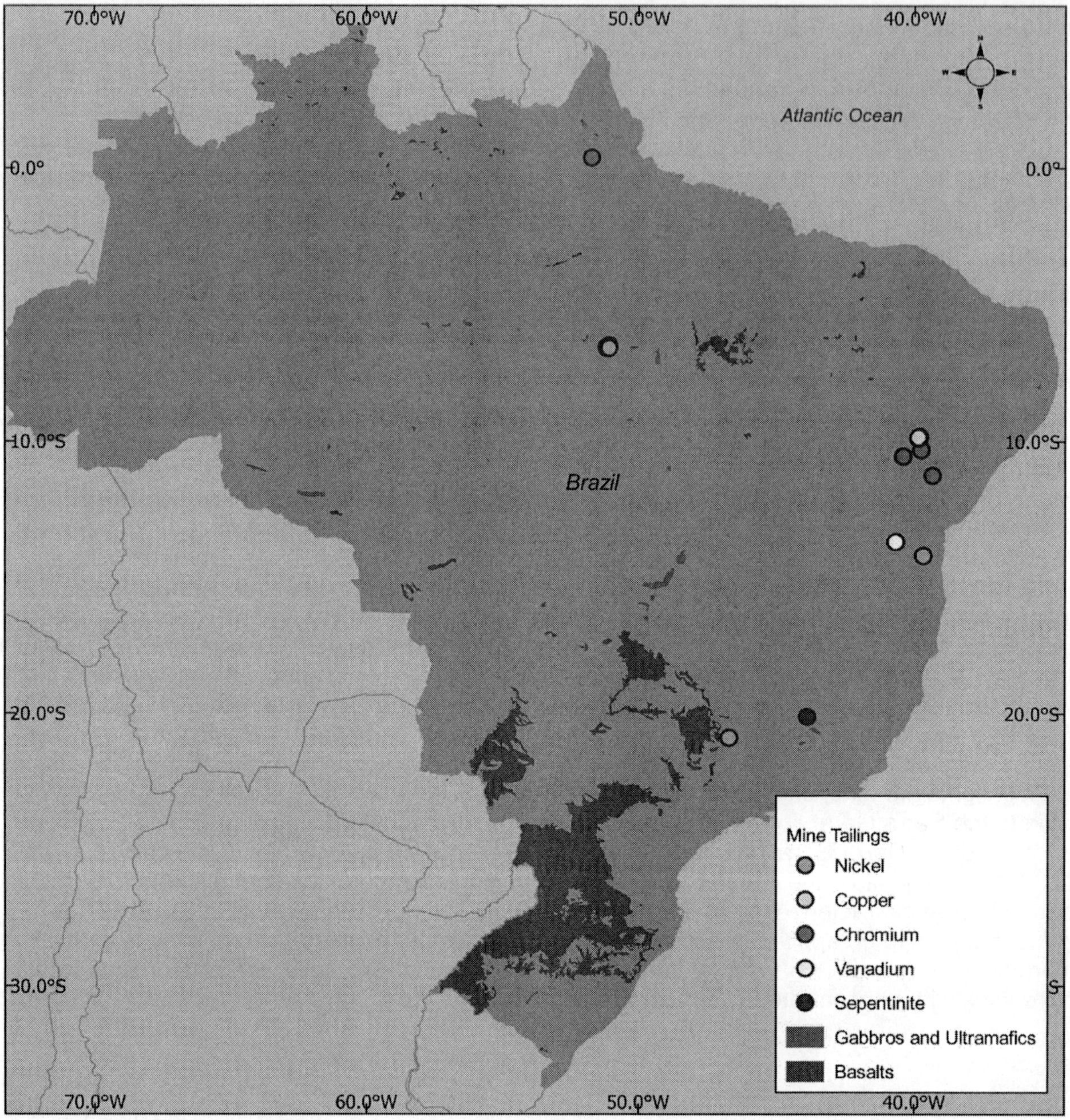

ST.034
N'tsi Tatay: A New Look at Windy Craggy

Bruce Downing[1], Rick Van Nieuwenhuyse[3], James Allen[2]
1. none, Langley, BC, Canada, 2. former chief of Champagne Aishihik First Nation, Haines Junction, YT, Canada, 3. CEO, Contango Ore Inc, Fairbanks, AK, USA

N'tsi Tatay is North America's largest resource of cobalt with excellent copper, zinc, gold, and silver grades hosted in an Upper Triassic Besshi-style volcanogenic massive sulfide. This deposit occurs within the Alexander Triassic metallogenic belt that has excellent potential for future discoveries.

The deposit is comprised of three vertical massive sulfide zones termed North, South, and Ridge. The 1992 historical resource estimate is 297,000,000 tonnes grading 1.38% Cu, 0.069% Co, 0.20 g/t Au, and .83 g/t Ag using a 0.5% copper cutoff grade. This estimate should be considered a minimum, as a new zinc-rich (Ridge) zone was discovered at the end of the last drilling campaign in 1990 and remains open. In addition, a supergene zone enriched in copper is capped by gold-silver-rich gossans.

Significant gold is associated with a chert-carbonate-sulfide unit. One drill hole (83-14) intersected 29.7 metres grading 14.72 grammes/tonne gold with microscopic native gold and electrum occurring as 7- to 8-micron-diameter grains.

Mineralization is comprised of chalcopyrite, pyrrhotite, pyrite, and sphalerite with lesser marcasite, galena, digenite, arsenopyrite, an unidentified bismuth telluride, cobaltite, cubanite, native gold, electrum, and native silver. Cobalt generally occurs within the sulphides (primarily pyrrhotite) as a solid solution. Primary textures and geologic relationships are well preserved for the most part. Gangue components include quartz, iron carbonates, magnetite, chlorite, and calcite.

From 1988 to 1991, approximately $50,000,000 in expenditures were spent completing 4,139 metres of underground development and 4,618 metres of diamond drilling from surface and underground drill sites as well as the completion of a preliminary feasibility study.

N'tsi Tatay was discovered in 1957 by prospecting under the supervision of Mr. James McDougall (an inductee into the Canadian Mining Hall of Fame in 2006). After years of highly successful exploration the project was shuttered by inclusion in a park (1993) by Provincial and Federal politics without due consultation and compensation for the land claims and ancestral territory of the rightful First Nation owners. This event has led to an unparalleled historical legacy of over one billion dollars invested and spent on bringing other mineral deposits to fruition but nothing for the impacted First Nations.

Recent innovations in recovery technology, mine development, block caving, backfill paste and dry stack tailings storage techniques, environmental management and mitigation strategies, combined with the Canadian recognition and implementation of the United Nations Declaration on the Rights of Indigenous Peoples (UNDRIP), necessitate that this high-quality resource be granted another look after 63 years.

There are numerous social and economic benefits from this development for all stakeholders. This is the time in which to showcase a vision of development with ownership and leadership by an Indigenous Nation entity that respects their culture, the environment, park sustainability, and economic realities. In a world that values a green economy requiring huge amounts of copper and ethically sourced cobalt, N'tsi Tatay warrants consideration.

The historical Southern Tutchone translation of N'tsi (windy) Tatay (mountainous) is Windy Craggy. This deposit is located in northwestern British Columbia, Canada.

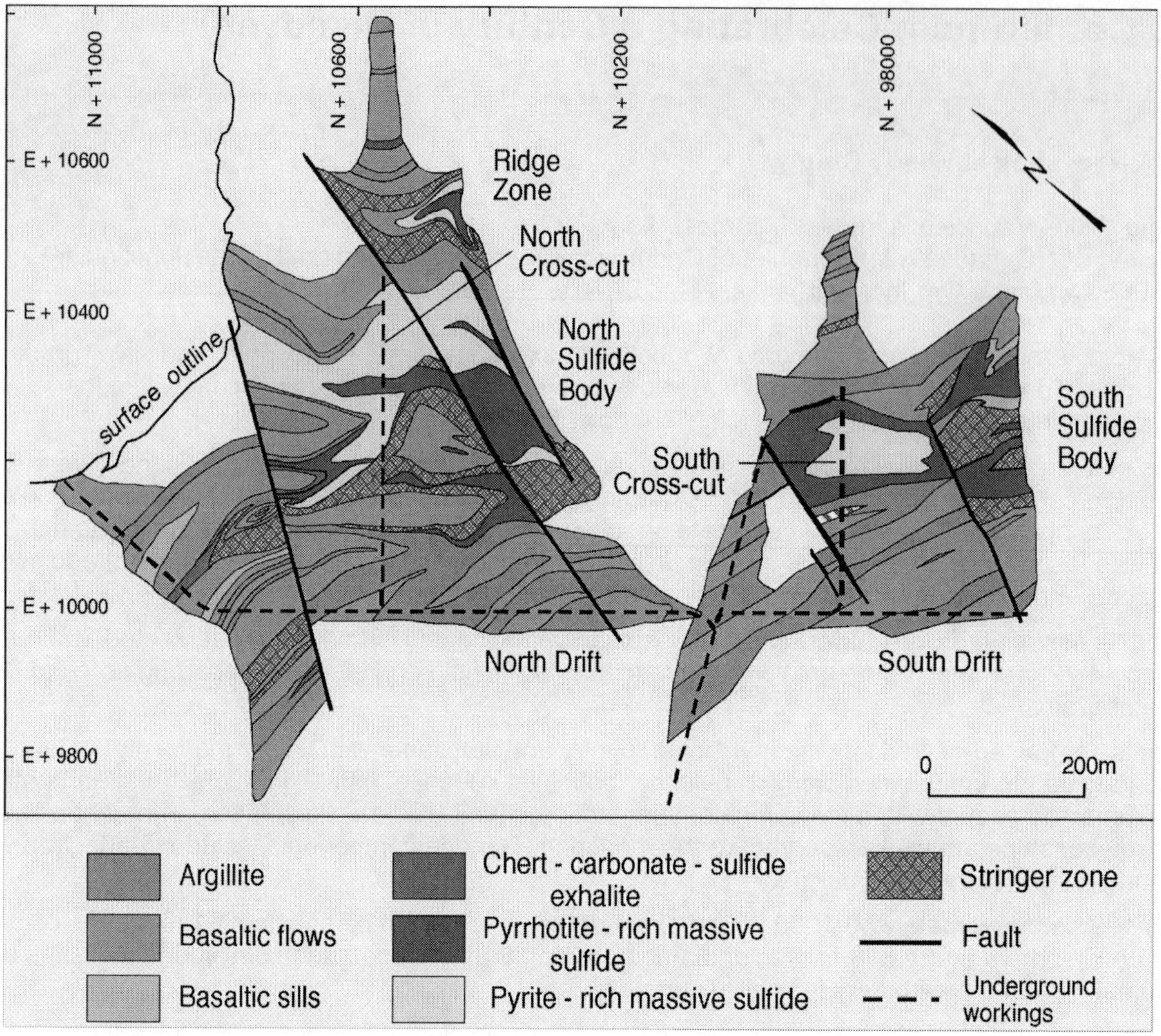

N + 11000
N + 10600
N + 10200
N + 98000
E + 10600
E + 10400
E + 10000
E + 9800
N
Ridge Zone
North Cross-cut
North Sulfide Body
surface outline
South Cross-cut
South Sulfide Body
North Drift
South Drift
0
200m
Argillite
Basaltic flows
Basaltic sills
Chert - carbonate - sulfide exhalite
Pyrrhotite - rich massive sulfide
Pyrite - rich massive sulfide
Stringer zone
Fault
Underground workings

ST.035
A New LA-ICP-MS Approach to Quantifying Variable Gold Deportment within Pyrite for Complex Orebodies

Angela J. Escolme, Leonid V. Danyushevsky
CODES - University of Tasmania, Sandy Bay, TAS, Australia

Variable Au content in pyrites from complex orebodies presents opportunities for mineral processing optimization. In this study we establish a routine method for laser ablation-inductively coupled plasma-mass spectrometry (LA-ICP-MS) analysis of pyrite to quantify the proportion of pyrite with high Au and the distribution of Au within pyrite grains (e.g., high Au-bearing pyrite rims) in mineral processing samples.

Pyrite can host a wide range of trace elements, including Au, in variable concentrations reflecting formation temperature, absorption properties, grain size, fluid composition, and availability of Fe and/or S. In overprinted porphyry systems, pyrite grains may reflect a complex history of pyrite precipitation and dissolution under changing conditions. As a result, the Au content of pyrite may be highly variable both between grains and within individual grains. Three end members are recognized, early porphyry-stage pyrite with low Au, late epithermal-stage pyrite with high Au, and composite pyrite grains with low-Au cores and high-Au rims (Sykora, 2018). This phenomenon provides opportunities to optimize mineral processing for refractory Au if the distribution of Au in pyrite can be quantified at various stages of mineral processing.

We present a new method for routine LA-ICP-MS analysis of pyrite in order to quantify variable Au deportment in pyrite using a case study of crushed and size-sorted samples from mineral processing samples from an overprinted porphyry system. In this method, two perpendicular ablation lines are run from core to rim on each pyrite grain/fragment using a 4-µm beam, moving at 3 µm/sec, firing at 20 Hz with a resulting in the line depth of ~2.5 µm. The ICP-MS signals for Fe (2 ms), As (5 ms), and Au (20 ms) were measured with a total sweep time of 37 ms. Line data was interpreted assuming a simple cuboid model to estimate the volume of Au-rich pyrite and low-Au pyrite as well as the proportion of Au contained in outer rims of pyrite.

Results were validated against diagnostic leach data for each size fraction. Data sets collected using the method described were interrogated for the following information: a) pyrite grain/fragment size, b) presence of Au-rich rims, c) Au concentration in high-Au pyrite, c) thickness of Au-rich rims, and d) variability in all the former by size fraction. We propose that this method can be applied to assess the characteristics of ore feed as well as gauge processing efficiencies in the treatment of refractory Au in pyrite and could be adapted to rapidly review the deportment of critical metals (e.g., Co, Te) in pyrite in crushed materials.

ST.036
The Distribution and Redox State of Arsenic in Low- to Medium-Temperature Massive Sulfides

Daniel D. Gregory[4], Anthony Chappaz[1], John Cliff[2], Daniel E. Perea[2], Sandra Taylor[2], Ivan Belousov[3]
1. University of Central Michigan, Mount Pleasant, MI, USA, 2. Environmental Molecular Sciences Laboratory, Pacific National Laboratory, Richland, WA, USA, 3. CODES, University of Tasmania, Hobart, TAS, Australia, 4. University of Toronto, Toronto, ON, Canada

Arsenic is an important element in economic geology, both in its co-occurrence with Au and in its potential adverse environmental impacts after mine closures. In addition to individual As minerals such as arsenopyrite, realgar, or orpiment, pyrite can often contain large amounts of As. Arsenic redox states have been argued to affect the uptake of several trace elements. When As substitutes for S(-II), potentially as As(-I), it is thought that incorporation of elements like Co and Ni are enhanced. Conversely, when As—potentially as As(+III)—substitutes for Fe(+II), it results in a dislocation in the pyrite structure that accommodates large cations like Au. Additionally, how the pyrite structure is altered by the presence of As displaying different redox states may affect its oxidation rate.

In this contribution we use a series of microanalytical techniques to document the variability of As content from the cm to nm scale. We first utilize LA-ICPMS spot analyses to track As content in different pyrite textures in samples from the Leicester pyrite member, New York, and the Black Bute SEDEX deposit, Montana. Areas representative of the most abundant textures were then selected for LA-ICP-MS mapping. These analyses highlight the variation in As and other associated trace elements across the cm-scale samples (with the spot data) and at the mm scale in the different maps. This is followed with NanoSIMS trace element mapping of pyrite textures of interest informed by the LA-ICP-MS analyses. The NanoSIMS analyses revealed microscale enrichments of As, often at the intersection of different generations of pyrite. These interfaces were targeted for atom probe tomography, which identified late As enrichments between the microcrystals that make up pyrite framboids. In addition to elemental concentrations, we also investigated As speciation using μ-XANES analyses. Taken together, our new findings suggest As spatial distribution is quite heterogeneous in the pyrite and that the redox state of the As may range from (-II) to (+III). This is likely to significantly affect both where less abundant elements, like Au, are to be found in the pyrite and how easily extractable these elements are likely to be. These findings will have important implications for the effectiveness of heap leach processes. Incorporating microanalytical techniques should be considered to implement new strategies to improve our management of different types of pyrite in tailings facilities.

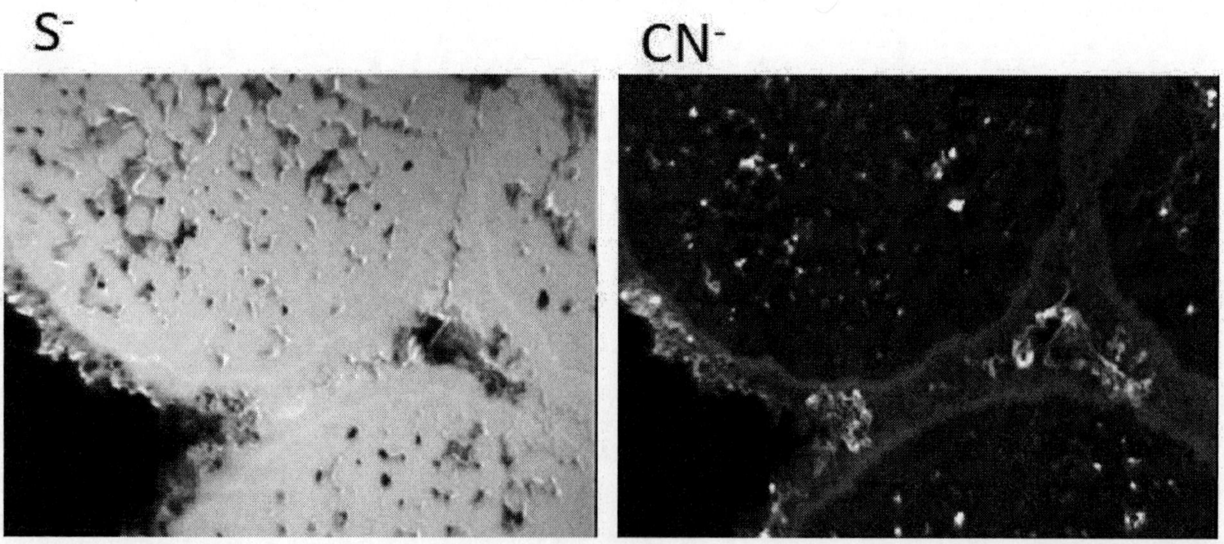

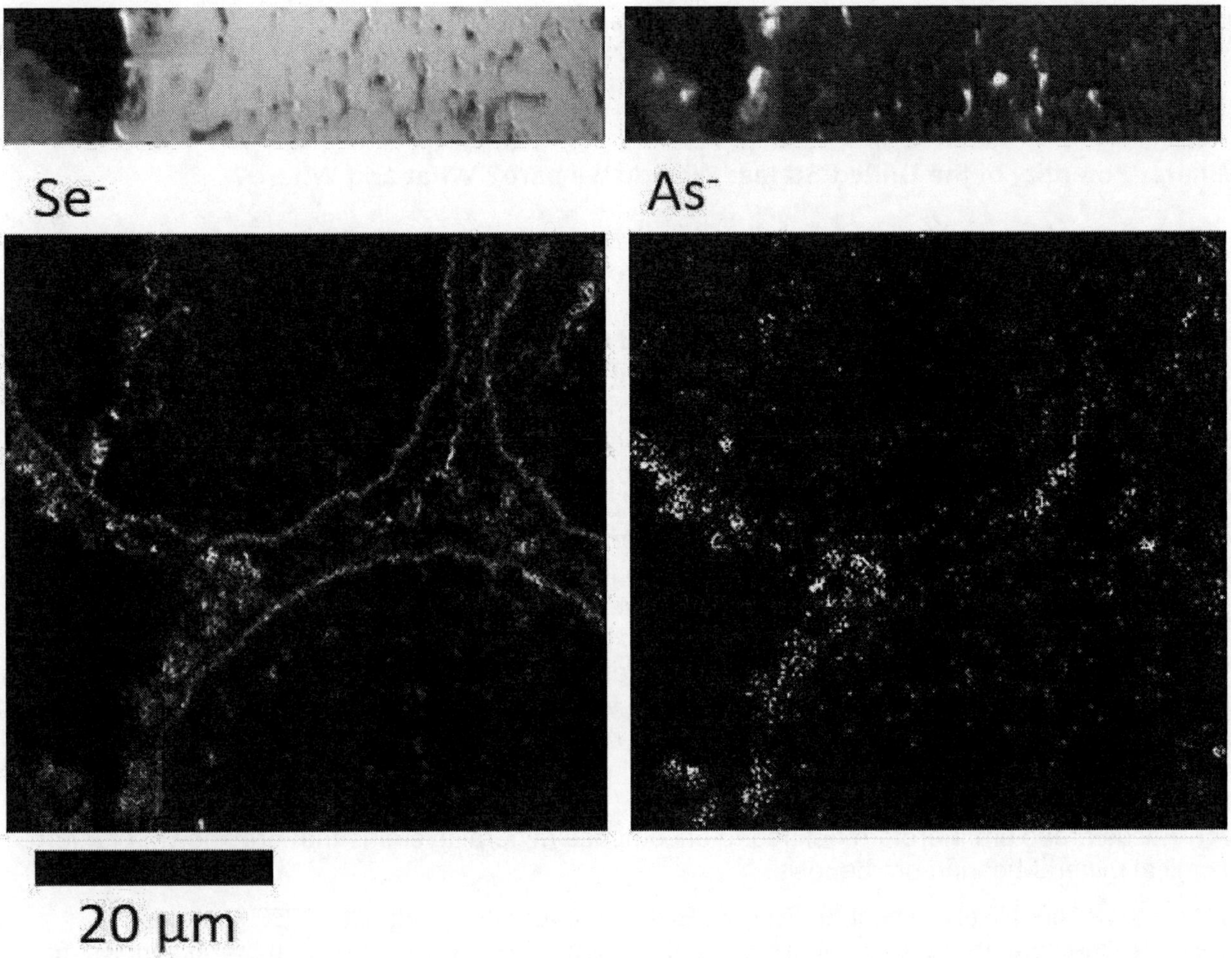

Figure 1: NanoSIMS trace element maps of pyrite from the Leicester pyrite member. Note that the As is primarily enriched on the outer rim of the pyrite framboid, similar to, but at a later stage than, the Se enrichment. Further, these trace elements are associated with organic matter (as indicated by the CN^- map).

ST.038

Critical Mineral Potential of the United States: Why do we care? What and Where?

Jane M. Hammarstrom[1], Albert H. Hofstra[2], Warren C. Day[2]
1. USGS, Reston, VA, USA, 2. USGS, Denver, CO, USA

The U.S. is 100% import reliant for graphite, niobium, and rare earth elements (REEs), and more than 50% reliant for cobalt, tin, tungsten, and other commodities that are increasingly in demand for emerging technologies. Interest in identifying domestic sources of critical minerals in the United States led to mandates from the government for the USGS to develop criteria for defining criticality, establish a list of critical mineral commodities, and partner with State Geological Surveys and others to acquire new framework geologic data (mapping, geophysics, lidar) for evaluating the potential for new and as-yet-undiscovered critical mineral resources. To guide selection of areas for new data collection, the Earth MRI (Mapping Resources Initiative) was established to outline focus areas that could contain critical mineral resources based on existing geologic data. The first phase of Earth MRI identified focus areas for REEs; the second phase considered focus areas for aluminum (bauxite, alunite), cobalt, graphite, lithium, niobium-tantalum, platinum-group elements (PGEs), tin, titanium, and tungsten. These commodities occur in various deposit types that form in diverse mineral systems; these systems operate in specific geotectonic settings to form a variety of genetically related ore deposits. Mineral systems are much larger than ore deposits and have features that are more likely to be detected by new framework geologic data. Identification of one part of a large mineral system raises the possibility that related ore deposit types may be present nearby or under cover. Focus areas are therefore defined to encompass mineral systems that are known, or inferred, to contain critical mineral-bearing ore deposits.

The mineral systems most likely to host future U.S. REE resources are magmatic REE systems (carbonatites, alkaline igneous rocks), placers and paleoplacers (monazite in heavy mineral sands), marine chemocline systems (REE-rich phosphorites), and deposit types associated with iron oxide-apatite and iron oxide copper-gold systems (Fig. 1). Focus areas for mid-Paleozoic phosphates that could host REEs were selected from formations identified on state-scale geologic maps. Generalized outlines of known mining districts or mineral belts, distributions of occurrences, and geochemical or geophysical anomalies are the basis for some focus areas. Mafic magmatic systems that host PGEs and cobalt, for example, are partly exposed in the mid-continent rift area of the central U.S. where new geophysical data are crucial for evaluating potential buried resources. Some focus areas contain known mineral deposits or occurrences of critical mineral commodities that are actively mined or may have produced in the past. Other focus areas are geologically permissive for the occurrence of critical minerals because the focus area has evidence of the presence of particular mineral systems. Priorities for selecting specific focus areas for new work consider the land status relative to exploration and development potential, the data needs for defining target areas, and the likelihood that new discoveries could reduce import reliance. National and regional maps represent a first step in selecting areas for new data acquisition to evaluate the critical mineral potential of the United States.

Fig. 1. Focus areas for REE-bearing mineral systems in the United States.

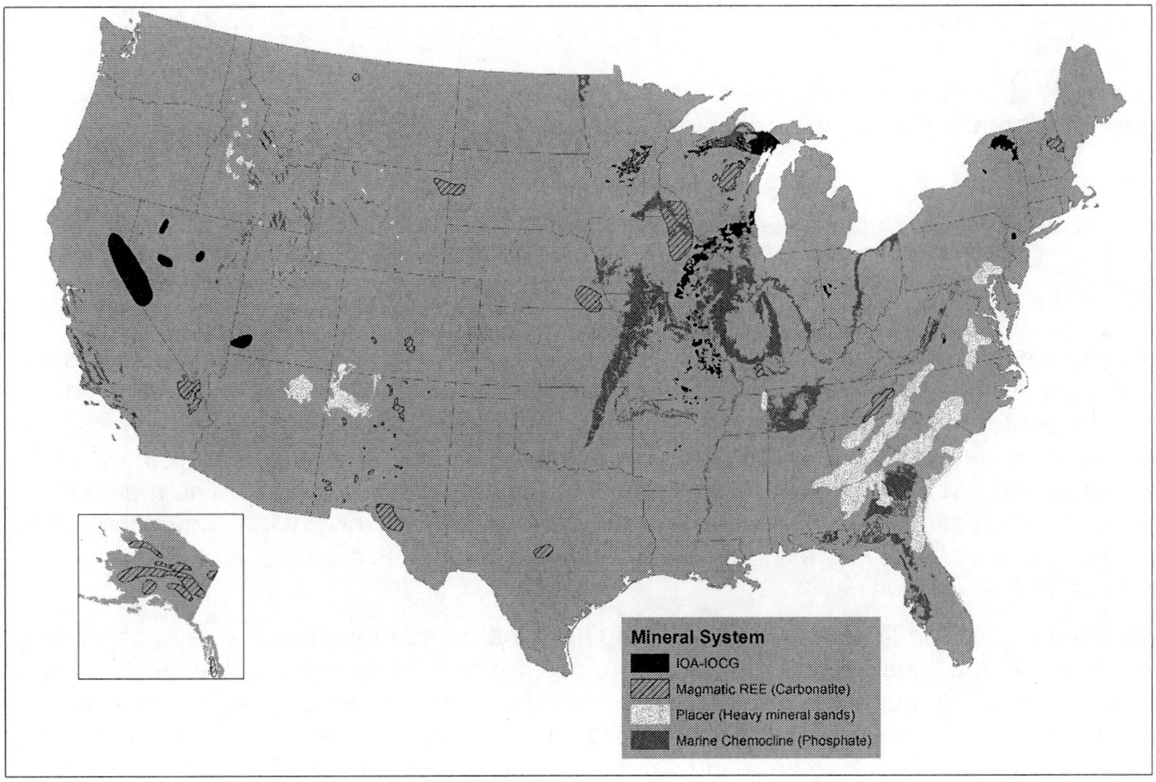

ST.039

From Deep Space to Shallow Crusts and Asteroids: Muons as One Possible Tool in the Future of Mineral Exploration, Planetary Missions, and Asteroid Mining

Marko Holma[1, 2, 3], Pasi Kuusiniemi[1, 2, 4], Timo Enqvist[1], Kai Loo[1, 4]
1. Muon Solutions Oy, Saarenkylä, Finland, 2. Arctic Planetary Science Institute, Rovaniemi, Finland, 3. Kerttu Saalasti Institute, University of Oulu, Oulu, Finland, 4. Virtual Muography Institute, Operates internationally, Finland

Many mineral deposits are associated with drastic density gradients in the form of lithological contacts, structural breaks, the excess accumulation of dense (or low-density) minerals, hydrothermal alteration, and/or weathering. Variations in rock densities can be geophysically followed by a variety of methods, the principles of most based on classic physics. One of the fields of physics not yet applied widely in mineral exploration is that of muography, although its principles were introduced already in 1955. The method is currently experiencing a renaissance due to new-generation semiconductors, increasing computing power, and advanced particle physics codes. However, in geosciences the method is currently mostly used in volcanology, not in mineral exploration.

Muons are heavy, charged electron-like particles originating from the upper atmosphere due to constant interactions between high-energy cosmic rays and air molecules. Some muons have such high energies that they travel in rocks for tens to hundreds of meters—the really rare ones even up to a few kilometers. While muons travel through dense materials, they lose energy until attenuated. This phenomenon is directly related to an average density of the rock column and can be exploited in muography for constructing X-ray-like density variation maps of the rock material between the ground and the muon detector(s). Muography applied to mineral exploration is thus based on extracting relative rock densities in various directions from the statistics of muons passing through the detector (or an array of detectors).

We anticipate muography to find its place it the plethora of geophysical methods in the following few decades. One challenge is that muography originates outside the realm of classic physics and hence most geoscientists are not yet familiar with its principles. However, as the technology matures it may become a competing option for detecting density variations in both greenfield and brownfield exploration. There already are solutions to apply muography via tunnels and drill holes, and we can foresee that method development will continue for many more decades. Preliminary analyses show that, theoretically, muography can be applied at least up to 2-km depth and brought to celestial objects like the Moon and Mars. If the atmosphere is thin (Mars) or there is no atmosphere whatsoever (Moon, asteroids), other natural muon generation mechanics may occur. At the Moon and Mars, muography could be used to locate resources (water ice and other raw materials for in situ resource utilization, ISRU) and caverns for human settlements, storage space, and the search for life. The imaging of the asteroids using muons, on the other hand, may provide a useful tool for screening which ones have the highest potential for mining. In the time frame of 100 years, many such activities could be automatized by robots and AI. It is also possible that will be able to manufacture transportable muon generators in the distant future. With those, all types of muon-based imaging techniques could be brought to the mainstream and as deep in terrestrial crusts as there are tunnels available.

ST.040

Exploration of Lunar In Situ Resources Can Be Conducted by Applying Density-Sensitive Cosmic-Ray-Based Geophysical Muon Imaging Method Called Muography

Timo Enqvist[1], Marko Holma[1, 2, 3], Pasi Kuusiniemi[1, 2, 4], Kai Loo[1, 4]
1. Muon Solutions Oy, Saarenkylä, Finland, 2. Arctic Planetary Science Institute, Rovaniemi, Finland, 3. Kerttu Saalasti Institute, University of Oulu, Oulu, Finland, 4. Virtual Muography Institute, Operates internationally, Finland

In the future, the mining industry becomes increasingly challenged by the need for applying its state-of-the-art techniques off the Earth. A permanent human presence on the Moon, for example, requires effective new technologies that will enable the in situ resource utilization (ISRU). On the Moon, the ISRU concept targets for developing methods for the exploration and characterization of a wide range of local resources, including water ice, helium-3, oxygen, regolith, and metals. We propose that many of these essential resources are related to density anomalies (e.g., high water content, high porosity) and can possibly be detected and mapped by a novel geophysical density imaging method called muography.

On Earth, muography is based on the detection of muons that are constantly produced in the interactions between air molecules and primary cosmic rays. The attenuation of muon flux is measured after the particles have passed the object, such as a volcano or a large body of rock (e.g., measurements are performed in caves or tunnels). The measured muon fluxes from different directions are finally translated to 2D or 3D density models based on the fact that high average densities reduce the flux faster than those of low. However, as the Moon has no atmosphere, lunar muography requires that muons are produced in the interactions between primary cosmic-ray particles and the uppermost parts of the lunar soil. Our preliminary results suggest that muons are indeed generated in the uppermost lunar soil. Hence, muography would work on the Moon.

Furthermore, and perhaps surprisingly, the energy appears to have only a small impact in the production depth of muons. This is illustrated in the figure, which shows the muon production depth distributions for three different selected primary cosmic-ray particle energies in petaelectron volts (PeV). Note the logarithmic scale in the vertical axis. The unit of the horizontal axis is mwe (meter water equivalent), which is a common reference in these types of particle physical simulations (1 mwe equals 1-m-thick column of water). For other materials, such as for high-porosity lunar regolith, the absolute depth is obtained by dividing the mwe depth with the density of the material (i.e., for the density of 2.65 g/cm^3, the depth of 5.0 mwe corresponds to 1.89 m). The differences in muon counts/mwe in the first meters can also be due to statistical uncertainties. The two arrows indicate the approximative depths where a half (50%) and every 10th (10%) of the produced muons are able to penetrate.

Our next aim is to determine the absolute muon flux as a function of the depth on the Moon. This is work in progress, and as the first step we have used, as shown, protons as primary cosmic-ray particles, water as the target material, and simulated three different primary energies: 1 PeV, 3 PeV, and 10 PeV. The number of simulated primaries were 1,000, 375, and 125, respectively. The zenith angle of incoming protons was selected to be uniform and limited to 75°. Simulations were carried out using the Fluka simulation package.

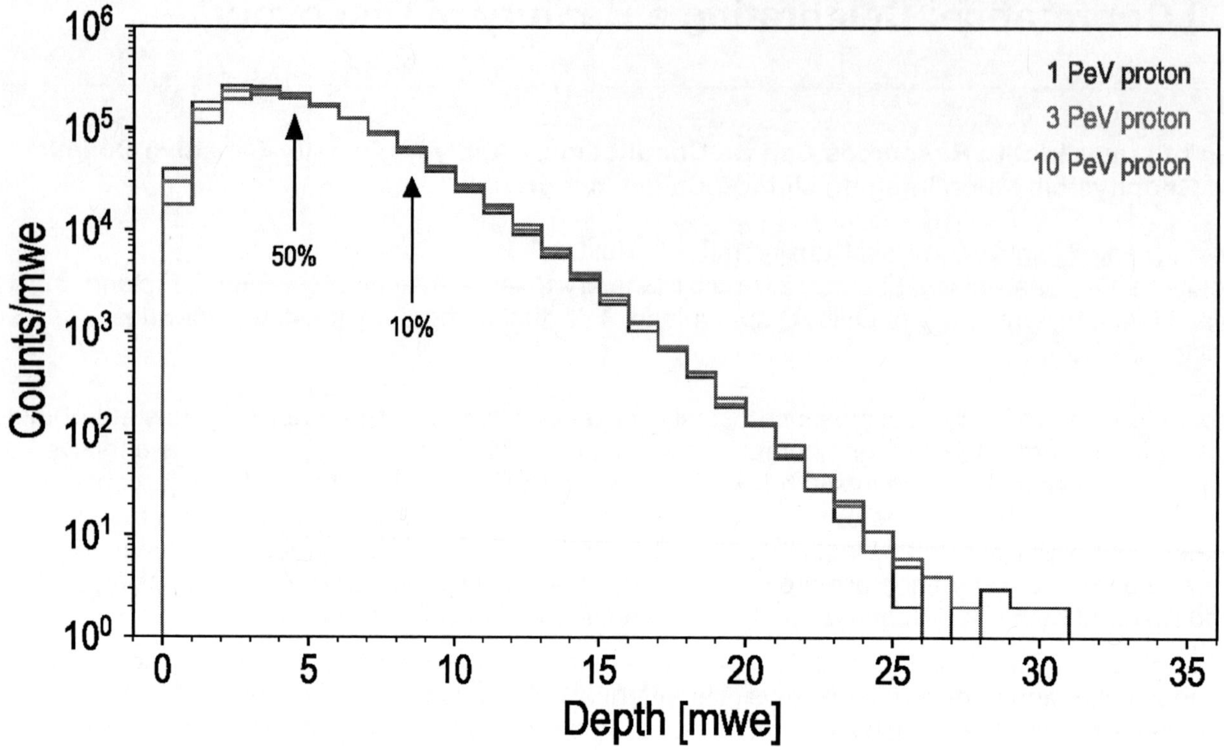

1 PeV proton
3 PeV proton
10 PeV proton
50%
10%
Counts/mwe
Depth [mwe]

ST.041
Increasing Efficiency Through Understanding Geological Controls on Grade Engineering

Julie Hunt[1], Nathaly Guerrero[1], Greg Wilkie[2]
1. CODES - UTAS, Hobart, TAS, Australia, 2. CRC ORE, Brisbane, QLD, Australia

CRC ORE (Cooperative Research Centre for Optimising Resource Extraction, www.crcore.org.au) works to improve the productivity, energy, and water signatures of mining operations and has developed Grade Engineering®, which is a range of technologies to reject nonvaluable material at a coarse particle size (~10–100 mm) early in the mining value chain. This reduces the amount of material sent for further size reduction and in doing so significantly reduces the amount of energy and water use per unit of metal produced.

Grade Engineering® is based on combinations of screening, sensor-based sorting, and heavy media separation related to five rock-based levers: preferential grade deportment by size, differential blasting for grade by size, sensor-based bulk and stream sorting, and coarse gravity separation. CRC ORE, in collaboration with CODES (Centre for Ore Deposit and Earth Sciences, www.utas.edu.au/codes), is working to identify relationships between geologic parameters (e.g., mineralogy, texture, hardness) and Grade Engineering®, particularly preferential grade deportment by size.

Preferential grade deportment by size is the propensity for some ores to exhibit preferential breakage leading to concentration of minerals in specific size fractions. This typically involves an increase of valuable mineral phases in finer size fractions. For example, at an Australian gold deposit ore screened into +50, 50 to 19, and -19 mm size fractions showed that >85% of the gold is in the -19-mm particles, and this size fraction makes up only ~25 to 35% of the total mass. Screening of subeconomic ores at a South American Zn-Pb-Ag operation into +70, 70 to 19, and -19 mm size fractions demonstrated that most of the value is in the -19-mm size fractions, which make up only ~25% of the mass.

Anecdotal information suggests that preferential grade deportment by size response is related to geologic textural parameters—for example, mineralization styles dominated by veins are likely to have a higher preferential grade deportment by size response than disseminated mineralization styles. Early research results indicate this is true for at least some ores. Vein and disseminated ore from a porphyry copper deposit shows clear differences in Response Ranking (RR), a measure of preferential grade deportment by size developed by CRC ORE (Fig. 1 a, b). Where veins are important, vein abundance, paragenesis, and alteration can be significant with respect to RR, as illustrated by samples from a gold porphyry deposit (Fig. 1c, d). The aim is to use various techniques (detailed logging, hyperspectral mineralogy, high resolution images, etc.) to identify geologic parameters that can be linked to RR. Ideally these will be measured (semi)-automatically in drill core. The parameters will be used to predict RR and allow it to be mapped and included in 3D models.

Fig. 1. Porphyry copper ore – a) vein-hosted (1.2% Cu), b) disseminated sulphides (0.7% Cu); Gold porphyry ore – c) multiple veins and significant potassic alteration (0.5 ppm Au); d) single vein and minor potassic alteration (0.3 ppm Au). Drill core is ~50 mm across. RR = response ranking.

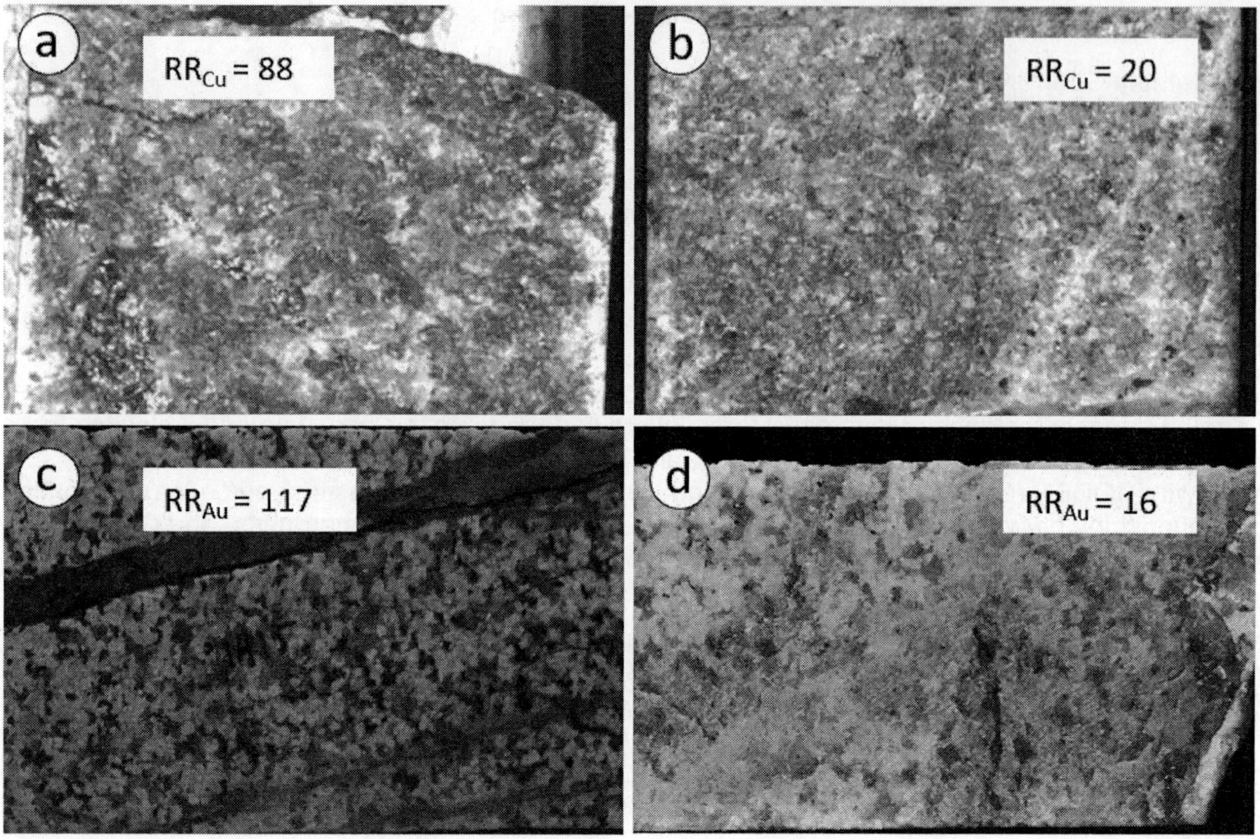

ST.042
Reading Minerals: Rare Element Enrichment, the Magmatic-Hydrothermal Transition, and Geochemical Exploration of Lithium Pegmatites in Ireland

David Kaeter[1, 2], Renata Barros[2, 3], Julian F. Menuge[1, 2], John Harrop[4]
1. iCRAG - SFI Research Centre for Applied Geosciences, Dublin, Ireland, 2. School of Earth Sciences, University College Dublin, Dublin, Ireland, 3. Geological Survey of Belgium, Brussels, Belgium, 4. Coast Mountain Geological Ltd., Vancouver, BC, Canada

The battery market for electric vehicles and renewable energy storage is dominated by rechargeable lithium-ion batteries, making lithium supply essential to climate action through decarbonization. In 2019, more than half of the world's lithium was sourced from lithium pegmatites of the Li-Cs-Ta (LCT) family, predominantly from Australia. Current global lithium supply involves long diesel-fueled maritime transport routes, which counteracts lithium's role in climate action. Responsible consumption and production require shorter supply chains from deposit to battery.

Reading the mineralogical record of LCT pegmatite deposits can help address the challenge of reducing the climate impact of lithium production, by informing deposit models, mineral exploration, and geometallurgy, therefore promoting local supply. Our research focuses on a belt of LCT pegmatites, which is located along the eastern margin of the late-Caledonian S-type Leinster Batholith, southeast Ireland. The LCT pegmatites are hosted by a major regional shear zone and are part of a tin-lithium province that stretches subparallel to the Iapetus suture from Europe through Nova Scotia to North and South Carolina.

We investigated crystal chemical zoning in muscovite, cassiterite, and columbite-tantalite using petrography, scanning electron microscopy, and LA-ICP-MS chemical mapping. The zoning patterns record that pegmatite rare element mineralization resulted from an interplay of magmatic crystallization, metasomatism, and hydrothermal processes. Late-stage metasomatic alteration led to partial resorption of early minerals including the lithium ore-mineral spodumene, followed by dispersion of lithium and other rare elements into country rocks, mostly within dark mica. Dispersion led to formation of geochemical halos around the LCT pegmatites with the potential to use country-rock lithogeochemistry and mica composition as geochemical vectoring tools.

Geochemistry of mica in the granite host analyzed by handheld LIBS has been found to exhibit coherent spatial patterns occurring adjacent to and above LCT pegmatites known at depth from drilling. These channels of mineral-specific geochemical information are distinct from soil geochemical patterns and are not influenced by the same secondary, surface processes such as dilution.

As outcrop is virtually absent in the study area, regional stream sediment geochemistry data (Geological Survey Ireland) was assessed as an LCT pegmatite exploration tool. After correcting for geologic background using a linear regression approach, catchments containing LCT pegmatites show high residuals for concentrations of both tantalum and tin. The mineralogy of stream sediment samples from a subsample of these catchments was subsequently analyzed to characterize the host minerals of tin and tantalum. Cassiterite and columbite-tantalite were identified, and both show geochemical and textural signatures that correspond to the zoning patterns mentioned above, which indicates that these minerals were derived from the local LCT pegmatites.

These results suggest that, when regional geology and tectonic setting are prospective, lithium pegmatite prospectivity can be further assessed for tin-tantalum associations in (often publicly available) regional stream sediment data. Following geospatial analysis, stream sediment samples could be obtained from individual prospective catchments and their mineralogy analyzed. Local-scale geochemical surveys could follow where stream sediments of prospective catchments contain tin and tantalum oxides with chemistries and textures indicative of a lithium pegmatite source.

ST.044
Cluster Analysis of Magnetite Geochemistry to Infer Subtypes of Magnetite

Timothy C. Lui, Daniel D. Gregory
Department of Earth Sciences, University of Toronto, Toronto, ON, Canada

The current classification of minerals is based on idealized crystal structures as determined by X-ray diffraction. In reality minerals form in many dissimilar environments, and differences in a variety of properties can provide insight into the subtypes of a mineral that its crystal structure cannot. For example, magnetites that form from igneous versus hydrothermal environments were found to contain differences in trace element abundance. However, more formation environments exist and many elements can be analyzed. The goal of the study is to determine subtypes of magnetite via applying machine learning techniques to elemental abundances.

This study compiles more than 25,000 geochemical analyses on magnetite from around 1,200 papers, mostly from electron microprobe analysis and laser ablation-inductively coupled plasma-mass spectrometry studies. We also add our own geochemical data into this data set. The resulting data set functions as a large and representative feature space to run unsupervised learning algorithms on. Then, the study compares multiple clustering algorithms on this data set. Different clustering algorithms use varying properties to define their clusters; hence, we will test multiple algorithms to observe which model performs best. K-means and partitioning around medoids are two algorithms that function as centroid models; hierarchical cluster analysis functions as a connectivity model, Gaussian Mixture Modeling as a distribution model, and density-based spatial clustering of applications with noise (DBSCAN) as a density model. After determining the most adequate clusters for our magnetite data set, the study interprets how these clusters give us information about the subtypes of magnetite that exist and their geochemical qualities.

Magnetite trace element concentration is currently used to determine formation environment for different magnetite samples. However, generally only a few trace elements are used, and as more data is collected the boundaries between different subtypes is becoming blurrier. Furthermore, because the classification scheme currently employed is completely empirical it is possible that important groups that do not fit the deposit type headings are being missed. By using machine learning algorithms on such a large and representative database we will be able to generate a more complete understanding of magnetite subtypes. Given that the approach is data driven, this study helps determine subtypes in a more unbiased manner.

The results of this study are significant to the scientific community since they provide comparisons for the efficacy of multiple unsupervised learning algorithms applied to large geochemical data sets. Results also help develop the understanding of magnetite deposits and the different subtypes and geochemical qualities of these.

ST.045
The Degree of Undercooling and Critical Metal Mineralization Potential of Granitic Pegmatites

Dalton M. McCaffrey, Simon Jowitt
University of Nevada, Las Vegas, Las Vegas, NV, USA

Granitic pegmatites are notorious amongst many other geologic systems for their ability to concentrate critical metals (e.g., Li, Ta, Be) to mineable abundances. However, the petrologic processes that concentrate these metals to economic abundances in granitic pegmatites have not been fully determined. Undercooling, or crystallization below the liquidus, has been shown experimentally to cause saturation and associated critical metal mineralization (i.e., form distinct critical metal phases) at much lower concentrations than in more typical (i.e., minimally or nonundercooled) haplogranite systems at 700°-750°C. This process can theoretically explain the formation of pegmatites enriched in critical metals, but undercooling is difficult to demonstrate in natural systems. Previous research has demonstrated undercooling with textural evidence, such as unidirectional growth textures, the presence of graphic granite, and other skeletal growth forms, in addition to a number of case studies that used a variety of geothermometers to quantify undercooling. However, comparing results obtained from the latter is difficult as a result of the variable zones that were sampled, the different thermometers used, and heterogeneity between pegmatite types and classifications.

This study presents a compilation of crystallization temperatures from over 100 granitic pegmatites classified via the Černý and Ercit (2005) pegmatite classification scheme. These temperatures were estimated using several different geothermometers, including fluid and melt inclusion microthermometry, two-feldspar solvus and other mineral exchange thermometers, Ti-in-mineral, O-isotope, and mineral saturation thermometry, and phase-assemblage relationships. In addition to the pegmatite type and thermometer used, the pegmatite zone sampled was also noted to determine the effect of sampling location on temperature of crystallization. Preliminary data from this study show that more than 75% of pegmatites with reported crystallization temperatures have a lithium-cesium-tantalum (LCT) affinity, the rare element class has the most reported crystallization temperatures by more than twice any other pegmatite class, and fluid inclusion microthermometry is the most widely utilized thermometer in these systems. Plotting all of the data from individual thermometers together (ignoring the effect of pegmatite class and zone) yields ranges of primary crystallization temperatures of ~275°C for Ti-in-quartz thermometry, ~200°C for two-feldspar thermometry, and ~500°C for primary fluid inclusion microthermometry. On average, pegmatite classes that host significant critical metal mineralization (i.e., rare element and miarolitic classes) experienced ≥100°C of undercooling, whereas barren pegmatites (i.e., abyssal and muscovite classes) experienced little to no undercooling. LCT pegmatites yield crystallization temperatures that average ~100°C lower than the niobium-yttrium-fluorine (NYF) pegmatite family. Different pegmatite zones (i.e., border-wall, intermediate, and core zones) yield variable crystallization temperatures that both increase and decrease from the outer zones inward. The range of primary intrapegmatite crystallization temperature, or the crystallization temperature range between pegmatite zones, may act as a significant control on critical metal mineralizing potential. This is exemplified by the Tanco pegmatite, one of the most well-documented highly evolved granitic pegmatites known, which records a 300°C range in crystallization from margin to center, much larger than any other pegmatite in this study. This suggests that intrapegmatite crystallization temperature may be a useful criterion for identifying granitic pegmatites with high critical metal mineralizing potential.

ST.046
Identifying Critical Mineral Resources in New Mexico

Virginia T. McLemore
New Mexico Bureau of Geology and Mineral Resources, New Mexico Institute of Mining and Technology, Socorro, NM, USA

Critical minerals are mineral resources that are essential to our economy and whose supply may be disrupted; many critical minerals are 100% imported into the U.S. Many of these critical minerals have been identified in New Mexico. Both uranium and potash are important commodities in NM and are considered critical minerals. New Mexico is the leading state in potash production and has produced significant uranium in the past. Recently, exploration has occurred for rare earth elements (REEs) in several areas in NM. REE-Th-U veins are found in the Gallinas, Capitan, and Cornudas Mountains and Laughlin Peak-Chico Hills; all are associated with Tertiary alkaline igneous rocks. Disseminated Y-Zr deposits in Proterozoic nepheline syenite are known at Pajarito Mountain (Mescalero Apache Indian Reservation). Tellurium production has occurred from the Lone Pine mine, Wilcox district, and beryllium has been produced from pegmatites in NM. Other critical minerals are associated with various mineral deposits in NM. For example, vanadium and molybdenum, by-products of uranium mining, as well as selenium and REEs, are associated with sandstone uranium deposits in the Grants uranium district. Rhenium is found in porphyry copper and porphyry molybdenum deposits in NM. Coal deposits are abundant in the state and could be source of several critical minerals (REEs, Se, V, Ge), but more work is needed to fully understand the distribution of critical minerals in NM coal deposits. Mining districts in NM that contain critical minerals have been previously identified. However, specific mines, occurrences, deposits, and favorable geologic terranes within these mining districts and elsewhere in NM have not been well defined for critical minerals. Therefore, geologists at NM Bureau of Geology and Mineral Resources (NMBGMR) are creating a database of critical minerals mines and occurrences in the state. We also are creating a database of existing chemical analyses, available drill core, and other subsurface information stored at NMBGMR. Legacy data as well as newly acquired data are being used. These databases will be available to government officials, mining companies, and the general public, in most cases online through the NMBGMR interactive map (see https://maps.nmt.edu/), open file reports, or by request. Not only are these data required in order to delineate favorable geologic terranes and priority areas containing potential critical minerals for the USGS mandate, but identification and examination of critical minerals is a high priority of the NMBGMR. These databases will aid in refinement of the geologic models of critical minerals, needed for exploration. This project also is important to the state of NM because mineral resources must be identified before land exchanges, withdrawals, or other land use decisions are made by government officials. Future mining of potential economic critical minerals will directly benefit the economy of NM. Another aspect of this project is the training of our future workforce since students at NM Tech will be hired to work on this project.

ST.047
Gold-Telluride Ore Deposits: Tellurium Mineral Economics, Resource Assessment, and Potential Value Add

Brian A. McNulty, Simon M. Jowitt
University of Nevada Las Vegas, Las Vegas, NV, USA

Tellurium is a key raw material required for manufacturing thin-film cadmium-telluride photovoltaic (CdTe PV) modules that will have a vital role in the transition to a carbon-neutral future. This reflects the fact that solar photovoltaics are currently one of the most rapidly deployed renewable energy technologies with global grid-connected capacity achieving some 707.5 gigawatts by the end of 2020. Of this, CdTe PVs contributed ~31 gigawatts, although annual installation predictions suggest this will increase to >250 gigawatts per year by the year 2050, indicating a likely rapid increase in demand for tellurium. However, global tellurium resources and production chains remain unclear, indicating that significant uncertainties exist over whether current supplies are sufficient to meet this forecasted increase in demand.

With the exception of the Kankberg gold-silver-tellurium mine in Sweden and unconfirmed production from tellurium-bismuth mines in China, over 90% (~400 tonnes) of annual tellurium production is sourced as a by-product of copper mining, typically from porphyry copper and volcanic-hosted massive sulfide deposits. However, the latter represents by-product tellurium that is only recoverable from copper anode slimes associated with pyrometallurgical metal production rather than hydrometallurgical approaches. In addition, of the 124 copper refineries worldwide, less than 10 recover tellurium. This reliance on copper production increases uncertainty around whether increased demand for tellurium can be met by current mining and refining practices.

One possible solution to a potential tellurium shortage is by-product tellurium derived from active gold mines. Accessory, precious metal-bearing tellurides are common in many gold deposits although tellurium concentrations are not routinely measured, making tellurium deportment difficult to quantify. Here we present the results of an analysis of 518 mineral deposits that are known to contain tellurium, of which 64 are active gold mines (Fig. 1). The tellurium potential of these active gold mines was assessed using three proxies based on publicly available tellurium assay data, telluride mineralogy, and tellurium concentrations in sulfide. These data suggest an unrealized by-product tellurium potential of 167,273 tonnes or ~300 years worth of current global tellurium production. These initial estimates require further research to verify their accuracy as well as to assess mine-specific recovery potential. Assuming a tellurium price of $70,000 per tonne, annual recoveries of 1.4 or 14.3 tonnes could yield 1% of an operation's cost of $10 or $100 million. This indicates that the potential recovery of this critical element may be economically beneficial. A recent example of this value add is the Kennecott mine where Rio Tinto is investing $2.9 million to construct a 20 tonne/year plant that will extract ~$1.4 million worth of tellurium that would otherwise report to waste. Tellurium exemplifies the opportunity to add value to active mines by considering the recovery of by-product critical elements. For some active gold mines, recovering tellurium that would otherwise report to waste could provide increased economic value, global resource stability, and the necessary raw materials for the technologies needed to achieve carbon neutrality.

Fig. 1. Key geological and geographical results of the identified 518 tellurium-bearing mineral deposits.

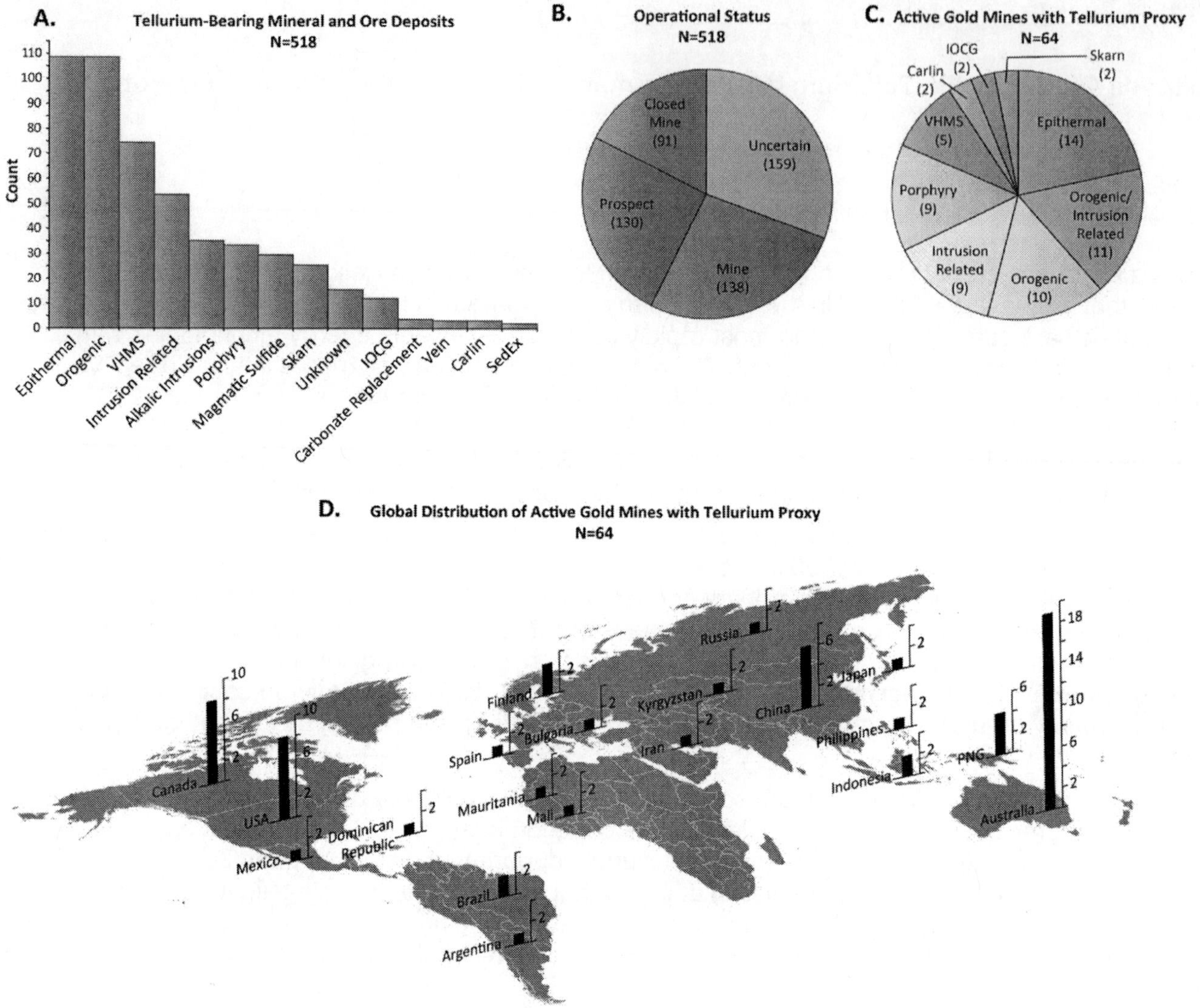

ST.049
Drilling Down into Geophysics to Uncover British Columbia's Next Porphyry Deposits

Dianne Mitchinson[1], Dominique Fournier[2], Craig Hart[1], Thibaut Astic[3], Devin Cowan[3]
1. Mineral Deposit Research Unit, University of British Columbia, Vancouver, BC, Canada, 2. Mira Geoscience Ltd., Vancouver, BC, Canada, 3. University of British Columbia - Geophysical Inversion Facility, Vancouver, BC, Canada

Rapid, near-surface mineral discovery is becoming less frequent in well-explored parts of the world. Investigating the deeper, undercover, or more difficult terrane within known mineral districts presents the next best chance for discoveries going forward. This requires a commitment to new data collection and application of multidisciplinary teams and approaches to exploring that data. The Quesnel Terrane of central British Columbia (B.C.) represents such a district, where excellent porphyry mineral potential is suggested by geologic trends but bedrock exposure is limited due to overburden. The full potential of this region will ultimately be realized when the wealth of geophysical data available is interpreted hand-in-hand with geologic knowledge.

Several important porphyry copper and gold deposits, including the Mount Milligan and Mount Polley mines, sit amongst hundreds of porphyry copper and gold showings along a terrane-parallel trend in B.C.'s Quesnel Terrane (Fig. 1). A significant 250-km gap in the trend of porphyry occurrences coincides with a region covered extensively by glacial till. From geologic mapping and geophysical survey data, the Quesnel Terrane volcano-magmatic arc package that hosts known porphyry deposits is contiguous along strike through the gap. Recognizing the opportunities of the region, two large geophysical surveys were deployed under the Quesnel Exploration Strategy (QUEST) program in 2007. These gravity and electromagnetic surveys provided significant new data and data coverage where geologic mapping has been obviously hindered. Fourteen years after the collection of these data, the results continue to provide opportunities to better understand the geology and mineral deposit settings in the Quesnel Terrane.

Recent work in the gap takes existing and new conductivity, magnetic susceptibility, and density geophysical inversion models and combines them with geologic map information, field data, and drilling data to extract details about the central Quesnel Terrane's hidden bedrock geology that are of interest to porphyry explorers. This 3D integrated analysis aims specifically to identify alkalic porphyry deposit host intrusions hidden beneath the overburden. Alkalic porphyry deposits are found throughout the Quesnel Terrane and are exciting targets due to their often elevated gold and platinum group element abundances. Intrusions related to alkalic porphyry deposits in the southern Quesnel Terrane have distinct geophysical signatures and commonalities in their spatial distributions. Magnetic, gravity, and electromagnetic data and models aided us in seeking similar rocks in the central Quesnel Terrane and also provided valuable understanding of the amount of cover material that could be obscuring them. In addition to highlighting areas of new porphyry potential in the central Quesnel Terrane, interpretations advanced during this work help extend our more developed knowledge of porphyry deposit settings in the northern and southern Quesnel Terrane into the covered region between.

An understanding of the value of a mineral district by geoscience organizations, a commitment to data collection, a passion by geoscientists to persevere in unravelling geologic and mineral settings, and crucially, an integrated approach to interpretation of data are the ingredients that will allow explorers to uncover the next porphyry deposit in central B.C.

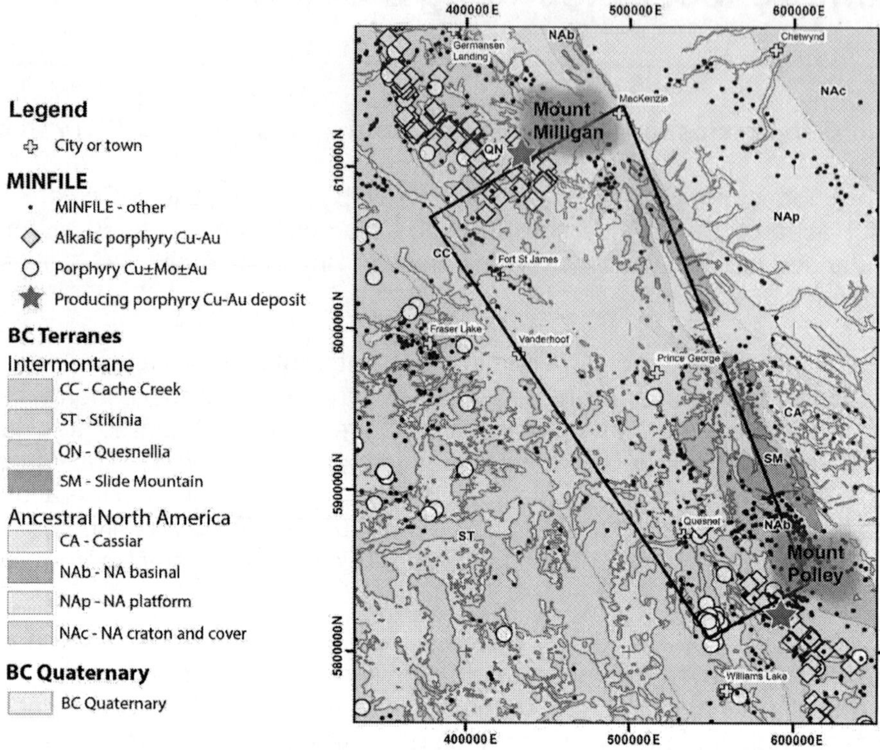
Legend
City or town
MINFILE
MINFILE - other
Alkalic porphyry Cu-Au
Porphyry Cu±Mo±Au
Producing porphyry Cu-Au deposit
BC Terranes
Intermontane
CC - Cache Creek
ST - Stikinia
QN - Quesnellia
SM - Slide Mountain
Ancestral North America
CA - Cassiar
NAb - NA basinal
NAp - NA platform
NAc - NA craton and cover
BC Quaternary
BC Quaternary
400000 E
500000 E
600000 E
6100000 N
6000000 N
5900000 N
5800000 N
Mount Milligan
Mount Polley
Germansen Landing
MacKenzie
Chetwynd
Fort St James
Fraser Lake
Vanderhoof
Prince George
Quesnel
Williams Lake
NAb
NAc
NAp
QN
CC
CA
SM
ST

ST.050
Machine Learning for Geochemical Exploration: Classifying Magma Fertility in Arcs and Insights into Porphyry Copper Deposit Formation

Chetan Nathwani[1, 2], Jamie Wilkinson[1, 2], Robin Armstrong[1], Daniel Smith[3], Christian Ihlenfeld[4]
1. Natural History Museum, London, United Kingdom, 2. Imperial College London, London, United Kingdom, 3. University of Leicester, Leicester, United Kingdom, 4. Anglo American, London, United Kingdom

Machine learning algorithms provide a powerful new approach to help in the discovery of ore deposits that are not exposed at the Earth's surface. A current exploration focus is the development of tools, based on large geochemical data sets, that will enable the identification of magmatic districts predisposed to host porphyry copper deposits. In this study, we train and test four common supervised machine learning techniques: logistic regression, support vector machines, artificial neural networks, and Random Forest™ to classify magma fertility in arc rocks using a global data compilation. Several properties of geochemical data sets, such as class imbalance, high multicollinearity, sparsity, missing values, and compositional data effects, preclude the application of traditional statistics and many machine learning techniques. Application of machine learning techniques on geochemical data therefore requires critical preprocessing steps. We evaluate the classification accuracy of each supervised machine learning technique using cross-validation, showing high accuracy for each model. A test data set, comprising three deposits of varying size, metal endowment, and tectonic setting, shows similar classification performance in all cases. We further illustrate the efficacy of our approach by demonstrating that our models outperform those derived from bivariate fertility indicators currently used by the exploration industry. This highlights the value in utilising the high dimension space available in geochemical data for classification problems. Feature analysis of our models reveals that the most important parameters for classification reflect early and abundant fractionation of amphibole and/or garnet and suppression of plagioclase during lower crustal, hydrous magma evolution. We cannot exclude that some signatures may additionally reflect relatively subtle hydrothermal alteration effects. Empirically, our models underscore the potential of using supervised machine learning approaches in mineral exploration, and that the "black box" nature of these techniques can be overcome.

ST.051
The Occurrence of Gold Mineralization in (Eastern and Western) Dharwar Craton, Karnataka India: Avenues for Looking Beyond 2021

Sunder Raju V. Perumala[1], Prabhakar Sangurmath[2]
1. CSIR-NATIONAL GEOPHYSICAL RESEARCH INSTITUTE, HYDERABAD, TN, India, 2. Hutti Gold Mines Company Limited, Bengaluru, KA, India

Gold occurrences spread over a vast geological time span from Archaean to Recent in Indian cratons and the world-class lode gold Kolar deposit in EDC were studied. Principal types of deposits in order of decreasing bulk reserves are in ancient conglomerates, greenstone belts, epithermal veins, metamorphosed turbidities, and carbonate-rich sediments. In India, a fair understanding of gold prospects in the 2.7 Ga megametallogenic epoch, featuring oxide-sulphidic banded iron formations (BIFs) of different stratigraphic horizons and profusely developed intense auriferous quartz veins hosting lode gold deposits in greenstone belts and high-grade terrains of the Dharwar craton, has emerged in recent years from studies on Sargur, Chitradurga, Gadag, and Shimoga schist belts. During the COVID-19 pandemic, gold demand in India suffered a setback, with demand for April-June 2020 quarter at 63.7 tonnes compared to 213.2 tonnes in the same quarter the previous year (World Gold Council). The average grade produced in India has fallen from 47.5 g/t in 1891-1900 to 3 g/t in 2020. It is thus imperative that a high degree of accuracy is to be attained in the evaluation of the lean orebodies. In view of the high degree of variation in width (pinch-and-swell structure) of lodes and heterogeneity of distribution of gold in the host rock, borehole data by itself is not a reliable mining tool. Therefore, a judicious integrated programme of exploratory drilling and exploratory mine development with close-spaced sampling is considered to be the reliable method of assessment of gold lodes beyond 2021. About 70 tonnes of gold metal is categorized as reserves (economically mineable), and gold resources are estimated at 585 tonnes of gold metal. These resources await upgrade by substantial deep diamond drilling >2 km and RC drilling, resource modeling, exploratory mining, and mining feasibility studies before they reach the status of mineable reserves. To look beyond 2021, we have to look for gold mineralisation and deposits in unexplored areas (concealed) by application of sophisticated methodology/equipment and technology such as 1) conceptual exploration by geological modeling, 2) use of remote sensing techniques with satellite imageries and air photos, 3) ground geophysical mapping using MT and seismic data integrated with traditional geophysical methods and airborne geophysics, 4) use of on-site portable analysers such as hand held XRFs, trace element, and isotopic geochemistry, 5) several sample preparation protocols using automated mineralogy tools and mass spectroscopic analyses, 6) geological and structural mapping, 7) data management using AI and machine learning using drone-based technology, 8) data integration, 9) shortwave infrared (SWIR) and laser petrology and mineralogical studies using XTSCAN, and 10) ore beneficiation and geometallurgical studies. In the next century, the Dharwar Craton will be a gold hub with another new Kolar deposit in India beyond 2021.

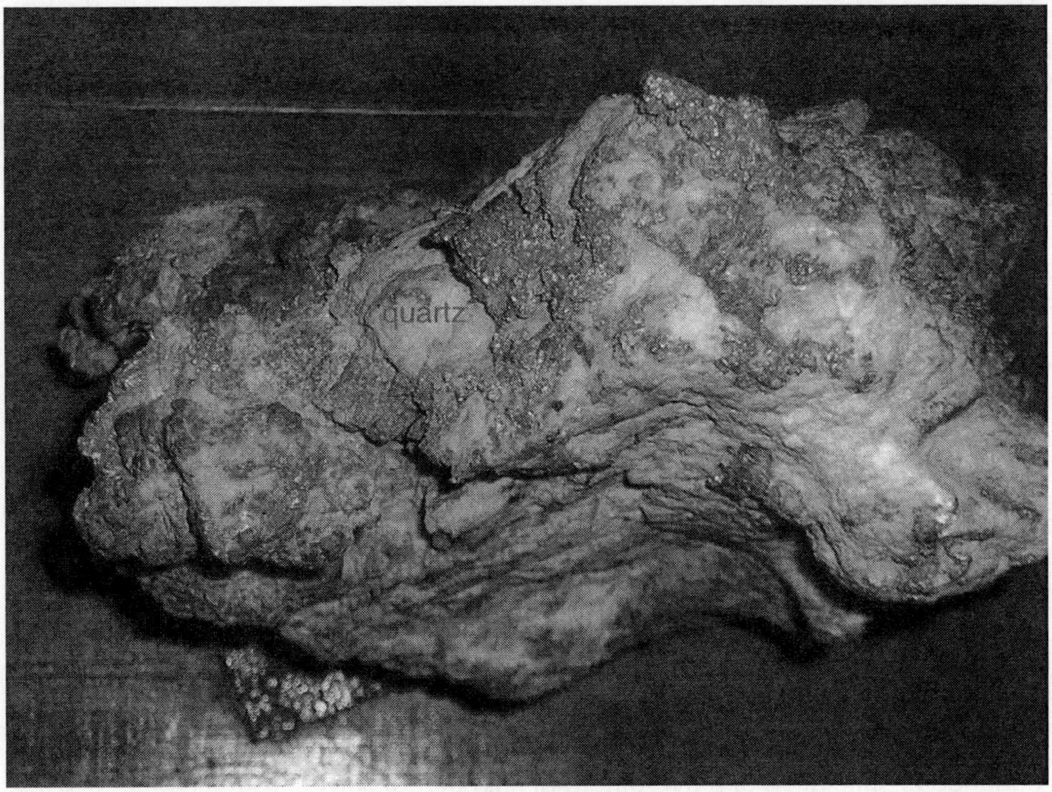
quartz

ST.052

The Hydrothermal Mobility of Vanadium: Insights from Modeling with Implications for the Vanadium Enrichment of Iron Oxide-Apatite Deposits

Maria A. Rodriguez-Mustafa[1], Jihua Hao[2], Mark R. Frank[3], Adam C. Simon[1]
1. University of Michigan, Ann Arbor, MI, USA, 2. School of Earth and Space Sciences, University of Science and Technology of China, Hefei, China, 3. Department of Geology and Environmental Sciences, Northern Illinois University, DeKalb, IL, USA

Vanadium is a critical element, as it is necessary in steel alloys and in V redox batteries, both of which are essential for the development of a low-carbon energy infrastructure. Vanadium is traditionally a by-product of the extraction of Fe and Ti in magmatic deposits and of U in calcrete-hosted U-V deposits. Its current production is limited, and its demand is forecasted to increase significantly, so new sources of this element will be needed in order to achieve the required supply. Iron oxide-apatite (IOA) deposits are dominated by magnetite, an iron oxide mineral. They are major sources of Fe, and recent studies have shown that they can contain substantial V hosted in magnetite as well. This project explores the favorable physicochemical conditions for the formation of V-rich magnetite in these deposits.

Magnetite in IOA deposits is thought to result from magmatic-hydrothermal processes, but the exact conditions of the hydrothermal fluid from which it precipitates are still unclear. We present the thermodynamic modeling of aqueous speciation of a hydrothermal fluid that exsolves from a melt of intermediate composition. Our model evaluates the behavior of V species in the fluid as a function of temperature, pressure, pH, and fO_2. Our preliminary results indicate that the solubility of V is higher at higher pH, temperature, and fO_2 and that V is most favorably incorporated in magnetite as V^{3+} at high temperature and low pH and fO_2 (Fig. 1).

Our results indicate that purely hydrothermal magnetite can host considerable amounts of V in its structure and that the parental melt for the hydrothermal fluid needs to be of intermediate composition. These results have implications within the tectonic setting that is suitable for the formation of IOA deposits, which will allow for improved exploration campaigns to more efficiently find this type of deposit in order to increase the resources of V needed for the energy transition.

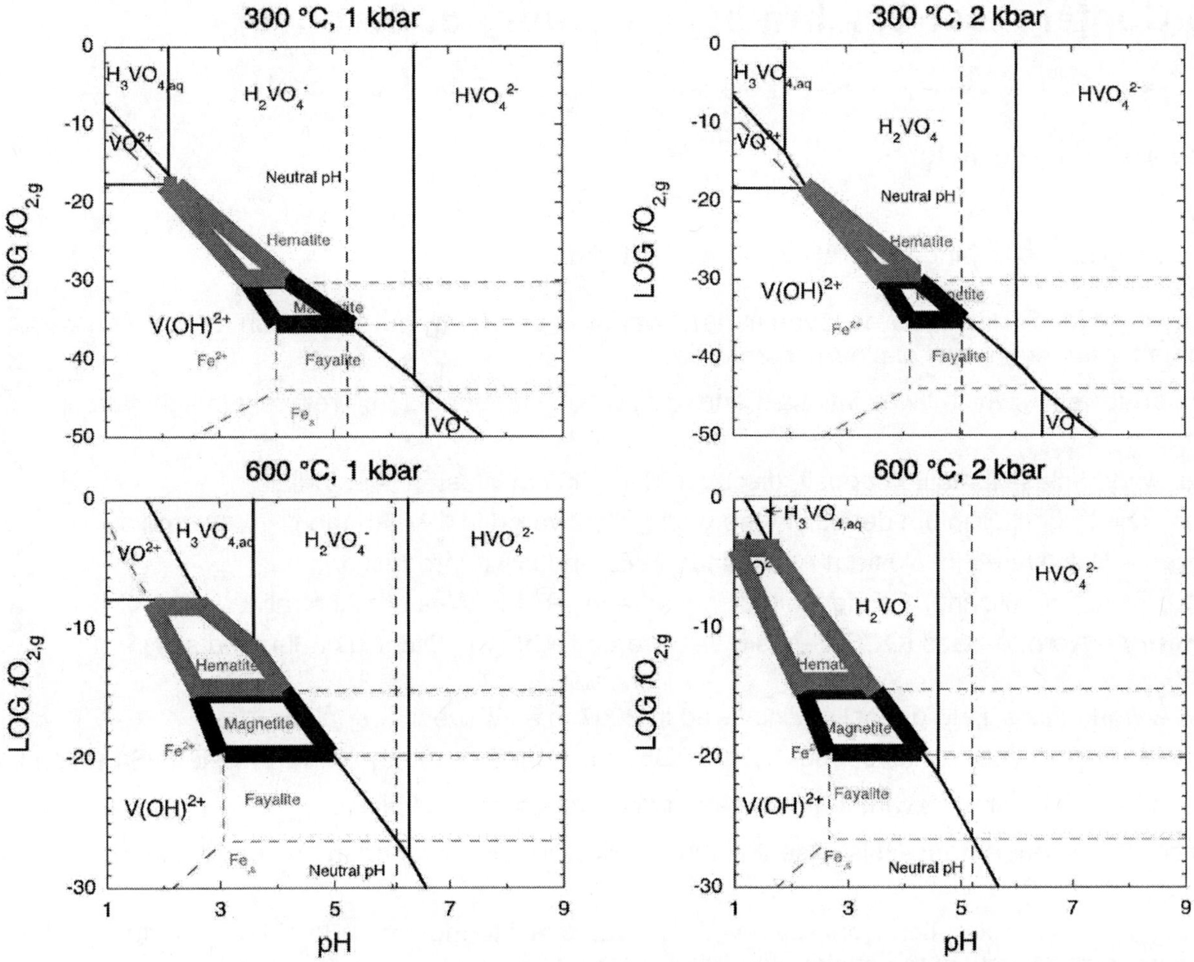

Fig. 1: Phase diagrams showing the stability of iron oxides and V species in pH and fO2 space for different temperatures and pressures. Highlighted are the fields where hematite (red) and magnetite (black) incorporate V3+ in their structure. The favorable conditions for V incorporation in magnetite are high temperature and acid pH.

ST.053
Seven Discoveries - Seven Insights

John C. Rowntree
Hunter Minerals Pty Limited, Kangaroo Valley, NSW, Australia

I have been involved in the discovery of seven mineral deposits during my career. I wish to share my seven relevant insights for future mineral exploration activities.

The **seven discoveries** in which I was involved and from which I gleaned some relevant insights are as follows:

- **Wodgina** –World-class tantalite deposit, discovered in 1969 in Western Australia.
- **Jabiluka** – The largest uranium deposit in the world, discovered in 1971 in the N.T., Australia.
- **Paddington** – Discovered in Western Australia in 1982 and still in production.
- **Munni Munni** – A significant PGE deposit, discovered in 1984 in Western Australia.
- **Ernest Henry** – A world-class IOCG deposit, discovered in 1991 in Qld, Australia, and still in production.
- **Isortoq** – A world-class TVM deposit, discovered in 2007 in southwest Greenland.
- **Maniitsoq** – A world-class nickel/copper/cobalt/PGE camp, discovered in 2011 in western Greenland.

The **seven insights** which are relevant to these seven discoveries are as follows:

Efficient Corporate Management - It is essential that public and private company management understand the following:

- The country in which exploration work will be carried out, considering their rule of law, Aboriginal rights, infrastructure, and security in the exploration project area;
- The management requirements for corporate stability, integrity, and financial development.

The Rule-of-Law - Mineral exploration is a high-risk endeavour and risks should not be increased by exploration in countries that do not have a true and effective Rule of Law and legal system, including the following:

- Genuine legal rights for exploration and exploitation, as finance can be compromised without such rights;
- An assessment of aboriginal and environmental rights, surface rights and access to infrastructure.

Finance - Exploration requires essential finances.

- Using Joint Venture (JV) funding can result in the incoming JV party managing the field exploration and ongoing exploitation.
- The financiers providing funding must understand that they are part of the discovery team.

Exploration Team - The exploration team must have the unique skills required to carry out requisite exploration.

- Team and corporate management must communicate effectively and respect and inspire all team members.
- Annual review of projects gives team members an opportunity to give their advice for ongoing teamwork and project selection.

Pattern Recognition - An assessment is required of the best geological mode of occurrence for a commodity prior to obtaining exploration rights.

- Ensure that the exploration team utilizes effectively the exploration techniques in geology, geochemistry, geophysics, and drilling to implement the requisite field work.

Lateral Thinking and Positive Exploration -

- Continue to review the selected target area. Understand that **Lateral Thinking** of the mode of occurrence is required to adapt and expand the field work to find alternate targets.
- All projects are always uniquely different and therefore exploration should not be carried out simply to ensure that there is no discovery of the targeted mode of occurrence.

Full Exploration Rights - Once the target area for exploration has been selected, the company should acquire 100% of the exploration rights for early-stage exploration.

SEVEN DISCOVERIES	SEVEN INSIGHTS						
	Efficient Corporate Management	Rule-of-Law	Finance	Exploration Team	Pattern Recognition	Lateral Thinking & Positive Exploration	Full Exploration Rights
Wodgina	✓✓	✓	✓	✓✓	✓✓✓	✓	✓
Jabiluka	✓✓		✓✓	✓✓✓	✓✓✓	✓✓	✓
Paddington	✓✓	✓	✓✓	✓✓	✓✓	✓✓✓	✓
Munni Munni	✓✓	✓	✓	✓✓	✓✓✓	✓	✓
Earnest Henry		✓	✓	✓	✓✓✓	✓✓	
Isortoq		✓	✓	✓	✓	✓✓✓	✓
Maniitsoq	✓	✓	✓✓	✓	✓✓	✓✓	✓✓

Seven Discoveries, Seven Insights

ST.054
Using ASTER Mineral and Vegetation Maps in the U.S. Basin and Range to Evaluate and Assess Lithium-Rich Playas

John C. Mars, Federico Solano, Bernard E. Hubbard
U.S. Geological Survey, Reston, VA, USA

Critical resources such as lithium will play an increasing role in the world economy over the next decade. Lithium-rich brines in playas are the main source for current lithium resources. Assessment criteria to identify new potential lithium playa deposits are incomplete but include the presence of diagnostic saline minerals, high-silica volcanic rocks, hydrothermally altered clay and silicic mineralogy, and the presence of green vegetation indicative of a saline playa. In addition, current studies suggest that hydrothermally altered silica-rich volcanic tuffs are a source for lithium in brines. This study compiles and uses Advanced Spaceborne Thermal Emission and Reflection Radiometer (ASTER) mineral and vegetation maps of playas in the Basin and Range to look for mineralogical similarities between a known lithium brine deposit at Clayton Valley, Nevada, and other playas.

The ASTER vegetation and mineral map of Clayton Valley shows a suite of silica-rich, hydrothermally altered rocks surrounded by volcanic rocks near the playa (Fig. 1). In addition, halite and the presence of green vegetation indicate that there is water present to support an underground brine reservoir. These vegetation and mineralogical properties are mapped and described at other playas in the region, such as Railroad Valley, Nevada, to discuss and evaluate their potential for lithium-rich brine resources in the study area.

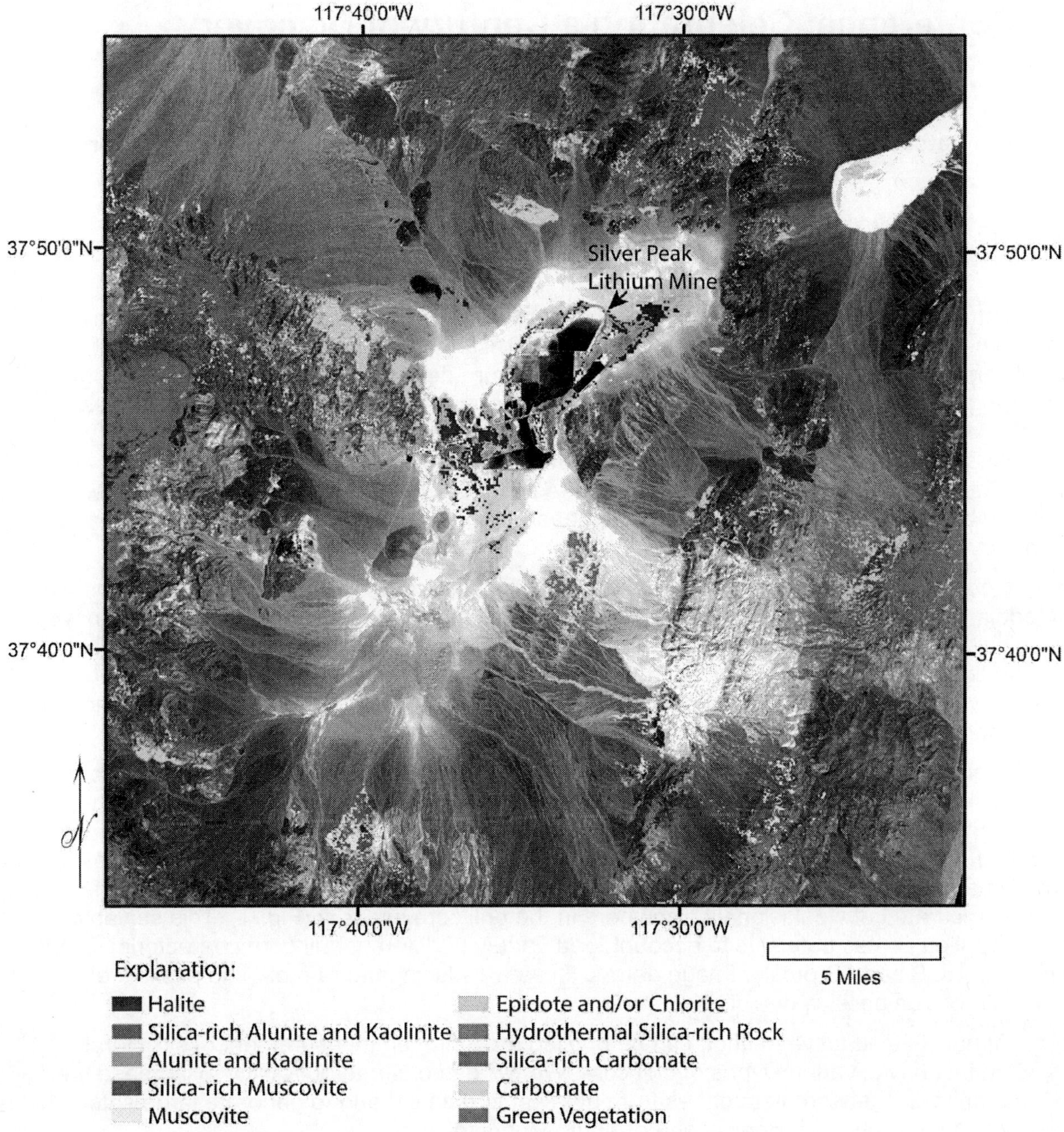

Fig. 1 ASTER mineral and green vegetation map of Clayton Valley, Nevada

ST.055
Adaptive Feature Detection for Improved Measurement of Structural Orientations Using Core Imagery

Laurel A. Stothers, Cassady L. Harraden
Corescan Ltd, Vancouver, BC, Canada

High-quality structural measurements play a key role in building effective geological models and predicting the geotechnical behavior of an orebody. Measurements of fracture, joint, vein, foliation, and fault orientations all contribute to the evolving understanding of an ore deposit and are key to making informed geological and engineering decisions. As the number of high-quality input measurements increases, so does the accuracy of the output model. Since these measurements are typically collected manually, recent efforts have focused on the generation of larger, high-quality data sets by obtaining structural measurements from drill core imagery in an automated fashion. Doing so has the potential to expedite the otherwise lengthy and expensive process of manual measurements, and to output larger quantities of data with greatly improved consistency.

Automated methods offer a decisively convenient way to collect structural measurements and enhance geological models; however, they must be robust in order to maintain measurement accuracy even with diverse geological inputs. Parameters such as drill core texture, size, shape, and colour, along with competency and the image resolution in which data is captured, can change significantly across projects and drill hole intervals. If these factors are not considered upon deployment of an automated method, the accuracy of structural detection and measurements can be affected.

To maximize the accuracy of structural measurements acquired automatically from drill core imagery in a generalized capacity, a statistical technique was developed for automated feature detection, which can adapt the detection method to suit the input image data. This method examines image statistics localized to probable target structures and amplifies signal of the detected feature accordingly. In doing so, statistical outlier removal of a relative magnitude can be applied on a localized basis to clean structural image data while preserving accurate but weak signals elsewhere in the drill core imagery (Fig. 1). This scalable solution can be applied across data sets to account for a variety of different drill core characteristics and pixel resolutions in RGB and 3D profiler image data, ultimately reducing the rate of false positive errors and increasing the rate of true positive detections.

Here, we present how this adaptive method can be applied for improved measurement of structural orientations when faced with various types of drill core. We will demonstrate our ability to increase the accuracy of automatically measured structures from data set to data set and to generate large sets of high-quality measurements for robust geological and/or engineering solutions.

Fig. 1. Example of adaptive feature detection applied to the measurement of fracture orientations where (a) is an RGB image in which fractures and orientation line (black marker) are displayed; (b) automatically detected fractures (shown in yellow and blue representing high and low detected signal, respectively) before adaptive feature detection is applied; and (c) results of adaptive feature detection showing two fractures accurately detected and oriented in space, despite their variable signal strength (as shown in image b).

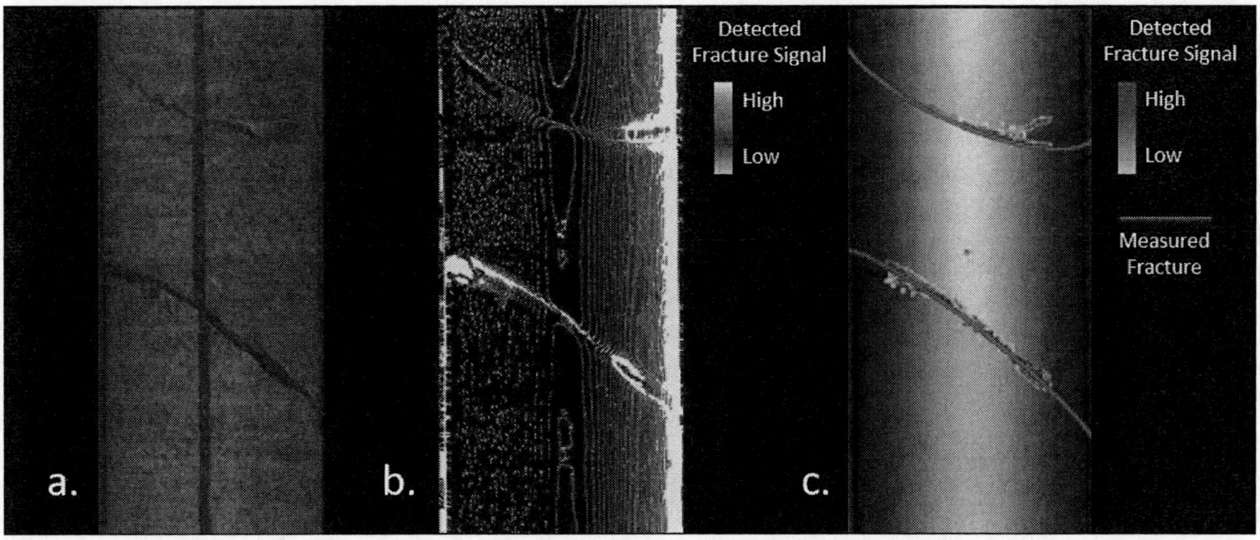

ST.057
Geoscientists: Explorers and Architects of the New Energy Economy

Edith Newton Wilson
Rock Whisperer LLC, Tulsa, OK, USA

What a difference a year makes. Pandemic-driven changes in energy usage have accelerated demand for low-cost renewable resources as the world's appetite for fossil fuels shrinks. The percentage of global electricity generated by renewables continues to rise at rates exceeding 25% year-on-year. The old energy business model of high-risk, high-return, capital-intensive, and infrastructure-dependent production of fossil fuels is giving way to a world of abundant clean energy characterized by rapid deployment, low risk, low up-front capital investment, and distributed, lean infrastructure. Electrification of civilization carries with it the potential for long-deferred global energy equity. Renewable energy projects are delivering an order of magnitude higher return than shares of fossil fuel companies, while providing more reliable and affordable energy options for all global citizens—and with lower carbon cost. The new energy economy is here, and it rests on a foundation of economic geology.

The energy transition is synonymous with a fundamental breakthrough in energy storage. Since the discovery of fire, humans have relied on storage of energy through photosynthesis, geologic processes to concentrate energy more efficiently, and the subsequent breaking of hydrogen-carbon bonds in wood, coal, oil, or natural gas to release energy. This process not only produces carbon dioxide and other greenhouse gases as a biproduct, but it also destroys the energy storage device forever, requiring that new quantities of hydrocarbons be extracted and combusted for every joule or watt hour of energy consumed. Renewable energy storage devices—batteries—require a similar amount of carbon for extraction and manufacturing but emit no carbon dioxide on use. Moreover, the extracted components used to make batteries (or solar panels or wind turbines) are not destroyed to produce energy, but can be reused, repurposed, or fully recycled. Not only is the new energy economy here, but the circular economy is just around the corner.

The talents of economic geologists will translate effectively to all aspects of the rapidly developing sustainable energy supply chain. First and foremost, we bring our geologic expertise, along with our exploration and mining experience, to the problem of filling the rising demand for metals and other conductive materials used to power the Internet of Things, electric vehicles, distributed energy resources, and utility systems. The extractive industries that produce materials for energy storage will rely on sustainable development, environmental management, and proper social license now more than ever. Mineralogy, crystallography, material science, and nanotechnology will be increasingly essential skills as we onshore industrial processing as well as electrode design and manufacturing. Optimization of battery chemistries and recycling metals from waste will also call on metallurgical and refining expertise. As we move from a society that extracts and uses up our energy resources to one that optimizes energy as a service, the most valuable talents we bring to the table may well be our creativity and thirst for discovery, as well as our willingness to embrace change.

Energy storage...

yesterday...

and tomorrow.

Hydrocarbons

Store energy from the sun in H-C bonds

Emit ~400 g CO_2 eq/kWh to make/use

Are finite on the human time-scale

Are destroyed to give up stored energy

Batteries

Store energy from the sun in electrons

Emit ~40 g CO_2 eq/kWh to make/use

Store energy from renewable resources

Preserve components through end-of-life

D1

Crustal Architecture of a World-Class Mineral District: Transcrustal Upflow Zones and Metal Endowment

Taus R. Jørgensen[1], Harold L. Gibson[1], Eric Roots[1], Rajesh Vayavur[1], Graham Hill[2], David Snyder[3], Mostafa Naghizadeh[1]
1. Mineral Exploration Research Centre, Harquail School of Earth Sciences, Laurentian University, Sudbury, ON, Canada, 2. Institute of Geophysics of the Czech Academy of Sciences, Prague, Czech Republic, 3. Geological Survey of Canada, Ottawa, ON, Canada

Lithospheric-scale crustal growth processes are first-order controls on metal source, transport, and concentration. However, the crustal architecture and geologic features that characterize mineral districts and the processes responsible for district-scale metal endowment are poorly understood. Here we present a whole-of-crust analysis that provides insights into belt- and district-scale features and processes, which influenced the metal endowment of the Neoarchean Noranda District (Abitibi greenstone belt, Canada) as documented by its ~20 volcanogenic massive sulfide deposits (VMS), ~19 orogenic Au deposits, and minor synvolcanic intrusion-hosted Cu-Mo ± Au ± Ag mineralization. Much of the metal budget (17 of 20 VMS deposits; 104 Mt of 130 Mt of VMS ore) is confined to a smaller area characterized by a magmatic center with the greatest combined surface area of synvolcanic plutons and felsic volcanic rocks and the highest density of major synvolcanic structures. Crustal-scale integration of seismic, gravity, and magnetotelluric imaging with surface geology shows that the more endowed area is marked by increased felsic plutonism at depth and located above subvertical pipe-like corridors that connect to relatively low resistivity middle-lower crust. The largest and most Au-rich VMS deposits are situated where three major synvolcanic faults correspond to one of the most predominant structural corridors at the locus of the magmatic center. This new crustal reconstruction explains the clustering of ore deposits on both belt and district scales by ancestral transcrustal structures that are interpreted to have localized, optimized, and sustained magmatic and ore-forming processes. The results highlight a strong magmatic control on metal and in particular Au endowment in VMS systems and indicate that the VMS hydrothermal system was not restricted to a near-surface (~<5 km) convective subseafloor seawater system but is part of a larger vertically extensive but areally localized, deep crustal to mantle magmatic system. Overprinting by ca. 30 m.y. younger orogenic Au deposits suggests that VMS Au enrichment may be a proxy for SCLM Au and base metal enrichment.

D2
Geochemical Signatures of Felsic Rocks in Modern Intraoceanic Settings and Implications for Archean Greenstone Belts

Marc L. Fassbender[1], Mark D. Hannington[1, 2], Margaret Stewart[3], Philipp A. Brandl[2], Alan T. Baxter[1], David Diekrup[1]
1. Department of Earth and Environmental Sciences, University of Ottawa, Ottawa, ON, Canada, 2. GEOMAR Helmholtz Centre for Ocean Research Kiel, Kiel, Germany, 3. Department of Earth and Evnironmental Sciences, Mount Royal University, Calgary, Calgary, AB, Canada

Felsic volcanic rocks are abundant in ancient greenstone belts and important host rocks for volcanogenic massive sulfide (VMS) deposits. Dacites and rhyolites are the immediate hosts for about half of all VMS deposits, related to anomalous heat flow during rifting, fractional crystallization, and partial melting of basaltic crust. For over 30 years, geochemical signatures of these rocks (e.g., F-classification of Archean rhyolites) have been widely used to identify possible hosts for VMS deposits in greenstone belts, based on processes of melt generation and fractionation and inferred geodynamic setting. Until recently, it has not been possible to rigorously compare these findings to modern oceanic settings owing to the limited sampling of felsic volcanic rocks from the oceans. In particular, it remains unclear to what extent the differences between modern and ancient oceanic crust may be related to the lower mantle temperatures, thinner crust, and subduction-related processes in present-day settings. In this study, we have compiled high-quality analyses of more than 2,200 unique samples of modern submarine felsic volcanic rocks (>60 wt % SiO_2) from mid-ocean ridges, ridge-hotspot intersections, intraoceanic back-arc basins, island arcs, and ocean islands. This new compilation shows significant geochemical diversity, spanning the full range of compositions of silicic lavas in VMS-forming environments. In the F-classification, FI-like felsic volcanic rocks are mostly related to calc-alkaline magmatism in arc-related assemblages, and FII signatures are interpreted to be from intracontinental arc-backarc environments. FIIIa signatures are mainly found in tholeiitic rocks from rift-related intraoceanic island arc and back-arc assemblages, and FIIIb-like dacites and rhyolites are from mid-ocean ridges and back-arc spreading centers. LREE-depleted FIV signatures are interpreted to be mainly from forearc and intra-arc rift settings and are largely absent in Archean greenstone belts. However, in modern oceanic settings, we show that felsic volcanic rocks ranging from FII to FIV dacites and rhyolites can occur within a single arc-backarc system. We interpret the diversity of modern arc-backarc assemblages to reflect highly variable i) stress regimes, ii) crustal thicknesses, iii) presence or absence of slab-derived fluids (dry melting versus wet melting), and iv) mantle anomalies. These differences are revealed by a combination of principal components analysis, unsupervised hierarchical clustering, and supervised random forest classification of the compiled data. We suggest that the melting conditions recorded by the felsic volcanic rocks are related to short-lived microplate domains, specifically where multiple overlapping spreading centers have contributed to basin opening, as in the northern Lau Basin. The felsic volcanic rocks in modern arc-backarc systems share strongly depleted mantle signatures and well-known subduction-related chemistry (strong LILE enrichment in combination with strong negative Nb-Ta anomalies and HREE depletion). This contrasts with felsic volcanic suites in Archean greenstone belts, a high proportion of which show HFSE and HREE enrichment due to a less depleted mantle, a lack of wet melting, and highly variable crustal contamination. We suggest that thickened oceanic crust in the Archean may have resulted in widespread buoyant microplate domains that were obducted during amalgamation, preserving the felsic volcanic rocks with HFSE and HREE enrichment.

D3

Mineralogy and Stable Isotope Geochemistry of Carbonatite Dikes in Southern Ravalli County, Montana, USA

Christopher H. Gammons
Montana Tech, Butte, MT, USA

Approximately 20 small carbonatite "vein-dikes" have been documented in southernmost Ravalli County in southwest Montana. Dozens of similar occurrences have also been described in the neighboring Mineral Hill district of Idaho. The deposits are steeply dippling, tabular bodies several centimeters to several meters wide, and up to 150 m long. The deposits are parallel to the surrounding foliation in Precambrian amphibolite and quartzo-feldspathic gneiss, and locally have internal fabrics that appear deformed. The deposits are highly enriched in Nb and light REE, and some of the larger deposits were mined at a small scale in the 1950s. Previous workers have suggested either a magmatic or a hydrothermal origin for these deposits. However, very little academic or mineral exploration work has been done since the 1960s.

Three carbonatite deposits in the Sheep Creek area were sampled in this study, only one of which could be examined in outcrop. The bulk of the carbonatite bodies consists of a granular mosaic of calcite, Fe-dolomite, barite, apatite, and quartz. REE minerals include monazite, ancylite and allanite, with subordinate REE-F-carbonates. REE also occur with Nb-Ti minerals, including aeschynite, pyrochlore, fersmite, and chevkinite. Other Nb phases include columbite and Nb-rich rutile. Concentrations of tantalum are negligible in these minerals. The carbonatites are often banded, which could be a primary igneous feature or perhaps a metamorphic overprint. Fe-oxides (magnetite and specular hematite) are common accessory minerals, and trace sulfides have been found, including pyrite, pyrrhotite, chalcopyrite, galena, molybdenite, siegenite, and cobaltite. Hornblende in the surrounding amphibolite is altered to phlogopite near the veins. However, no alteration resembling fenite is apparent.

Samples of matrix calcite and dolomite (n = 21), as well as a single sample of ancylite, were submitted for stable isotope analysis. The data are consistent with a primary igneous origin for the carbonates (Fig. 1), although there appears to have been some modification through interaction with high-temperature meteoric water. The S-isotope composition of pyrite and chalcopyrite (n = 4) is unexpectedly light, ranging from -7.9 to -6.9 per mil. Coarse barite from a different carbonatite body has $\delta^{34}S$ = +5.0 to +6.4 and $\delta^{18}O$ = +6.7 to +7.7. This barite is isotopically distinct from hydrothermal barite veins further north in Montana that have $\delta^{34}S$ = +13 to +35 and $\delta^{18}O$ = +10 to +15 per mil. Overall, the stable isotope data are consistent with a magmatic origin for the carbonatites, although more work is needed to unravel trends in the S-isotopes. Also, an important question remains as to the age of the carbonatite vein-dikes. Some early Pb-alpha dates of 90 to 99 Ma were reported by the USGS for monazite samples from the Mineral Hill district. However, these dates should be interpreted with caution considering recent studies showing that U-Pb systematics in monazite can be reset during overprinting metamorphic events. It is intriguing to speculate that the carbonatites straddling the Idaho-Montana border could be related to Nb-REE carbonatites in the British Columbia Alkalic Province.

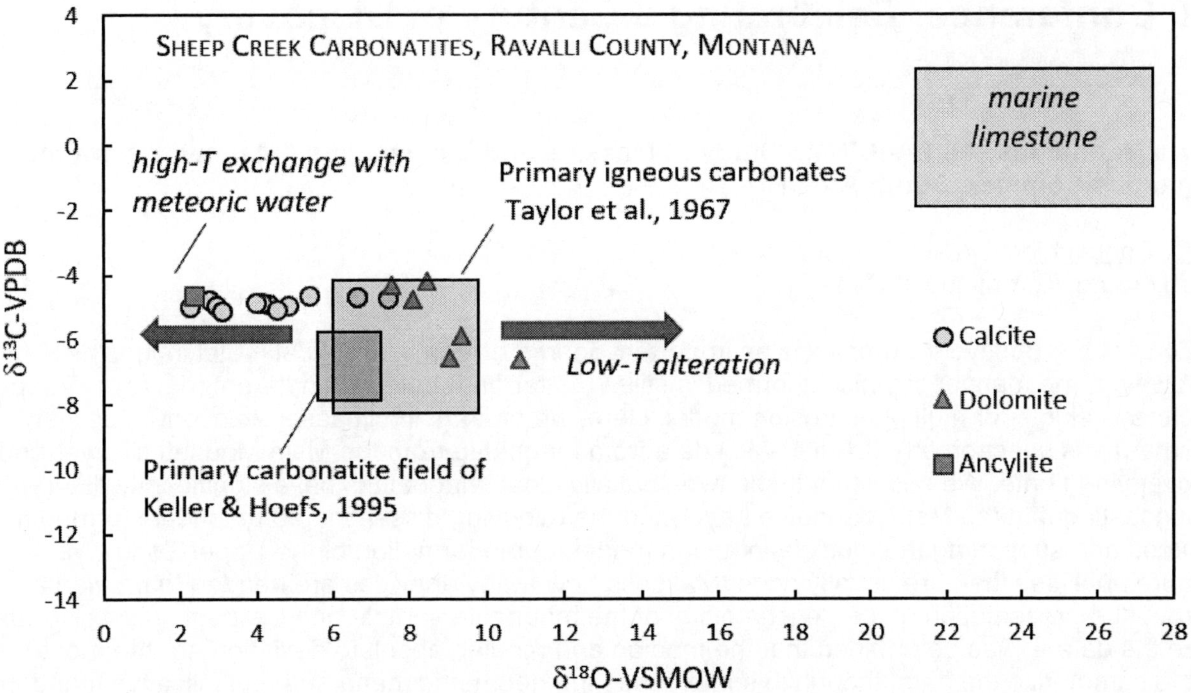
Sheep Creek Carbonatites, Ravalli County, Montana
marine limestone
high-T exchange with meteoric water
Primary igneous carbonates Taylor et al., 1967
Low-T alteration
Primary carbonatite field of Keller & Hoefs, 1995
Calcite
Dolomite
Ancylite
δ13C-VPDB
δ18O-VSMOW
4
2
0
-2
-4
-6
-8
-10
-12
-14
0
2
4
6
8
10
12
14
16
18
20
22
24
26
28

D4

A Tale of Two Magnetites: An LA-ICP-MS Study of Massive and Disseminated Magnetite from the Bushveld Igneous Complex, South Africa

Daryll Bien C. Concepcion, Daniel Gregory
University of Toronto, Toronto, ON, Canada

The Upper Zone of the Bushveld Complex is an important source of V, a vital critical metal in today's economy. However, the manner in which it formed is still a matter of debate with different research groups arguing for either a single- or multiple-intrusion model. Here, we present in situ laser ablation-inductively coupled plasma-mass spectrometry (LA-ICP-MS) data from magnetite from the Main Magnetite Layer and the directly overlying norite. We show that these two spatially close magnetites have significantly different chemistry, suggesting that the Main Magnetite Layer and the overlying disseminated magnetite formed from different sources and supporting the multiple-intrusion model for the formation of the Upper Zone. We performed more analyses than are usually done to assess how many analyses are needed to provide a statistically robust representation of the geochemistry of the magnetite sample and the best statistics to use to summarize the data. It was determined that the median and median absolute deviation are the most robust ways to summarize the data, though in some cases the geometric mean and multiplicative standard deviation can be used. Additionally, it was determined that for the Bushveld Complex, 10 analyses are sufficient for characterizing the trace element chemistry of the massive magnetite layer, but more are needed for the disseminated magnetite in the norite overlying it.

Figure caption: A summary of some of the median element geochemistry of the massive and disseminated magnetite analyzed by this study.

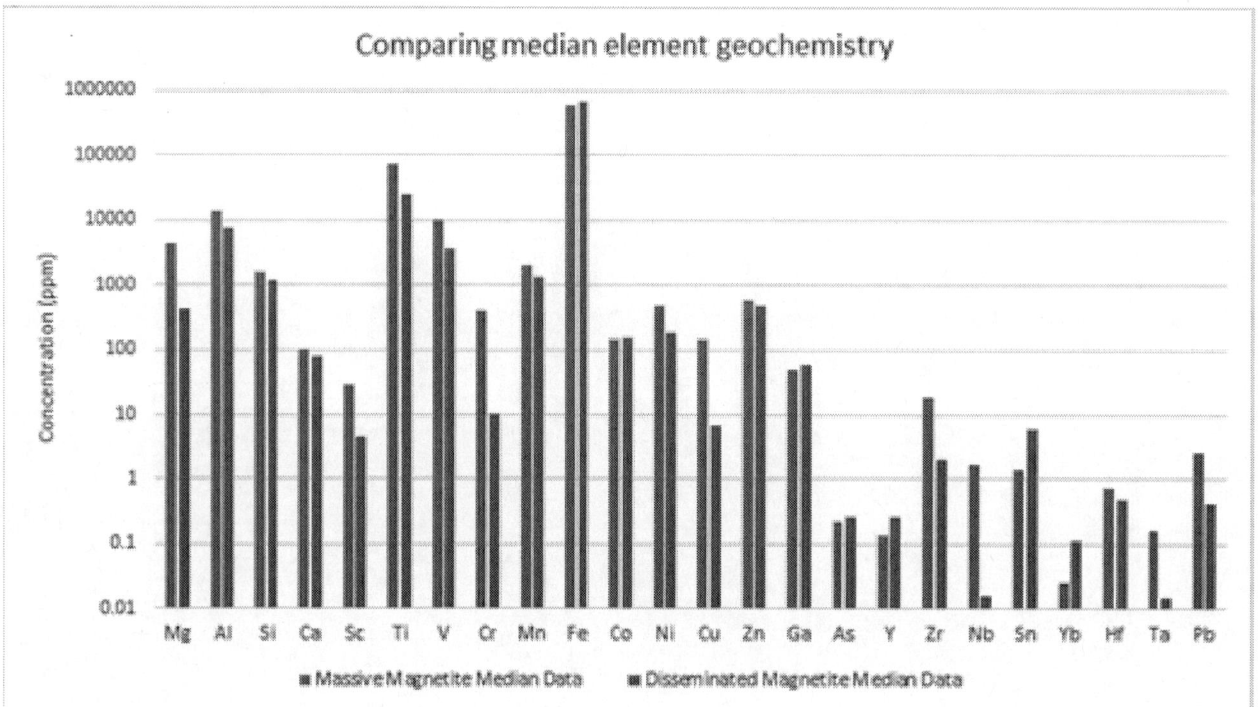

D5

Titanium Isotopes in Magnetite as Tracers of Ore-Forming Processes in Iron Oxide Apatite (IOA) and Iron Oxide Copper Gold (IOCG) Systems

Christopher Emproto[1], Ryan Mathur[2], Adam C. Simon[1]
1. University of Michigan, Ann Arbor, MI, USA, 2. Juniata College, Huntingdon, PA, USA

Iron oxide apatite (IOA) and iron oxide copper gold (IOCG) deposits collectively are major sources of Fe and Cu and minor or potential sources of Au and the rare earth elements (REE)—resources crucial for modern and emerging technologies and whose demand is expected to increase significantly in the next few decades. IOA and IOCG deposits occur globally and comprise structurally controlled Fe oxide (± Cu sulfide in IOCG systems) bodies hosted within highly metasomatized country rocks and exhibiting characteristic element enrichment suites. Mounting evidence suggests that IOA deposits represent deeper portions of IOCG systems, however, a universal connection remains tenuous. These deposits are geologically diverse in their ages, petrologic associations, mineralization styles, and particular elemental enrichments but are believed to represent a common set of respective processes. Additional uncertainties as to the source of their metallic resources have posed difficulties in establishing consensus for a holistic formation model. Here, we present Ti isotope compositions for magnetite, an oxide phase common in both IOA and IOCG deposits, and apply these results to better understand the formation of Fe oxide resources in these and other magnetite-bearing mineralized systems including Fe-rich skarns, layered mafic intrusions, and Cu-Au porphyries. The utility of Ti isotopes in deconvolving ore-forming processes is premised in the suggested exclusivity of Ti isotopic variations in crustal materials reflecting the precipitation of primary magmatic Ti-bearing oxide phases while resisting hydrothermal overprinting.

Ti isotope compositions were determined via multi-collector inductively coupled plasma mass spectrometry (MC-ICP-MS). Ranges in δ^{49}Ti of magnetite from IOAs (5.55‰), IOCGs (5.49‰), and layered mafic intrusions (3.78‰) each eclipse the variation of bulk rock δ^{49}Ti observed in terrestrial igneous rocks (c. 0–0.19‰) and in meteorites (c. 0–0.3‰), and variation within individual deposits (e.g., 2.43‰ variation at El Laco IOA in Chile) far exceeds the $\Delta^{49}Ti_{oxide\text{-}melt}$ (0.39‰) predicted from earlier work by Johnson et al. (2019) for magnetite monotonically precipitating (i.e., not resorbed) from a fractionating magmatic reservoir (Figure 1). The large overlap (from c. -2 to 3‰) and similar range in δ^{49}Ti observed in IOA and IOCG magnetite strongly suggests that similar processes occur in the formation of magnetite in both of these deposit types and is consistent with models suggesting a genetic relationship between them. Further work is necessary to understand the mechanism responsible for such large degrees of Ti isotope fractionation in individual ore deposits and to better characterize the ranges of δ^{49}Ti in magnetite from different types of iron oxide ore deposits.

Figure 1. Measured $\delta^{49}Ti$ compositions for banded iron formation (BIF), Cu-Au porphyry, Fe-Ti-V deposit, iron oxide apatite (IOA) deposit, iron oxide copper gold (IOCG) deposit, layered mafic intrusion (LMI), and Fe skarn magnetite. Analytical uncertainties are smaller than the symbol size.

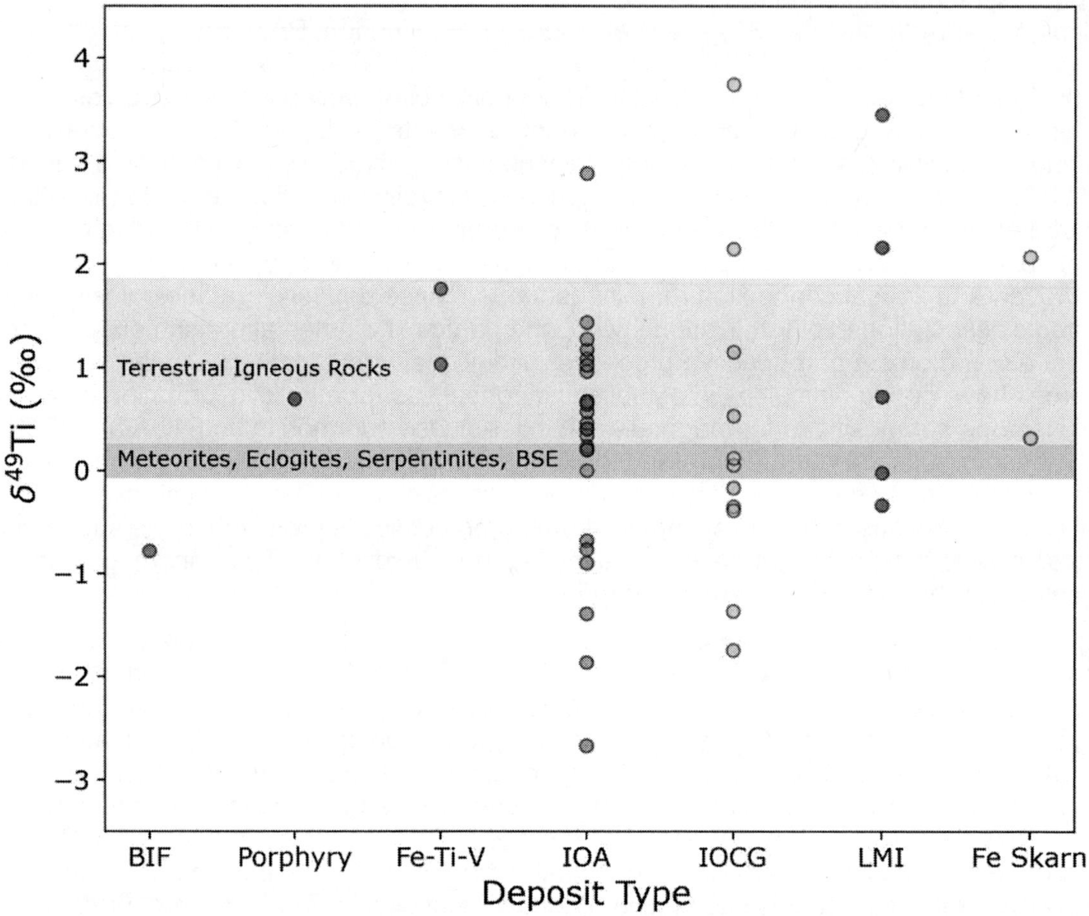

D6

The El Alacrán Cu-Au Deposit: A Hybrid Iron Oxide-Copper-Gold and Carbonate-Replacement Deposit in the Colombian Andes

Julian Manco[1], Craig J. Hart[1], Hildebrando Leal-Mejía[1], Janet Gabites[2], Robert Creaser[3]
1. Mineral Deposit Research Unit (MDRU), The University of British Columbia (UBC), Vancouver, BC, Canada, 2. Pacific Centre for Isotopic and Geochemical Research (PCIGR), The University of British Columbia (UBC), Vancouver, BC, Canada, 3. Canadian Centre for Isotopic Microanalysis (CCIM), University of Alberta, Edmonton, AB, Canada

The El Alacrán Cu-Au Deposit in NW Colombia is the most explored prospect within the San Matias District (SMD). The genesis of the deposit is controversial because it exhibits mixed characteristics of different deposit models (e.g., IOCG, VMS, and Skarn deposits), besides an evident spatial relationship with nearby porphyry Cu-Au prospects (e.g., Montiel E, Costa Azul). Understanding the origin of the El Alacrán Cu-Au Deposit and its relationship with porphyry Cu-Au mineralization in the regional metallogenic context is critical for the exploration of similar deposits in the western Colombian Andes.

The El Alacrán Cu-Au Deposit lies within a Late Cretaceous intra-oceanic arc formed in an extensional to transpressive tectonic regime upon the Caribbean Large Igneous Province (CLIP). It is hosted in an intermediate to mafic volcano-sedimentary sequence of the Barroso Fm. and intruded by a series of calc-alkaline phreatomagmatic breccias and tonalites (e.g., El Alacrán Oeste Tonalite). The Cu-Au mineralization extends over 1 km north-south and comprises different styles, including stratabound replacement, dissemination, and minor veins. Based upon a detailed petrographic characterization, at least thirteen alteration assemblages have been recognized and interpreted to be formed through three different pathways (A, B, and C; Figure 1): Path A (>79.5 Ma) developed an early barren calc-sodic alteration (albite-apatite-quartz-chlorite) spatially associated with sill intrusions followed by Path B (ca. 79.5 Ma) with abundant magnetite (mushketovite)-quartz-apatite-Fe-rich chlorite-carbonate ± titanite pyrite ± chalcopyrite, providing a Fe-Au ± Cu ± light REE (Ce-La) IOCG geochemical signature. The incursion of an additional fluid (Path C) between 73 and 68 Ma under a syn-deformation regime produced a calc-potassic alteration assemblage dominated by V-mica (roscoelite composition)-quartz-actinolite-ilmenite-pyrrhotite-chalcopyrite-pyrite-Au, followed by sericite-Mg-rich chlorite-carbonate-quartz-chalcopyrite-pyrite-titanite±molydenite and later by a carbonate-base-metal (CBM) vein alteration producing a Cu-Au-Mo-Zn-Ag-As geochemical signature.

Pb isotopes in sulfides and feldspars suggest a relationship of Path B with a non-porphyry source (e.g., El Alacrán Oeste Tonalite). In contrast, disseminated sericite of Path C yielded a ^{40}Ar-^{39}Ar age of ca. 68 Ma, slightly younger with respect to U-Pb zircon ages obtained for intrusions associated with porphyry Cu-Au mineralization in the San Matias District (ca. 70 Ma). S isotopes in sulfides (avg $\delta^{34}S$ ~11‰) suggest an important seawater contribution to the ore-forming fluids. Fluid in Path B is consistent with IOCG mineralization, whereas CRD mineralization associated with Path C seems to explain better the variety in mineralogy, isotopic composition, sericite age, and the multiple Au generations identified in the El Alacrán Cu-Au Deposit.

El Alacrán Cu-Au Deposit mineralization is interpreted as the product of a hybrid process between an iron oxide-copper-gold (IOCG) spectrum and a carbonate replacement deposit (CRD) mineralization possibly associated to a porphyry Cu-Au source (Figure 1). The mineralogical and compositional features are comparable to those of well-known Andean IOCG deposits in the Coastal Cordillera province of Peru and Chile. However, dominant seawater ore fluid, a juvenile crust involvement, and the proposed relationship with a porphyry Cu-Au source make it unique respect to other Andean IOCG analogues.

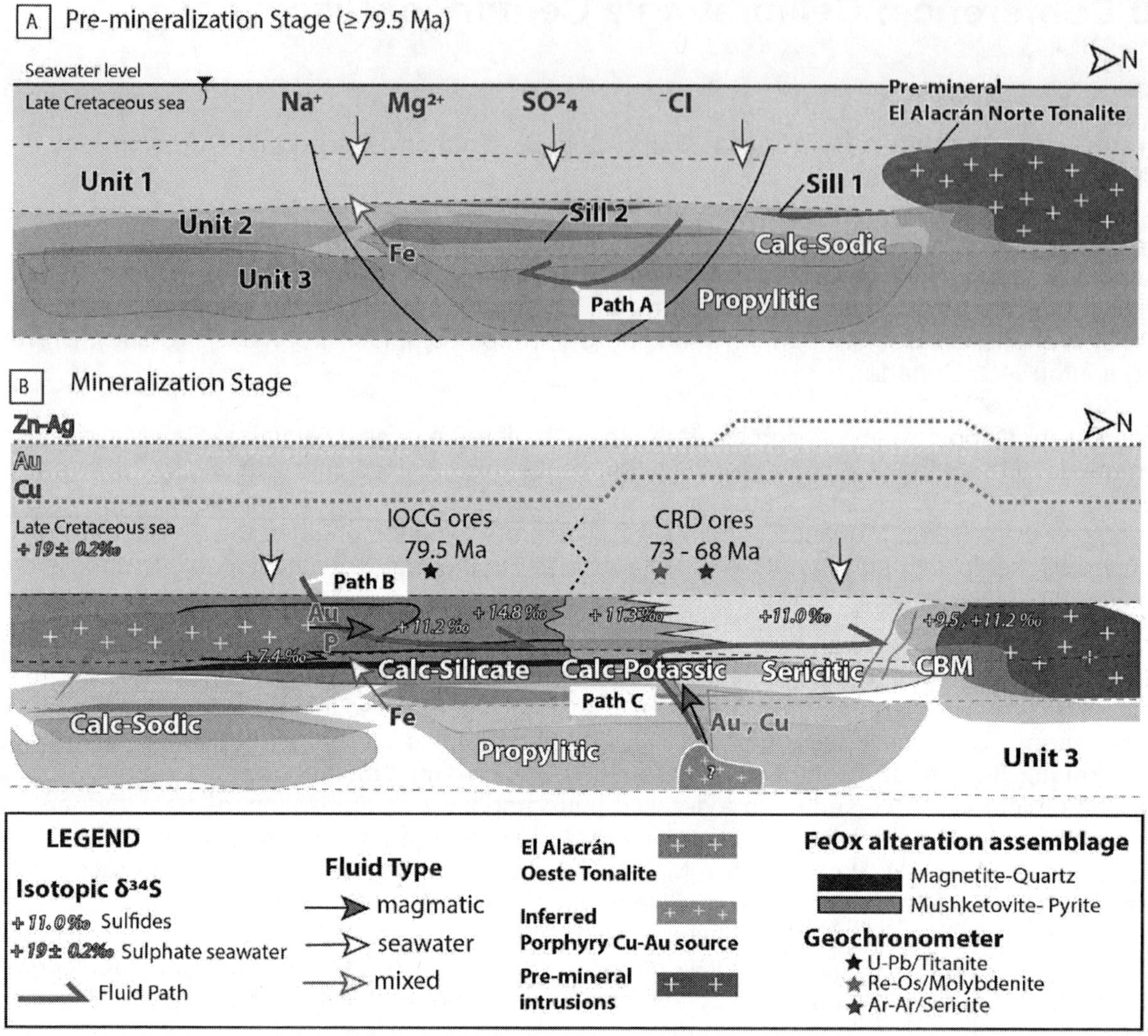

Figure 1. *Fluid evolution of the El Alacrán Deposit. A. West-looking schematic diagram of the initial configuration of the El Alacrán Deposit. The schematic diagram shows the El Alacrán Deposit sequence units (Unit 1, Unit 2, and Unit 3), the pre-mineral location of the El Alacrán Norte Tonalite magmatism, and the beginning of the Sill 2 magmatism that led to the formation of the cooling Path A alteration assemblages. B. Mineralizations and alteration assemblages formed from seawater interaction with at least two magmatic fluids; one fluid sourced from the El Alacrán Oeste Tonalite (in Path B) and other sourced from a inferred concealed porphyry source (Px2) consistent with the cooling Path C.*

D7
Fluid Inclusions and Origin of Carlin-style Mineralization at the Cove Deposit, Nevada

Sarah R. Shapley[1], John Muntean[2], Matthieu Harlaux[2]
1. University of Nevada, Reno, Reno, NV, USA, 2. Nevada Bureau of Mines and Geology, Reno, NV, USA

Carlin-style gold deposition provides up to 5% of the global gold budget and is the leading gold source in the Unites States. Despite its significance, there is not a consensus on the source of gold or fluid for these deposits. The fine-grained nature of ore assemblages as well as lack of associated quartz veins and suitable/workable fluid inclusions (FI) makes it difficult to ascertain the conditions of mineralization, including the temperature, pressure, salinity, and source of the hydrothermal fluids. The Cove deposit, located in north-central Nevada, grades from deep base metal vein type mineralization (BMVT) to Carlin-style mineralization. Both types formed during the Eocene between 39.2 and 34.3 Ma. Past research provides a geologic framework, including identifying magmatic signatures for sulfur isotopes of BMVT mineralization. The association of BMVT with Carlin-style mineralization affords a unique opportunity to investigate fluid sources of Carlin-style mineralization. Additionally, oxygen isotopes in carbonates have been proposed as a vector for Carlin-style deposits to evaluate the extent of fluid-rock interactions. Current work tests this vector at Cove.

This study combines petrography, microthermometry, and Raman spectrometry analysis of FIs to characterize BMVT mineralization fluids. Petrographic study identified three categories of FIs. Group (I) are rounded, lobate, two-phase FIs with 20-40 volume % vapor bubbles. Preliminary microthermometry for group (I) indicates moderate homogenization temperatures (236°-362°C, avg = 330°C) and low salinities (1.7-4.9 wt % NaCl eqiv, avg = 3.7 wt % NaCl eqiv). These conditions are consistent for "E-veins"- an emerging classification of base metal veining distal to or late in porphyry systems, when coupled with previously obtained $\delta^{34}S$ values at Cove. Group (II) crosscuts (group I) and shows reequilibration textures, prohibiting further work. Group (III) consists of rounded, lobate to irregularly shaped FIs. Multi (5-7)-phase hypersaline brine FIs locally coexist in the same fracture planes with vapor FIs in low proportions, indicating that they may be trapped during phase separation, while other FI planes trap only brines. The presence of multiple fluids at similar modern depths at the Cove deposit may indicate overprinting of the deposit during its formation or multiple pulses of fluid emerging from depth. Next steps will evaluate the potential for BMVT fluids to evolve to Carlin-style mineralization by quantifying the metal content of individual FIs by LA-ICP-MS to search for specific geochemical signatures such as Au, As, Hg, and Tl enrichments.

A challenge in locating Carlin-style mineralized systems is the scarcity of exploration vectors. Oxygen isotope analysis in carbonates has been invoked as a possible method of measuring the extent of fluid-rock interactions, with depleted $\delta^{18}O$ values correlating to higher degrees of fluid-rock interaction and elevated gold grades. Preliminary results show depleted $\delta^{18}O$ values in Au-bearing samples relative to barren ones, supporting that $\delta^{18}O$ values in carbonates are a potentially effective vector into Carlin's cryptic alteration halos. The knowledge gained from characterization of fluids, geochemistry, and $\delta^{18}O$ values of carbonates could result in more advanced metallogenic models for distal-disseminated and Carlin-style gold deposits in the McCoy district, Nevada, and elsewhere in the world.

D8

The Vergenoeg Strato-Volcano – IOCG-like Mineralization Associated with Felsic Magmatism in the Bushveld Magmatic Province, South Africa

Laurence Robb[1], Teimoor Dehkordi[2]
1. University of Oxford, Oxford, United Kingdom, 2. CIMERA, University of Johannesburg, Johannesburg, South Africa

The Vergenoeg Igneous Complex (VIC) defines a strato-volcano that erupted as the terminal phase of the bimodal Rooiberg Group (RG), the latter representing the extrusive component of the 2057-2054 Ma Bushveld Magmatic Province. The RG was extruded immediately prior to emplacement of both the mafic (Rustenburg Layered Suite - RLS) and felsic (Lebowa Granite Suite - LGS) intrusive phases of the Complex. The lowermost portion of the RG comprises the basaltic-to-andesitic Dullstroom Formation, much of which occurs beneath the RLS and therefore predates the main layered mafic intrusion. The upper portions of the RG, including the VIC, comprise the dacitic-to-rhyolitic, Damwal, Kwaggasnek, and Schrikkloof Formations, which are closer in composition to the highly fractionated A-type granites of the LGS. The VIC hosts world-class fluorite mines associated with a Fe-F-(Cu-REE) mineralization style that has marked IOCG affinities.

Numerous polymetallic, magmatic-hydrothermal mineral deposits occur throughout the various felsic phases of the Bushveld Magmatic Province, and these are typically represented by a three-stage paragenetic sequence: early magmatic Sn-W-Mo-F ores (600° > T > 400°C), followed by a Cu-Pb-Zn-As-Ag-Au paragenesis (400° > T > 200°C) and then late-stage Fe-F-U mineralisation (<200°C). Borehole core from mineralized intervals of the VIC reveals a sharp, erosional contact (~35°) between the uppermost RG volcanics (i.e., the Schrikkloof Formation) and pyroclastic rocks of the VIC, suggesting that the latter postdates the RG and may be synchronous with subjacent LGS emplacement. A well-defined vent occurs in the centre of the complex and is also the site of the world-class F-Fe-REE deposit of the Vergeneog Mine. The vent preserves a coarse volcaniclastic breccia comprising mainly hematite (goethite/siderite)-fluorite-magnetite-fayalite—minor minerals include apatite, cassiterite, monazite, titanite, and REE-carbonate phases. It is zoned vertically and records a complex interplay between volcanic eruptive processes, magmatic-hydrothermal remobilization of ore constituents, and later supergene overprinting. Prominent sulphide mineralization occurs in the form of pyrite and chalcopyrite with lesser arsenopyrite and sphalerite. Away from the vent, subaerially-deposited volcaniclastic material defines a strato-volcano (ca. 10 x 5 km in extent) comprising two distinct units, becoming finer grained with increasing distance from the vent. The lower portion of the VIC is built of siliceous ignimbrite with a minor rhyolitic component. This is overlain by volcanic hematite-fluorite breccia intercalated with laminated, locally reworked, fan-delta sediments, all deposited in a maar-like structure. The recently developed Nokeng fluorite mine is hosted in these bedded VIC volcaniclastic sediments, distal to the vent.

Genesis of the polymetallic mineralization in the VIC, and other felsic phases of the Bushveld Complex, remains poorly understood. The association between intrusive and extrusive pulses of felsic magmatism, together with evidence for widespread magmatic-hydrothermal Fe-F-(Cu-REE) mineralization processes, define a Bushveld metallotect that has similarities to other iron-oxide-copper-gold (IOCG) provinces in, for example, Brazil and Sweden. An IOCG footprint in the felsic phases of the Bushveld event would render them more prospective than previously thought.

D9

Tungsten Mineralization in the Eastern Alps – Tracking Ore-Forming Processes Using Scheelite Trace Element Chemistry and Micro-textures

Florian Altenberger[1], Johann G. Raith[1], Joachim Krause[2], Kai Bachmann[2], Jasper Berndt[3], Julia Weilbold[4], Holger Paulick[4]

1. Montanuniversität Leoben, Department Applied Geosciences and Geophysics, Chair of Resource Mineralogy, Leoben, Austria, 2. Helmholtz-Zentrum Dresden-Rossendorf, Helmholtz Institute Freiberg for Resource Technology, Freiberg, Germany, 3. Westfälische Wilhelms-Universität Münster, Institute for Mineralogy, Münster, Germany, 4. Geological Survey of Austria, Department of Raw Materials, Vienna, Austria

The increasing demand for raw materials used in high-technology applications, and their economic importance, is reflected in efforts by the European Union to achieve strategic autonomy on key raw materials. Tungsten is one of the prominent raw materials highlighted by EU criticality assessments. Discovery of the world-class Felbertal W-deposit (Austria) in 1967, which is still Europe's largest producing tungsten mine, triggered substantial greenfield exploration during the 1980s. Many tungsten occurrences were discovered in the Eastern Alps and are currently being investigated with the aim to define differences in mineralization processes. The goal is to establish assessment criteria ("fingerprint") to identify regions with high potential for significant tungsten endowment that can be used in mineral exploration.

Scheelite ($CaWO_4$) is the most common tungsten mineral in the Eastern Alps, occurring in different geological settings and mineralization styles, including stratabound mineralization in metabasites, meta-carbonate and calc-silicate rocks, skarns, orogenic Au-W veins, and scheelite-bearing metamorphic veins etc. Scheelite from these different settings is studied using a combination of cathodoluminescence (CL), electron probe micro-analyzer (EPMA), SEM-based automated mineralogy (MLA), and laser ablation-inductively coupled plasma-mass spectrometry (LA-ICP-MS) techniques.

In the metamorphosed Felbertal vein-stockwork deposit, several stages of scheelite (re)crystallization are recorded. In addition to magmatic-hydrothermal scheelite, there is metamorphic scheelite, which formed during polyphase regional metamorphic overprint. CL-documented micro-textures and in situ trace element data allow us to define individual scheelite generations and track changing conditions during ore formation. Based on the Felbertal trace element signature of scheelite, a comparison with scheelite from other ore-forming environments will be carried out.

The results of this study show that Na, Mg, Sr, Y, Nb, Mo, Ba, REE, Hf, Pb, and U are sensitive to specific mineralization processes and, hence, useful to discriminate scheelite formed in different environments. In combination with CL imaging, in situ trace element analysis of scheelite distinguishes metamorphic scheelites from those of other provenance. Metamorphic scheelite exhibits homogeneous micro-textures, whereas magmatic-hydrothermal scheelite commonly shows primary and oscillatory zoning. Secondary zoning and replacement textures develop in scheelite affected by subsequent metamorphism, resulting in formation of several generations of scheelite with different trace element signatures. Molybdenum, Sr, and, particularly, REEs experience significant redistribution between these generations. Total REE concentrations and REE patterns may vary considerably; e.g., LREE enrichment or depletion and negative or positive Eu-anomalies. Furthermore, fractionation of REE between scheelite and coprecipitating minerals (e.g., apatite, allanite-Ce, epidote, titanite, etc.) is also important.

D10

Characterization of Polymetallic Vein-Type Occurrences in the Meguma Terrane: A Lesser-Known Gold Deposit Type for Nova Scotia?

Naomi Welt[1], Erin Adlakha[1], Joshua Jackman[1], Jacob Hanley[1], Mitchell Kerr[1], Geoff Baldwin[2]
1. Saint Mary's University, Halifax, NS, Canada, 2. Nova Scotia Department of Energy and Mines, Halifax, NS, Canada

This study focuses on three Au-bearing, polymetallic (Sb-Pb-Zn-Co-Ni-Cu-Bi-Ag-Au) vein occurrences in the western Meguma Terrane with a suspected, but unproven, genetic link. The aim is to understand how the occurrences formed, including their timing, sources of metals, and relationships to other Meguma Au deposits and metasedimentary-hosted polymetallic vein deposits globally (e.g., Sb-Au deposits of the Bohemian Massif, Variscan Orogen, Europe).

At the Lansdowne and Cape Saint Mary's (CSM) occurrences in Digby County, Bear River Formation, metapelites are intruded by gabbro (~440 Ma; apatite U-Pb) and crosscut by quartz ± carbonate veins, some of which host multi-stage, sulfide-sulfosalt mineralization. At Lansdowne, mineralization crosscuts the early quartz veins and consists of an early Fe-As stage (pyrite-arsenopyrite-pyrrhotite), a Zn-Cu stage [(Cd-rich) sphalerite-chalcopyrite], and a later Pb-Sb stage (boulangerite-jamesonite-galena) with multiple generations of quartz-calcite-chlorite gangue. Arsenopyrite thermometry indicates temperatures of 425°-450°C, while chlorite of the Zn-Cu stage provides a lower temperature of 350°-390°C. The latest chlorite provides a temperature of 120°-160°C. Isocon diagrams generated using least- and most-altered samples from the mafic intrusions and metapelites indicate variable depletion in the above metals (except Sb, which was gained), potentially indicating these rocks as metal sources.

At the CSM Stibnite Occurrence, siderite infilled-breccia hosts mineralization and crosscuts earlier, brittle-ductile deformed quartz veins in Bear River Formation metapelites. Mineralization consists of an early As ± Co-Ni stage [arsenopyrite-(Co-rich) gersdorffite], followed by a Cu-Sb-Ag stage [tetrahedrite-(Ag-rich) freibergite], a Cu-Sb-Bi stage (chalcopyrite, Bi-Sb alloy), and late REE-P alteration (florencite-pyrite). At the Mavillette Beach Occurrence, Pb-Sb mineralization (boulangerite-jamesonite-galena) is observed in quartz-pyrite veins.

At the Nictaux Falls Dam Occurrence (NFDO), variably mineralized quartz veins and quartz-infilled breccia cut Kentville Formation metapelites near their contact with the South Mountain Batholith (370 Ma), diabase dykes, and a gabbro intrusion (~380 Ma; apatite U-Pb). Mineralization consists of early pyrite, a zoned Fe-Co-Ni-As stage (arsenopyrite-cobaltite-gersdorffite) with chlorite (~280°C) and rutile, and late-stage electrum.

The mineralogy, paragenesis, metal associations, and ore textures are distinct from typical Meguma Au deposits. The occurrence of mineralization in breccias and mineral thermometry suggest their formation in relatively shallow crustal environments at moderate to high temperature. Striking similarities between Lansdowne and CSM with the West Gore Sb-Au deposit (Hants Country, N.S.) in terms of mineralogy and paragenesis are noted. Future work includes Re-Os dating of arsenopyrite and trace element work on ore minerals to compare with similar deposit types.

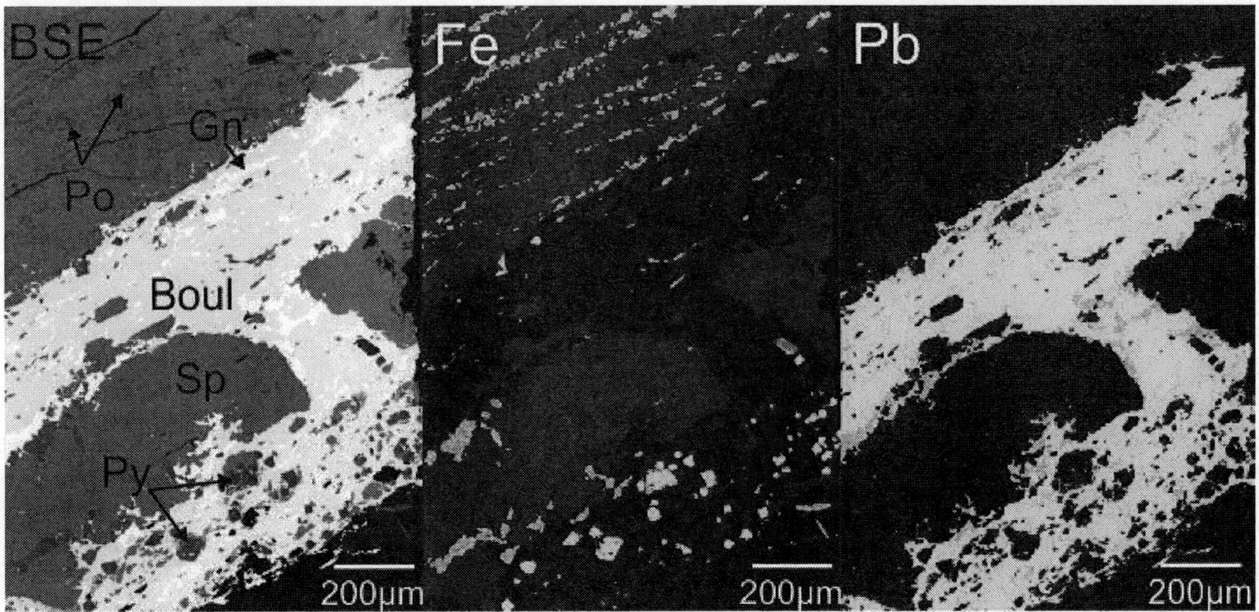

Electron Probe Micro-analyzer elemental maps for core sample of mineralization at Lansdowne (sample LAN2-2). BSE image with labelled minerals: sphalerite (Sp), pyrite (Py), pyrrhotite (Po), boulangerite (Boul), and galena (Gn). Fe map differentiates pyrite (green) and pyrrhotite (orange). Pb map differentiates boulangerite (green) and galena (orange).

D11

Remote Sensing-based Mapping of Zn-Pb-Carbonate Hosted Ore Deposits Using Sentinel-2 and PRISMA Satellite Imagery: The Jabali Test Site (Western Yemen)

Rita Chirico[1], Nicola Mondillo[1, 2], Carsten Laukamp[3], Giuseppina Balassone[1, 4], Diego Di Martire[1], Alessandro Novellino[5]
1. University of Naples Federico II, Naples, Italy, 2. Natural History Museum, London, United Kingdom, 3. CSIRO Mineral Resources, Kensington, WA, Australia, 4. INVG-OV, Naples, Italy, 5. British Geological Survey, Keyworth, United Kingdom

In this study, we have developed an integrated methodology, based on hyperspectral and multispectral imagery, for mapping the spatial distribution of the hydrothermal and supergene alteration associated with the Zn-Pb(-Ag) Jabali deposit (Yemen). The Jabali deposit is located in a desert area northeast to the capital of Yemen, Sana'a, along the western NW-SE-trending margin of the Sab'atayn basin. The Zn-Pb mineralization consists mainly of nonsulfides of supergene origin, mainly represented by smithsonite, minor hydrozincite, and hemimorphite (+ acanthite and greenockite in small amounts). The nonsulfide ore is hosted by a partly hydrothermally dolomitized Jurassic limestone. The supergene ore derived from oxidation of the primary sulfide mineralization (sphalerite, pyrite, and galena) that has been interpreted as a Mississippi Valley Type deposit emplaced via fluid migrations in extensional tectonic regime. Satellite data enabled the identification of the hydrothermal dolomitization, as well as the outlining of the gossans and secondary mineralized products, which extensively outcrop in the test area. The area was analysed by means of Sentinel-2 multispectral (European Space Agency) and PRISMA* hyperspectral (Italian Space Agency) imagery. These two missions provide to the scientific community and to other users free-of-charge data and were chosen for sensing the different materials' spectral responses in the visible near infrared (VNIR) and the shortwave infrared (SWIR) regions, respectively. PRISMA provides images at a spectral resolution of 12 nm in a continuum of 240 bands in the VNIR to SWIR wavelength region of the electromagnetic spectrum (400-2,505 nm). This satellite imagery allows the clear distinction between dolomite and limestone in the SWIR region by enhancing the spectral differences between these two carbonate phases in bands 148 and 150 (2,320 and 2,335 nm, respectively). The distribution of gossans and alteration products, instead, was tested on 10-m resampled Sentinel-2 products, which offer a full set of 12 bands, with several bands in the VNIR region (B2, B4, B6, B7, and B8A) covering the Fe-bearing minerals spectral features. Spectral signatures of representative host-rock samples and Fe- and Zn-bearing mineralized alteration products were collected and validated via XRD analyses in order to build up a spectral library needed for image classification, which has been produced with the support of band combinations, spectral indices, and statistical analyses. Despite the great number of bands provided by PRISMA, a prompt visualization of the known Fe and alteration minerals outcrops with the 30-m spatial resolution was not achieved yet. However, the availability of multiple bands in the SWIR region and the hyperspectral resolution make PRISMA a suitable tool for mapping the dolomitization footprint, while the use of multispectral, Sentinel-2 "easy-to-handle" products suits best for the identification of the gossan outcrops associated with the Jabali ore type.

This work proves that an integrated satellite-ground-based mapping approach using multiple sources of spectral data with different spectral resolution is a promising exploration strategy for the Jabali ore type at regional scale in the Sab'atayn basin of Yemen, transferable to other sedimentary basins worldwide.

*Project carried out using PRISMA Products, © of the Italian Space Agency (ASI)

D12

Petro-Geochemical Characterization Of Supergene Copper Mineralization In Atacama Desert (Northern Chile): U-Pb Chronometric Potential And Formation Conditions

Steven Kahou[1], Stéphanie Brichau[1], Stéphanie Duchene[1], Marc Poujol[2], Eduardo Campos[3], Rodrigo Riquelme[3], Sébastien Carretier[1]
1. Géosciences Environnement Toulouse (GET), Université de Toulouse, Toulouse, France, 2. Univ Rennes, CNRS, Géosciences Rennes, Rennes, France, 3. Departamento de Ciencias Geologicas, Universidad Catolica del Norte, Antofagasta, Chile

Supergene copper mineralizations (SCM) are nowadays the economic viability of many porphyry copper deposits worldwide. These mineralizations are derived from supergene processes, defined as sulfide oxidation and leaching of ore deposits in the weathering environment, and any attendant secondary sulfide enrichment. For supergene copper mineralization to form, favorable tectonics, climate, and geomorphologic conditions are required. Tectonics control the uplift needed to induce groundwater lowering and leaching of sulphides from a porphyry copper deposit. Climate controls copper leaching in the supergene environment and groundwater circulation toward the locus where supergene copper-bearing minerals precipitate. In the Atacama Desert of northern Chile, SCM seem to take place during specific Tertiary climatic periods and relief formation. But many uncertainties remain regarding the genesis and the exact timing for SCM formation.

We present mineralogical and chemical data on supergene copper-bearing minerals sampled from in situ and exotic SCM from the Atacama Desert of northern Chile. Although northern and southern areas of the Atacama Desert have experienced different geological evolution, they both underwent similar geological and climatic controls to form and preserve SCM.

Chrysocolla and pseudomalachite are the most common copper-bearing minerals found in SCM from the Atacama Desert of northern Chile. This led us to test the potential of both minerals as dating materials to place the SCM formation in a geochronological context. However, variation of U content from one mineral to another, common Pb contribution, and the absence of matrix-matched-standard for both minerals made U-Pb LA-ICP-MS dating of chrysocolla and pseudomalachite challenging. A mixed approach combining texturally-controlled in situ LA-ICP-MS U-Pb dating and multi-elements mapping, together with SIMS oxygen isotope analyses, was applied to the Mina Sur, Damiana, and El Cobre exotic copper deposits from the Atacama Desert.

Regardless of location, the results obtained demonstrate the important role played by the local geological parameters on the control of the U-Pb chronometer in supergene copper minerals. Initial high U content of the porphyry source can promote U-rich supergene copper minerals as is the case at Mina Sur. Additionally, a long-term supergene alteration will favor long-term water circulation in the depositional environment, which can leach U from the porphyry source and allow the formation of U-low supergene copper minerals during recent times.

The promising results obtained represent a new tool to understand the physicochemical, climatic, and geological conditions that prevailed during the formation of supergene copper deposits, as well as a proxy for their prospection around the world.

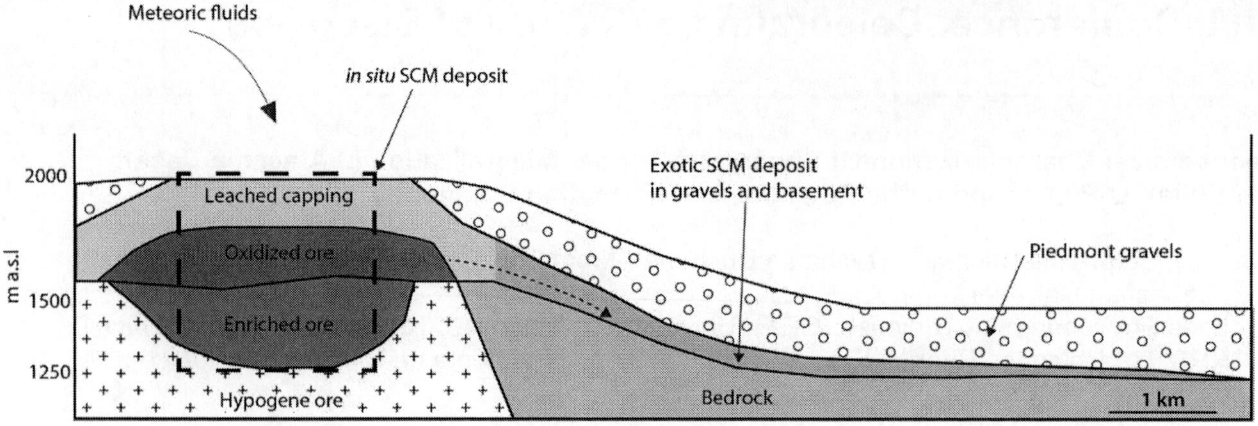
Meteoric fluids
in situ SCM deposit
Exotic SCM deposit
in gravels and basement
Piedmont gravels
2000
m a.s.l
1500
1250
Leached capping
Oxidized ore
Enriched ore
Hypogene ore
Bedrock
1 km

ST.058
Potential Field Imaging of the Pembine-Wausau Terrane Wisconsin-Michigan: Implications for Volcanogenic Massive Sulfide Deposit Exploration

Eric D. Anderson[1], Klaus J. Schulz[2], Christopher S. Holm-Denoma[1]
1. US Geological Survey, Denver, CO, USA, 2. US Geological Survey, Reston, VA, USA

The Pembine-Wausau terrane represents a major Paleoproterozoic belt of metavolcanic and intrusive rocks that hosts several volcanogenic massive sulfide (VMS) deposits including the Back Forty deposit. The island-arc complex accreted to the continental margin of the Superior craton along the Niagara fault zone during the Penokean orogeny and was subsequently intruded by syn- to postorogenic granitoids. Limited outcrop makes bedrock mapping difficult. In addition, Phanerozoic sedimentary rocks cover the eastern extent of the belt. A high-resolution aeromagnetic survey over parts of the Pembine-Wausau terrane was conducted to better understand the geology and potential mineral resources. The data were collected along north-south flight lines spaced 150 m apart at a nominal height of 80 m. These data, along with in-fill gravity stations and targeted new radiometric age dates, help improve our understanding of the VMS potential.

The complete Bouguer gravity anomaly map shows a west-northwest trending linear high that correlates with mapped mafic-ultramafic rocks along the Niagara fault zone delineating the northern extent of the Pembine-Wausau terrane. A moderate-amplitude high continues south for about 15 km where a strong east-west trending gradient is observed. The gradient can be traced more than 150 km to the west, and most known VMS deposits occur north of it. Gravity anomaly lows south of the gradient correlate with felsic volcanic rocks of the Beecher Formation, which hosts the Back Forty deposit. New zircon U-Pb age dates indicate that rhyolite in the Beecher Formation was deposited at 1835.9 ± 4.5 Ma, significantly younger than the ca. 1870 Ma age for VMS deposits in the belt. The Back Forty deposit occurs on the northern edge of a gravity high that continues for more than 25 km to the south. A previously unmapped granitoid at the southern extent of the gravity high has been dated at 1837.3 ± 3.7 Ma.

The reduced-to-pole magnetic anomaly map shows detailed changes in rock magnetization. North-south-trending magnetic lineaments interpreted as dikes are observed near the Back Forty deposit. The dikes appear to be truncated at the strong gravity gradient, indicating a significant change in magmatic setting across the gradient. Several low-amplitude magnetic highs with diameters around 1 to 4 km are observed around the Back Forty deposit. Field observations indicate that the anomalies are sourced in previously unmapped granitoids that may have been the heat source for VMS formation. East-northeast-trending lineaments highlighted with the tilt derivative align with mapped fold axes in the volcanic rocks that presumably overlie the heat source. A high-amplitude magnetic anomaly over the granitoids to the south of the district indicates significantly greater magnetite content than in the granitoids in the district. The granitoids to the south are separated from the granitoids in the district by east-west- to east-northeast-trending tilt derivative lineaments that are interpreted to delineate a shear zone. Taken together, the geophysical data and the new radiometric age dates suggest VMS potential south of the gravity gradient that is younger than the ca. 1870 Ma age of VMS deposits in the Pembine-Wausau terrane.

ST.059
An Innovative Tectono-Metallogenic Model for Revitalised Targeting of IOCG-related Deposits in the Gawler Craton

John A. Anderson
Austrike Resources, Glenalta, SA, Australia

A new tectono-metallogenic model for the Olympic Metallogenic Event (OME) provides the framework for a significant step-change in ore vectoring for a spectrum of IOCG and coeval deposit styles.

Minerals exploration resulted in the recognition of the Nankivel advanced argillic cap in the nineties and the discovery of the 42 Moz Paris silver deposit in 2011. Paris is interpreted as an intermediate sulphidation epithermal associated with a porphyry system at Nankivel and the Helen copper-gold-silver magnetite skarn, all within a 100-km^2 sericite pyrite lithocap. The mineralisation is hosted in part by 1620 Ma subduction monzodiorites, with the Paris-Nankivel mineralisation dated within 2 Ma of the 1590 Ma Olympic Dam mineralisation.

The revised model proposes the Paris-Nankivel epithermal-porphyry belt was rapidly formed at the same time as Olympic Dam, other IOCG deposits, and many of the gold deposits in the Hiltaba/Gawler Ranges SLIP. Both the SLIP and speed, size, and diversity of the OME are explained by precursor double subduction under compression followed by a rapid extension and formation of a super caldera filled with Gawler Range Volcanics (GRV). Northerly subduction associated with the St. Peters suite intrusives and southerly subduction by the colliding North Australia Craton resulted in super metasomatism of the mantle during 1640 to 1605 Ma. A change to extension from 1597 Ma resulted in two subparallel belts of low-eNd and high-F Hiltaba granites above the fossil subduction zones, the formation of the 90,000-km^2 super caldera, and highly fractionated (Zr/Hf = 20) cupolas telescoped into the rising shoulders of the craton. Differing precursor subduction conditions resulted in epithermal-porphyry deposits on the southern margin of the caldera, whereas IOCGs formed on the northern and eastern margins with haematite- or magnetite-dominated systems respectively forming on the shoulder or in hotter, more reduced conditions under the GRV blanket within the caldera.

Variants of a mid-GRV strato-tectonic marker of the caldera collapse and OME are interpreted as the Bitalli Rhyolite at Paris and Nankivel palaeosurface on the southern shoulder, and as lateral ferruginous sediments and volcanogenic conglomerate facies that collapsed into the IOCG systems on the northeastern shoulder (Fig. 1). Drill logs indicate the overlying Pandurra (Sandstone) Formation is transitional with the OME marker, so the cover role of that unit needs reassessment with implications for deposit preservation and exploration.

The new model provides an improved 4-D guide to selecting target areas with preferential structural preparation and favourable intrusives around the super caldera margin and at prospective levels beneath the OME marker. This revitalises previously mature IOCG districts with better prioritisation of gravity targets and offers a new paradigm of targeting other hydrothermal deposit styles both in old and new priority areas. The application of new magneto-telluric and pathfinder geochemical techniques to extensively map hydrothermal systems is agnostic to deposit styles and will expand the discovery space beyond the successful but tiring dogma of IOCG gravity targeting.

Figure 1: Regional Long Section looking west across the proposed Gawler Super Caldera with different copper-gold deposit styles determined by precursor subduction preparation.

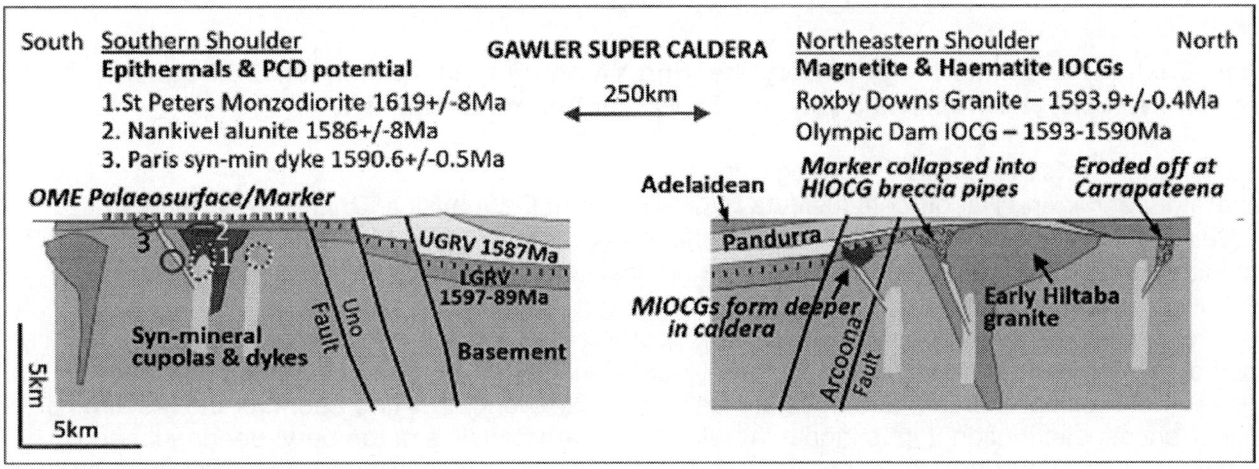

ST.060

Occurrence, Texture, and Mineralogy of Beryl-bearing Yamrang Pegmatite in Ikhabu Pegmatite Field, Eastern Nepal: New Beryl-Columbite Subtype REL Pegmatite from Himalaya

Sushmita Bhandari[1, 2], Kezhang Qin[1, 2], Qifeng Zhou[3]
1. Key Laboratory of Mineral Resources, Institute of Geology and Geophysics, Chinese Academy of Sciences, Beijing 100029, China, 2. University of Chinese Academy of Sciences, Beijing 100049, China, 3. Institute of Mineral Resources Research, China Metallurgical Geology Bureau, Beijing 101300, China

Rare metals, especially beryl-aquamarine, have been sporadically reported from Nepal Himalaya. However, occurrences of rare-element pegmatites (REL) from Nepal Himalaya, their distribution, mineralization characteristics, host-rock properties, and ages are unknown. Our work aims to present for the first time the detailed occurrences, distribution, types, and mineralization characteristics of the beryl-bearing Ikhabu pegmatite field from the Taplejung district of Eastern Nepal. We accessed the Yamrang, Sangsabu, and Wairung beryl-bearing pegmatite dikes in the Ikhabu area and presented the detailed study from the biggest Yamrang pegmatite. Yamrang pegmatite is an intensely beryl-mineralized, zoned pegmatite dike on Yamrang hill with 25- to 30-m width and 130-m visible extension, lying at 27.50° N, 87.73° E. Based on field geology, microscopic observation, and major and trace element analysis of beryl, garnet, and tourmaline, mineralogy, texture, and zonations of Yamrang pegmatite are characterized, classification of Ikhabu pegmatite field is done, evolution degree is identified, and possible exploration implications for REL mineralization in the Himalaya are discussed. The Ikhabu pegmatite field belongs to the beryl-columbite subtype, rare-element pegmatite, but also consists of aquamarine-bearing miarolitic pockets formed by late-stage hydrothermal activities. Yamrang pegmatite shows a pattern of axial symmetric zonation and can be differentiated into 5 mineralogical-textural zones, as follows: Zone 1) Saccharoidal albite+quartz+muscovite zone (wall zone); Zone 2) Blocky perthitic microcline zone (outer intermediate zone); Zone 3) Perthite+quartz+muscovite+albite+beryl (inner intermediate zone); Zone 4) Quartz+beryl core zone; and Zone 5) Miarolitic beryl pocket zone. Beryl occurs in zones 3, 4, and 5 in association with quartz and k-feldspar. Beryls are homogeneous to weakly zoned, green to blue, prismatic to columnar, and belonging to the sodic beryl type. They consist of low total alkali contents (<1 wt %), Na ~0.5 wt %, Li and Cs <0.5 wt %, and high Na/Li (i.e., >3.5) and high Fe content (0.7 wt %) with low REE, mostly below detection level (bdl) but elevated Be (59898.82 ppm) and Ga (25.06 ppm) content compared with world's evolved pegmatites.The garnets are reddish-brown, euhedral, and belong to the almandine-spessartine series (Xsps ~50) with high HREE (217 ppm) and Y vs. ∑REE (1.9). Tourmalines are black to greenish colored, Mg-Ca rich dravite-schorl type from outer zone to Fe-Mn rich schorl variety from inner zone with overall high Ti (0.4 wt %) and low Pb, Bi, and REE (bdl). Sodic beryl with low REE contents, spessartine garnet with Xsps ~50, high HREE and Y vs. ∑REE, and schorl tourmaline with high Ti and low Pb, Bi, and REE contents indicate that Yamrang pegmatite and thus the Ikhabu pegmatite field is modestly fractionated REL pegmatite. The higher evolution of Yamrang pegmatite than other pegmatites suggests the regional evolution direction from North to South in the Ikhabu Pegmatite field. Along with high alkali contents of beryl and high Xsps contents of garnet, high Li, Ga, Nb, Mn, Fe, and low Mg, Ti contents of schorl tourmaline are found as evolution indicators in the Ikhabu Pegmatite field and possibly are useful indicators in modestly fractionated pegmatites elsewhere.

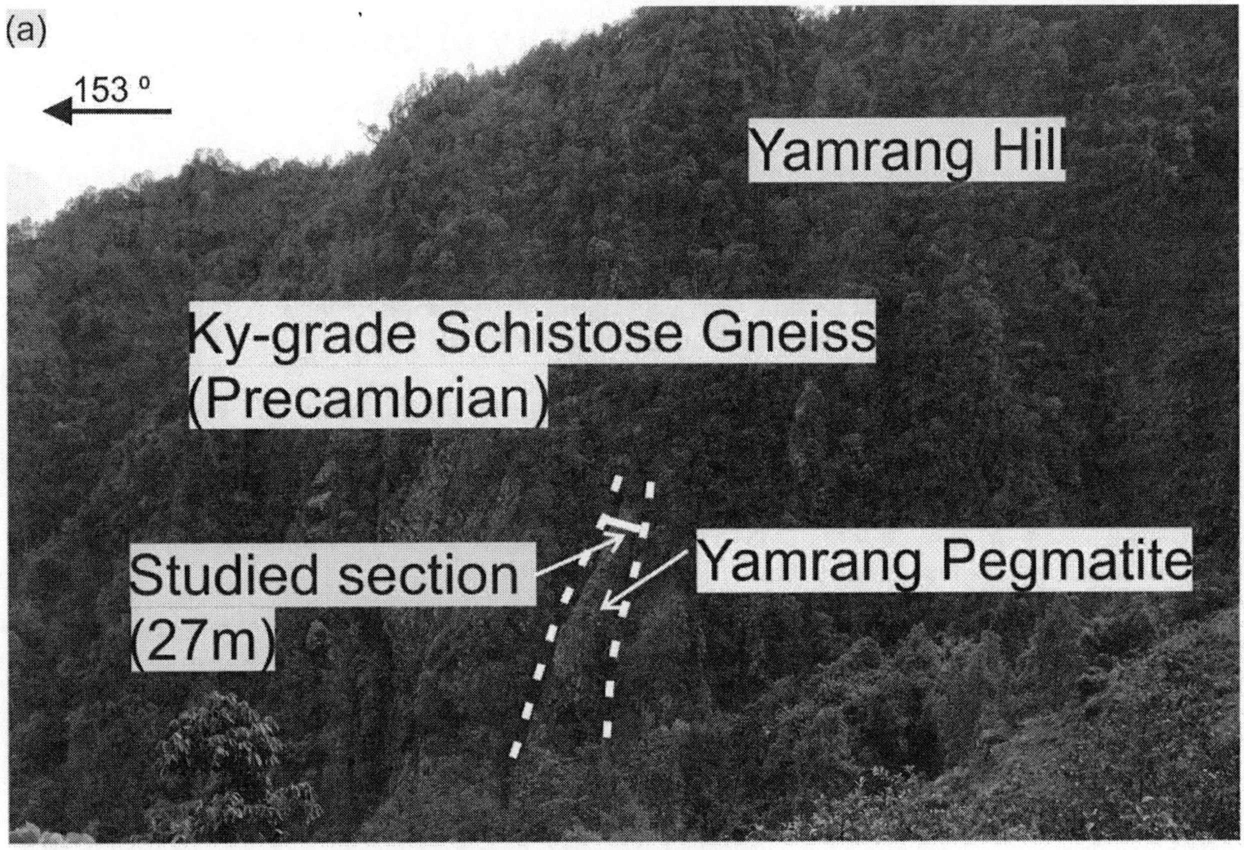
(a)
153 º
Yamrang Hill
Ky-grade Schistose Gneiss
(Precambrian)
Studied section
(27m)
Yamrang Pegmatite

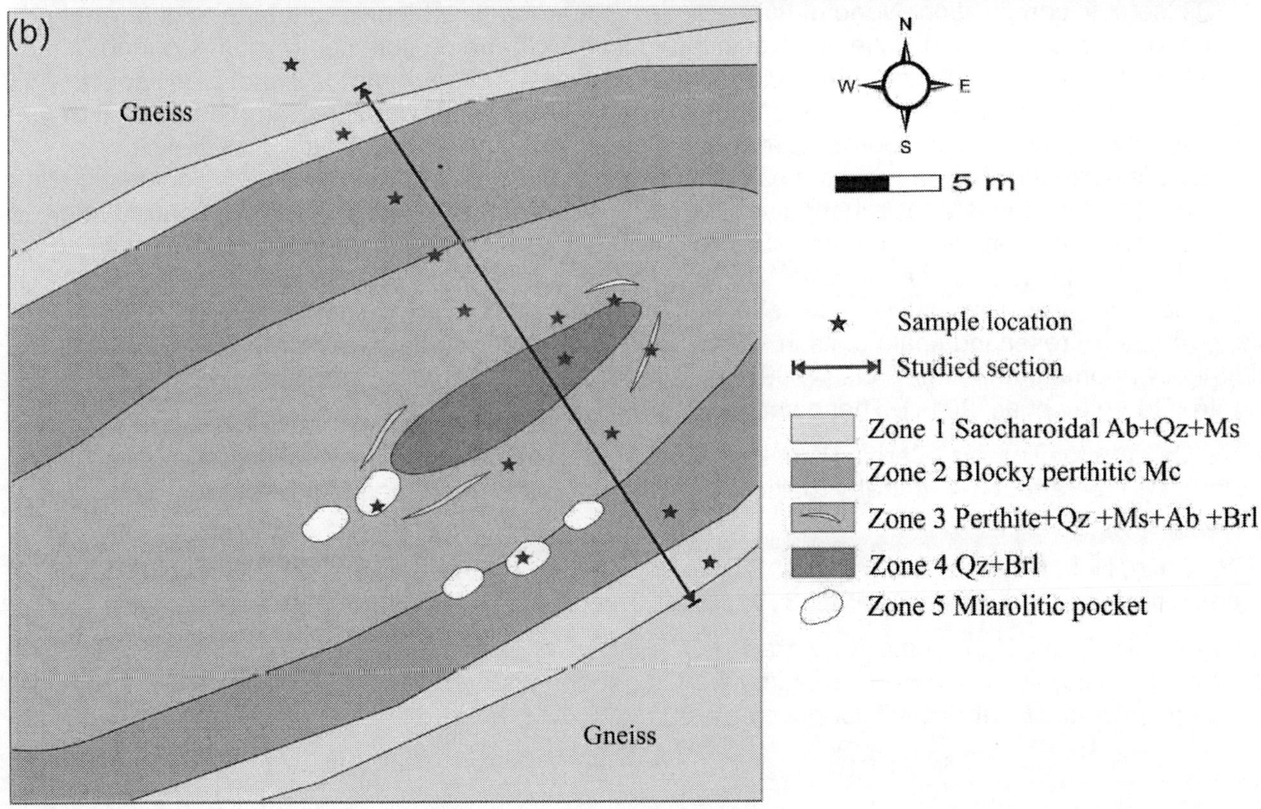
(b)
Gneiss
Gneiss
N
W
E
S
5 m
Sample location
Studied section
Zone 1 Saccharoidal Ab+Qz+Ms
Zone 2 Blocky perthitic Mc
Zone 3 Perthite+Qz +Ms+Ab +Brl
Zone 4 Qz+Brl
Zone 5 Miarolitic pocket

ST.061
Re-Os Isotopic Analyses of Wulfenite (PbMoO4) in the Oxidation Zone of the Alpine Zn(Pb) Deposits

Maria Boni[1,4], Holly Stein[2,3], Giuseppina Balassone[1], Gang Yang[2], Nicola Mondillo[1,4]
1. University Napoli Dipartimento Scienze della Terra dell'Ambiente e delle Risorse, Napoli, Italy, 2. AIRIE Program, Colorado State University, Fort Collins, CO, USA, 3. Institute of Geosciences, University of Oslo, Oslo, Norway, 4. Earth Sciences Department, The Natural History Museum, London, United Kingdom

The lead molybdate mineral, wulfenite ($PbMoO_4$), is a relatively rare mineral that usually occurs in oxidized zones of Zn(Pb) deposits. Wulfenite is generally observed to be one of the last mineral species formed in the oxidation zone, on top of other non-sulfides (e.g., cerussite, anglesite, smithsonite, hemimorphite, mimetite, pyromorphite, and Fe- and Mn-(hydr)oxides). Wulfenite is found in several mines in Triassic carbonates of the Alpine realm in Austria and Italy (Bleiberg, Raibl, Gorno) and is particularly abundant at Mežica (Slovenia). Wulfenite mines were a source of Mo through WWII.

Wulfenite forms xenomorphic aggregates that can replace host carbonates or surround altered galena crystals. Cavities formed in the oxidation zone may contain zoned wulfenite laths of up to 30 mm. Crystals vary from colorless, yellow, yellow-red, and red-brown to brown-black. Trace element analyses show As, Hg, V, W, Bi, and U in wulfenites, especially in earlier-precipitated, darker varieties. Since these elements are not known in the precursor sulfide minerals nor in the host carbonates, oxidizing fluids external to Triassic carbonates may be involved.

In this study, we analyzed pure handpicked wulfenite crystals from three localites (Bleiberg, Mežica, Gorno) for their Re-Os isotopic composition. None of the data was isochronous, which we attribute to different and varying Os isotopic compositions in wulfenite-precipitating fluids. Changing fluid compositions during wulfenite deposition are readily observed in zoning at the macroscale (see photo). All analyses show that 80 to 97% of the total Os is common Os. For two of the deposits (Bleiberg, Mežica), Re ranged from 1.6 to 4.9 ppb, whereas the two analyses from Gorno were much higher, 48.3 and 48.5 ppb Re. Extremely low $^{187}Re/^{188}Os$ for Bleiberg and Mežica preclude calculation of model ages, but with so little radiogenic ^{187}Os, it can be assumed that wulfenite ages must be very young (post-Alpine). A much higher Re content and $^{187}Re/^{188}Os$ for Gorno coupled with Os that is 95% common Os also suggests a very young wulfenite age, perhaps <10 Ma. For all samples, measured $^{187}Os/^{188}Os$ ratios are very high (Mežica ~5, Bleiberg ~14, Gorno ~26). Therefore, the initial $^{187}Os/^{188}Os$ ratios for fluids responsible for wulfenite deposition are highly radiogenic, perhaps representing shale units in much older (Precambrian) basement. More likely, however, shales with highly anomalous $^{187}Re/^{188}Os$ (>2000), as are known in the upper Permian, can explain the wulfenite data (Georgiev et al., 2011). These shales also have very high Mo.

Based on Re-Os data for Triassic (Carnian) shales (Xu et al., 2014), present day measured Os isotopic ratios ($^{187}Os/^{188}Os$) are about 3-4, thereby making derivation of wulfenite-depositing fluids from Triassic shales unlikely.

Georgiev, S., Stein, H.J., Hannah, J.L., Bingen, B., Weiss, H.M., and Piasecki, S. (2011) Hot acidic Late Permian seas stifle life in record time: *EPSL*, 310: 389-400.

Xu, G., Hannah, J.L., Stein, H.J., Mørk, A., Vigran, J.O., Bingen, B., Schutt, D., and Lundschien, B.A. (2014) Cause of Upper Triassic climate crisis revealed by Re-Os geochemistry of Boreal black shales: *Palaeogeography-Palaeoclimatology-Palaeoecology*, 395: 222-232.

Mežica, Slovenia
zoned wulfenite crystals
2 cm

ST.064
Multivariate Statistical Studies in the Faina Deposit: Insights on Element Association Fluid Composition and Vectoring Within Ore-Bearing Hydrothermal Systems

Gabriel M. da Silva[1], Atlas V. C. Neto[1], Mariana B. Soares[2], Felipe E. Alves[1], Piero A. B. de Sampaio[1]
1. Federal University of Rio de Janeiro, Rio de Janeiro, RJ, Brazil, 2. State University of Rio de Janeiro, Rio de Janeiro, RJ, Brazil

The Quadrilátero Ferrífero (QF) is one of Brazil's the most prolific mineral provinces, which hosts several important gold and iron deposits. Northwest of the QF, the Pitangui Greenstone Belt (PGB) hosts the Turmalina orogenic gold deposit comprising several orebodies with contrasting features with gold deposits in the central QF, such as higher metamorphic grade up to amphibolite facies and a distinct geochemical signature. Near Turmalina, there are a number of deposits, including the Faina deposit. This work provides new information on the stratigraphy of Faina and insights on the main pathfinders for gold in the region of the PGB. Because the gold deposits in the NW QF are complex, there is a need for new prospective criteria. We combined drill core description and multi-element geochemistry and multivariate statistical analyses to identify geochemical correlations as a proxy for gold and geochemical gold vectoring. Cores of four drill holes were described, and a total of 770 samples from 9 drill cores were assayed for 49 elements. Faina shows a total of 7 lithotypes from top to bottom. These includethe following: a) Saprolite (oxidized shallow rocks), b) biotite-actinolite-chlorite-schist, c) talc-chlorite-schist, d) amphibole-chlorite-schist, e) carbonaceous phyllite, f) amphibole-rich rocks, and g) carbonate-rich hydrothermal rocks. Bulk geochemical results were used to divide the samples into two major alteration zones, an oxidized zone (n = 138) and a hypogenic zone (n = 632). A subset of 15 elements was chosen (Au, Ag, As, Bi, Cr, Cu, Mo, Ni, Pb, S, Sb, Sn, Tl, W, and Zn) as representative of the geochemical signature; those elements can be related to gold mineralization depending on the deposit model. Assays results were then transformed by centered log-ratio (CLR) before applying Principal Component Analyses (PCA), Hierarchical Cluster Analyses, and Factorial Analyses. Two high-grade (>0.684 ppm) intervals with different behaviors were identified, one at the oxidized zone hosted by the saprolite, which displays a distinctive Au-As-W-S association, and a second at the hypogenic zone hosted by amphibole-chlorite-schist, which presents an Au-As-W-Sb-S association. The two geochemical associations suggest at least two different mineralization styles at Faina. At the hypogenic zone, drill core description revealed the presence of arsenopyrite at high-grade zones, which may explain the observed relationship between Au, As, and S. Moreover, the element association of the hypogenic zone is attributed by some authors to a typical fluid chemistry consisting of S-rich reduced fluid that selectively transport this group of elements, although further analyses are necessary to test this hypothesis. The multivariate statistical methods are hence suitable for reducing complexity of large and complex geochemical datasets to identify a correlation between elements and provide vectors towards gold mineralization in a context of mineral exploration.

ST.066
Cesium Deposits

Tom Richards, David Trueman, Bruce Downing
none, Langley, BC, Canada

Cesium (Cs) is a little-known element, the bulk of which is used in oil and gas well servicing in the form of cesium formate, but it also has diverse uses in high tech specialty applications. Historically, cesium has been won from the minerals pollucite and lepidolite from "giant" pegmatites like Tanco in Manitoba and Bikita in Zimbabwe, or the Yichun granite, PRC. These sources are now problematic, production having been stopped at Tanco by a fall of loose rock and at Bikita by exhaustion of stockpiles. The small Sinclair deposit in Australia was mined out in 2019 and the ore shipped to China. Several small cesium deposits have been identified in Australia, Africa, the U.S., and Canada, but none have reached systematic production. A second type of cesium is found in the form of epithermal geyserites which have been identified in Tibet, India, and Argentina. Although of lower Cs grades and widely varying mineral assemblages, these may be equal in their contained cesium to the large pegmatite sources. In turn, their differing mineralogy may lend to more easily processible and less costly metallurgy. The Taron deposit (Argentina) is a recently discovered cesium-bearing geyserite. A third type of cesium deposit is formed through greisening. The Yichun rare metal granite is an example of Li-Cs endogreisening formed in situ and at the expense of its protolith. In Yellowstone National Park in Wyoming, Li-Cs exogreisening is currently occurring, and mineral Li-Cs assemblages are forming at very low temperatures and pressures in sedimentary rocks. The source of Cs-bearing pegmatites is probably that of fractionation of S-type granites, and the source of Cs-bearing geyserites is probably similar to that at Yellowstone in which large volumes of rhyolites over the Yellowstone mantle plume are being leached by circulating epithermal fluids. Cesium is now on the U.S. Government's critical elements list, and there is burgeoning exploration for Cs-bearing pegmatites. At present, Albemarle Corp., in China, and the China's Sinomine Resource Group control the cesium industry. This study was undertaken to characterize the importance of geology, geochemistry, mineralogy, and geometallurgy of these types of cesium deposits and to offer some insight as to the origins of novel deposits.

ST.067
Miocene IOCG Deposits Associated with the Trans-Mexican Volcanic Belt

Edith Fuentes[1, 2, 3], Eduardo González-Partida[4, 1], Antoni Camprubí Cano[1], Geovanny Hernández-Avilés[3], Janet Gabites[5], Giovanni Ruggieri[6], Alexander Iriondo[4, 1], Margarita López-Martínez[7]
1. UNAM universidad Nacional Autonoma de Mexico, MEXICO CITY, CMX, Mexico, 2. Laboratorio Nacional de Geoquímica y Mineralogía, UNAM, MEXICO CITY, CMX, Mexico, 3. Posgrado en Ciencias de la Tierra, UNAM, MEXICO CITY, CMX, Mexico, 4. Centro de Geociencias, UNAM, MEXICO CITY, CMX, Mexico, 5. University of British Columbia, Vancouver, BC, Canada, 6. Istituto di Geoscienze e Georisorse, Consiglio Nazionale delle Ricerche, Pisa, Italy, 7. Centro de investigación científica y de educación superior de Ensenada, Baja California, CICESE, Ensenada, BCN, Mexico

The Cu- and Au-rich Tatatila-Las Minas IOCG skarn deposits are circumscribed to the earliest stages of the Trans-Mexican Volcanic Belt (TMVB). These deposits were formed as skarns between Mesozoic carbonate rocks and Miocene intermediate to acid hypabyssal rocks. New U-Pb zircon and $^{40}Ar/^{39}Ar$ ages provide evidence for four epochs of magmatic activity in the area: (1) Triassic, in association with the Paleozoic basement to the Mesozoic series, (2) late Oligocene to early Miocene suite of pre-TMVB intrusives, (3) middle to late Miocene suite of TMVB-related intrusives, and (4) Pliocene intrusive and extrusive rocks of the TMVB. The obtained ages range between 24.60 ± 1.10 and 19.04 ± 0.69 Ma for stage 2 and between 17.80 ± 0.50 and 13.80 ± 1.40 Ma for stage 3. Stage 2 corresponds to a previously unknown magmatic stage in the area. Only stage 3 rocks are associated with the IOCG skarn mineralization, whose retrograde stages have been dated at 12.44 ± 0.09 and 12.18 ± 0.21 Ma. Therefore, the ages of stage-3 intrusive rocks are interpreted as part of the prograde skarn associations (~15.4 to <14 Ma). The petrogenetic affinity of stage-2 and stage-3 rocks is about the same—their main difference has to do with higher Y and Yb contents in stage-3 rocks, which is suggestive of an interaction of their parental magmas with alkaline magmas that are likely to belong to the conterminous and contemporaneous Eastern Mexico Alkaline Province. Petrological indicators in Cenozoic rocks point to intermediate to acid granitic rocks emplaced in a subduction-related continental arc with high-silica adakitic signatures due to deep-seated magmas that underwent crustal contamination to some degree.

ST.069

Triple-Halogen (Cl, Br, I) Systematics of Scapolite from the Tietangdong Fe-Au Skarn Deposit, North China Craton

Wensheng Gao[1], Xiaodong Deng[2], Lei Chen[3], Jianwei Li[1, 2]
1. School of Earth Resources, China University of Geosciences (Wuhan), Wuhan, China, 2. State Key Laboratory of Geological Processes and Mineral Resources, China University of Geosciences, Wuhan, China, 3. MLR key laboratory of Metallogeny and Mineral Assessment, Institute of Mineral Resource, Chinese Academy of Geological Sciences, Beijing, China

Halogens, especially chlorine (Cl), bromine (Br), and iodine (I), are important complexing agents for various metals and have been used as indicators for the sources and evolution of hydrothermal systems. However, such a use has been limited due to the challenge in performing precise analysis of halogens of fluid inclusions trapped in minerals, notably for Br and I. Experimental studies have shown that the partition coefficients of halogen ratios (Br/Cl and I/Cl) between scapolite and the ore-forming fluids are close to 1. As such, the halogen ratios in scapolite-group minerals approximate to values of the fluids from which the minerals precipitated. Here, we present results of fs-LA-HR-ICP MS (femtosecond laser ablation high-resolution inductively coupled plasma mass spectrometry) halogen analysis of scapolite from the Tietangdong Fe-Au skarn deposit in the North China Craton to better understand triple-halogen behavior during ore-forming processes.

Scapolite at Tietangdong occurs as coarse-grained crystals (0.2–1.5 cm) closely associated with garnet, diopside, and magnetite. Both scapolite and diopside contain vapor-rich and daughter crystal-bearing fluid inclusions that form petrographically defined fluid inclusion assemblages, which indicate fluid unmixing during the skarn formation. The scapolite has high and relatively homogeneous halogen concentrations, with 1.84–3.41 wt % Cl, 388.8–806.2 ppm Br, and 8.4–21.4 ppm I, corresponding to high Br/Cl ratios of (6.1–14.7) × 10^{-3}, and I/Cl ratios of (90.8–302.0) × 10^{-6}. Besides, the scapolite displays a positive correlation between Br/Cl and I/Cl molar ratios (Fig. 1).

The halogen ratios are an order of magnitude higher than the values typical of magmatic hydrothermal fluids ([0.65–2.47] × 10^{-3}; for Br/Cl and [9.34 121.00] × 10^{-6} for I/Cl). This contrast cannot be explained by any mixing trends because no end-member fluids with higher halogen ratios exist (Fig. 1). We suggest that the high Br/Cl and I/Cl ratios in the Tietangdong scapolite most likely reflect the preferential partition of Br and I into brine during fluid phase separation. This view is partly supported by the halogen compositions of daughter mineral-bearing fluid inclusions in quartz of the Tsushima magmatic-hydrothermal system in Japan, which have Br/Cl ratios ranging from 0.49 × 10^{-3} to 5.5 × 10^{-3} (Kurosawa et al., 2016). Recent experimental study has also revealed preferential partition of Br into brine phases formed by fluid unmixing so that the brine has high Br/Cl ratios (Liebscher et al., 2006). Their experiments also showed that the partition coefficient of Br between the brine and vapor phases is pressure- and temperature-dependent and positively correlated with the extent of fluid unmixing. Using a Rayleigh trace element fractionation model, we found that the brine would have Br/Cl and I/Cl ratios of 18 × 10^{-3} and 500 × 10^{-6}, respectively, if the fluid unmixing occurred at temperature lower than 450°C and pressure of ~300 bar (Fig. 1), conditions similar to those under which the scapolite at Tietangdong formed.

Fig. 1. Molar I/Cl vs. Br/Cl ratios of scapolite samples and the modeling results of halogen partition during fluid unmixing using a Rayleigh fractionation model. Various halogen reservoirs are also shown for references.

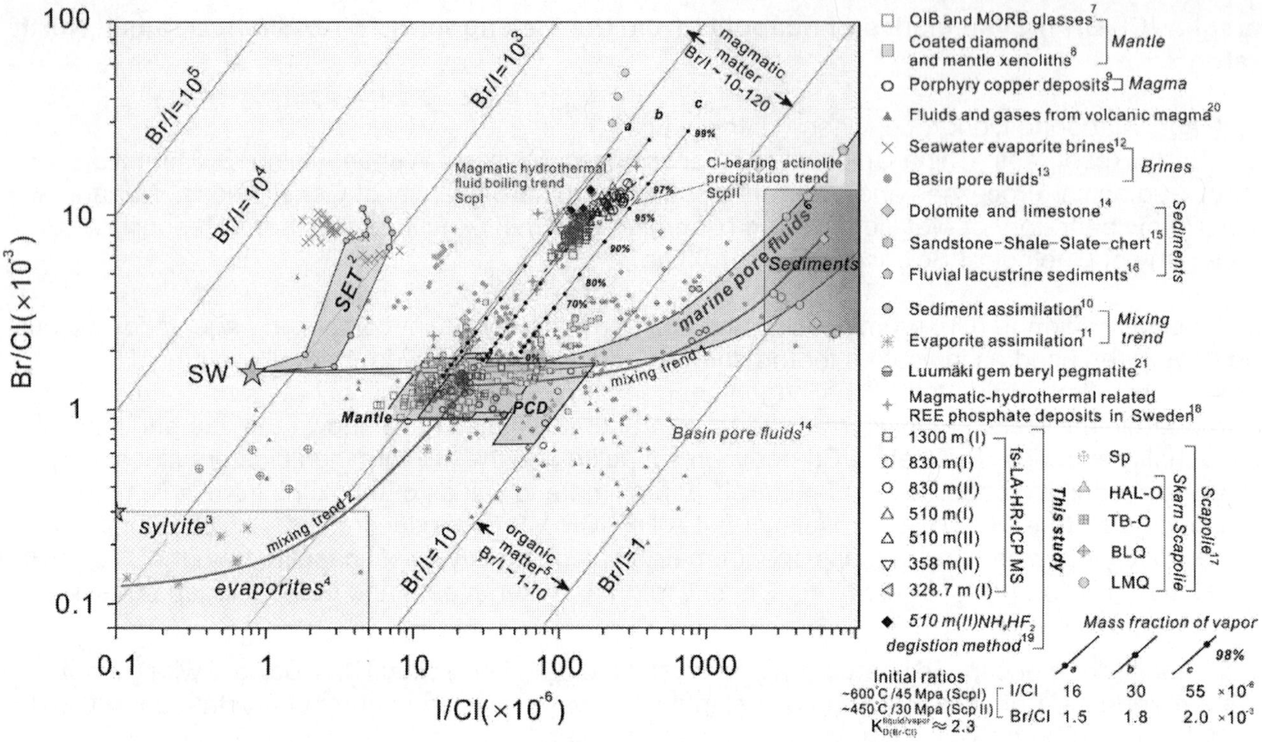

ST.072

Trace Element Compositions and Cathodoluminescence Textures of Apatite from Neoarchean and Paleoproterozoic Fe Oxide Cu-Au (IOCG) Deposits of the Carajás Province, Brazil

Raphael B. Hunger[1], Carolina P. Moreto[1], Roberto P. Xavier[1], Poliana I. Toledo[1], Gustavo Henrique C. de Melo[2], Xin-Fu Zhao[3]
1. University of Campinas, Campinas, SP, Brazil, 2. Federal University of Ouro Preto, Ouro Preto, MG, Brazil, 3. China University of Geosciences, Wuhan, China

Apatite is a common accessory mineral in either hydrothermally altered or mineralized zones of several iron oxide copper-gold deposits (IOCG) of the Carajás Province, northern Brazil. It mainly occurs as individual prismatic- to ovoid-shaped crystals of variable sizes (up to 10 cm wide), which are paragenetically related to sulfide-rich ore zones. Additionally, apatite may be associated with early sodic-calcic alteration, eventually accompanying the formation of massive magnetite-rich bodies. Based on in situ LA-ICP-MS data of apatite, combined with cathodoluminescence (CL) imaging, we investigate morphological and compositional variations among apatite from six IOCG deposits of Carajás formed during the Neoarchean (ca. 2.7 and 2.5 Ga) and the Paleoproterozoic (ca. 1.8 Ga). Apatite from all deposits contains low amounts of Na_2O (<0.05 wt %), Mn (<360 ppm), and Ba (<35 ppm), and low to moderate Fe (<8400 ppm), Mg (≤1400 ppm), and Sr (<635 ppm) contents. Ore-related apatite from the ca. 1.88 Ga Sossego orebody (Sossego deposit) displays brownish-yellow luminescence and has the most fractionated REE+Y (REY) distribution, with significant LREE enrichment (Ce/Yb_{CN} = 3.084–60.430) and moderate negative Eu anomalies (Eu/Eu* = 0.423–0.569). Similarly, apatite from the 2.7 Ga Sequeirinho orebody (Sossego deposit) is LREE-enriched (Ce/Yb_{CN} = 1.408–16.896) and shows orange-brown luminescence; it demonstrates lower total REY contents (206.043–1423.255 ppm) but larger negative Eu anomalies (Eu/Eu* = 0.311–0.436). In contrast, apatite from the ca. 2.57 Ga Igarapé Cinzento (GT-46) and ca. 1.88 Ga Alvo 118 deposits is characterized by a distinct HREE enrichment (Ce/Yb_{CN} = 0.323–1.351 and 0.316–0.765, respectively) and exhibits positive Y anomalies. Under CL light, apatite crystals at GT-46 are predominantly dull yellowish-green and may show sectorized darker blue zones, whereas apatite from the Alvo 118 deposit is generally bluish-grey. Within the main orebody of the ca. 2.57 Igarapé Bahia deposit, apatite displays bright greenish-yellow rims with darker cores and have overall higher Sr (318.244–631.401 ppm) contents. Chondrite-normalized fractionation trends indicate that these grains are mostly MREE-enriched and have weak to moderate positive Eu anomalies (Eu/Eu* = 1.151–1.803), which could collectively be an indication of alkaline, CO_2-rich fluid conditions. Lastly, apatite from the 2.5 Ga Grota Funda deposit is commonly patchy, zoned, and displays LREE-enriched domains (Ce/Yb_{CN} = 2.067–4.667) identified by a purplish luminescence, which are surrounded by relatively REY-depleted, yellowish-green zones. Such compositional variation in apatite could be attributable to coupled dissolution-reprecipitation mechanisms caused by later interaction with metasomatic fluids. Notably, Eu/Eu* and Ce/Ce* values suggest that apatite crystallized under relatively distinct oxidation states, from moderately reduced (Grota Funda and GT-46) to moderately oxidized (Sequeirinho, Sossego and Alvo 118) and predominantly oxidized conditions (Igarapé Bahia). The results from this study indicate that the processes that controlled the REY and trace element incorporation into apatite in these deposits do not seem to be linked to a specific metallogenic epoch recorded at Carajás. Thus, the observed compositional and morphological variations in apatite most likely reflect the history of fluid-rock interaction and hydrothermal evolution at a deposit scale, regardless of its age.

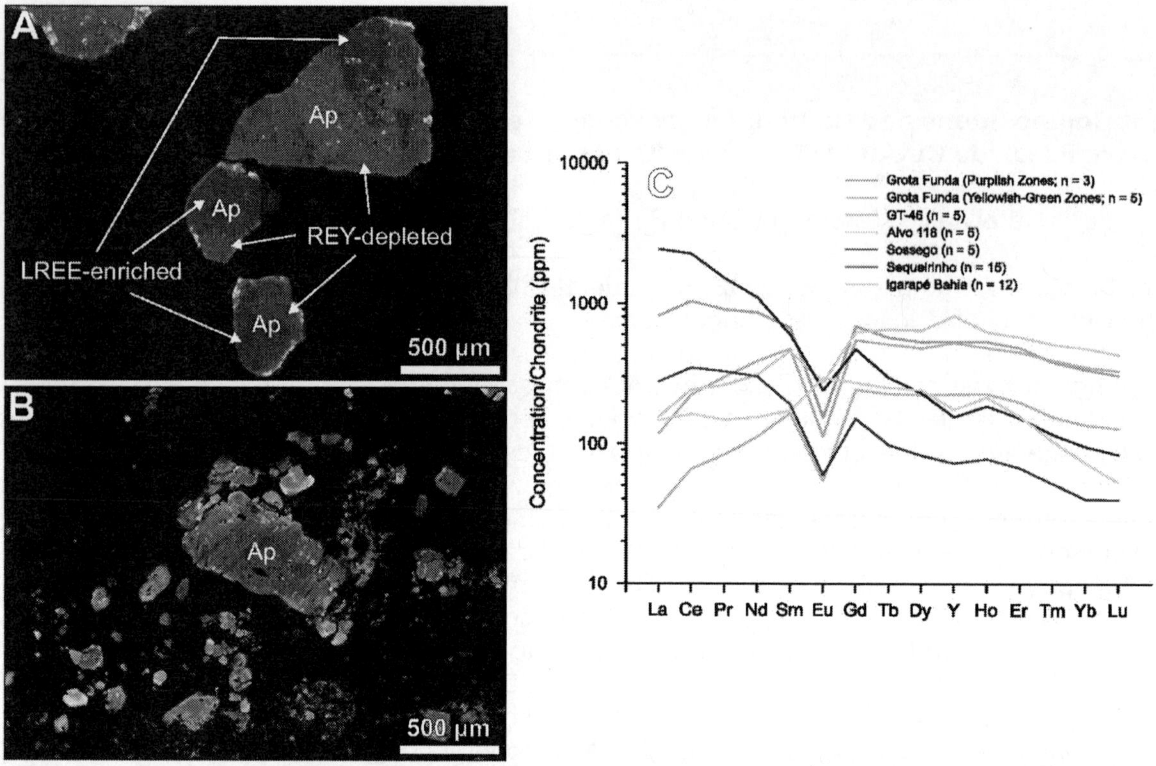

Figure 1. **A** and **B** Cathodoluminescence images of yellowish and greenish-yellow apatite crystals from ore zones of the Grota Funda and Igarapé Bahia deposits, respectively. Note that apatite from the Grota Funda deposit displays purplish LREE-enriched domains, which are surrounded by relatively REE + Y (REY)-depleted yellowish zones. **C** Average rare earth element (REE) distribution patterns of apatite from iron oxide copper-gold (IOCG) deposits of the Carajás Province. Chondrite values are from McDonough and Sun (1995)

ST.073
In-Situ LA-ICP-MS Trace Elements and U-Pb Geochronometry of Melanite Garnet: The Eocene Mengya'a Skarn Pb-Zn Polymetallic Deposit in Gangdese Belt, Tibet of China

Xiaojia Jiang, Youye Zheng, Shunbao Gao, Xin Chen
China University of Geosciences, Wuhan, Wuhan, China

Ti-enriched garnet such as melanite occurs in a variety of geological environments, participating in Si-unsaturated igneous rocks, high-temperature metamorphic rocks, and contact-metasomatized skarns, and it is stable throughout a wide range of pressure-temperature conditions. Despite containing fair amounts of U, there is a lack of detailed research on its use as a U–Pb geochronometer. Abundant skarn Pb-Zn deposits in which ore-related pluton is not exposed have been discovered along the eastern Gangdese metallogenic belt, western China. The mineralization age of these skarn Pb-Zn deposits remains controversial. Therefore, the present paper first investigates the petrography, geochemistry, and in situ U-Pb geochronology of melanite coupled with the Re-Os geochronology of molybdenite. These data are then used to explore the potential of melanite dating and to constrain the age links between skarn alteration and mineralization.

Mengya'a Pb-Zn polymetallic deposit is one of the largest Pb-Zn deposits in Gangdese metallogenic belt. Its unique feature is that the ore-forming pluton is not exposed and the melanite is well developed. Therefore, the determination of the forming time of the melanite can build up the bridge and link for the diagenesis and mineralization of the Mengya'a deposit. The large Mengya'a Pb-Zn deposit is used as case study. Two melanite samples from the Mengya'a deposit are TiO_2-rich (mean of 3.47 wt %) and of brown-black color. They span wide ranges of Pb (0.02-4.40 ppm), Th (0.48-4.22 ppm), and U (0.60-6.43 ppm) contents and have U/Pb values between 0.57 and 81.82. Laser ablation inductively coupled plasma mass spectrometry dating showed that these melanites have a U-Pb age range of 54.9-53.8 Ma, which is consistent with the ages of two adjacent molybdenite Re-Os models (53.36-52.86 Ma). This indicates that skarn alteration occurred temporally, spatially, and genetically associated with a Pb-Zn mineralization event during main-collisional orogeny.

The positive linear Ti vs. Si, Ti vs. Fe^{3+}, REE vs. Y, and REE vs. U correlation indicate that Ti-enriched garnets may have high U content, and the incorporation of U and Ti is largely controlled by crystal chemistry of garnet in near closed system. Ti-rich garnets (e.g., melanite) are also widely distributed in Si-unsaturated alkaline rocks, carbonatite complexes, and high-temperature metamorphic rocks. Thus, this study confirms the reliability and accuracy of the U-Pb dating of melanite and highlights that Ti-rich garnets may have potential to become an invaluable geochemical and geochronological tool in magmatic, metamorphic, and hydrothermal mineral assemblages.

ST.074

Unraveling the Genesis of Fe-Ti Oxide-bearing Ultramafic Intrusions in the Duluth Complex, Minnesota, USA

Jackie M. Kleinsasser[1], Adam C. Simon[1], Amartya Kattemalavadi[1], Kyle LaChance[1], Dean Peterson[2], George Hudak[2]
1. University of Michigan, Ann Arbor, MI, USA, 2. Natural Resources Research Institute, Duluth, MN, USA

The Duluth Complex is a 1.1 Ga large mafic intrusion that was emplaced as part of the Midcontinent Rift System. Located in northeastern Minnesota, the Duluth Complex hosts numerous types of mineral deposits, including Cu-Ni, PGE, Mn, Ti, and others, yet it remains largely undeveloped. Significant Fe-Ti ± V mineralization has remained particularly unstudied over the past 20 years and represents one of the United States' most promising domestic resources of these energy- and infrastructure-critical metals. Fe-Ti ± V mineralization is hosted in 14 oxide-bearing ultramafic intrusions along the western margin of the Duluth Complex. Although all are broadly similar in mineralogy, two contrasting genetic models have been proposed to explain the source of Fe and how the deposits formed. There exists incredibly scarce geochemical evidence to support or refute either of the models, and the main objective of this project is to unravel the genesis of Fe-Ti oxide-bearing ultramafic intrusions.

Focusing on the Longnose and Titac intrusions, two well-drilled intrusions ~40 km apart hosted in the Partridge River and Western Margin intrusions, respectively, we used detailed optical and microbeam methods to compare the textures and compositions of Fe-Ti oxides (ilmenite, titanomagnetite, and magnetite) and olivine to gain insight into their origins. Both intrusions show similar textures despite differing proportions of ilmenite and titanomagnetite, with Longnose having higher amounts of ilmenite. Olivine in both intrusions is serpentinized and fractured, and magnetite is present as thin, secondary veins that formed via serpentinization reactions, being more pervasive in the Titac Intrusion. Major and minor element compositions of titanomagnetite and ilmenite are similar between Longnose and Titac. Microtextural observations of ilmenite and titanomagnetite indicate complex temperature and redox trends during emplacement. Hematite lamellae in ilmenite, multiple types of pleonaste exsolution in titanomagnetite and ilmenite, and granule, sandwich, and trellis type ilmenite exsolution in titanomagnetite are found in both intrusions. These textures indicate that the original Fe-Ti oxide liquid underwent extensive subsolidus cooling and oxidation-reduction reactions starting at >900°C and was most likely assisted by deformation and secondary fluid incursions during and immediately following emplacement. Olivine from the Longnose Intrusion increases in forsterite content with depth (Fo_{60}–Fo_{68}), while olivine from the Titac Intrusion neither increases or decreases with depth and is typically ~Fo_{60}.

The initial results of this study point towards a similar complex, slow cooling history for Fe-Ti oxide liquids and a relatively primitive magma for both the Longnose and Titac intrusions, despite being emplaced in different rock types. Future work will involve measuring stable Fe and Ti isotopes of the ilmenite and titanomagnetite to fingerprint the metal sources and directly assess a key question in the genesis of the deposits.

ST.076
Indium Mineralization in the Magmatic-Hydrothermal Deposits of Herberton Mineral Field, Northeast Queensland, Australia

Avish A. Kumar, Ioan Sanislav, Paul Dirks
James Cook University, Townsville, QLD, Australia

The increasing demand for indium is related to its high usage in the manufacture of LCD, touchscreens, and solar panels. Indium does not have a primary mineral that occurs in economic quantity and is sourced as a by-product from copper and zinc ores. Indium can occur in a variety of mineral systems including VMS, porphyry, skarn, SEDEX, polymetallic veins, and magmatic-hydrothermal deposits. In this study, the genesis of indium mineralisation within the magmatic-hydrothermal domains of Herberton Mineral Field was investigated. The largest indium-bearing deposits in the region are the Baal Gammon and Isabel. The Baal Gammon deposit consists of a multistage tin and copper-zinc-indium mineralization. The tin mineralization is hosted almost exclusively by a granitic porphyry dyke that is overprinted by the copper-zinc-indium mineralization. The deposit has a total resource of 2.8 Mt at 1% copper and 39 g/t indium. The Isabel deposit consists of a series of lead-zinc polymetallic veins hosted by brecciated sediments along the contact with a quartz-feldspar porphyry dyke. The deposit has a total resource of 83 kt at 15% zinc and 370 g/t indium. Field relationships suggests that indium mineralization at both deposits was introduced prior to the emplacement of unmineralised, crosscutting porphyry dykes at 288 Ma. The main host for indium is sphalerite, followed by chalcopyrite, and high indium concentrations were also found in stannite at the Baal Gammon deposit. So far, no indium minerals have been identified. Although sphalerite is the main host for indium at the scale of the deposit, there is a very good correlation between indium and copper, suggesting that the incorporation of indium into sphalerite is dependent on the availability of copper. Thermodynamic modelling suggests that the colligative properties of the mineralising fluids play an important role in concentrating and precipitating indium-rich sulphides while the physicochemical parameters dictate its mobility in the hydrothermal environment.

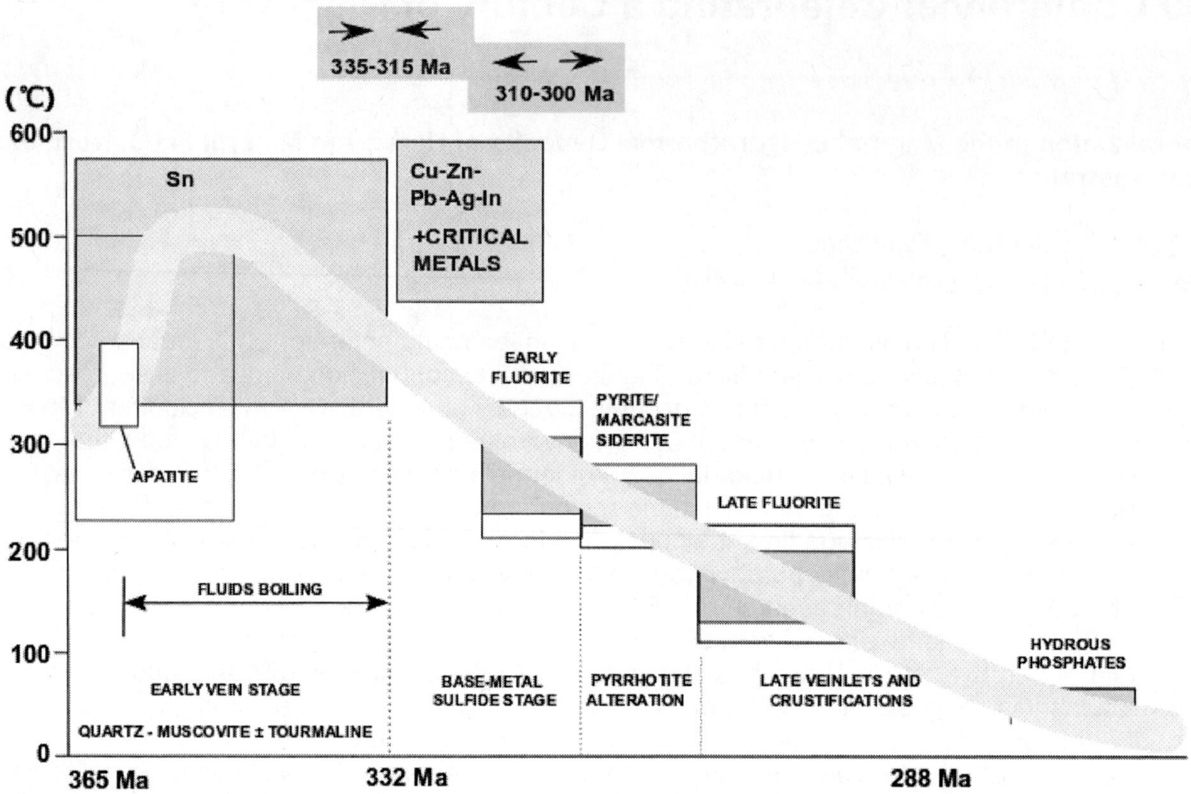

335-315 Ma
310-300 Ma
(℃)
600
500
400
300
200
100
0
Sn
Cu-Zn-
Pb-Ag-In
+CRITICAL
METALS
APATITE
EARLY
FLUORITE
PYRITE/
MARCASITE
SIDERITE
LATE FLUORITE
HYDROUS
PHOSPHATES
FLUIDS BOILING
EARLY VEIN STAGE
QUARTZ - MUSCOVITE ± TOURMALINE
BASE-METAL
SULFIDE STAGE
PYRRHOTITE
ALTERATION
LATE VEINLETS AND
CRUSTIFICATIONS
365 Ma
332 Ma
288 Ma

ST.077
Evaluation of Magnetite in Stream Sediments at the Casino Porphyry Copper-Gold-Molybdenum Deposit, Central Yukon as an Indicator Mineral and Vectoring Tool

Martin W. McCurdy[1], Jan M. Peter[2], M. B. McClenaghan[2], Michael G. Gadd[2], Dan Layton-Matthews[1], Matthew I. Leybourne[1], Robert G. Garrett[2], Scott G. Casselman[3]
1. Queen's University, Kingston, ON, Canada, 2. Geological Survey of Canada, Ottawa, ON, Canada, 3. Yukon Geological Survey, Whitehorse, YT, Canada

The Casino porphyry Cu-Au-Mo deposit is hosted in Late Cretaceous quartz monzonite of the Casino suite intruding Mesozoic granitoids of the Whitehorse suite in the unglaciated Dawson Range, west-central Yukon, Canada. It is one of Canada's largest and highest-grade porphyry deposits. Research is being carried out on relationships between the composition of magnetite in stream sediments and magmatic and magmatic-hydrothermal magnetite in this well-preserved deposit. Magnetite recovered from bedrock samples of mineralized, hydrothermally altered rocks from the Casino deposit and unmineralized granodioritic and monzonitic host rocks was compared to magnetite in sediment from two streams directly draining the deposit. For comparison, magnetite from sediments in four streams not draining the Casino deposit was also compared.

Optical and scanning electron microscopy, electron probe microanalysis (EPMA), and laser ablation-inductively coupled plasma mass spectrometry (LA-ICP-MS; "LA") were used to determine physical and chemical properties of igneous and magmatic-hydrothermal magnetite in polished thin sections, thick sections, and epoxy grain mounts. Data were acquired from 90 EPMA spot analyses from thick sections, 608 LA analyses of thick sections and grains from disaggregated bedrock samples, and 915 LA analyses of magnetite grains (0.25-2.0 mm) in grain mounts from six stream sites (1,523 total LA analyses). Spot analyses of magnetite from two bedrock samples containing igneous magnetite and one sample containing magmatic-hydrothermal magnetite, supported by petrographic observations, were compared with magnetite grains recovered from stream sediments using a variety of statistical and graphical methods.

Significant geochemical variability in magmatic-hydrothermal magnetite exists within the potassic alteration zone of the Casino deposit. Relatively low Ti, Al, and Mg concentrations in magmatic magnetite from the Casino deposit, compared with recent studies of similar deposits, are attributed to sub-solidus re-equilibration during cooling. A plot of Ti versus Ni/Cr (Fig. 1) is used to discriminate magmatic-hydrothermal magnetite from the potassic alteration zone and igneous magnetite from granodiorite and quartz monzonite hosting the Casino deposit.

A small number of studies have considered the chemistry of magnetite recovered from glacial sediments as a vector for porphyry copper deposits, but few, if any, have examined the chemistry of magnetite in stream sediments to detect porphyry copper deposits in Canada. The elements Mg, Al, Ti, V, Mn, Co, Cr, Ni, Cu, Ga, and Ge are used to discriminate between magmatic-hydrothermal and igneous magnetite using linear discrimination analysis and binary discriminant diagrams. Magmatic-hydrothermal magnetite from the potassic alteration zone at Casino is identifiable in streams draining the deposit. A detectable level of Cu in the magmatic-hydrothermal magnetite, present as inclusions or substituting in the cubic lattice, is a strong indicator of Cu mineralization. This approach may be of practical use in the exploration for porphyry systems in Canada.

Fig. 1. Ti versus Ni/Cr plot effectively separates igneous magnetite in granodiorite (red) and quartz monzonite (green), and magmatic-hydrothermal magnetite in contact breccia (blue) (modified from Dare et al., 2014). Point attributes (colour and shape) are based on linear discriminant analysis classifications. Re-equilibration of granodiorite during cooling reduces Ti concentrations in magnetite.

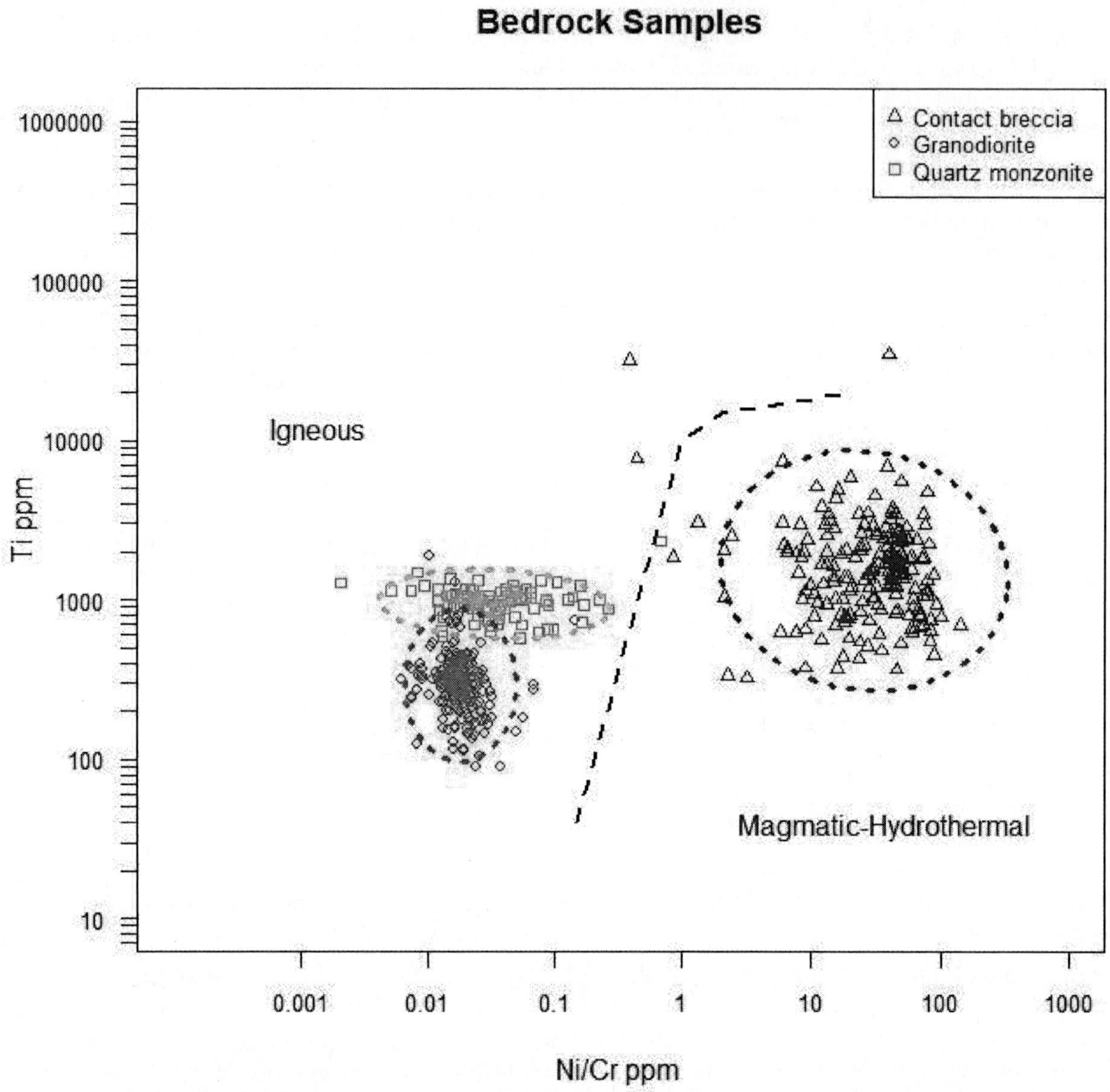

ST.078
Geology of the REE Mineral Deposits in the Gallinas Mountains (Gallinas District), Lincoln County, New Mexico; Update from New Mapping and Geochemical Sampling

Virginia T. McLemore, Shari Kelley, Matthew Zimmerer, Evan Owen, Lynn Heizler, Alexander Gysi
New Mexico Bureau of Geology and Mineral Resources, New Mexico Institute of Mining and Technology, Socorro, NM, USA

Rare earth elements (REE) are critical minerals that are essential in the electronics, automotive, and metallurgical industries. REE deposits are found in New Mexico, but they have not been important exploration targets in past years because supply has been met elsewhere. However, with a projected increase in demand, New Mexico deposits are being re-examined for their REE potential. Current, new geologic mapping and petrographic, mineralogical, and geochemical analyses in the Gallinas Mountains, New Mexico are being conducted, where past production of REE, as bastnaesite, occurred in the 1950s. Copper, silver, lead, zinc, iron, and fluorite also have been produced from the district. Host rocks include Proterozoic granite and gneiss overlain by Permian sedimentary units belonging to the Abo, Yeso, and Glorieta formations. The mineral deposits are also hosted by several textural varieties of 28-30 Ma, alkaline to alkalic-calcic, peraluminous trachyte and syenite sills and dikes. Additional igneous rocks include older andesite and Cougar Mountain trachyte/trachyandesite (~38 Ma) and a younger, fluorite-bearing rhyolite laccolith (28-30 Ma). The Red Cloud Canyon fault forms a major north-trending fault system that is downthrown to the west. No significant mineral deposits are found along the Red Cloud Canyon fault, but along smaller fault spays and fracture zones. New mapping and chemical analyses have allowed defining six types of mineral deposits in the district distinguished by mineralogy, chemistry, and style: REE-F veins, Cu-REE-F (± Pb, Zn, Ag, Au) veins, F veins, intrusive breccia pipes, carbonate breccias, and iron skarn-contact replacement deposits. The REE-F, Cu-REE-F, and F veins are hydrothermal, with variable fluorite and barite, low in sulfur, and controlled by minor faults, fractures, and bedding planes. Most veins are <20 m long and typically <1.5 m wide. New veins have been mapped that have expanded the mineralized area for exploration. The magmatic, intrusive breccia pipes form a north-east-trending belt approximately 2 km long. Breccia pipes are matrix-supported, pipe-like bodies that intrude the trachyte and Yeso and Glorieta formations and consist of angular to subrounded fragments of sandstone, shale, limestone, granite, granitic gneiss, and trachyte/syenite that are as much as 1 m in diameter. The contacts are covered and are not well exposed, but the pipes appear to be roughly elliptical in shape. Most of the breccia pipes are cemented by quartz, feldspar, fluorite, and hematite along with small crystals of other minerals and rock fragments, and locally host REE-F and F veins and anomalously high gold values (as much as 222 ppb). Electron microprobe studies indicate that the predominant matrix is albite, with local trachytic or porphyritic texture. Carbonate breccias are found in the vicinity of thin limestone and gypsum (?) beds in the Yeso Formation and many contain fluorite, local trace gold (as much as 178 ppb), and variable amounts of REE. Iron skarn-contact replacement deposits consist of hematite and magnetite, with local elevated REE, replacing Yeso limestone and Yeso and Glorieta sandstone. Different alteration styles are associated with the veins, breccia pipes, skarns, and trachyte/syenite sills/dikes and are under study.

ST.080
The Naka-Osaka Magnetite Ore Deposit is the First IOA (Kiruna) Type in Japan

Atsushi Miyashita[1], Hiroyasu Murakami[2], Koushi YAGI[3], Wataru Fujita[1], Takuto Kanai[4]
1. Faculty of Science and Technology, Seikei University, Tokyo, Japan, 2. School of Education, Waseda University, Tokyo, Japan, 3. Hiruzen Institute for Geology and Chronology Co., Ltd., Okayama, Japan, 4. Research Institute of Cultural Properties, Teikyo University, Fuefuki, Japan

The Naka-Osaka Mine in Shimonita, Gunma Prefecture, Japan, is the geosite of the Shimonita Geopark. This mine produced high-grade magnetite ore. It is also an industrial heritage site where iron manufacturing was carried out in the 19th century. We confirm the Naka-Osaka magnetite ore deposit is the IOA (Kiruna) type.
The Naka-Osaka Mine is located on the north side of the Ookitano-Iwayama tectonic line (the Median Tectonic Line of Japan). The iron ore deposit is composed of small but high-grade magnetite orebodies, and it is embedded in both the Nanjai Formation and the Name Granite. The Nanjai Formation is correlative with the Jurassic Tanba-Mino accretionary complex. The 70 Ma Name Granite is correlative with the Cretaceous Ryoke I-type granite.
The magnetite orebodies are lenticular and/or vein-shaped with wide alteration zones which are composed of actinolite or hydro-biotite + chlorite with carbonate stock-work veinlets. Magnetite contains Si up to 3 wt % and often contains euhedral Cl-apatite micro inclusions (Fig. 1A). The Fe-Si-Al-O amorphous phase (quenched melt?) is found among magnetite and actinolite (Fig. 1B). The minor amounts of sulfides (pyrrhotite, pyrite, lollingite-arsenopyrite composite grains) are associated with the magnetite ore suggesting a low sulfidation state. The filling temperatures of primary fluid inclusions in fine fluorite veins are above 500°C, which is a higher-temperature as compared with common hydrothermal deposits.
The Naka-Osaka magnetite ore deposit has the feature that the main ore mineral assemblage (Mgt + Apa) with a neutral to alkaline alteration zone at high temperatures above 500°C is close to the large-scale tectonic line. This feature of the Naka-Osaka Mine strongly suggests that the magnetite ore deposit classified as the IOA (Kiruna) type. The formation age of the ore deposit assumes 62.5 Ma - 2.2 Ma, which is one of the youngest IOA types in the world.

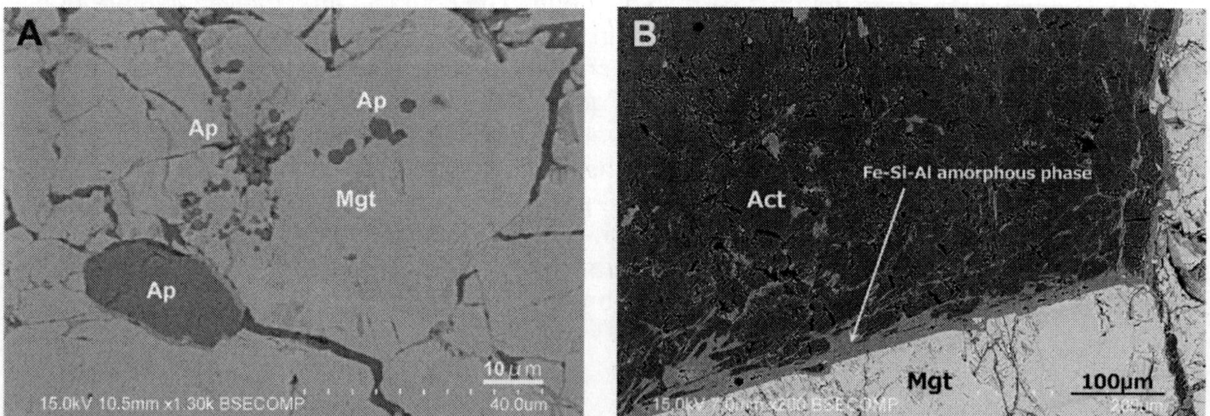

Fig. 1. Back Scattered Electron Images (BEIs) of magnetites of the Naka-Osaka Mine.: A: NO18. Apatite (Ap) inclusions in magnetite (Mgt); B: NO76, Fe-Si-Al amorphous phase occurs among euhedral magnetite and actinolite (Act).

ST.081
Nonsulfide Zinc-Lead-Vanadium Ores in the Kihabe and Nxuu Prospects (Aha Hills, Botswana) - A Mineralogical Insight

Nicola Mondillo[1], Francesco Putzolu[1], Giuseppina Balassone[1], Nigel Forrester[2], Maria Boni[1]
1. Dipartimento di Scienze della Terra, dell'Ambiente e delle Risorse - University of Naples, Naples, Italy, 2. Mount Burgess Mining N.L., East Victoria Park, WA, Australia

The Kihabe and Nxuu Zn-V (>Pb>Cu-Ag-Ge) prospects are located at the boundary between Namibia and Botswana (Aha Hills, Ngamiland District). The prospects, currently under exploration by Mount Burgess Mining N.L., contain total resources (2004 JORC compliant) of 25.3 million tonnes ore at 3.00% zinc equivalent, a silver amount of 3.3 million ounces, and notable V+Ge credits that are still not evaluated at a resource level. The orebodies, consisting of mixed sulfide-nonsulfide minerals, occur in a strongly deformed Proterozoic terrain representing the eastern extension in Botswana of the Damara belt of Namibia, and are hosted by quartzwackes at the contact with dolostones. Sulfides have been formed before or during the Panafrican orogeny, similarly to those occurring in the Namibian Otavi Mountainland.

Several drillcore samples from the Kihabe and Nxuu partly oxidized ores were analyzed for conventional optical and scanning electron microscopy, X-ray diffraction, and bulk-rock geochemistry. Sulfide minerals, mainly consisting of sphalerite, Ag-bearing galena, and minor pyrite, appear to be finely intergrown with quartz and mica within the host rock. This mineral assemblage is overprinted by secondary oxidized minerals, mostly consisting of willemite, various Zn-bearing phyllosilicates, smithsonite, and vanadates. Willemite has been detected in patches and/or as vein filling, and is As-rich. Zn-bearing phyllosilicates are finely dispersed in the quartzwacke, texturally growing on preexisting micas and locally replacing illite and K-feldspar; they can also occur as aggregates of newly formed crystals directly precipitated in cavities. Specific X-ray diffraction analysis on clay minerals allowed determining that they mostly consist of fraipontite (a Zn-bearing species of the kaolinite-serpentine group), in contrast with previous studies that prevailingly identified the most abundant phyllosilicate as baileychlore. The Zn-phyllosilicates at Nxuu are characterized by contents ranging from 36 to 46 wt % ZnO, MgO variable between 1 and 4 wt %, Al_2O_3 between 11 and 15 wt %, and SiO_2 between 25 and 30 wt %. Smithsonite occurs as concretions in cavities or it replaces willemite. Locally the Zn-carbonate is associated with cerussite, which directly replaces galena. The most abundant V-mineral is descloizite, which has been widely detected in the most surficial zones of the orebodies, in association with calcretes within the Kalahari sedimentary cover.

In analogy with the Zn deposits of the Otavi Mountainland, willemite and fraipontite at Kihabe and Nxuu could be genetically related to a hydrothermal mineralization stage, which shortly followed or was contemporaneous with the Panafrican orogeny. The other secondary minerals, i.e., the Zn-Pb carbonates and vanadates, likely represent more recent weathering-derived products.

ST.082

Alunite Supergroup Minerals from an Acid-Sulfate Alteration Assemblage in the Southern Atacama Desert as Indicators of Paleo-Hydrothermal and -Supergene Environments

Jorge E. Morales-Leal[1], Eduardo A. Campos[1], Kalin Kouzmanov[2], Rodrigo Riquelme[1]
1. Departamento de Ciencias Geológicas, Universidad Católica del Norte, Antofagasta, Chile, 2. Department of Earth science, University of Geneva, Geneva, Switzerland

Hydrothermal activity genetically related to pluton intrusion during the Incaic tectonic phase affected the Potrerillos district in the Chilean Precordillera at the southern edge of the Atacama Desert which produced acidic fluid-driven alteration lithocaps, later overprinted by supergene processes. Outcrops with advanced argillic (AA) alteration commonly contain alunite supergroup minerals (e.g., alunite, jarosite, aluminum-phosphate-sulfate [APS] minerals) of both hypogene and supergene origin, indicating overprinting of multiple alteration events over the system's lifespan. Morphology, textural patterns, and composition of alunites, jarosite, and APS minerals related to hypogene and supergene processes in the Potrerillos district were investigated as proxies for physicochemical parameters as well as composition and source of the fluids producing AA alterations in the area. Field observations, optical microscopy, scanning electron microscopy (SEM), X-ray diffraction, short-wave infrared spectroscopy, in situ mineral chemistry (μ-XRF, EPMA), and sulfur isotope analyses define three different types of fluids, each related to a specific depositional environment (Fig. 1): The first is a magmatic-hydrothermal environment dominated by acidic fluids of magmatic origin, with sulfur species derived from SO_2 disproportionation at depth. This environment is characterized by a mineral assemblage consisting of tabular-shaped alunite, quartz, and traces of Ti oxides, indicative of precipitation at temperatures >200°C. In shallow and oxidized environments, the magma-derived fluids mix with meteoric water, producing fluid containing SO_4^{2-} originating from both disproportionation at depth and H_2S-oxidation from the vadose zone, thereby triggering the precipitation of tabular-shaped alunite, quartz, kaolinite, and small amounts of smectite. The second is a steam-heated, acidic fluid-dominated environment at a temperature of 150° to 200°C with highly oxidized and high-PO_4^{3-} activity solutions (pH ~1 to <5) containing sulfate generated upon oxidation of distilled H_2S in the vadose zone. This environment is characterized by fine-grained tabular alunite associated with quartz, kaolinite, traces of dickite, and APS minerals with svanbergite-woodhouseite composition. The third is an oxidized supergene environment with acid-sulfate meteoric waters (pH ~1), characterized by an assemblage of jarosite, pseudocubic alunite, kaolinite, halloysite, Fe-oxides, smectite, and low amounts of APS minerals. The acidic solutions therein originate from sulfuric acid produced upon oxidation of sulfide minerals and dissolution of other minerals such as alunite and APS minerals at the time of rock leaching, resulting in enrichment of the supergene fluid in Al^{3+}, Fe^{3+}, Fe^{2+}, H^+, K^+, SO_4^{2-}, Ca^{2+}, Sr^{2+}, and PO_4^{3-}. The textural relationships between alunite, jarosite, and APS minerals and SEM analyses suggest variations in physical and chemical characteristics of the fluid in the steam-heated and supergene environments. Multiple events responsible for AA alteration in the studied area have induced complex evolution of an epithermal system that grades from an early magmatic-hydrothermal stage to late leaching processes driven by meteoric waters. Furthermore, the capability of alunite supergroup minerals to record environmental physicochemical changes renders them suitable for the evolution of epithermal systems to be studied, including both hypogene and supergene processes.

Figure 1: Acid environments in the studied area. A) Magmatic-hydrothermal and steam-heated environments. B) Supergene environment shows the extinct epithermal system being leached by acid-meteoric fluids during supergene alteration.

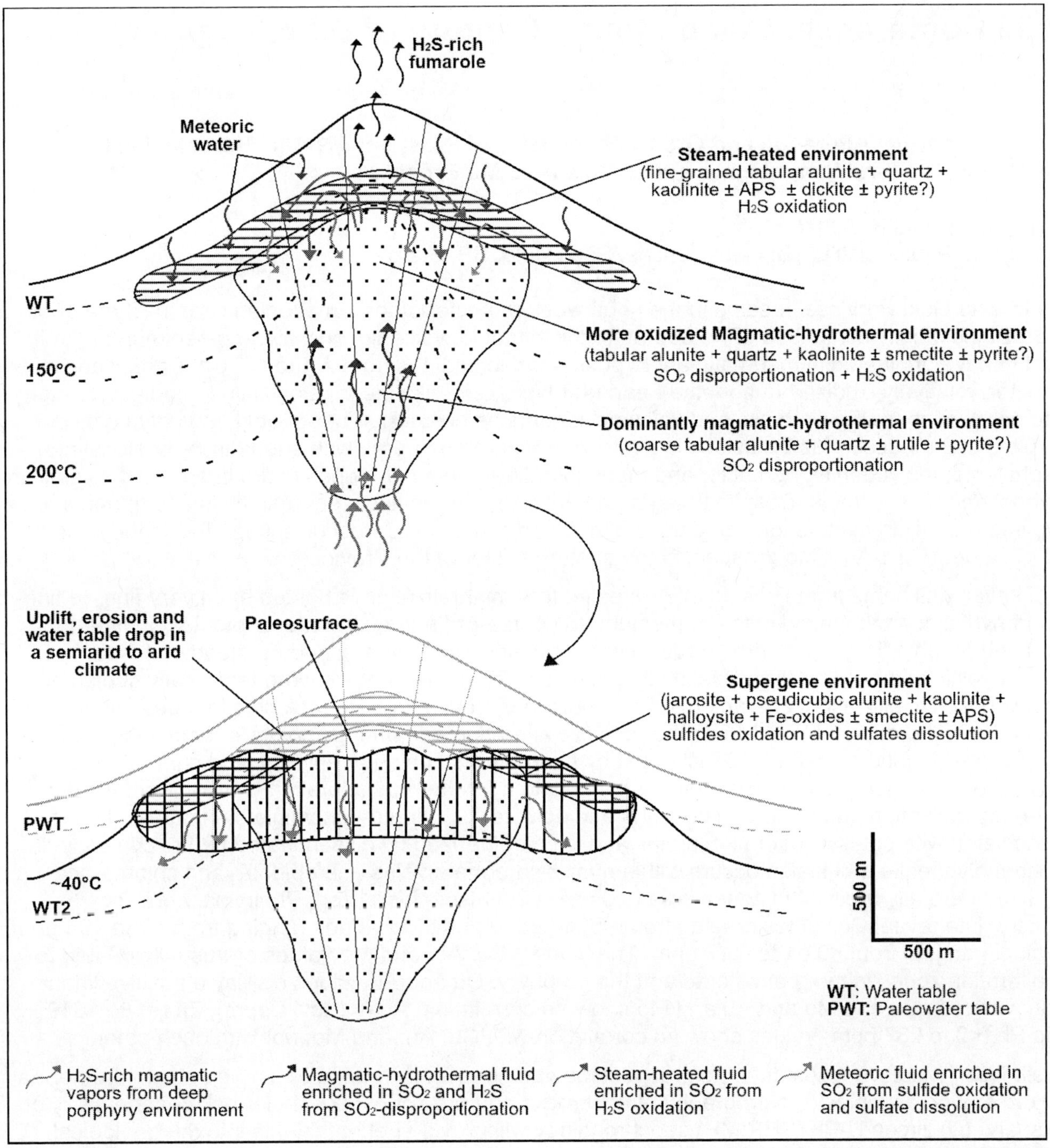

H2S-rich fumarole
Meteoric water
Steam-heated environment
(fine-grained tabular alunite + quartz + kaolinite ± APS ± dickite ± pyrite?)
H2S oxidation
WT
More oxidized Magmatic-hydrothermal environment
(tabular alunite + quartz + kaolinite ± smectite ± pyrite?)
SO2 disproportionation + H2S oxidation
150°C
Dominantly magmatic-hydrothermal environment
(coarse tabular alunite + quartz ± rutile ± pyrite?)
SO2 disproportionation
200°C
Uplift, erosion and water table drop in a semiarid to arid climate
Paleosurface
Supergene environment
(jarosite + pseudicubic alunite + kaolinite + halloysite + Fe-oxides ± smectite ± APS)
sulfides oxidation and sulfates dissolution
PWT
~40°C
WT2
500 m
500 m
WT: Water table
PWT: Paleowater table
H2S-rich magmatic vapors from deep porphyry environment
Magmatic-hydrothermal fluid enriched in SO2 and H2S from SO2-disproportionation
Steam-heated fluid enriched in SO2 from H2S oxidation
Meteoric fluid enriched in SO2 from sulfide oxidation and sulfate dissolution

ST.083
Geology, Hydrothermal Alteration and Ore Geochemistry of Prospect-47, Alta Floresta Gold Province, Brazil: The First Insights into a New Cu ± Au ± Mo Prospect

Danilo Pineschi, Rafael R. Assis
University of São Paulo, Institute of Geosciences, São Paulo, SP, Brazil

The Alta Floresta Gold Province, located in the southwestern sector of the Amazonian Craton (State of Mato Grosso, Brazil), is mainly composed of Paleoproterozoic plutonic, volcanic, and volcano-sedimentary units interpreted to have originated in magmatic arc to postorogenic settings from 2.1 to 1.7 Ga. Particularly in its eastern sector, relatively oxidized (magnetite-bearing) I-type, calc-alkaline to subalkaline, medium- to high-K, meta to peraluminous granitic rocks and volcanic sequences host a number of gold and polymetallic deposits. Gold usually occurs as inclusions in pyrite, the main ore mineral, with subordinate chalcopyrite, galena, sphalerite, molybdenite, hematite, and magnetite. Most gold-rich deposits are interpreted as similar to porphyry-epithermal systems. Despite being a primarily Au-rich province, several mining companies have put efforts into greenfield exploration campaigns aimed at discovering Cu occurrences. This is the case of Prospect-47, a new Cu ± Au ± Mo prospect in the eastern sector of the Province.

Logging of seven drill cores from Prospect-47 revealed that mineralization is hosted by (i) very fine- to fine-grained porphyritic quartz monzodiorite, (ii) medium- to coarse-grained quartz monzodiorite, (iii) fine- to coarse-grained porphyritic granodiorite (intrusive in the previous rock types), and (iv) quartz-feldspar porphyry of rhyolitic composition (intrusive in all rock types). These rock types were temporally subjected to (1) pervasive potassic alteration, characterized by K-feldspar, biotite, or both replacing igneous minerals; (2) pervasive sericitic alteration that overprints the potassic alteration and affects mostly igneous plagioclase; and (3) pervasive chloritic alteration characterized by hydrothermal chlorite, epidote, and carbonate overprinting previous hydrothermal alterations. All rock and hydrothermal alteration types are crosscut by syn- to late-mineralization quartz-carbonate veins and epidote- or chlorite-rich veinlets. Mineralization is mainly associated with potassic alteration zones and is characterized by disseminated pyrite, chalcopyrite, bornite, and molybdenite, but it also occurs within quartz-carbonate veins and epidote- and chlorite-rich veinlets. Hydrothermally brecciated zones are uncommon but also mineralized. Preliminary ore geochemistry data reveal Cu values ranging from <50 to 9500 ppm, Mo values ranging from <1 to 996 ppm, and Au values ranging from <0.01 to 0.17 ppm. The highest Cu, Au, and Mo values are usually related to potassic alteration zones in the granodiorite and the porphyry. Cu and Au values display a positive linear correlation, whereas Cu and Mo and Au and Mo show no correlation. Ag (<1 to 14 ppm), Zn (<1 to 1619 ppm), and Pb (<3 to 237 ppm) values show no correlation with Cu, Au, and Mo, nor with each other.

These preliminary results suggest that the prospect may be genetically related to a magmatic-hydrothermal system. Further work, such as (i) petrography of host rocks and ore zones; (ii) whole-rock geochemistry of host rocks; and (iii) zircon U-Pb (SHRIMP) geochronology will provide better insights into the geological, tectonic, and metallogenetic history of the prospect.

Fig. 1. Location map of Prospect-47 and nearby mineral deposits, Amazonian Craton.

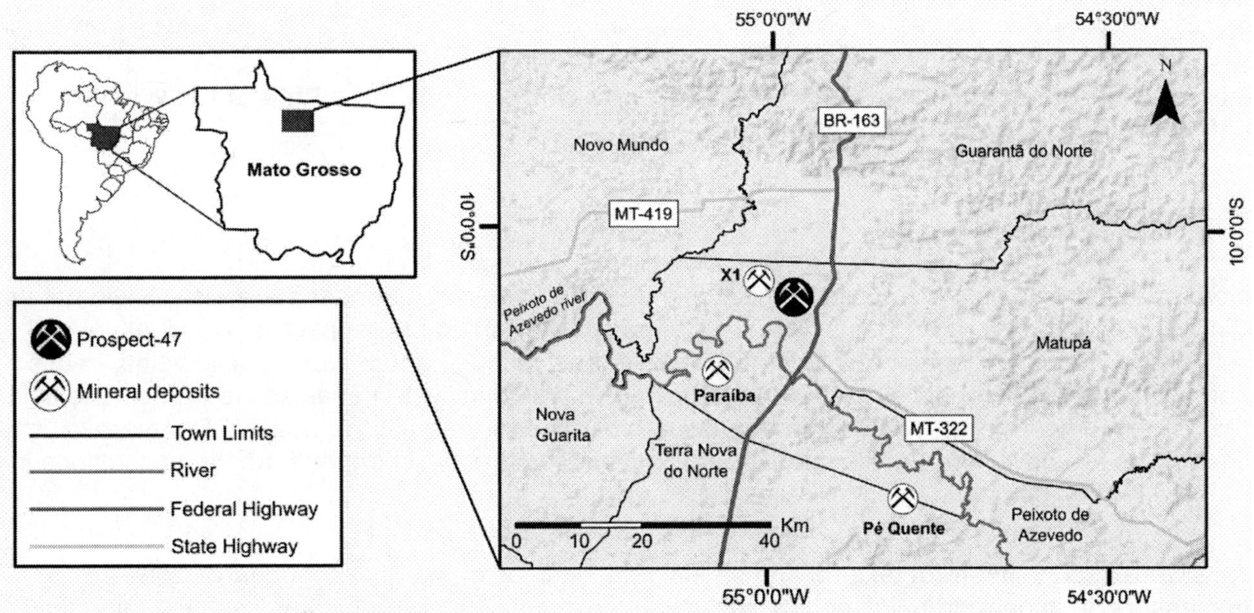

ST.087
Constraints on Conditions of Mineralization in the Quartz-Hübnerite Breccia Pipe of the Western Dunmore Vein System, Silver Gulch, Colorado

Diego Schutz
Fort Lewis College, Durango, CO, USA

The Dunmore system contains mineralized breccia chimneys and veins in a zone that trends ~3 kilometers east-west in Silver Gulch, Colorado. This system has been poorly studied but represents a well-exposed snapshot of mineral deposition analogous to Red Mountain breccia pipes and Silverton style veins. In this investigation, field and petrographic studies were combined with fluid inclusion analyses to unravel the fluid evolution, constrain temperatures, and estimate pressures of mineralization in the western Dunmore system. The results are novel and provide additional constraints of breccia chimney formation conditions in the region.

Field surveys were used to interpret textural and spatial relationships of the quartz-hübnerite zone and to collect samples for use in petrographic and fluid inclusions analyses. Petrographic analyses revealed the tungsten-breccia zone contains three dominant quartz generations: (1) Euhedral quartz with no growth zones (5 to 10 millimeters); (2) anhedral quartz with hübnerite and pyrite (1 to 2 millimeters); and (3) euhedral quartz with growth zoning (4 to 6 millimeters).

Fluid inclusions within the first quartz generation represent high temperatures (~300°C) but are highly deformed from rapid pressure and temperature fluctuations and system instability. No inclusions in the quartz-hübnerite generation are preserved, suggesting a transition from higher to lower temperatures. Late vuggy euhedral quartz with growth zoning hosts fluid inclusions with homogenization temperatures between ~240° and 250° C. Pressure estimations based on topographic differences and paleosurface assumptions suggest system formation at ~1,400-meter depths at ~60 bars of pressure.

Surrounding breccia dikes and pipes represent a volatile-rich magmatic system with intense brecciation. The quartz-tungsten zone in the west Dunmore system shows minor brecciation, representing a later transitional stage from a mesothermal magmatic system into a cooling epithermal stage of mineralization.

Breccia pipes are commonly important mineralized components found with numerous types of economic mineral deposits. This study contributes to the developing genetic model of the Dunmore system along with other breccia pipe occurrences in the northern San Juan Mountains.
The Dunmore system illustrates a complex evolution of fluids and gases and the development of different but spatially related mineralized bodies in breccia systems.

ST.088
Linking All Valuable Primary Resources to Volcanic Glass

Peter B. Schwann
Australian Institute of Geoscientists, Cottesloe, WA, Australia

The greatest rock building component is seriously discounted in geology at all levels. The almost complete lack of turbidite-ash fall tuffs and Bouma sequences in the mineral industry geological naming of rocks, especially in the Archaean and Paleoproterozoic shields, has led to a profusion of intrusive sills that do not make sense.

The examination of modern volcanic arc provinces will show that the majority of the rocks are subaerial-submarine ash-derived sediments. The huge land and submarine basins of Australia and North America are made up of large amounts of ash derived sediments, which are correctly named by oilfield geologists but change as soon as the mining geologists look at them. The definition of oil and gas by studying LIPs has been critical in identifying hydrocarbon deposits in many basins, notably in the North American, where vanadium identifies the hydrocarbon temperature and formation.

A simple look at any volcano will tell the observer that ash is very important to the shape and makeup of the cone. When these ingredients are deposited in water, fresh or salty, there is a fundamental bonus to the start of all systems that yield hydrothermal orebodies, and that is interstitial water. The other critical ingredient is volcanic glass, which is not silica, as most geologists think, but is basalt-andesite in composition. When this hydrates, it goes to the ferrous clay chlorite, and, from there, the journey to mineral and hydrocarbon deposits is well recorded.

I will demonstrate that the formation of gold, BIF and pisolie iron ores, and nickel and copper ores are related to this process. The ability of the mineral shere to adopt oil field processes will also be a great boon to exploration, and the work by Ross Large and others will peice this together in leading to new discoveries.

ST.089

Geology, Mineralogy, and Geochemistry of the Intrusion-Related Polymetallic Quartz Veins at Laodikino, Serbo-Macedonian Massif, N. Greece

Christos L. Stergiou[1], Vasilios Melfos[1], Panagiotis Voudouris[2], Lambrini Papadopoulou[1], Paul G. Spry[3]
1. Faculty of Geology, Aristotle University of Thessaloniki, Thessaloniki, Greece, 2. Faculty of Geology and Geoenvironment, University of Athens, Athens, Greece, 3. Department of Geological and Atmospheric Sciences, Iowa State University, Ames, IA, USA

Polymetallic Cu-Au-Bi-Pb-Ag-Te ± Au-bearing quartz veins occur at Laodikino in the Serbo-Macedonian Massif in northern Greece. They are shear-related and hosted in metamorphic rocks, including two-mica gneisses and schists, biotite gneisses, and amphibolites. Barren aplites, pegmatites, and quartz veins are parallel or intrude along the main schistosity of the metamorphic rocks. Magmatic rocks are not exposed in the broader area. Ductile deformation episodes that affected the area include the pre-Alpine to Alpine D1 to D3 phases, and the post-Alpine D4 phase, which was followed by brittle deformation. The D3 deformation event is associated with Jurassic to Early Cretaceous amphibolite facies metamorphism, while the D4 deformation event is related to Late Cretaceous to Eocene retrograde greenschist facies metamorphism. Brittle deformation is mainly characterized by NW- and NE-trending normal to oblique faults. The D4 and the brittle deformation phase resulted in the Late Cretaceous to Miocene extensional collapse of the Serbo-Macedonian Massif and the onset of the main Cenozoic magmatic and mineralizing event that affected the region during the Oligocene-Miocene. The polymetallic quartz veins are generally <1 m wide and crosscut D3 folds, and are subparallel to or crosscut the D4 folds. Locally the metallic mineralization appears discontinuous and forms lenses. Exploration drilling during the early 1990s revealed that these veins reach a depth of 200 m and locally exceed 3 m in width. They exhibit typical hydrothermal characteristics including massive, comb, and breccia textures (Fig. 1A-C). The main metallic assemblage includes pyrite, arsenopyrite, chalcopyrite, sphalerite, tetrahedrite, and galena, and minor amounts of pyrrhotite, magnetite, ilmenite, rutile, cobaltite, Bi-tellurides, electrum, and native bismuth (Fig. 1D-G). Restricted disseminations, aggregates, and pods of magnetite, pyrite, chalcopyrite, rutile, ilmenite, along with minor galena, arsenopyrite, xenotime, and bastnäsite occur in the host rocks (Fig. 1H, I). The alteration mineral assemblage includes quartz, sericite, chlorite, and minor calcite and barite. Hematite, goethite, malachite, chalcocite, covellite, cuprite, and native copper formed due to supergene oxidation. The polymetallic quartz veins are enriched in Ag (<2,433 ppm), Au (<3 ppm), Co (<166 ppm), Bi (<83 ppm), Se (<58 ppm), and In (<53 ppm), while the disseminated assemblage is relatively more enriched in rare earth elements, including Ce (<57 ppm), Nd (<31 ppm), La (<30 ppm), Sm (<7 ppm), and Dy (<5 ppm), as well as in Ga (<10 ppm). Tetrahedrite, chalcopyrite, and electrum are carriers of silver and gold. Bismuth and Te enrichment is mainly related to inclusions of pilsenite hosted in chalcopyrite. In the disseminated mineralization, Dy occurs in xenotime, while Ce and Nd are hosted mainly in bastnäsite. The Laodikino mineralization shares in common mineralogical, geochemical, and structural characteristics with other Cu-Au-Bi-Pb-Ag-Te-enriched intrusion-related deposits in northern Greece (e.g., Koronouda, Drakontio). These deposits constitute future exploration targets for rare and critical metals.

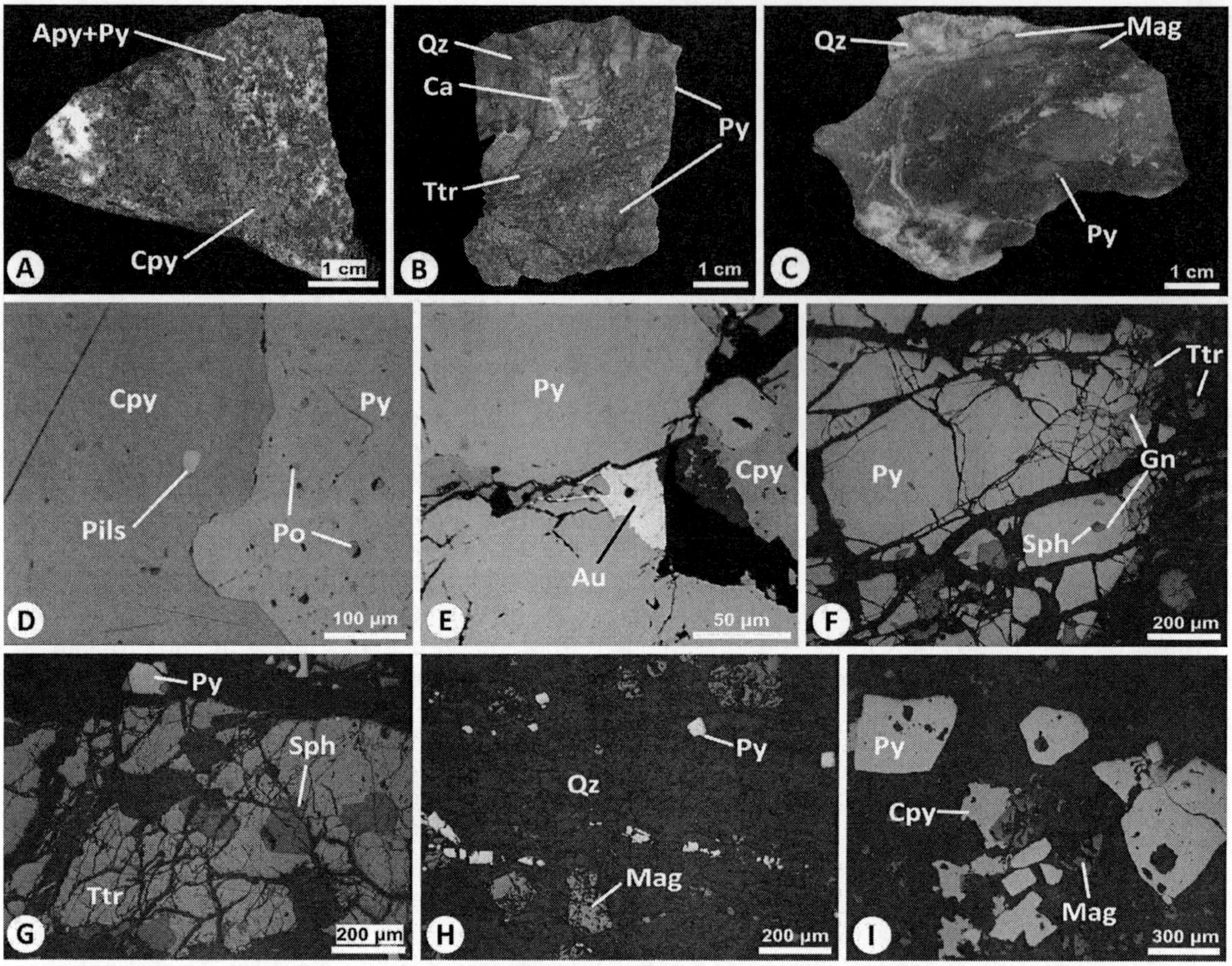

Fig. 1. Photographs of hand specimens of the studied mineralization styles (A-C) and photomicrographs of the metallic assemblages (D-I: plane reflected light). (A) Pyrite (Py) and arsenopyrite (Apy) from a polymetallic quartz vein exhibiting massive texture. (B) Pyrite (Py) and tetrahedrite (Ttr) rimmed by quartz (Qz) in a mineralized quartz vein. Calcite (Ca) is found in quartz interstices. (C) Quartz (Qz), magnetite (Mag), and tiny pyrite (Py) disseminations in a pod hosted by the metamorphic rocks. (D) Pilsenite (Pils) included in chalcopyrite (Cpy) and pyrrhotite (Po) inclusions in pyrite (Py) from a polymetallic quartz vein. (E) Native gold (Au) associated with pyrite (Py), and chalcopyrite (Cpy) from a polymetallic quartz vein. (F) Pyrite (Py) replaced by sphalerite (Sph), galena (Gn), and tetrahedrite (Ttr) in a mineralized quartz vein. (G) Sphalerite (Sph) in tetrahedrite (Ttr) and pyrite (Py) in a mineralized quartz vein. (H) Pyrite (Py) and magnetite (Mag) in quartz (Qz) from a pod hosted by the metamorphic rocks. (I) Pyrite (Py), magnetite (Mag), and chalcopyrite (Cpy) from a pod within the metamorphic rocks.

ST.090
Geology and Genesis of the Nam Xe Carbonatite-Associated REE Deposit in NW Vietnam

Yue Sun[1], Hao Hu[2], Jian-Wei Li[1, 2], Xuan Dac Ngo[3]
1. School of Earth Resources, China University of Geosciences, Wuhan, China, 2. State Key Laboratory of Geological Processes and Mineral Resources, China University of Geosciences, Wuhan, Wuhan, China, 3. Department of Mineral Resource Prospecting and Exploration, Hanoi University of Mining and Geology, Hanio, Vietnam

Carbonatite-associated rare-earth element (REE) deposits are the world's most important source of REE, as best represented by Bayan Obo Fe-REE-Nb deposit in Northern China and the Maoniuping REE deposit in Southwestern China, but their genesis remains hotly debated. The Nam Xe deposit in NW Vietnam (7.8 Mt of TREO) is hosted in carbonatite dykes and associated alkaline intrusions that were emplaced along the NW-striking Ailao Shan-Red River shear zone. However, the nature, age, and genesis of REE mineralization remain poorly understood. Here we present geological, mineralogical, and geochronological data to place new constraints on the formation of Vietnam's largest REE deposit.

Field relationships suggest an intrusive sequence from calciocarbonatite through magnesiocarbonatite to ferrocarbonatite. Both calciocarbonatite and magnesiocarbonatite have been extensively altered by carbonate-rich fluid, forming Ba- and carbonate-rich REE ores grading at 11.5-15.6% TREO. REE-minerals in the ores mainly consist of monazite, REE-fluorocarbonate, and minor donnayite-(Y), which are associated with gangue minerals dominated by Sr-Ba-Mg-Ca carbonate and barite. Ferrocarbonatite has been extensively altered by sulfate-rich fluid to form Sr- and sulfate-rich, high-grade REE ores containing 14.0-15.6% TREO, with REE-fluorocarbonate and baric-celestite being the major minerals.

Ferrocarbonatite contains 7,624-14,903 ppm REE, which is slightly higher than precursor calciocarbonatites (3,400-11,898 ppm). This observation indicates that differentiation of the parental magmas had limited contributions to REE accumulation in residual melts. Rather, hydrothermal alteration of carbonatites has played a critical role in the formation of high-grade REE ores at Nam Xe. Monazite grains from Ba-rich REE ores yield a Th-Pb age of 34.82 ± 0.41 Ma (2σ). This age is consistent with REE mineralization ages (34–11 Ma) of Maoniuping and other REE deposits that are part of the 270-km-long Mianning-Dechang REE metallogenic belt in SW China. This age consistency suggests that carbonatite-associated REE deposits in NW Vietnam and those of the Mianning-Dechang metallogenic belt in SW China likely represent different portions of an extended REE metallogenic belt generated under a Cenozoic post-collisional extension setting, related to the Indo-Asian collision.

ST.091

Identifying Cotectic and Xenomelt Magmatic Sulfides Using R-factor and S-isotope Variability in the Mount Keith MKD5 Komatiite-Hosted Nickel Deposit

Anne B. Virnes[1], Marco Fiorentini[1], Stephen J. Barnes[2], Laure Martin[3], Nicolas Thebaud[1], Quentin Masurel[1]
1. Center For Exploration Targeting, The University of Western Australia, Crawley, WA, Australia, 2. CSIRO Mineral Resources, Perth, WA, Australia, 3. Centre for Microscopy, Characterisation, and Analysis, The University of Western Australia, Crawley, WA, Australia

Komatiitic melts are inherently sulfur-undersaturated and must assimilate external sulfur in order to segregate a sulfide liquid that can equilibrate with the silicate melt and form nickel-sulfides. Multiple sulfur isotopic signatures of orthomagmatic sulfides, specifically the mass-independent signature (MIF-S, denoted $\Delta^{33}S$), inform us of the nature of external sulfur reservoirs and elaborate on the relative amount of assimilated sulfur. Predicting the S-isotope composition of external sources can be used in exploration targeting for favourable host rocks. However, internal isotopic variations due to magma-sulfide liquid interactions must be considered in order to use magmatic S-isotope compositions to infer the isotopic composition of the external source.

Recent work on samples from the world-class Mount Keith MKD5 komatiite-hosted nickel deposit discovered that sulfides in komatiite adcumulates can be classified between two distinct endmembers: 1) small, cotectic sulfide blebs, which form from in situ nucleation in equilibrium with crystallising olivine. Upon crystallisation, these blebs are trapped by the crystallising olivine pile. As a result, the blebs have equilibrated with small amounts of silicate magma (low R-factors) and display relatively low Pd content; 2) large blebs of sulfide melt, called xenomelts, which are formed due to direct melting of sulfide minerals in assimilated sulfidic country rocks. These xenomelts can be transported in the magma over varying distances, resulting in high-Pd tenors strongly dependent on R-factors. Both types will display mixed, non-zero S-isotope signatures between that of the mantle and contaminant reservoirs. But, as cotectic sulfides crystallise directly out of the magma, the S-isotope signature of this type will reflect that of the magma, independent of R-factors. In contrast, the S-isotope signature of assimilated xenomelt blebs will be determined by the amount of silicate melt they can equilibrate with and are, as such, highly dependent on R-factors. Hence, the relationship between PGE tenor (R-factor-dependent) and S-isotope composition can discriminate between these genetic processes. Identifying the xenomelt sulfides with R-factor-dependent $\Delta^{33}S$ values informs us on the amount of assimilant and its original S-isotope signature (Fig. 1A).

Our work presents new in situ multiple sulfur isotope data and sulfide metal contents from samples of the MKD5 nickel deposit from the Archean Agnew-Wiluna Greenstone Belt, Western Australia. We employ detailed petrography, mineral chemistry, and multiple sulfur isotope analyses to identify distinct cotectic and xenomelt populations (Figure 1B). R-factor for each sample is estimated using Pd in pentlandite analysed by LA-ICP-MS as a proxy for metal tenor, and S-isotope signatures are measured using the in situ SIMS technique. The resulting relationship between R-factor and S-isotope signatures is used to discriminate between cotectic and xenomelt sulfide blebs and to estimate the original isotopic signatures of the contaminant. Accurately predicting the contaminant isotopic signature can aid camp-scale targeting for favourable host rocks to komatiites.

Figure 1: A) Relationship between final (measured) and initial sulfide melt isotopic composition as a function of R-factor and silicate melt sulfur composition, for an assimilated xenomelt equilibrating at varying R values. B) Polished mount and photomicrograph of sample MKD153-618, displaying both cotectic and xenomelt sulfide blebs.

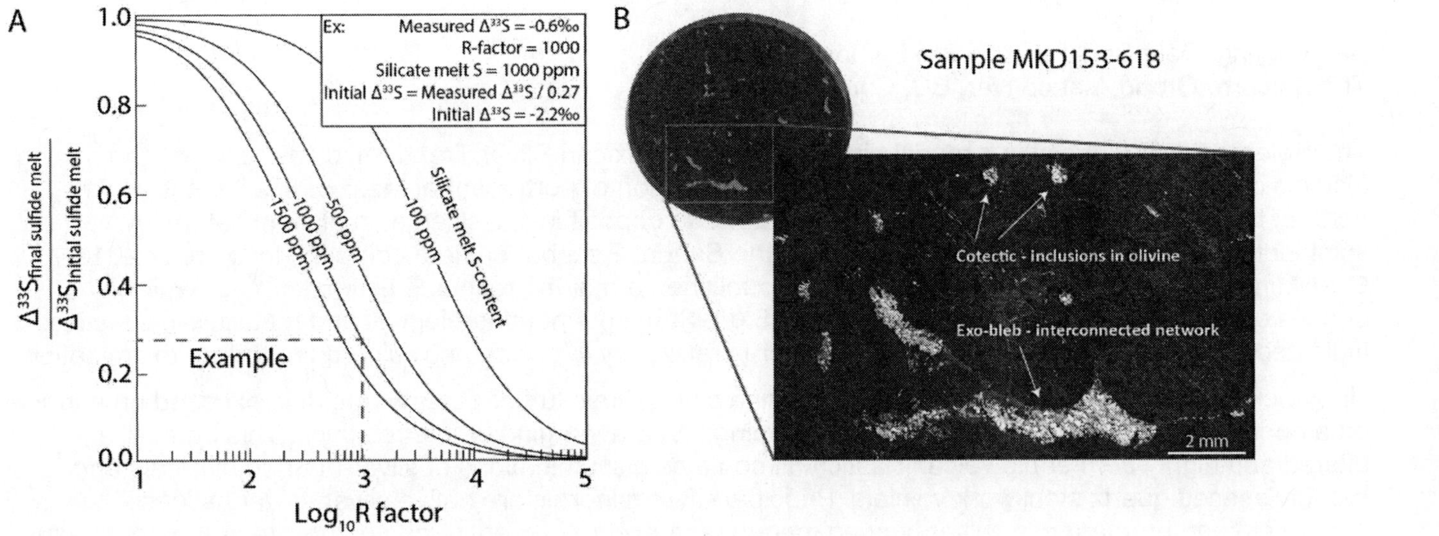

ST.092

Geology of the Boleras Deposit: A Tertiary, Stratiform, Disseminated Silver Deposit in the Mesa Central, Durango, Mexico

Ben Whiting, Dale Brittliffe, Rob van Egmond, Art Freeze
The Belcarra Group, Vancouver, BC, Canada

The Boleras Silver Deposit is a recent discovery in the "Mexican Silver Trend," midway between the mining districts of Tovar and Guanacevi in the Mesa Central region of north-central Mexico. It is located 45 km north of the town of Tepehuanes in Durango State. The deposit is classified as a disseminated, low-sulphidation epithermal deposit and is a part of the Sandra-Escobar caldera complex. In October 2016, Orex Minerals Inc. and Canasil Resources Inc. published a maiden resource estimate of 33.3 million ounces of silver in 9.8 million tonnes grading 106 g/t. Subsequent mineralogical and metallurgical testing indicated that a considerable portion of the silver is refractory in nature and will require further examination.

Host rock for the Boleras silver deposit is a rhyolite crystal lithic tuff of Tertiary age. It is exposed on surface on a gentle hillside as a stratiform, shallowly dipping layer, averaging 20 metres thick. This potassically altered and highly permeable volcaniclastic unit contains disseminations of silver-bearing minerals and broadly spaced quartz stockwork veinlets. Principal silver minerals are native silver, silver halides - cerargyrite and bromargyrite, silver bearing manganese oxides - argentiferous pyrolusite and aurorite, plus fine silver in silicates.

Plans are underway to explore the Sandra-Escobar mineral district through a joint venture between Orex Minerals Inc. and Pan American Silver Corp. in order to evaluate the various styles of precious and base metals mineralization in the caldera complex.

Figure 1. Dale Brittliffe reading the rocks on the Boleras Deposit. Photo by: Ben Whiting.

ST.094
Antimony Isotope Fractionation in Hydrothermal Systems and its Potential Exploration Application

Degao Zhai[1], Jiajun Liu[1], Ryan Mathur[2]
1. China University of Geosciences Beijing, Beijing, China, 2. Juniata College, Huntingdon, PA, USA

To date, few studies report antimony (Sb) isotope compositions in natural systems and the dominant factors controlling Sb isotope fractionation are still unclear. The literature has focused on the environmental and geological applications in some detail, while no studies have focused on one location that possesses concentrated amounts of Sb. Antimony mainly occurs as stibnite (Sb_2S_3) in a variety of intermediate-low temperature hydrothermal ore deposits worldwide and has been the simplest matrix to analyze. Scant Sb isotope data from stibnite exists in the literature described above, yet the key controls of its isotope fractionation in these systems are still largely unknown

In order to characterize antimony (Sb) isotope fractionation in hydrothermal systems, we present Sb isotope compositions of primary stibnite ores from a large Sb deposit in south China. A total number of 39 analyses reveals a large δ^{123}Sb range of -0.27 to +0.86‰, representing an up to 1.13‰ variation in this hydrothermal system. A gradual increase of Sb isotope ratios from the proximal to the distal parts was observed in stibnite ores. Rayleigh distillation models the systematic variation trend of the Sb isotope values, demonstrating that ore fluids are preferentially enriched in heavier Sb isotopes during the precipitation of isotopically light stibnite. In this way, we consider separation of stibnite from an Sb-bearing fluid related to reaction kinetics as a cause for Sb isotope fractionation.

The modeled data constrain a Sb isotope fractionation factor ($\alpha_{fluid\text{-}stibnite}$) between hydrothermal fluid and stibnite at approximately 0.9994. The model also constrains an initial Sb isotope value of ~0.45‰, which indicates metal sources in the basement rocks of the study area. Stibnite ores precipitated further away from their fluid source will gradually have higher Sb isotope compositions, which indicates that Sb isotopes can be used to fingerprint hydrothermal fluid flow directionality and may provide a vector to mineral mineralization.

ST.095
Identifying Potential Porphyry Copper Deposits at the Zhunuo Ore-Cluster District in Western Gangdese, Tibet: Insights from SWIR Spectrometry and Geochemical Anomalies

Shunli Zheng[1], Song Wu[2], Youye Zheng[1], Zaiying Yu[2]
1. China University of Geosciences (Wuhan), Wuhan, China, 2. China University of Geosciences(Beijing), Beijing, China

The Zhunuo ore-cluster district in the western Gangdese belt, southern Tibet, China, is characterized by a significant geochemical anomaly, where the Zhunuo super-large porphyry copper deposit has been discovered through geochemical methods. However, due to the extended coverage of Linzizong volcanic rocks, it is quite difficult to find concealed porphyry copper deposit only by geochemical methods. Thus, the application of new technical methods is urgently needed to guide exploration breakthroughs in the Zhunuo district. Recently, more studies have shown that the wavelength variation of diagnostic absorption peaks of white mica and chlorite measured by shortwave infrared (SWIR) spectrometry is closely related to the mineralization temperature, which can be footprints to trace the hydrothermal/mineralization (H/M) centers of porphyry ore systems. In this study, 3,666 surface rock samples from 20 km southwest of the Zhunuo deposit were systematically collected from an area of about 130 km^2, and 10,022 spectral data were obtained by SWIR. In general, fifteen kinds of alteration minerals, mainly consisting of white mica, chlorite, epidote, and kaolinite, have been identified. The white mica and chlorite were widely distributed in the study area, and the Al-OH wavelength of white mica and the Fe-OH wavelength of chlorite varied between 2,191-2,219 and 2,239-2,260 nm, respectively. Moreover, there is an increasing trend of Al-OH wavelength of white mica and Fe-OH wavelength of chlorite from porphyry centers to distal locations. When combined with geochemical anomalies of stream sediments and geological characteristics in the Zhunuo district, three potential M/H centers have been determined that are dominated by short wave anomalies (Al-OH, 2,191-2,206nm; Fe-OH, 2,243-2,250 nm), namely Cimabansuo, Rimujucuo, and Luobugangmu. Significantly, we found that a Dongshibu copper prospect previously identified by geochemical methods didn't show any porphyry mineralization-associated alteration information as reflected by SWIR, consistent with the barren copper results of later verification through drilling exploration. This decoupling may be caused by the long-term migration and weathering of stream sediment samples, during which the obtained geochemical anomaly cannot represent their in situ information. Therefore, caution must be taken when using stream sediment data to delineate prospecting targets. Instead, SWIR is a potentially powerful vectoring tool to aid exploration for porphyry systems, especially in the western regions where the level of prospecting work is extremely low.

*** Formatting Issue - Best Available Paper/Graphic ***

Title: Genesis and Oxygen Isotope Signature of Hematite and Magnetite from Magmatic-Hydrothermal BIF-hosted Iron Ore

Theme: Diversity in Ore Deposits

Presentation Type: Oral

Session:

Authors: Flávia C. Silveira Braga[1], Carlos A. Rosière[2], Andreas Pack[3], Steffen G. Hagemann[4], João O. Santos[5]
1. UEMG (Universidade do Estado de Minas Geraos), Belo Horizonte, MG, Brazil, 2. UFMG (Universidade Federal de Minas Gerais), Belo Horizonte, MG, Brazil, 3. Georg-August-Universität, Göttingen, Germany, 4. UWA (University of Western Australia), Perth, WA, Australia, 5. UEA (Universidade do Estado do Amazonas), Manaus, AM, Brazil

Abstract Text:

The formation of the high-grade iron orebodies in the Ediacaran-Cambrian Araçuaí/Brasiliano orogenic belt in the southeastern part of the São Francisco craton exemplifies a new and unusual type of iron ore developed during the tectonic reworking of the consolidated sialic crust associated with anatectic melting in the course of a West Gondwana continental collision.

Massive, coarse-grained, granular magnetite-hematite and schistose hematite iron orebodies (>60 wt % Fe) comprise several small deposits (approximate resource between 5 and 30 Mt for each one) bordering the Araçuaí orogenic belt in the southeastern São Francisco craton in Brazil. These deposits are hosted by tectonic slivers of Orosirian-Statherian amphibolite facies metamorphic iron formation (i.e., itabirite) enclosed in Statherian granites and pre-Rhyacian gneissic rocks of the Guanhães Complex.

The iron formation slivers (BIF) have undergone multistage hydrothermal alteration and Fe mineralization associated with a large hydrothermal system activated by crustal reworking and melting of the crust during the Brasiliano orogeny. In the syn- to late-collisional stage (ca. 580-560 Ma), partial Fe enrichment developed along shear zones, which control the location of the schistose, high-grade Fe orebodies. During the syn- to post-collisional stage (ca. 560-530 Ma) massive magnetite bodies formed by contact metamorphic-metasomatism with pegmatite intrusions. The post-tectonic stage (ca. 530-490 Ma) is characterized by the oxidation of magnetite to kenomagnetite/maghemite and martite and by crystallization of granular, massive hematite ore.

A characteristic variation in the oxygen isotope signature is associated with the changing hydrothermal alteration and Fe mineralization, with a progressive decrease of the $\delta^{18}O$ signature from the BIF (1.7 to 8.1‰) to high-grade iron ore (-1.6 to 2.6‰). In the magnetite-hematite and schistose hematite iron orebodies, the magmatic-hydrothermal fluids (calculated $\delta^{18}O_{fluid}$ between 4.6 and 13.1‰) from the pegmatite bodies have imprinted a relatively "heavier" $\delta^{18}O$ signature than observed in most hypogene BIF high-grade iron ores worldwide. Hematite and magnetite crystallized in the quartz and pegmatite veins, at the contact zone with the high-grade Fe ores, yielding $\delta^{18}O$ values between 1.8 and 5.0‰, well in the range of magmatic-hydrothermal systems, indicating a high magmatic fluid/rock ratio. Lower $\delta^{18}O_{mineral}$ values are observed in hematite and magnetite from the high-grade Fe bodies (-1.6 to 2.6‰). These values are probably related to the significant influx of low-temperature meteoric fluids that descended along extensional shear zones during the collapse phase of the Brasiliano orogeny.

*** Formatting Issue - Best Available Paper/Graphic ***

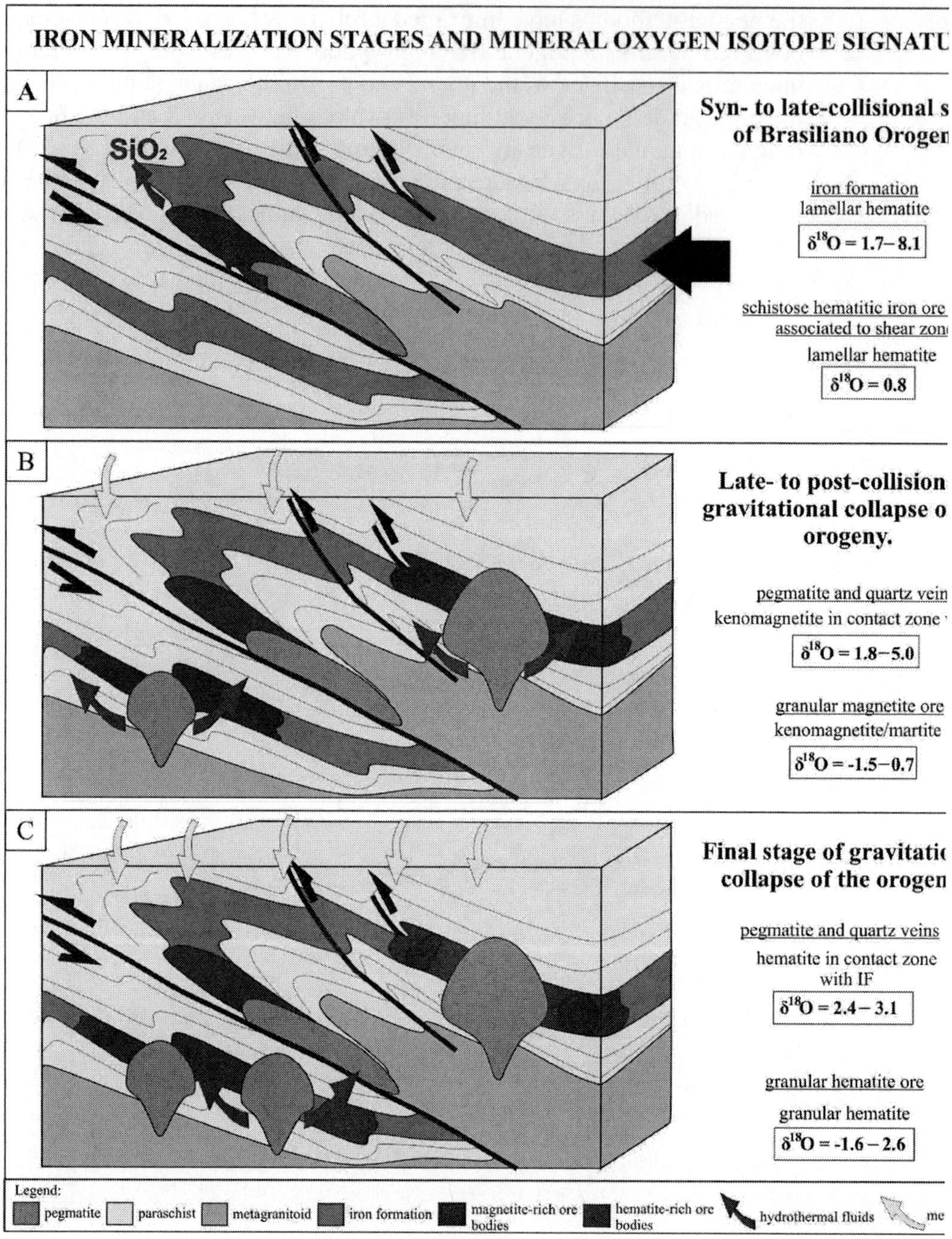

BC.2
History and Geopsychology of the Tintina Gold Belt

Curtis J. Freeman
Avalon Development Corp, Fairbanks, AK, USA

The following is a summary of the history and psychological impacts of the Tintina Gold Belt in Alaska and the Yukon Territory by a person who was there on the day of its birth. During the late 1970s and early 1980s, Alaska was emerging from a decades-old lode gold exploration and mining hiatus dating back to World War II. As the gold price rose through the 1970s and peaked in 1980 at over $800 per ounce, the mining industry was developing new ways to find and mine gold deposits. Heap lech technology, new and better mining equipment, computer technologies, and advances in geochemical and geophysical exploration helped create a rapidly changing environment for exploration and mining. And perhaps most important of all, a new army of geologists and engineers trained in these new technologies began to take over exploration, development, and mining all over the world, creating a brain trust that held the seeds of new discoveries. This coalescing of mining science and technology created the critical mass that was needed to set off a wave of new gold discoveries in the 1980s and 1990s, particularly in the western U.S., Australia, and South America. Alaska lagged behind those front-runners, but by the mid-1980s, Alaska started to reap the benefits of this mining industry revolution. Fort Knox was the first of the "'new breed" of gold deposits discovered in Alaska. Many others soon followed, and Alaska's gold resources grew from a few million ounces in 1985 to over 200 million ounces by 2021. Alaska and the adjacent Yukon Territory experienced a wave of new gold discoveries that covered a broad arc from the western Yukon through east central Interior Alaska into southwestern Alaska. During the 1990s and early 2000s, deposits like Fort Knox (14 Moz), True North (1.3 Moz), Donlin (+40 Moz), Dolphin (6.5 Moz), Ryan Lode (2.3 Moz), Vinasale (1.9 Moz), Pogo (10 Moz), Livengood (20.5 Moz), Dublin Gulch (+6 Moz) and others were discovered and advanced, some to production. Although each of these deposits had some characteristics that were unique to itself, they contained enough similarities to suggest they might share a common genetic ore deposit model. Academic research at a number of universities around the globe began to quantify these common denominators and, by the mid-1990s, a new genetic model known as the Intrusive-Related Gold (IRG) deposit model was being assigned to many of these gold deposits. It was in this environment that the Tintina Gold Belt was "created" in the spring of 1997 by an Alaskan exploration geologist and three junior mining company Presidents while sitting at an outdoor café in Montreal, Quebec. This presentation will detail the circumstances leading up to the creation of the Tintina Gold Belt, the personalities involved, and the impact its creation had on the mining industry's technical and financial communities.

BC.3
The Bre-X Fraud -- Impact on Geoscience in Exploration

Anne Thompson
PetraScience Consultants Inc., Vancouver, BC, Canada

On March 20, 1997, the geologist in charge of the Busang Gold project plunged to his apparent death from a helicopter in East Kalimantan, Indonesia on route to camp. What led to the moment that shattered dreams, devastated gold markets, and forced the industry to confront major fraud? What were the lessons learned and what changed?

Much of the description of the Bre-X debacle focused on the characterization of gold grains and resource calculations based on data sets provided by Bre-X employees. On site process and sampling is also considered, but what is largely missing from the story is the geology.

Throughout the exploration and promotion of the property, geologic terms and models were freely used to frame the project as one with substantial potential. "Maar-diatreme" was a popular model, particularly applied in the SW Pacific to describe a subset of epithermal/magmatic-hydrothermal deposits. Bre-X referenced this model in press releases and promotional materials (Fig. 1) and used the example of the Kelian gold deposit, 200 km away as an analogy.

I was hired by Bre-X in 1996 to characterize the mineralization, document gold occurrences, and create a more detailed model for the deposit formation. The work included a 10-day site visit to train staff in spectroscopy, review whole drill core, and carry out analysis on skeleton core. Follow up included an extensive petrographic study.

Alteration is characterized by selectively pervasive grey to white zones of mica-clay along well-defined fractures that host minor visible base metal sulfides. Kaolinite dominates the spectral results, persisting to depth in the fractures. Petrographic work identified quartz, carbonate, and white mica alteration with minor anhydrite and tourmaline at depth. Notable are extensive zones of epidote and chlorite alteration both in core and outcrop. Reported gold correlated with the kaolinite alteration, based on SWIR spectroscopy. However, during the PetraScience petrographic study, no gold was found.

Several key observations on site were concerning, but were not noted by other visitors – likely on short, scripted trips. As a consultant at the time, there were no processes for addressing concerns or reporting discrepancies. Could we have known? Humans tend to believe what we are told, yet we are trained as geoscientists to observe and ask questions. It is worth remembering that even with that training, unconscious bias plays a big role in how we value exploration projects.

Subsequent to the collapse of the exploration market due to the fallout from Bre-X, the 43-101 National Instrument was introduced in Canada to create standards for disclosure. This was a game changer for exploration. Two key factors are important: 1) the instrument applies to all disclosures; and 2) it created a structured format for reporting all data including geologic information.

The lessons from the fraud at Busang continue to impact our business and the way we operate. Among them is the importance of geoscience, including accurate descriptions of rocks and relationship to mineralization, rather than reliance on models. Quality geoscience is critical to investor confidence and a responsible mining industry.

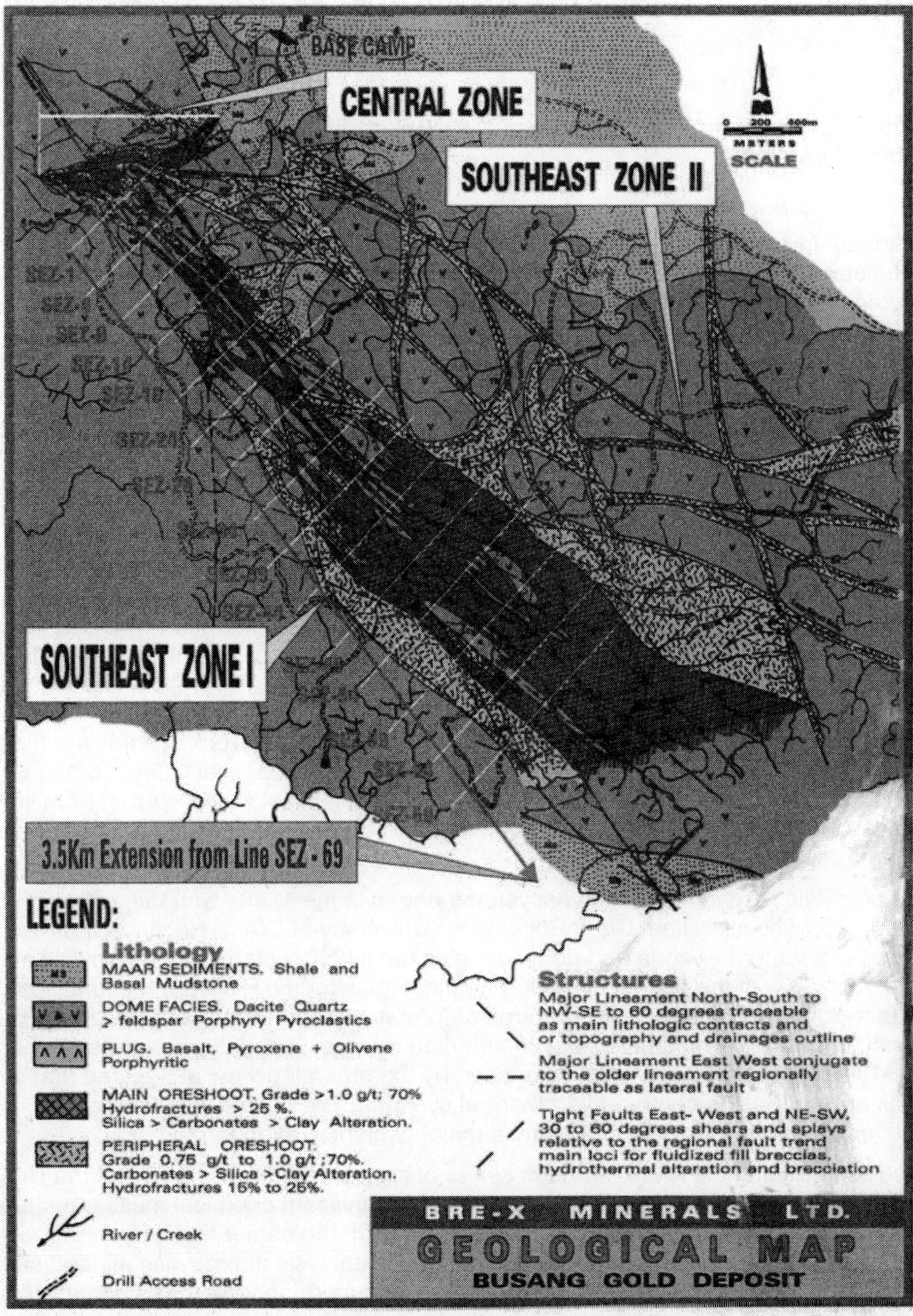

NOTE: Main Oreshoot based upon surface and drill results.
Peripheral Oreshoot subject for update after drill results.

BC.4
Propylitization 1865–2020: Historical Perspective and New Paradigm for the Origin of Propylitic Alteration in Porphyry Systems

Jamie J. Wilkinson[1, 2], Adam Pacey[2, 1, 3], Lisa A. Hart-Madigan[1, 2], Jacob Longridge[2, 1], David R. Cooke[4], Michael J. Baker[4], Adrian J. Boyce[5], Clara C. Wilkinson[1]
1. Natural History Museum, London, United Kingdom, 2. Imperial College London, London, United Kingdom, 3. Rio Tinto, Melbourne, VIC, Australia, 4. University of Tasmania, Hobart, TAS, Australia, 5. Scottish Universities Environmental Research Centre, East Kilbride, United Kingdom

The first recorded use of the term "propylite" by Ferdinand von Richthofen in 1865 was applied to greenish porphyritic rocks with chlorite and epidote surrounding the Comstock intermediate sulfidation lode, Nevada. The term came from the Classical Greek "propylon," meaning "gateway," because von Richthofen believed these rocks were a unit at the stratigraphic base (i.e., gate) of the Tertiary stratigraphy. It was subsequently realised that they were simply chlorite-, epidote-, and calcite-altered andesites and diorites, directly associated with mineralization. In tandem with the development of the concept of hydrothermal alteration came the understanding that metal zonation also existed around magmatic-hydrothermal deposits, extending well into the propylitic domain (e.g., Emmons, 1927). Until the 1950s, the term "propylitic" was just used for andesitic protoliths, but once the widespread occurrence of the alteration type was recognised in a variety of ore systems, the term was broadened to refer to a hydrothermal alteration assemblage of epidote, chlorite, albite, and calcite, often with pyrite, in any original lithology. Propylitic alteration was given a theoretical foundation by the seminal work of Meyer and Hemley in 1964, but at this point, a classification dichotomy developed in which the term propylitic was applied either broadly to rocks containing the typical mineral assemblages of the alteration type, or just to the weak, mainly isochemical, subset of these assemblages. The view that propylitic alteration is essentially an isochemical process apart from limited hydrogen metasomatism, hydration, and the addition of minor carbonate and sulfide, has since become entrenched in the literature.

The "modern" era of propylitic research in porphyry systems began in the 1980s with the visionary Ph.D. thesis of the "father" of propylitic alteration, Geoff Ballantyne (University of Utah). He studied Silver Bell, Safford, and Christmas in southern Arizona and derived the essential characteristics of propylitic alteration that we are really just refining with the benefit of modern technologies today. His work recognized the diverse mineralogy that exists, constrained temperatures of formation and the gradational relationship with potassic alteration, and implicated a magmatic fluid contribution using oxygen isotopes. Later stable isotope studies in the 1980s-1990s appear to have been influenced by the emerging view at the time that meteoric waters were predominant in intrusion-centred hydrothermal systems. The inferred dominance of groundwater as a driver of propylitic alteration remains a majority opinion even today.

However, recent studies, including the AMIRA-funded series of projects led from CODES at the University of Tasmania, have begun to change this paradigm. Increasingly, significant chemical mass transfer is recognized in the propylitic zone of porphyry deposits, consistent with the earlier findings of Emmons, and addition of magmatic components is supported by detailed isotopic analysis of propylitic mineral separates from the Northparkes system, New South Wales. Thus, there is an emerging view that magmatic fluids are pervasive throughout this domain, helping to account for the extensive trace element dispersion patterns now recognised in the major propylitic minerals, epidote and chlorite, that are being developed as powerful exploration tools for porphyry deposits under cover.

BC.6
Who Moved The Goalposts? Game-changers in the Bushveld Complex, South Africa – The First Hundred Years

Judith A. Kinnaird
University of the Witwatersrand, 2050 Wits, South Africa

One hundred years on, the Bushveld Complex is still one of the greatest ore deposits on Earth and still has undeveloped and unexplored resources.

The Bushveld intrusion was first shown on a map in the 1860's by Karl Mausch, although the first use of the term Bushveld Complex was by Molengraff in 1901. In 1906, William Bettel produced PGMs from gold-bearing conglomerate concentrates of the Witwatersrand Basin, but as far back as 1885, Dick Hart, a prospector, displayed samples containing "platina" in Church Square in Pretoria according to a newspaper of the time. By 1923, platinum was being mined from quartz veins near Mookgopong (formerly Naboomspruit) and, in 1923, the prospector Adolf Erasmus identified platinum on two farms near Mokopane (formerly Potgietersrus).

However, it was the discovery in 1924 of the platinum-bearing horizon by Andries Lombaard and confirmed by Merensky, at Maandagshoek in the Eastern Bushveld, that was really the beginning of the Bushveld bonanza. Within three years, Merensky had traced the reef that was named after him all the way around the eastern, western, and northern limbs of the Complex. The main Ni-Cu-PGE host of the northern limb was discovered by Merensky in 1925 and originally termed the Merensky Platinum Horizon, but termed the 'Platreef" by Van der Merwe in 1976. Platinum mining began in the late 1920's, but the Great Depression resulted in cessation of operations until 1992 when world-class opencast operations resumed. Meanwhile the western limb had become the focus of PGE mining.

Chromium, first noted by Mausch in 1865 near Rustenburg and then reported by Hall and Humphrey in 1908, wasn't mined until 1924 in the Steelpoort area of the eastern limb. Three groups were defined with decreasing chromium content upward from the LG's through the MG's to the UG's, with the LG6 being the thickest and most amenable chromitite for mining. A game changer in the late 1970's was the development of the methodology for extracting PGE from chromitite layers. As the UG2 was the richest, and had been left unmined because of its low Cr content, it has now become a greater PGE resource than the Merensky Reef.

Just as the fortunes of the Bushveld Complex might have appeared to be waning over the last decade, the discovery of two Tier 1 deposits in the northern limb, the sub-horizontal Flatreef mineralization located downdip of the exposed Platreef and the palladium-dominant Waterberg Ni-Cu-PGE deposit under cover north of the Hout River Shear Zone, have been game-changers, especially as the mineralization in the latter is at different stratigraphic levels to that in other limbs.

What is poorly appreciated is that the Bushveld Complex is a world-class producer of gabbronorite and granite dimension stone, andalusite from its metamorphic aureole, and fluorite from Vergenoeg. Cassiterite mineralisation has been of importance in the past, whilst vanadium from the magnetitite layers of the Upper Zone is increasingly important for vanadium batteries for energy storage for the future. Satellite intrusions are also the focus on ongoing exploration.

C1
Exploration and Mining – a World of Change

John F. Thompson
PetraScience Consultants, Vancouver, BC, Canada

Over the last one hundred plus years there have been remarkable changes in how and where we explore for minerals, and how we mine in an effective and responsible manner. For an industry not widely recognized for innovation, progress has been extraordinary albeit punctuated by periods of stasis. Three broad periods of change are discussed.

Rapid industrialization in the early part of the 20th century led to increasing demand for metals. Mining responded with the development of bulk mining – open pits, large equipment, and the flotation revolution. During this time the difference between geologists, miners and metallurgists was minimal – eminent men and women transgressed these roles although societal norms restricted women and those without stature. Mineral deposits were studied as critical geological environments related to physical and chemical processes. While there were several important contributors, Waldemar Lindgren was a leader.

Following the Second World War, metal demand surged again. Scientific and engineering advances during the war were transferred to the mining industry, for example solvent extraction to recover and process uranium, and the resulting development of SX-EW for copper. Researchers pursued new levels of experimental and theoretical thermodynamics contributing to the understanding of ore deposits. Simultaneously, earth science underwent the plate tectonic revolution, fundamentally changing our views on metallogeny, prospective terrains, and large-scale ore forming processes.

This period culminated (~1970-1990) with new comprehensive ore deposits models, some with revolutionary implications for processes and exploration (e.g., VMS, magmatic sulfide, porphyry and epithermal deposits). Exploration in these settings was supported by a wide range of new geochemical and geophysical tools. Studies on active systems – e.g., seafloor, geothermal, and weathering – contributed to our understanding of physical, chemical and biological processes related to ore formation. The fascination with new processes had a downside. At times, exploration pursued a panacea approach driven by the latest model or discovery – seafloor processes were widely invoked, while sinters, diatremes, shear zones and other phenomena were overused and promoted.

Post 1990s, societal scrutiny of mining led to an increased focus on efficiency, environmental and social performance, and innovation. At the same time, the exponential increase in the demand for metals ran counter to these efforts with increased production from massive low grade operations resulting in "less metal from more rock" and reliance on economies of scale. Exploration played its part with new discoveries, and systematic evaluation of mine environments. Significant technological advances at rock to micro-scales changed geochronology and improved analytical precision and detection with implications for research and exploration. The digital revolution supported rapid data collection, data analytics and modeling. Multidisciplinary, diverse research and exploration teams utilized numerous approaches, and while insight still came from individuals and new discoveries, panaceas disappeared.

Over the latter period, the world has seen upheaval from economic, societal and health challenges, combined with radical changes in data use and communication. The full ramifications for exploration remain to be seen, but as we enter a new period, there will be increasing demand for high quality mineral deposits that will deliver benefits across society.

C2
Game Changers: The First Hundred Years: Analytical Techniques

Dominique Weis[1], Corey J. Wall[2], Maghaleray Amini[1]
1. Pacific Centre for Isotopic and Geochemical Research, Department of Earth, Ocean and Atmospheric Sciences, University of British Columbia, Vancouver, BC, Canada, 2. Boise State University, Boise, ID, USA

Geochemistry develops chemical principles allowing us to understand how the Earth formed and how it works. It can also be used directly to help discover mineral resources and keep the planet habitable. The potential applications of these tools over the last century have been highly dependent on the development of analytical techniques. We have only been able to get an absolute age for the Earth in 1956. Nowadays, we can have precise U-Pb ages with precision down to 0.2 myrs for Archean rocks and less than 10 kyrs for Tertiary rocks, including in systems that were considered earlier on to be undatable (e.g., layered intrusions' source of large reserves of PGEs, Cr, and V and other metals such as Cu, Ni, and Au). The amount of material that is required for analysis has decreased from grams to milligrams and more recently to very tiny amounts as low as a few picograms. Another major analytical game-changer over the last decade has been the development of in situ techniques allowing for analyses of trace element concentrations simultaneously with isotopic compositions (Pb, for instance, in thin sections, or Hf in zircon grains) for provenance determination or/and U-Pb age dating. The split-stream techniques ((MC)-ICP-MS) are fundamentally transformative to unravel complicated metal distribution patterns and to define their association with mineral phases, the chemistry of which can also be determined in situ. They complete high throughput methods for measuring almost every geologically relevant element to below-average crustal concentrations routinely and rapidly. Simple developments such as pXRF allow instantaneous determination of elemental compositions on the field, now with a precision that competes with that of analyses in the lab – a game-changer for mineral exploration. Community initiatives such as EARTHTIME are key for high-precision dating used in porphyry systems, timescale calibration, etc. With increased sensitivity, these techniques can now be applied to use U-Pb to date carbonates or accessory phases such as monazite, titanite, rutile, and scheelite. Proper QA/QC requires access to different approaches, such as Raman microscopy, microdrilling, and thermoionization instruments. Another completely new field is the application of multi-collector ICP-MS for the analyses of transitional metal isotope compositions such as Fe, Cu, Zn, and Mo, among others that allow the deciphering of the processes leading to the concentrations of these metals in ore deposits or at the other end the process to determine their paths in the environment.

This keynote will review the significant analytical developments over the last decade and show some new game-changing applications.

C3

The First and Second Opening of Pandora's Box – How Re-Os Changed Resource Geology

Holly J. Stein[1, 2]
1. AIRIE Program, Colorado State University, Fort Collins, CO, USA, 2. Institute of Geosciences, University of Oslo, Oslo, Norway

Throughout the geosciences, the only unifying parameter for correlating and organizing earth processes is absolute time. As a budding young economic geologist working with the mineral industry in the 1980s, I set my sights on dating sulfides, reading with wonderment how two German-Swiss physicists in the 1960's recognized that $^{187}Re \rightarrow ^{187}Os$ formed a radiometric clock for molybdenite. That recognition lay dormant for 30 years as the ability to precisely measure isotopic concentrations for Re and Os was beyond reach, and the ^{187}Re decay constant remained poorly determined until 1996. In the world of ore geology, argon-based dating of non-sulfide (gangue) minerals was fraught with prolonged thermal histories in ore-forming environments, and the human error in picking plateaus to achieve an unjustifiably precise age from a mineral's sometimes complex outgassing spectrum. We needed a way to put the stethoscope directly on the sulfides.

The reducing environment offered by sulfides and organic material provides a haven for chalcophile-siderophile-organophile Re and Os. The Re-Os clock provides dates from Early Archean to Pleistocene. It is not temperature sensitive. It retains primary ages even with granulite-facies overprints. On oxidation, however, whether in the geologic past or in present-day weathering, Re and Os (and other metals) become readily unmoored and the Re-Os radiometric clock is compromised.

Not surprisingly, the critical first step is to turn geologic observation into a thoughtful sampling strategy. Relative time is useful for describing crosscutting relationships, but Re-Os gives us the immediate ability to separate crosscutting events in real time. Re-Os teaches us that crosscutting relationships cannot simply be extrapolated beyond the outcrop scale, or between drill holes. Re-Os shows that all veins of a certain type in a deposit are not necessarily the same age. In some ore systems, Re-Os revealed more than one episode of mineralization with a large intervening time gap. Re-Os gave us our first sense of duration, or time required to build a giant ore system. In a well-known Archean terrane, Re-Os showed that not all gold mineralization fell within a narrow orogenic window. Re-Os upset dogma in the first opening of Pandora's box, and was unpopular and even rejected by some where long-standing ore deposit models were confronted with a new kind of data. Indeed, there were early challenges and hurdles in establishing the credibility of Re-Os within the ore geology community.

Re-Os isotope geochemistry spread quickly beyond the ore geology community as the oil industry realized its potential and was willing to invest financially, long-term, in developmental work. The timing of maturation and migration, and the sorting out of multiple hydrocarbon charges from different source rocks presented new challenges, but surmountable with strategic sampling anchored in geologic relationships. Even at the center of hydrocarbon work, we turn to sulfides again and again, as metals (and sulfur) are integral to petroleum systems. Geologists who can work across the metals-hydrocarbon divide, and who can visualize and model metal sources before the creation of an ore deposit, will be invaluable to mineral exploration in the future.

U-Tube: The Myth of Pandora's Box The Gift of Curiosity

C4
Changing the Game with Geological Modeling: Then and Now

Rebecca Montsion[1, 2], Mark Jessell[2], MW Lindsay[2], Stéphane Perrouty[1], EA de Kemp[3], Laurent Ailleres[4]
1. Mineral Exploration Research Centre, Harquail School of Earth Sciences, Goodman School of Mines, Laurentian University, Sudbury, ON, Canada, 2. Centre for Exploration Targeting, School of Earth Sciences, The University of Western Australia, Crawley, WA, Australia, 3. Geological Survey of Canada, Central Canada Division, Ottawa, ON, Canada, 4. School of Earth, Atmosphere and Environment, Monash University, Monash University, VIC, Australia

From the early days of bedrock mapping, geoscientists have sought to capture and communicate geological features as models (2D and 3D). This modeling process involves gathering observations, applying assumptions, and generating interpretation realisations. However, as geological knowledge increases and computation power grows, new opportunities for data-focussed modeling and interrogation are revolutionizing how geological features are captured and communicated. The progression from hand-drawn interpretations such as cross-section and block models to fully computed 3D implicit model ensembles has facilitated a shift to more quantitative and repeatable modes of inquiry. With 3D modeling, it is now possible to pose complex questions related to geochemical or ore distributions, crustal-scale hypothesis testing, and uncertainty analysis. The process of 3D modeling also forces internal consistency checks between interpreted map features from all datasets, which is an informative process rarely achieved through 2D modeling alone. Looking forward, advancements in 3D modeling should integrate geological knowledge and principles with modern data analytics to produce interrogatable and geologically reasonable realizations. Cutting edge 3D modeling includes capturing increasingly complex geological feature geometries, respecting convoluted geological histories, and enhancing resource estimation.

Figure 1: Progression of modeling innovations from ancient Egypt quarry records in 1160 BC, 2D geological mapping by William Smith (1815), conceptual 3D block modling by Andreson (1884), and 3D graphical framwork modeling by Stirewalt and Henderson (1995). Novel and comprehesive implimentartions for the last decade include 3D prospectivity modeling by Payne et al. (2015) and uncertainy esitmation from Pirot and Lindsay (2021).

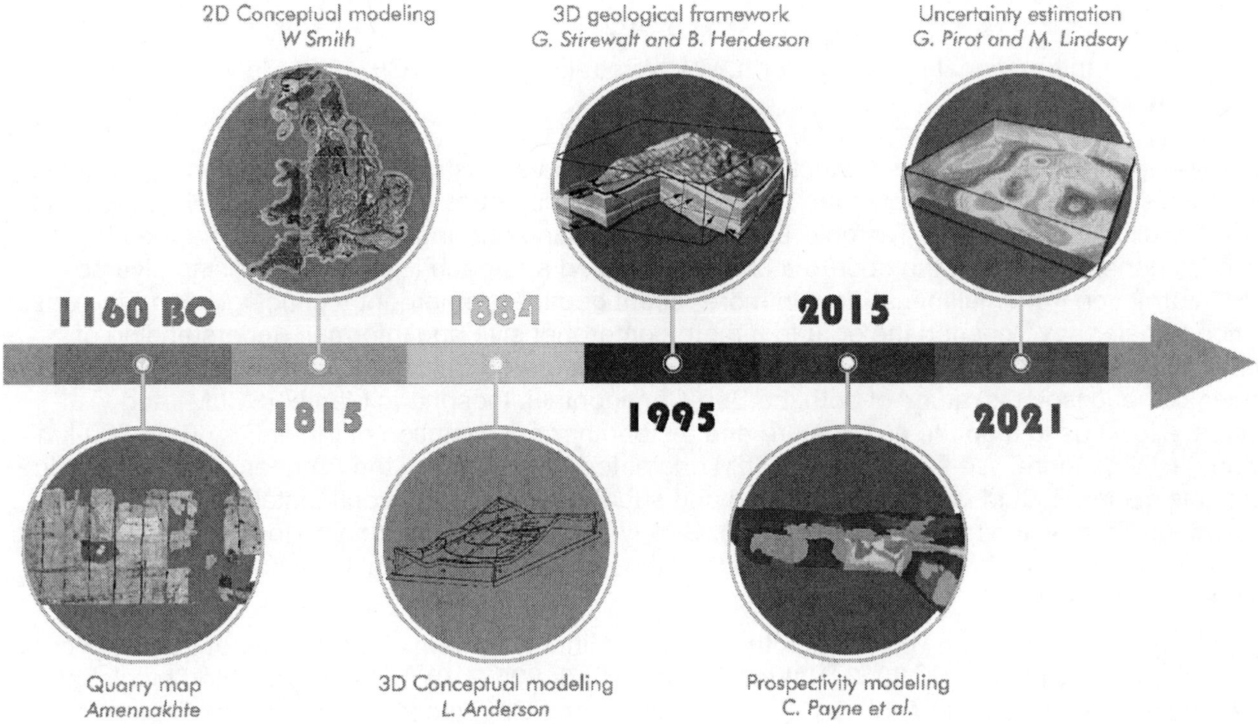

C5
Ore Deposit Science: The Emergence of Context

Graham C. Begg[1, 2], William L. Griffin[2], Suzanne Y. O'Reilly[2]
1. Minerals Targeting International PL, West Perth, WA, Australia, 2. CCFS/GEMOC, Macquarie University, Sydney, NSW, Australia

The early years of Earth Science were restricted to surface observations and were dominated by geological mapping. As a result, our view was very much “top-down,” and ore deposit science was largely conducted at the deposit to district scale. Plate Tectonic Theory revolutionised our understanding of both the planet and ore deposit genesis, opening the door to a prediction-based approach to targeting utilising diverse geoscience information from multiple scales. In more recent decades, emergent technologies and the ease of information transfer have opened the door to a more comprehensive and informed understanding of continents and Earth processes. Multi-disciplinary (geological, petrographic, geochemical, geophysical, geochronological, and isotopic) study of both the Sub-Continental Lithospheric Mantle (SCLM) and overlying crust allows us to map the architecture and tectonothermal evolution of the entire 60- to 280-km-thick continental lithosphere. We find that the SCLM originated dominantly in the Archean, and that outside the cratons, this ancient SCLM is strongly modified and subdivided into microcontinental blocks generally bounded by steep Trans-Lithospheric Faults (TLFs). Geodynamic processes, dominated by Plate Tectonics, have resulted in physical and chemical changes to the lithosphere, fundamentally influencing ore deposit formation and preservation.

Mineral deposit genesis is now viewed through the prism of Mineral Systems that operate at the scale of the entire lithosphere (e.g., Griffin et al., 2013, Nature Geoscience 6, 905-910) and are influenced by global geodynamic processes (e.g., Begg et al., 2018, In Processes and Ore Deposits of Ultramafic-Mafic Magmas through Space and Time; Mondal & Griffin (eds), Elsevier). The latter dictate lithospheric architectural and tectonic settings and availability of metal sources, and act as temporal triggers for the ore-forming stage(s) of the Mineral System. For example, magmatic Ni-Cu systems are controlled by supercontinent cycles, global patterns of whole-mantle convection, lithospheric architecture (including craton margins and variations in lithospheric thickness), and Plate Tectonic-driven stress states, including stress switching. Whilst Porphyry Copper systems are dependent on convergent margin processes, it is the spatial and temporal arrangement of zones of most intense shortening, and their moments of relaxation, that activate the ore-forming stages. Otherwise ordinary sub-arc magmas are transformed into Cu-fertile melts that ascend into the upper crust. Gold deposits arise from a two-stage process. The first, through the action of low-degree partial melting induced by changing physiochemical conditions, is the creation of zones of metasomatized mantle with enhanced frequency of gold-bearing veinlets. The second involves selective melting of such veinlets, through the agency of changing P-T conditions induced by tectonic triggers. Fluids exsolved from such melts at deep crustal levels migrate to shallower levels where deposits form at sharp P-T-X gradients.

Economic geology has evolved from its early deposit- and upper-crustal-centric focus to embrace structure and Mineral System processes at lithospheric-scale in 4D, and to an intimate connection with the global geodynamic system. This has raised a whole new set of questions and fertile areas for future research. The door to the next century of investigation and discovery is open.

C6
Game Changers in the Exploration of Submarine Hydrothermal Systems

Mark D. Hannington
University of Ottawa, Ottawa, ON, Canada

The discovery of black smokers, and especially their link to chemosynthetic life, was one of the most compelling scientific advances of the last century. Its impact on the model of volcanogenic massive sulfide (VMS) deposits is expressed in a quote from the 100th Anniversary Volume: "... the extraordinary intensity of marine research that followed the discovery of [black smokers] rapidly built our understanding of seafloor mineralization to such an extent that ore deposition in the submarine environment is now arguably one of the most clearly understood of all mineralizing systems ..." (Skinner, 2005). The origins of VMS deposits at seafloor hydrothermal vents and the connection to plate tectonic theory were already established concepts decades before the first images of black smokers appeared. In the early 1900s, research centered around the link to submarine volcanism – a relationship that is obvious today, but its acceptance was hard won at the time. The current model for VMS deposits was introduced in the 1950s, and unequivocal evidence of metal deposition directly associated with seafloor volcanism was uncovered at mid-ocean ridges in the 1960s. When black smokers were eventually found, critical aspects of the model were validated by sampling of actively venting hydrothermal solutions, including the details of water-rock reactions and the behaviour of different end-member fluids. However, the sheer numbers of black smokers and the scale of the global metal flux were unexpected. In a blur of discovery beginning in the late 1970s, more than 700 sites of submarine hydrothermal activity were located throughout the world's oceans, and new sites are still being found every year. The rush to explore the deep sea provided unparalleled opportunity to observe processes that could only have been supposed from the study of fossil deposits and many that would never have been predicted from our early understanding of black smokers.

Among the "game changers" in the exploration of submarine hydrothermal systems was the new partnership between ore deposit geologists and marine scientists who were already investigating the biology, chemistry, and physical oceanography of deep-sea vents. The powerful interdisciplinary research that followed had a direct impact on the changing VMS model, in particular aspects related to marine geodesy and geodynamics, oceanic heat flow, the diversity of submarine magmatic-hydrothermal systems, especially in subduction zones, the dynamics of melt and fluid flow in submarine volcanoes, the associated volcanic geochemistry, and the role of macro- and microfauna at hydrothermal vents, including the limits to life in ore-forming systems. These advances were all made possible by increasingly sophisticated technology for ocean exploration, such as satellite altimetry and navigation, marine gravity and magnetics, high-resolution seismic reflection and refraction, deep-sea drilling, underwater robotics, and autonomous vehicles. These tools enabled unprecedented documentation of submarine hydrothermal activity in the most remote regions of the oceans, revealing the complex geological make-up of different settings for mineralization and building the foundation for new concepts in marine metallogeny.

C7
Porphyry Copper Deposits - from Empirical Models to Mineral Systems

David R. Cooke[1, 2]
1. Centre for Ore Deposit and Earth Sciences (CODES), University of Tasmania, Hobart, TAS, Australia, 2. Transforming the Mining Value Chain (TMVC), an Australian Research Council Industrial Transformation Research Hub, University of Tasmania, Hobart, TAS, Australia

The porphyry copper model is well-established, robust, and long-lived. Its evolution over the past hundred years has occurred through incremental advance of deposit-scale knowledge gained from numerous characterization studies punctuated by "light bulb" moments where step-changes in understanding occurred due to insights from critical thinkers and the application of new technology. Collectively, these advances have facilitated thinking across scales and the development of a porphyry system model.

Deposit-scale characterization studies of porphyry deposits dominated the early 20th Century, providing the foundations for an empirical understanding of the hypogene and supergene characteristics of porphyry deposits. During this period, the recognition of the spatial and temporal association of ores with porphyritic intrusions eventually led to acceptance of an orthomagmatic genetic model for porphyry mineralization in the 1960s.

What many consider to be the classic porphyry deposit model was published in 1970 by David Lowell and John Guilbert. It is based on an exploration model developed by Lowell at San Manuel, Arizona, which led to the discovery of the Kalamazoo orebody. The Lowell and Guilbert model synthesized key deposit features – sulfide and alteration zoning patterns and vein distributions – that could guide explorers towards mineralization. The Lowell and Guilbert model stands as one of the most successful exploration models of the past 50 years – it profoundly influenced porphyry deposit discoveries.

The 1970s was a golden period of porphyry research. The Lowell and Guilbert model was refined by Gustafson and Hunt in 1975, who recognized that overprinting of multiple magmatic and hydrothermal events causes significant "variations on a theme" in the porphyry clan. Pioneering isotopic and fluid inclusion studies, coupled with hydrological and geochemical modelling, led to critical advances in the porphyry genetic model during this period, with Burnham's 1979 model a landmark advance.

The porphyry system model's foundations were laid in 1972 when Dick Sillitoe published "A plate tectonic model for the origin of porphyry copper deposits." This step-change in understanding led to a wide acceptance of the critical role of tectonic processes in porphyry mineral systems. Dick Sillitoe, perhaps more than any other individual, has profoundly shaped the porphyry model, effectively working across scales from crustal to regional to district to deposit. His porphyry district exploration model, published in 1989 and refined many times, most notably in 2010, has helped explorers to understand the links between porphyry and other deposit styles (epithermal, skarn, CRD), again leading to significant exploration success.

Generation of high-quality geological, geophysical, and geochemical datasets across all scales has facilitated significant, rapid advances to the porphyry system model over the past three decades. Critical insights have included the recognition of tectonic triggers for mineralization, the ingredients for fertile magma generation, the importance of trans-lithospheric structures, the role of mid-crustal magma chambers, and the influence of brines, vapors, and supercritical fluids on mineralization and alteration. The model continues to be refined, with a priority being to provide explorers with tools to aid discovery in greenfields terrains and under cover so they can meet society's urgent, ever-growing copper demand.

C8

The Discovery and Mining of the Olympic Dam Deposit: Impacts on the Optimization of Exploration and Mining/Processing of Complex Ores

Kathy Ehrig
BHP Olympic Dam, Adelaide, SA, Australia

Drilling of RD1 in July 1975 marked the discovery of the Olympic Dam breccia-hosted Fe-oxide Cu-U-Au-Ag deposit. The discovery was the culmination of approximately two decades of exploration for copper deposits across Australia by the former Western Mining Corporation. It is one of the renowned exploration success stories of the twentieth century, highlighted by the use of a mineralogy-based, holistic source-transport-trap system approach for generating conceptual exploration models and exploring "undercover" through flat lying, unaltered sedimentary rocks varying in thickness from at least 250 m in remote areas of South Australia. The Olympic Dam (OD) discovery ultimately ignited the recognition of a new class of deposits, iron oxide Cu-Au (IOCG). The global significance of OD is the amount of recoverable contained metals; it is the world's single largest accumulation of U and one of the largest Cu and Au deposits.

The deposit (~6x3 x 0.8 km deep, locally extending to >2.3 km depth) is hosted within the Olympic Dam Breccia Complex which is derived from weak to intense brecciation and texturally destructive iron metasomatism of the ~1594 Ma Roxby Downs Granite and lesser abundant lithologies (i.e., bedded clastic facies, felsic to mafic to alkaline ultramafic lavas and dykes). OD represents preservation of a post-mineral disrupted, shallow system, hence its somewhat unique character amongst IOCGs. Copper ± Fe sulfide, uranium, and gold and silver mineralization dominantly occurs as <200-micron-sized grains typically disseminated throughout the breccia matrix. Vein-style mineralization is rare. There is an unequivocal spatial relationship of Cu-U-Au-Ag with Fe at the macro- to nano-scale. The deposit also contains geochemically anomalous concentrations of a wide range of elements, including, but not limited to, REEs and other critical metals. Over 100 minerals are present, each with its own unique response (positive, benign, or deleterious) to each part of the metal recovery process. However, less than 25 sulfide, uranium, gangue, and As-, Se-, Te-, Bi-, Sb-bearing minerals account for >99.5% of the deposit mass and are critical to processing.

Metal production at OD commenced in 1988. Mining occurs via sublevel open stoping where ore is hoisted to the surface and processed onsite through a fully integrated metallurgical processing facility to produce Cu-cathode, gold and silver bullion, and uranium oxide concentrate. The presence of uranium and its associated short half-life radionuclides requires Cu-Au-Ag final products to contain extremely low concentrations of U and radionuclides prior to shipping offsite. A high level of complexity in processing OD ores is thus introduced because Cu-Au-Ag and U require multiple processing streams (i.e., milling, sulfide flotation, tailings leach, solvent extraction, solid-liquid separation, concentrate leach, flash and electric furnaces, sulfuric acid plant, ER and EW Cu refineries, and precious metals refinery) for economic recovery. This necessitates game-changing geometallurgical characterization. More than 30 elements (and density and mineralogy) of ore and waste across the deposit, along with fifty geometallurgical predictors of processing performance, are populated into each of the 20M blocks in the resource model, hence serving as the fundamental input into immediate and future planning of the OD asset.

C9

The Role and Development of Government Geological Surveys in making a Difference for Earth Resources Exploration, Discovery and Confident Evidence-based Decisions

Steve Hill, Marina Costelloe
Geoscience Australia, Canberra, ACT, Australia

Government geoscience agencies across the world have a diverse range of roles and objectives, largely dependent upon the needs and values of their jurisdiction. A significant number of these have a focus on using geoscience data and information to help support and stimulate the sustainability and operation of resource industries. This presentation looks at the development and impacts of government geoscience agency programs for the resources industry and then considers some of the emerging trends and future opportunities for government geoscience.

Many government geoscience agencies were initiated as trusted custodians of the geoscience data and knowledge of a jurisdiction. These were typically driven by systematic geological mapping and associated data acquisition such as geochronology, resource geology, geophysics, and geochemistry. For regions with some prospectivity for mineral or energy resources, the geoscience programs had a strong economic geology focus, including studies of mines and known mineralisation as well as data and information that stimulated exploration towards mineral discovery. This was heralded as the provision of "pre-competitive" geoscience data and information, driven by the philosophy that Government investment in geoscience data would help, if not drive, sustenance of or growth in private investment and industry and, therefore, provide return on the government investment. An important array of economic and benefit analysis studies have been conducted to help support the philosophy that pre-competitive geoscience investment provides a strong return from the initial government investment. This is particularly strong in the situations where pre-competitive geoscience can be directly (and even indirectly) attributable to the discovery of new mines. In some jurisdictions the pre-competitive programs have been extended to also include "collaborative" exploration programs where industry-government partnerships have been applied to regional data acquisition such as geophysical and geochemical surveys and drilling.

Government geoscience programs have been constantly looking for opportunities to redefine themselves, in many cases hoping to provide a competitive attractiveness for investment from private industry, but also to further accelerate or increase investment confidence in a particular jurisdiction. Some examples of new approaches have included a greater emphasis on open access data and greater alignment with FAIR data principles, as well as the use of crowd-sourcing geoscience data competitions. Another important development has been the recognition of the value of geoscience data across the resources value chain from exploration and mining to resource use and destiny, as well as the embrace of a broader definition of geoscience and its broader context within environmental, economic, and cultural issues. An example here is providing government guidance for issues of social license such as land access and community outreach.

C10
Game Changers in Understanding and Exploring for Magmatic Ni-Cu-PGE Deposits

C Michael Lesher
Laurentian University, Sudbury, ON, Canada

Ni-Cu-PGE sulfides are the type example of orthomagmatic mineralization. Although the origins of magmatic Cr, Fe-Ti-V, and PGE deposits and their exploration implications are still hotly debated with no consensus yet in sight, our better understanding of the genesis of magmatic Ni-Cu-PGE deposits was transformed by several game-changing discoveries in the late 1970s and 1980s. During the first half of the 20th century, most models involved sulfide exsolution with no explanation for the very high abundances of sulfides in some deposits. In the 1960s and 1970s, the first evidence for crustal S appeared, but as late as the mid-1970s, the leading model involved formation of sulfides in sulfide-rich parts of the mantle with transport to the surface in olivine-rich magmas. In the late 1970s, it was discovered that the partitioning of platinum-group metals between immiscible sulfide and silicate melts is orders of magnitude stronger than previously assumed and that the metal tenors of immiscible sulfide melts are strongly dependent on the effective magma:sulfide ratio. In the early 1980s, this led to the revelation that high-tenor magmatic Ni-Cu-PGE deposits cannot form from sulfide-saturated magmas, but must form from sulfide-undersaturated magmas. In the mid-1980s, it was shown that magmatic Ni-Cu-PGE deposits do not form from magmas containing intratelluric olivine phenocrysts, but that the host units represent dynamic lava/magma conduits, in which olivine crystallized in situ, that were capable of thermomechanically eroding wall rocks. This, increasing S-isotope evidence for non-mantle S in a wider range of deposits, and new geological and fluid dynamic constraints led to models in which sulfide melts do not come from the mantle, but S is incorporated during thermomechanical erosion of crustal rocks. The 1990s and 2000s provided additional geological, stratigraphic, mineralogical, geochemical, isotopic, fluid dynamic, and thermodynamic evidence for crustal incorporation models. These developments fundamentally and forever changed how we explore for magmatic sulfide deposits, a process that now involves exploring for magma conduits (lava channels, channelized sills, chonoliths, channelized dikes) that were able to access crustal S (sulfidic sediments, sulfidic volcanic rocks, evaporites). The debate in the 2010s turned to if and how sulfides can be transported vertically, with some models suggesting significant upward transport of sulfide droplets from the mantle or deeper "staging chambers" and other models requiring generation at more-or-less the same stratigraphic levels and only subhorizontal transport (see Figure: 1 = prospective channels with access to S, 2 = non-prospective channels without access to S, 3 = non-prospective, access to S but not channelized). During the next 10 years, we will develop more sophisticated fluid dynamic models of multiphase transport of silicate melt – sulfide melt ± phenocrysts ± xenoliths ± xenocrysts ± xenovolatiles, leading to advances in our understanding of how sulfides are localized in geometrically complex dynamic magmatic systems. We will also determine how to more precisely target mineralized magmatic plumbing systems by integrating more sophisticated understanding of tectonomagmatic controls with high-resolution seismic, magnetotelluric, and gravity data. MERC contribution number 2020-002.

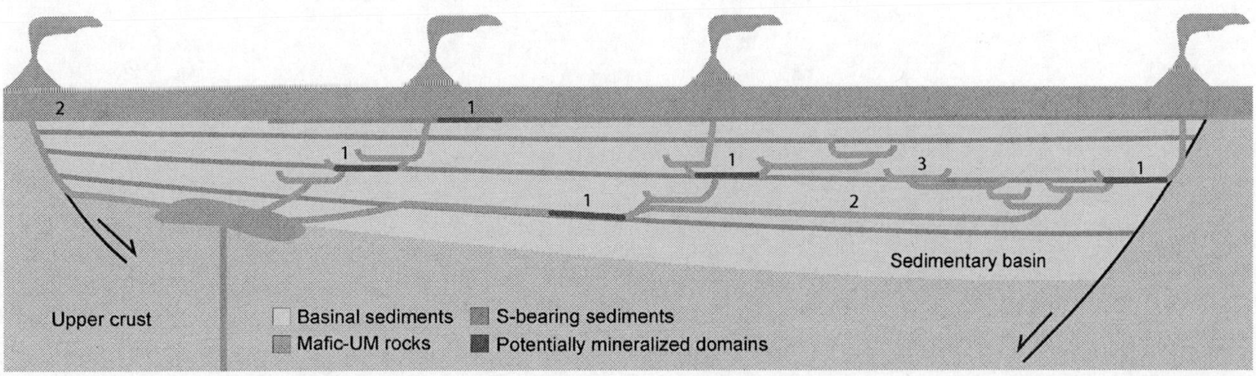

L1
Early Ore Deposit Observations and Advances by Waldemar Lindgren, A Rock-solid Foundation for Our Society

Jean S. Cline
University of Nevada Las Vegas, Las Vegas, NV, USA

During the 1990s and early 2000s, when Carlin researchers were trying to understand how Carlin deposits formed, difficulties, particularly the fine-grained nature of the mineralization, made ore characterization challenging. Carlin-type ores are distinct from other ores because they form replacement bodies, contain primary ionic gold in generally microscopic arsenian pyrite crystals, and exhibit alteration that is subtle but dominated by decarbonatization of silty calcareous rocks.

Ore mineralogy, textures, fluid inclusion studies, and numerical models indicate that gold did not precipitate in response to boiling or fluid cooling, but instead gold ions were captured by precipitating arsenian pyrite that formed rims on older pyrite crystals. Although a few studies determined pressure and temperature conditions during ore formation and sources of ore-fluid components, these studies did not converge on a genetic model and, instead, led to a proliferation of models that could be sorted into three classifications: (1) epizonal plutons contributed heat and possibly fluids and metals (Sillitoe and Bonham, 1990; Henry and Boden, 1998; Henry and Ressel, 2000); (2) meteoric fluid circulation resulting from crustal extension scavenged and precipitated metals with or without heat from widespread magmatism (Ilchik and Barton, 1997; Emsbo et al., 2003); and (3) metamorphic fluids from deep or mid-crustal levels, possibly with magmatism, transported and precipitated metals (Seedorff, 1991; Hofstra and Cline, 2000).

The difficulties in sorting out the genesis of Carlin-type deposits were related to the complex geologic history of northern Nevada and deposit features. Main ore-stage minerals (quartz, pyrite, illite, and dickite) are fine grained, typically microscopic, and volumetrically minor. Additionally, northern Nevada has undergone multiple diagenetic and hydrothermal events that also produced many of these same minerals, and these events were overprinted by or superimposed on the main-ore stage. Further complications include supergene alteration that oxidized some orebodies, contributing to misinterpretations about deposit genesis during the early years of mining. Bulk analyses of mineralized samples simply produced a mixture of several events. Analysis of mineral separates and microanalysis of pyrite, quartz, and fluid inclusions produced results related to the main-stage minerals; however, such analyses required time-intensive petrography to unravel mineral parageneses and to distinguish gold-related pyrite, quartz, and clay minerals from similar pre- or post-ore minerals. Perseverance and the continued study of newly discovered deposits eventually led to the recognition that these deposits formed at temperatures around 220°-240°C and that the deposits are genetically related to magmatism.

In hindsight, a read of early writings by Waldemar Lindgren dating to 1907, provide a clear origin for these deposits. In Lindgren's textbook *Mineral Deposits* he provides drawings of photomicrographs of jasperoid alteration identical to that typical of Carlin deposits. He further includes in this text his Classification of Mineral Deposits, first presented in 1911. We now know that Carlin-type deposits clearly fit the category of deposits with an "origin dependent on the eruption of igneous rocks......by hot ascending waters of uncertain origin, but charged with igneous emanations" at intermediate depths and temperatures of 200°-300°C. Lindgren called these deposits "mesothermal" deposits.

ST.098
Targeting Porphyry Copper Deposits at an Orogenic Belt Scale Using Zircon Trace Element Composition: Impact, Application, and Lessons Learned

Robert G. Lee, Alan J. Wainwright, Fabien Rabayrol, Eoin Cross, Farhad Bouzari, Craig J. Hart
MDRU - University of British Columbia, Vancouver, BC, Canada

Multiple studies have investigated zircon chemical composition to elucidate the temporal and chemical characteristics of magmas. Zircon has high potential as an indicator of magmatic fertility due to preferential inclusion of high field strength elements and rare earth elements such as Ce and Eu, which are used as proxies for the oxidation state of the parent magma. Zircons from magmatic rocks related to porphyry copper deposits typically have elevated Ce and Eu anomalies as well as anomalous trace element concentrations, which indicate increased fertility or the potential that the magma can produce mineralizing fluids. There is debate as to the magmatic physicochemical cause for these anomalies, whether they are related to the oxidation state, water content, or apatite-titanite fractionation. Additionally, the functionality of zircon chemistry as a method to indicate fertility and the direct application to belt-scale mineral exploration may not yet be fully utilized.

Ninety-eight samples collected from the western Tethyan orogenic belt across ~2,500 km from the Balkans to the Lesser Caucasus were selected to test the application of zircon composition as a method for targeting at the belt scale. Samples were collected throughout the belt from barren to copper-gold mineralized porphyry as well as high-sulfidation and intermediate-sulfidation epithermal deposits, including Rosia Montana, Romania; Bor, Serbia, Skouries, Greece; Kişladağ, Copler, and the Artvin district, Turkey, and the Adjara-Trialeti Belt, Georgia in addition to several pre-arc Jurassic to Paleozoic rocks. The composition of these samples includes calc-alkaline to alkaline diorite to monzonite intrusive and andesite to rhyolite extrusive rock types.

Zircons from these samples were analyzed using the LA-ICP-MS method and yielded >2,400 trace element and age analyses. Trace element compositions within individual samples typically show a high degree of variability that can lead to difficulty with interpretation. Using mean values for each of the samples as a first step, however, similar characteristics for mineralized centers in Cretaceous, Paleogene, and Neogene arc rocks are evident. Regardless of age, rock composition and location, the samples that are directly related to mineralization have mean Eu/Eu_N^* values >0.61, Ce/Ce_C^* >1,400, $(Ce/Nd_N)/Y$ >0.03, Yb/Gd_N >35, and Hf/Y >10. Calculated mean zircon temperatures for the mineralized samples tend to be low, from 730° to 650°C. Zircon from the barren Tethyan arc rocks and the older basement rocks generally have mean Eu/Eu_N^* values <0.40, Ce/Ce_C^* <1,000, $(Ce/Nd_N)/Y$ <0.01, Yb/Gd_N <30, and Hf/Y <10. Calculated zircon temperatures and total REE concentration are higher in the barren samples compared to the fertile ones. Mean zircon compositional data show clear differences between barren and mineralized populations. This method provides the first step to potentially identifying mineralized districts, which can be followed up through a more detailed assessment of the zircon composition and petrologic properties.

This project represents one of the most significant regional studies to apply the zircon method to assess the fertility potential of magmatic-hydrothermal porphyry-epithermal mineralized systems across an orogenic belt. Based on three to four samples collected at each site, compositional differences in zircon composition can be used as a fertility indicator.

ST.099

High Resolution vs. Standard Resolution: How an Increase in Spectral Resolution Using a Field Portable Spectrometer Affects Quality of Data – A Case Study on Nickel Exploration

Lori S. Pieniazek
Spectral Evolution, Haverhill, MA, USA

Standard resolution of a field-portable spectrometer at 3 nm (UV), 8 nm (VIS), and 6 nm (NIR) has been used successfully in the geological remote sensing and mining industry for a variety of applications. Alteration mapping has been key for finding different mineral assemblages that indicate an area of interest. Using a higher-resolution field spectrometer has proven to be beneficial in understanding and identifying new features in a variety of different mineral groups and subgroups.

A study was conducted using three different field spectrometers, each with different resolutions. As you increase resolution, features that are not visible with a standard resolution spectrometer appear in all three ranges (UV, VIS, and NIR), thus yielding a better understanding of alteration changes and geochemical conditions.

ST.100

Age and Rate of Accumulation of Volcanogenic Massive Sulfide Deposits: Case Study from the Lucky Strike Vent Field, Mid-Atlantic Ridge

Dennis Sánchez-Mora[1], John W. Jamieson[1], Mathilde Cannat[2], Thibaut Barreyre[3], Javier Escartín[4]
1. Memorial University of Newfoundland, St. John's, NL, Canada, 2. Institut de Physique du Globe de Paris, Paris, France, 3. Department of Earth Science/K.G. Jebsen Centre for Deep Sea Research, University of Bergen, Bergen, Norway, 4. Centre National de la Recherche Scientifique, Paris, France

One of the fundamental questions in ore deposit research is the rate at which mineral deposits form. Often, studies rely on geochronological methods to answer these questions. However, uncertainties of the analytical methods often exceed the timespan of deposit formation (e.g., application of U-Pb zircon dating).

Seafloor massive sulfide deposits offer an opportunity to study the rate of an ore-forming process in situ. Estimates of massive sulfide deposit volumes/tonnages that form at the seafloor are possible where high-resolution bathymetry has been collected. Volume estimates coupled with application of U-series dating to determine the ages of deposits provide constraints into the rate of formation of volcanogenic massive sulfide deposits.

In this study, we present U-series geochronology (^{226}Ra/Ba) of hydrothermal barite and deposit tonnage estimates from the mafic-hosted Lucky Strike vent field, an actively forming volcanogenic massive sulfide (VMS) deposit on the Mid-Atlantic Ridge. Radium-226/Ba dating suggests that venting at this site has been likely continuous for at least 6,500 years and evolved from a widespread, lower-temperature venting to more focused and higher-temperature fluid flow that resulted in the formation of sulfide/sulfate mounds and chimney structures at an average rate of ~200 t/year. This rate, combined with vent fluid composition and flux data, is used to calculate the efficiency of metal accumulation at the seafloor.

The accumulation rate estimates at Lucky Strike are comparable to other well-characterized mid-ocean hydrothermal sites (e.g., TAG and Endeavour). However, it is concentrated within a relatively small area of <2.5 km^2. The emplacement of a segment-centered volcano is likely a key process that has focused fluid flow and metal precipitation in a relatively small area. Results from Lucky Strike suggest that these mafic-dominated deposits can form in under 30,000 years.

G2
Gold Deposits of the Archean Abitibi Greenstone Belt, Canada

Benoît Dubé, Patrick Mercier-Langevin
Geological Survey of Canada, Québec, QC, Canada

The Neoarchean Abitibi greenstone belt in the southern Superior province has been one of the world's major gold-producing regions for almost a century with >6,100 tonnes (t) Au produced and a total endowment, including production, reserves, and resources (measured and indicated), of >9,400 t Au. A significant part of that gold is synvolcanic and/or synmagmatic and was formed during the volcanic construction of the belt between ca. 2750 and 2695 Ma. However, >60% of the gold is hosted in late, orogenic quartz-carbonate vein-style deposits that formed between ca. 2660 and 2640 ± 10 Ma, predominantly along the Larder Lake-Cadillac (LLCF) and Destor-Porcupine (DPF) fault zones. This ore-forming period coincides with the D_3 deformation, a broad north-south main phase of regional shortening that followed a period of extension and associated crustal thinning, alkaline to sub-alkaline magmatism, and development of fluvial-alluvial sedimentary basins (ca. 2680–2670 Ma). These sedimentary rocks are referred to, in the southern Abitibi, as Timiskaming-type. The tectonic inversion from extension to compression is <2669 Ma, the maximum age of the D_3-deformed, youngest Timiskaming rocks. In addition to the quartz-carbonate, vein-style, stockwork-disseminated, veinlet, and replacement-style mineralization is hosted in and/or associated with ca. 2683-2670 Ma, early- to syn-Timiskaming alkaline to sub-alkaline intrusions along major deformation corridors, especially in southern Abitibi. The bulk of such deposits formed late- to post-alkaline to sub-alkaline magmatism and the largest deposits are early- to syn-D_3 (ca. 2670–2660 Ma), whereas the bulk of the quartz-carbonate vein systems formed syn- to late-D_3. At belt scale, these illustrate a gradual transition in ore styles in orogenic deposits throughout the duration of the D_3 deformation event along the length of the LLCF and DPF. The sequence of events, although similar in all camps, was probably not perfectly synchronous at belt scale, but varied/migrated with time and crustal levels both along the main deformation corridors and from north to south.

The presence of high-level alkaline intrusions, which are spatially associated with Timiskaming-type conglomerate and sandstone, large-scale iron-carbonate hydrothermal alteration, and numerous gold deposits and prospects along the LLCF and DPF, indicates that these structures were deeply rooted and tapped auriferous metamorphic-hydrothermal fluids and melts from the upper mantle and/or lower crust, late in the tectonic evolution of the belt. The metamorphic-hydrothermal fluids, rich in H_2O, CO_2, and H_2S, were capable of leaching and transporting gold to the upper crust along the major faults and their splays. Although most magmatic activity along the faults predates the bulk of the gold, magmas have, in some cases, contributed fluids and/or metals to the hydrothermal systems. The exceptional gold endowment along the structural corridor defined by the LLCF and DPF may also suggest that these faults have tapped particularly fertile upper mantle-lower crust gold reservoirs. The concentration of large synvolcanic and synmagmatic gold deposits along the LLCF corridor supports the idea of a broad, gold-rich source(s) that may have contributed gold to the ore-forming systems at different times during the evolution of the belt.

G3
Metallogeny of the Neoarchean Malartic Gold Camp, Abitibi and Pontiac Subprovinces, Canada

Stéphane De Souza[1], Stéphane Perrouty[2], Benoît Dubé[3], Patrick Mercier-Langevin[3], Robert Linnen[4], Gema R. Olivo[5]
1. Université du Québec à Montréal, Département des sciences de la Terre et de l'atmosphère, Montréal, QC, Canada, 2. Laurentian University, Goodman School of Mines, Harquail School of Earth Sciences, Mineral Exploration Research Centre, Sudbury, ON, Canada, 3. Geological Survey of Canada, Québec, QC, Canada, 4. Western University, Department of Earth Sciences, London, ON, Canada, 5. Queen's University, Kingston, ON, Canada

The Neoarchean Abitibi greenstone belt hosts most of the gold resources in the Superior Province. Many of the gold camps of the belt are located along its southern margin and contact with the Pontiac subprovince, i.e., along the Larder Lake-Cadillac fault zone (LLCfz, Fig. 1A). This includes the Malartic camp and the Canadian Malartic deposit (10.5 Moz Au in past-production: 1935–1984 and 2011–2019; 16 Moz in reserves and resources), which is currently mined as a low-grade open-pit operation (~1 g/t Au; Fig. 1B). Recent work in the Malartic camp has led to an in-depth documentation of the Canadian Malartic deposit geology and footprint and to an improved understanding of ore-forming processes in the southern Abitibi and northern Pontiac subprovinces.

The Canadian Malartic deposit comprises disseminated-stockwork replacement-style mineralization in greenschist facies 2685–2682 Ma turbidites of the Pontiac subprovince and ~2678 Ma subalkaline to alkaline porphyritic quartz monzodiorite and granodiorite (Fig. 1C). The emplacement of these intrusions was coeval with the Timiskaming episode of fluvial-alluvial sedimentation (<2679 to <2669 Ma), prior to main-phase regional deformation (D_2) and associated Barrovian metamorphism (ca. 2665–2650 Ma). The mineralization occurs along two trends subparallel to the NW-striking S_2 cleavage and the axial surface of F_2 folds, and to the south-dipping Sladen fault. The syn-D_2 ductile-brittle Sladen fault is mineralized along a strike length of ~5 km. The orebodies are zoned outward and most of the higher-grade ore (>1 g/t Au) was deposited as a result of iron sulfidation, Na-K metasomatism, and formation of carbonates. Mineralization was dated using gold-associated molybdenite (Re-Os) and titanite (U-Pb) that yielded ages of ~2664 and ~2661 Ma, respectively. These ages are within error of the ~2660–2658 Ma age for peak metamorphism in greenschist facies metasedimentary rocks of the Pontiac Group south of the mine. Mineralization occurred ~15 Ma after the crystallization of the intrusions that host part of the ore and postdates the Timiskaming episode of sedimentation.

Smaller deposits (<1.7 Moz) in the camp include auriferous disseminated-stockwork zones, quartz ±carbonate-pyrite veins, and breccias along faults in chemically and mechanically favorable rocks. The age of these deposits is poorly constrained, but U-Pb zircon dating of post-mineral dikes (~2694–2692 Ma) highlights an early phase of mineralization. Field relationships suggest that the formation of these small deposits was incremental and can be bracketed between the end of volcanism (~2698 Ma), and the waning stages of D_2 deformation (~2640 Ma). Nevertheless, the bulk of the gold in the camp is concentrated in the Canadian Malartic deposit, which formed along the LLCfz and second order faults. Mineralized structures are compatible with north-south shortening and regional metamorphism, and the Canadian Malartic deposit is therefore attributed to the orogenic class of gold deposits. Hence, the timing, location, and nature of the mineralized structures along the LLCfz have played a major role in concentration of the gold endowment.

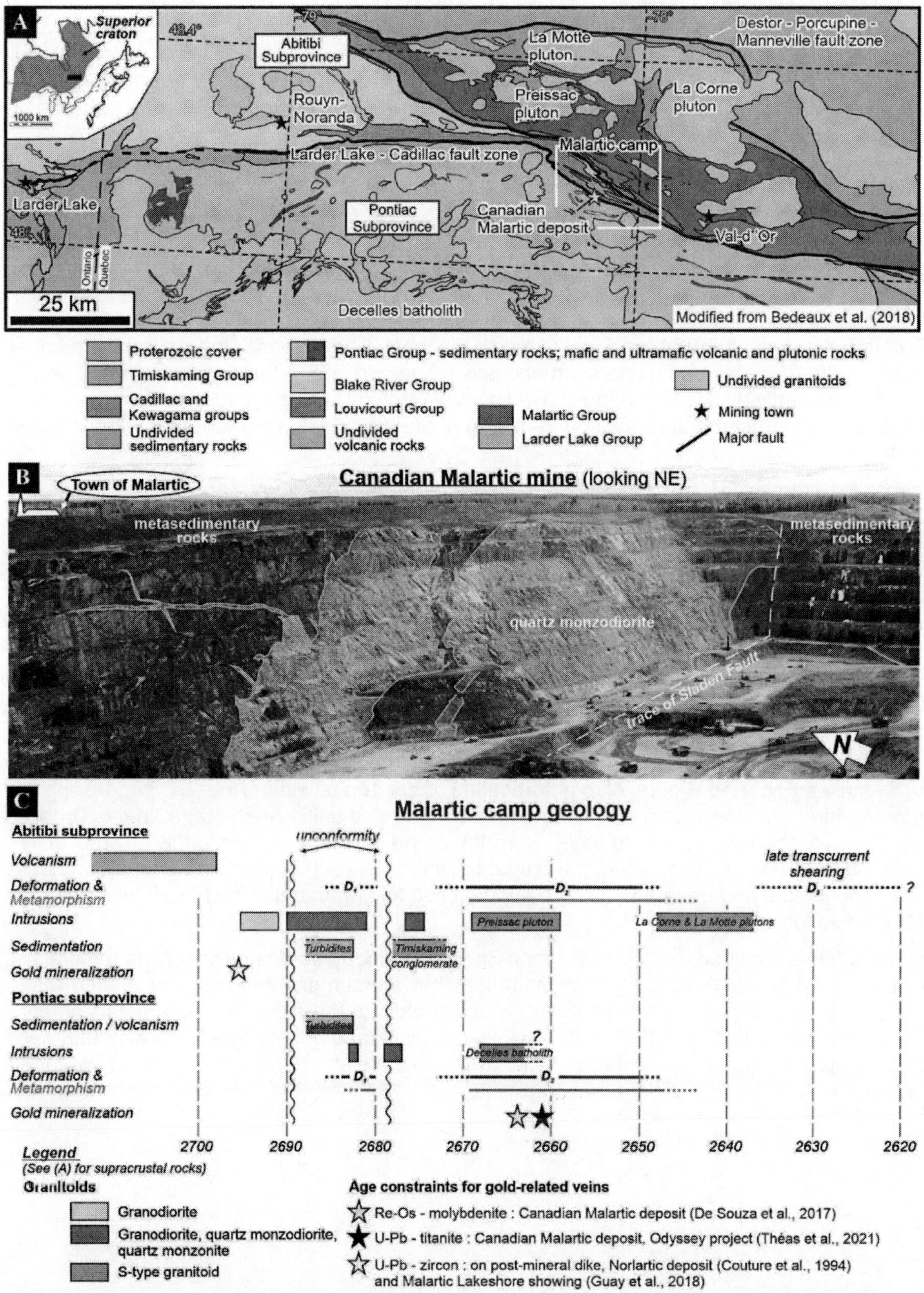

Figure 1: A) Map of the Abitibi and Pontiac subprovinces and location of the Malartic camp; B) Photograph of the Canadian Malartic mine; C) Geological summary of the Malartic camp.

G4
High-Grade Gold Mineralization at the Fenelon Deposit: Going Undercover to Study a Major Discovery in the Abitibi Greenstone Belt

Evan Slater[1], Anthony E. Williams-Jones[1], Sébastien Castonguay[2]
1. McGill University, Montréal, QC, Canada, 2. Geological Survey of Canada, Québec, QC, Canada

The Fenelon gold system is one of the most important mineral discoveries made in recent years in the Abitibi greenstone belt and this has re-ignited exploration in the northwestern Abitibi. The deposit lies north of the Sunday Lake Deformation Zone, a major east-west structure that controlled the emplacement of the world-class Detour Lake deposit (~25 Moz Au) ~70 km to the west. The belt and deposit area are covered extensively by till and, thus, continued exploration success hinges on understanding the controls on mineralization at this recently discovered deposit. Preliminary geologic and genetic models for Fenelon are presented here based on structural analysis and sampling of oriented drill core integrated with airborne geophysical data.

The deposit consists of three main ore zones that occur in shear corridors and narrow veins hosted in different types of host rocks. The Tabasco-Cayenne zone (e.g., 22.73 g/t over 48.01 m), hosted mainly within argillite and its contact with the 2697 Ma Jeremie diorite, defines the core of the known deposit. It consists of ESE-striking and steeply dipping, shear-hosted, sulfide-rich mineralization associated with silica flooding and intense sericite alteration. The Gabbro zones (e.g., 260.44 g/t over 7.02 m) have similar orientations but are restricted to a polyphase mafic sill of calc-alkaline affinity. The Area 51 zone is hosted within the Jeremie diorite, in which early centimetric auriferous quartz veins with local sulfides including rare molybdenite are overprinted by a second gold event consisting of chlorite and sulfides closely associated with shearing.

Oriented drill core show two foliations with different orientations. S_1 strikes ESE and is parallel to parts of the Tabasco-Cayenne and Gabbro zones, suggesting it was an important pre-existing control on shearing and fluid flow. S_2 strikes ENE and is parallel to gold-hosting shear zones within Area 51, suggesting it was also an important control on mineralization, at least for its chlorite and sulfide-rich stage. In the Tabasco-Cayenne zones, D_2 deformation appears to have overprinted and locally remobilized the S_1-parallel mineralization into high-grade ore shoots along the intersection of S_1 and S_2. Bedding orientations derived from airborne magnetic data and oriented drill core suggest two folding events which may have been responsible for the two foliations.

The gold mineralization is enriched in As, Ag, Cu, and Zn with chalcopyrite being the sulfide mineral correlating best with gold grade. Petrographic examination shows that native gold co-precipitated with pyrrhotite, chalcopyrite, arsenopyrite, and sphalerite, which speaks to the elevated base metal content of the ore fluid(s). Preliminary results suggest that Fenelon is a polyphase gold deposit with elevated base metals, which deposited in various types of host rocks over two main structurally controlled ore stages that may correlate temporally with multiple deformation events.

G5

Postsubduction Porphyry Cu-Au Emplacement During Transtension in Northwestern British Columbia (Canada): The Norm Rather Than the Exception?

Bram I. van Straaten[1], JoAnne Nelson[1], Lori Kennedy[2], Gayle Febbo[3], Emily Miller[2]
1. British Columbia Geological Survey, Victoria, BC, Canada, 2. University of British Columbia, Vancouver, BC, Canada, 3. Exploration Petrology Inc., Terrace, BC, Canada

Most porphyry Cu±Au±Mo systems are postulated to have formed under broadly contractional stress regimes marked by crustal thickening, surface uplift, and rapid exhumation in a synsubduction tectonic setting. We present evidence that most porphyry Cu-Au systems in the prolific Golden Triangle of northwestern BC were emplaced during transtension in a postsubduction tectonic setting. Such a structural and tectonic arrangement is probably more common globally than previously appreciated.

The Golden Triangle is a world-class mineral district containing seven >1 Gt porphyry Cu-Au deposits, five >1 Moz Au epithermal Au deposits, and numerous smaller deposits; it includes two operating mines. The district is in the Stikine terrane, one of the largest (2,000 x 300 km) island-arc terranes of the Canadian Cordillera that accreted to ancestral North America in the mid-Jurassic. However, significant preaccretionary mineral deposits within the terrane are restricted to a 300-km-long N-trending western belt (Golden Triangle) and a 100-km-long N-trending eastern belt (Toodoggone), formed during several distinct Late Triassic to Middle Jurassic metallogenic epochs. Most porphyry systems are coeval with the local onset of latest Triassic to Early Jurassic lower Hazelton Group volcanism, with magmatism hitherto ascribed to opposing subduction underneath Stikinia. The area north of these belts (Stikine arch) underwent latest Triassic to Early Jurassic uplift and erosion, and lacks lower Hazelton Group strata, their magmatic roots, and coeval mineral deposits.

We present detailed studies of Cu and/or Au systems at Bronson, KSM, Tennyson, Red Mountain, and Big Bulk (~204-190 Ma, latest Triassic to Early Jurassic). Our work shows that cogenetic intrusions are commonly tabular shaped and parallel to sheeted porphyry-style veins, indicating strong structural control during magmatic-hydrothermal deposit formation. The presence of fault-proximal megaconglomerates and rapid facies variations suggest lower Hazelton Group deposition was controlled by penecontemporaneous basin-bounding faults. Within the district, east-striking faults show normal movement, whereas north-striking faults show dextral strike-slip kinematics, suggesting porphyry and related deposits were emplaced into active transtensional pull-apart basins. Many of these nearly orthogonal faults likely represent reactivated pre-Devonian basement structures. Cretaceous Skeena fold-and-thrust-belt deformation overprints and inverts preexisting structures.

A postsubduction model for the Golden Triangle during the latest Triassic to Early Jurassic, starting ~207 Ma, is supported by 1) thin volcano-sedimentary successions exhibiting high structural control, 2) preceding contractional deformation (~215-208 Ma folds, thrusts, angular unconformity), 3) coeval uplift along the adjacent Stikine arch, 4) the Au-rich nature of mineralization and association with the onset of volcanism (both common features of postsubduction mineralization worldwide), and 5) geometric problems with long-lived opposing subduction zones. Transtension across preexisting translithospheric structures likely provided the mechanism and pathway for magma generated by remelting of previously subduction-metasomatized subarc lithosphere. Exploration programs should evaluate the potential of district-scale lineaments and possible dilational jogs or step-overs for hosting porphyry and related mineralization.

Globally, numerous postsubduction porphyry systems formed in extensional or transtensional settings (e.g., Cadia, Grasberg, Apuseni district), and we suggest that—particularly for porphyry deposits emplaced in postsubduction settings—a transtensional setting may be the norm, rather than the exception.

G6

The High-grade Gold Ore Paradox of the Brucejack Deposit: Insights from Nanoscale Imaging of Electrum and High-resolution Trace Element and Sulphur Isotope Analyses of Pyrite

Duncan F. McLeish[1], Anthony E. Williams-Jones[1], Olga V. Vasyukova[1], James R. Clark[1], Warwick S. Board[2], Richard A. Stern[3], Octavia E. Bath[2], Joel E. Ashburner[2]
1. McGill University, Montreal, QC, Canada, 2. Pretium Resources Inc., Vancouver, BC, Canada, 3. University of Alberta, Edmonton, AB, Canada

It has been long accepted that epithermal gold deposits form as a result of gold precipitation from hydrothermal fluids in which gold is present as dissolved species. A major weakness of this hypothesis is that it fails to explain the formation of ultra-high-grade or bonanza gold veins. Gold concentrations in the fluids responsible for epithermal mineralisation are typically on the order of 10-30 ppb, which are far too low to explain the occurrence of bonanza-grade epithermal deposits like Brucejack, where drilling has returned grades up to 41,582 g/t Au over 0.5-m intervals. Formation of such high-grade veins by direct precipitation of native gold or electrum from the ore fluids would require that fractures remain open for unreasonably long periods of time or that fluid fluxes be extraordinary.

A potential solution to the paradox of ultra-high-grade gold deposition, in geologically realistic time-frames, is offered by colloidal transport. We have studied the high-grade gold mineralisation from the Valley of the Kings (VOK) and Eastern Promises (EP) zones of Brucejack, a large and exceptionally high-grade, intermediate-sulphidation epithermal gold-silver deposit in British Columbia's Golden Triangle, to evaluate this possibility. Images obtained using transmitted electron microscopy show that: (1) gold commonly occurs as <1-10-nm spherical nanocrystals of electrum embedded within a calcite matrix; (2) larger (100-500 nm) particles of electrum, also embedded in calcite, are composed of hundreds of nanoparticles, each displaying distinct crystal lattice plane orientations; and (3) the margins of >1-μm-wide electrum masses surrounding the calcite matrix comprise nanoparticulate electrum partially crystallised to massive monocrystalline electrum. These images provide compelling evidence for the formation of colloidal suspensions of electrum and their flocculation. Analyses using LA-ICP-MS and EMP-WDS of the trace element composition of pyrite from phyllically-altered wallrock in the VOK and EP zones have identified a trend of increasing enrichment of Co, Ni, Se, and Cu with depth, documenting an otherwise cryptic transition to recently discovered propylitic and potassic alteration and associated Cu-Mo mineralisation in deep drill core from the adjacent Flow Dome Zone. Strong positive shifts in $\delta^{34}S_{pyrite}$ values (>20.0‰) from in situ SIMS analysis of VOK hydrothermal pyrite suggest that seawater mixed with hydrothermal fluids during the late stages of ore formation contemporaneous with the deposition of electrum. We present a model in which the high-grade epithermal gold mineralisation is due to the development of a boiling-induced colloidal gold suspension in a carbonate fluid originating from a porphyry source at depth and its flocculation during the incursion of seawater. According to this model, colloidal and flocculated gold particles are mechanically transported into networks of nano-veinlets that they block, producing millimetric to centimetric knots of electrum in centimetric carbonate veins. Our model offers a solution to the longstanding problem of high-grade gold transport and deposition in epithermal vein systems, as well as an explanation for the extraordinary and challenging grade variability, or "nugget effect," often encountered during the mining of these deposits.

G7

Magmatic Origins of Carlin-type Gold Indicated by NanoSIMS Sulfur Isotope and Trace Element Depth Profiling

Elizabeth A. Holley[1], Anne Fulton[1], Christie E. Jilly-Rehak[2], Michael J. Pribil[3], Craig A. Johnson[3]
1. Colorado School of Mines, Golden, CO, USA, 2. Stanford Nano Shared Facilities, Stanford, CA, USA, 3. Geology, Geochemistry, and Geophysics Science Center, U.S. Geological Survey, Denver, CO, USA

Nevada's Carlin-type deposits represent the largest gold accumulation in the northern hemisphere, but throughout 55 years of mining, their origins have been disputed. Did meteoric fluids scavenge and then redeposit the gold while circulating through the sedimentary host rock package? Or did the gold come from magmatic fluids during the southwest sweep of calc-alkaline magmatism through the region in the Eocene?

"Invisible" Carlin-type gold occurs most commonly in solid solution within micron to submicron-scale hydrothermal arsenian pyrite rims, overgrowing older sedimentary or magmatic pyrite grains. The lack of abundant alteration minerals and the small size of the ore grains have made it difficult to study the processes responsible for mineralization. Sulfur isotopes have long been considered the key to fingerprinting the source of the gold, but previous quantitative studies lacked the spatial resolution necessary to distinguish between the isotopic signatures of precursor pyrite and the thin, gold-rich hydrothermal pyrite overgrowths. We leverage advances in nano-scale secondary ion mass spectrometry (NanoSIMS) and laser ablation multi-collector inductively coupled plasma mass spectrometry (LA-MC-ICPMS) to produce and validate sub-micron, quantitative, standardized $\delta^{34}S$ and trace element data from five Carlin-type gold deposits in Nevada, as well as unaltered magmatic pyrites from Eocene dikes on the Carlin Trend.

At the Carlin, Getchell, and Turquoise Ridge deposits, ore pyrite is hosted in carbonaceous, realgar and orpiment-bearing limey mudstones and dolomitic limestones that have been decarbonatized during alteration. Most of the Au-poor precursor pyrite is isotopically heavy (22.0 to 54.4‰ $\delta^{34}S_{VCDT}$), consistent with formation during diagenesis of carbonaceous, iron-poor sediments beneath a marine basin that may have been closed. At Deep Star, the Au-poor cores of the ore grains are isotopically moderate (6.5 to 6.9‰) Jurassic pyrite from the hornfels aureole of the Goldstrike stock, similar to the mean $\delta^{34}S$ of Jurassic magmatic sulfur in the Great Basin. Unaltered Eocene magmatic pyrites hosted in dikes at Beast, Deep Star, and Betze Post contain minor Au and show variable $\delta^{34}S$ (7.9 to 15.2‰), within the range of Eocene magmatic sulfur in the Great Basin.

In the hydrothermal arsenian pyrite overgrowths, $\delta^{34}S$ varies systematically with Au. Submicron-scale zones with low Au concentrations show $\delta^{34}S$ values similar to the precursor pyrite hosting the overgrowths. In contrast, submicron-scale zones with high Au concentrations show $\delta^{34}S$ values as low as 4.2‰ at Deep Star, 2.5‰ at Carlin, 2.1‰ at Beast, 1.7‰ at Getchell, and 1.2‰ at Turquoise Ridge. These values are similar to those of porphyry-related sulfide minerals in the Battle Mountain district, the nearest location of definitively magmatic-hydrothermal Eocene mineralization.

We propose a model for Carlin-type gold formation wherein the gold was sourced from Eocene magmas isotopically and compositionally similar to the Beast dike. Fluids exsolving from the cooling magma bodies became enriched in Au. Fluid circulation in the magmatic-hydrothermal environment resulted in variable degrees of mixing with other sulfur sources in the host rock package. When these compositionally-evolving fluids ascended into favorable lithologic horizons, they precipitated the hydrothermal pyrite in sequential nano-scale zones on existing pyrite.

G8

Hydrothermal Alteration Mineralogy, Zoning and Paragenesis at the Low-Sulfidation Epithermal Cerro Blanco Deposit, Guatemala

Ekaterina Savinova[1, 2], Steffen G. Hagemann[1]
1. The University of Western Australia, Crawley, WA, Australia, 2. The University of Queensland, Indooroopilly, QLD, Australia

The Cerro Blanco deposit in southeast Guatemala is a classic hot spring-related low-sulfidation epithermal system with gold resource of 3 million ounces (61 million tonnes grading 1.5 g/t Au in Measured and Indicated categories), within a 400-m-wide and 900-m-long area. Hydrothermal alteration and mineralization of the Cerro Blanco deposit, mineral chemical composition, and textural and overprinting relationships are successfully characterized using a combination of hyperspectral imaging spectroscopy in the VIS-NIR-SWIR wavelength range, petrography, SEM, and EDS studies.

High-grade gold mineralization (>5 g/t Au) is primarily hosted in the Tertiary Mita Group sedimentary and volcaniclastic rocks within 60 currently defined veins that form two upward-flaring swarms converging downwards and merging into basal feeder veins. The overlying Salinas Group locally hosts low-grade (0.5–2 g/t Au), disseminated gold mineralization in silicified polymictic conglomerate and volcaniclastic rocks intercalated with siliceous sinter horizons formed by hot-spring activity. The Cerro Blanco deposit is located within a north-south striking graben. The East Fault separates the Au-Ag-bearing quartz-adularia veins from the Mita geothermal field with present-day active hot springs. The overall shape of the orebody is related to the geometry of pre-mineral faults, associated hydrothermal veins, and alteration zones.

Veins are dominated by crustiform, colloform, and amorphous quartz with associated adularia and calcite. Results of $^{40}Ar/^{36}Ar$ geochronology analysis on vein-hosted adularia indicate an age of 4.93 ± 0.47 Ma. The pristine nature of the veins at Cerro Blanco is explained by the young age of the deposit and lack of post-mineral faulting.

The main ore minerals are electrum (<1 mm) and gold as native inclusions (<500 μm) in pyrite, chalcopyrite, and acanthite. Selenides, sphalerite, and galena are sparse but also present. Two paragenetic stages of vein formation and hydrothermal alteration are defined. The main auriferous veins consist of crustiform and colloform bands that are characterized by paragenetic stage 1 equilibrium assemblage of quartz(chalcedony)-adularia-calcite-ankerite. Sulfides are located in ginguro bands consisting of fine-grained acanthite, with minor pyrite, chalcopyrite, and tetrahedrite. Stage 2 of the paragenesis is characterized by intense overprinting of the quartz-adularia-calcite-ankerite veins by montmorillonite and interstratified illite, and quartz replacement of bladed calcite. The bladed calcite veins are characteristically devoid of sulfides and gold. Hydrothermal alteration in the proximal zone of the sedimentary and volcaniclastic wall rocks is characterized by paragenetic stage 1 with quartz-adularia-illite-calcite and locally montmorillonite. The distal alteration zone is marked by quartz-illite-chlorite-calcite.

Wall rock-hosted illite suggests a temperature of formation >230°C. The crustiform and colloform-banded texture of quartz-adularia and carbonate veins, the location of native gold in stage 1 veins, lattice-bladed calcite and associated silica replacement are compatible with episodic fluid boiling and drop in the total confining pressure. The combination of these mineralogical and textural features suggests hydrothermal boiling is the principal precipitation mechanism for gold, silver, adularia, and bladed calcite.

The results of this hydrothermal alteration study carry significant implications for exploration, including targeting for extension of high-grade (>5 g/t Au) gold veins in the Cerro Blanco deposit and nearby concealed orebodies and new epithermal deposits in the region.

G9
Gold and Copper Fertility in Porphyry Systems: Insights from Sulfide Inclusions

Ariadni A. Georgatou[1], Massimo Chiaradia[1], Andreas Audétat[2], Frances Jenner[3], Jung-Woo Park[4]
1. Earth and Environmental Sciences,University of Geneva, Geneva, Switzerland, 2. Bayerisches Geoinstitut, Universität Bayreuth, Bayreuth, Germany, 3. School of Environment, Earth and Ecosystem Sciences, The Open University, Milton Keanes, United Kingdom, 4. School of Earth and Environmental Sciences, Seoul National University, Seoul, Korea

Although Cu and Au porphyry deposits show similar mineralisation and alteration styles, they occur in different geodynamic situations, with Cu porphyries being mostly associated with subduction arc-magmas and Au-porphyries with post-collisional settings. As a result, both porphyry types are characterised by fundamental differences regarding the petrogenesis, geochemistry, and fertility of the associated magmas. A recent study shows that while Cu-rich deposits linked to high calc-alkaline magmas require large volumes of hydrous magmas to form, Au-rich porphyry deposits associated with alkaline rocks form due to a more efficient Au precipitation. Moreover, magma fertility in porphyry deposits has also been attributed to remobilization, assimilation, and dissolution of sulfide-rich cumulates stored at depth, leading to the release of metals in the later exsolving hydrothermal ore fluid. In fact, studies looking at whole-rock PGE, Cu/Au, and Cu/Ag show that it is possible to determine the timing of sulfide saturation and thus to track the evolution of Cu and Au in the residual magmas.

We studied magmatic sulfides occurring in volcanic rocks from different geodynamic settings of barren and mineralised areas hosting Cu or Au porphyry deposits in order to investigate the role of magmatic sulfide saturation with respect to the fertility of the ore-forming systems. The volcanic areas investigated are the Ecuadorian volcanic arc including (among other volcanoes) Yanaurcu and Pichincha and hosting the Llurimaga Cu-rich porphyry, the Western Anatolian volcanic belt including Beydagi and hosting the giant Au-porphyry Kışladağ, and Konya volcanic belt, hosting the high-sulfidation Au epithermal deposit of Inlice and Doğanbey Cu porphyry. The findings are compared to Kula-Western Anatolia and Nisyros-Greece barren volcanic areas, representing intraplate and subduction geodynamic settings, respectively. Our results indicate that magmas from all areas underwent sulfide saturation independently of geodynamic setting, magma composition, and whether the system produced an economic deposit. In addition, while early saturating sulfides hosted by early crystallising phenocrysts are found in all study areas and are characterised by low Cu and Au abundances (EPMA-$Cu_{(med)}$ = 0.46 wt % and LA-ICP-MS-$Cu_{(med)}$ = 1.3 wt %, $Au_{(med)}$ = 0.089 ppm), late sulfides ($Cu_{(med)}$ = 32-56 wt %) hosted exclusively by magnetite are observed only in andesitic–dacitic and oxidised lavas of Yanaurcu, Pichincha, Beydagi, and Konya. $Cu/Ag_{(med)}$ values of sulfides (LA-ICP-MS) and whole rock (ID-ICP-MS) in Konya are 2,580-7,830 and 450-475, respectively, with the latter being representative of continental crust values (ca. 500). Lastly, PGE contents of whole rock confirm that all areas are extremely depleted, with $Pd/Pt_{(med)}$ of Konya and Beydagi being 1.61 and 1.53, respectively (compared to Pd/Pt = 7-60 for porphyry Cu-Au). These results suggest that all areas are characterised by similar and low initial Cu, Au, and PGE contents, with no pre-enrichment at the source needed, and show no correlation with the type and extent of mineralization. Regardless of the low metal contents caused by either early metal removal in sulfide-rich cumulates or metal-poor magma sources, "normal" magmas with "normal" chalcophile element contents are capable to form an economic deposit.

G10
Metallogeny of the Hod Gold Corridor, Eastern Pontides Belt, Turkey: Transitions from VMS to Porphyry-Epithermal Environments

Fabien Rabayrol[1, 2], Alan J. Wainwright[2], Robert G. Lee[2], Craig J. Hart[2], Robert A. Creaser[3], Alfredo Camacho[4]
1. Equity Exploration Consultants Ltd., Vancouver, BC, Canada, 2. Mineral Deposit Research Unit, The University of British Columbia, Vancouver, BC, Canada, 3. Department of Earth and Atmospheric Sciences, University of Alberta, Edmonton, AB, Canada, 4. Department of Geological Sciences, University of Manitoba, Winnipeg, MB, Canada

Porphyry and VMS deposits rarely form concurrently and in the same geological domain along convergent margin environments. Their temporal and spatial association indicates that contrasting ore-forming processes occurred in the evolving tectonic settings. The Artvin Au-Cu district is one of the major clusters of VMS bimodal-felsic, porphyry, and epithermal deposits in the Eastern Pontides Belt in northeast Turkey. Whereas ore-forming processes, timing, and tectonic setting of VMS mineralization are well defined, those for gold-rich porphyry and epithermal mineralization remain less constrained. The ~110-km-long Hod Gold Corridor in the Artvin district is defined by the northeast trend of several recent Au ± Cu mineral discoveries (~205 t Au; ~0.33 Mt Cu) including Hod Maden, Ardala-Salinbaş, and Taç-Çorak. These deposits include gold-rich porphyry, high- and intermediate-sulfidation epithermal, carbonate-replacement, and hybrid VMS-epithermal mineralization styles.

We present new U-Pb, $^{40}Ar/^{39}Ar$, and Re-Os geochronological results, interpreted with previously-compiled data, constrains Artvin district magmatic events to the Carboniferous (358-325 Ma), Jurassic (182-174 Ma), Late Cretaceous (92-78 Ma), Eocene (51-40 Ma), and Oligocene (30 Ma). Porphyry and epithermal mineralization along the Hod Gold Corridor peaked in the Early (~113 Ma; Berta prospect), Late Cretaceous (~88-82 Ma; e.g., Taç and Çorak deposits), and Eocene (~50 Ma; e.g., Ardala deposit), whereas VMS bimodal-felsic mineralization was restricted to the Late Cretaceous (~89-83 Ma). Therefore, we interpret that the Hod Gold Corridor was a long-lived, deep, crustal-scale structural feature along which the successive magmatic and mineralization events were emplaced.

The Late Cretaceous precious and base metal mineralization event is genetically associated with the emplacement of the felsic volcanic units of the Kızılkaya Formation (~89-83 Ma). Although this stratigraphic horizon is most well-known as host for VMS bimodal-felsic Cu-Zn mineralization along the Eastern Pontides Belt, we show that contemporaneous high- and intermediate-sulfidation epithermal Au ± Cu mineralization formed in it along the Hod Gold Corridor (e.g., Taç and Çorak deposits). Epithermal features and gold enrichments are increasingly pronounced and elevated eastwards across the Artvin district (Fig. 1). We interpret that the Cu-rich VMS (e.g., Cayeli and Murgul deposits) and Au-rich epithermal-style deposits represent two spatially and tectonically distinct end-members that formed during the same magmatic-hydrothermal event in the back-arc and arc settings, respectively. The eastward shallowing of the sea level and increased contributions from magmatic fluids are two possible first-order factors that could account for the contemporaneous variations of mineralization styles and metal associations.

This study examines the metallogenic responses to the spatial transition from back-arc to arc settings, contrasting geological domains, and implications for gold endowment in magmatic-hydrothermal systems in submarine environments.

Figure 1: Schematic, northwest-southeast, tectono-magmatic cross section of the Artvin district in the Late Cretaceous (~89-83 Ma) that shows the variation of mineralization style (i.e., from VMS bimodal-felsic to porphyry-epithermal) and metal association during the emplacement of the Kızılkaya Formation.

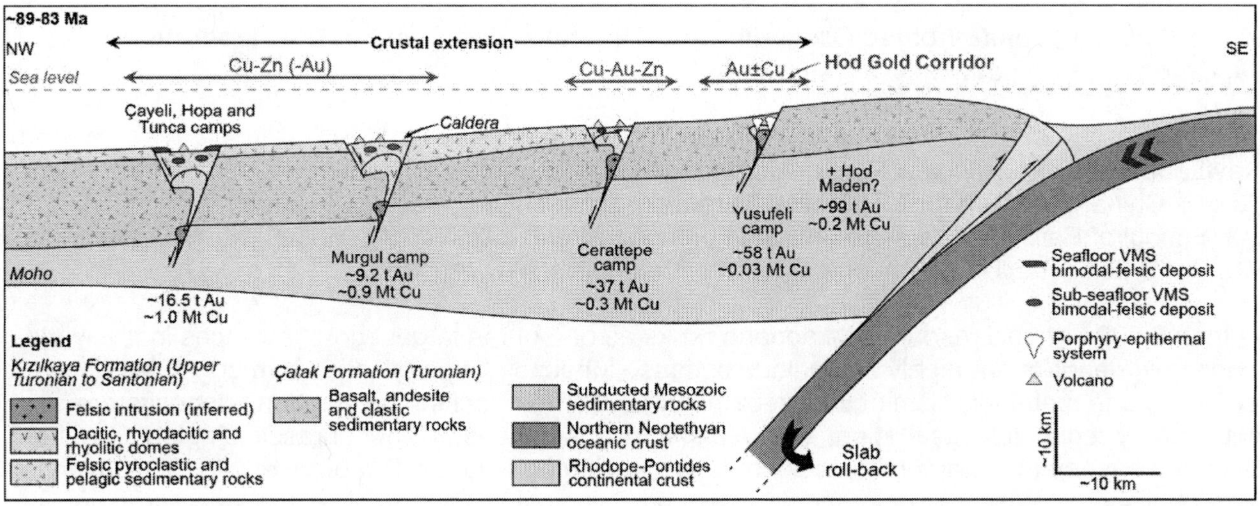

G11

A Focused ca. 120 Ma Granite-hosted Orogenic Gold Mineralization Event in the Jiaodong Peninsula, China

Liang Zhang[1], Li-Qiang Yang[1], David I. Groves[2], Roberto Weinberg[3], Yue Liu[4]
1. State Key Laboratory of Geological Processes and Mineral Resources, China University of Geosciences, Beijing, Beijing, China, 2. Centre for Exploration Targeting, University of Western Australia, Perth, WA, Australia, 3. School of Earth, Atmosphere and Environment, Monash University, Melbourne, VIC, Australia, 4. School of Earth Sciences, University of Melbourne, Melbourne, VIC, Australia

With more than 5,000 t of gold resources, Jiaodong ranks as one of the largest gold provinces in the world. The Jiaodong gold deposits are highly anomalous because Mesozoic mineralization formed ~1.8 billion years after high-grade metamorphism of the Precambrian basement rocks. The Jiaodong deposits are mainly controlled by regional faults that are sited along contacts between Upper Jurassic Linglong granites and Precambrian basement metamorphic rocks or Lower Cretaceous Guojialing granites. A critical overview of previously published isotope age data and new geochronological studies resolves the long-standing controversy concerning the temporal and genetic relationship between gold mineralization and gold-hosting (and other) granites. Most published reliable zircon U-Pb ages of the gold-hosting Linglong and Guojialing granites, representing the most voluminous magmatism, range from 166 ± 5.0 to 149 ± 2.0 Ma and 133 ± 3.0 to 126 ± 0.6 Ma, respectively. The Aishan granite, which is barren of gold-only deposits, yields zircon U-Pb ages of 120–108 Ma. The recent reliable hydrothermal muscovite $^{40}Ar/^{39}Ar$ and monazite U-Pb ages of ca. 120 Ma accord with previous age constraints of 123 ± 1–117 ± 3 Ma obtained mainly by robust $^{40}Ar/^{39}Ar$ dating of hydrothermal muscovite and U-Pb ages of hydrothermal zircon. These concordant ages constrain the timing of the major gold mineralization event across the Jiaodong province at 120 ± 2 Ma (arithmetic mean), although earlier or later insignificant gold mineralization events cannot be ruled out. This major ca. 120 Ma gold mineralization event is also consistent with constraints provided by crosscutting relationships between pre- and post-ore granitic and lamprophyric dikes and the gold deposits. A genetic relationship between the gold-hosting Linglong and Guojialing granites and mineralization is excluded because of the age gaps in their timing and by crosscutting relationships of auriferous quartz veins with the granites. The majority of Aishan granites are significantly younger than the gold mineralization event, and, although several of their U-Pb ages overlap the 120 ± 2 Ma age of gold mineralization, the fact that they show no spatial relationship to the gold deposits argues strongly against a genetic relationship. The Jiaodong gold deposits are structurally controlled, lack high-T alteration assemblages and metal zonation, and share geochemical and isotopic compositions with orogenic deposits, indicating that they are orogenic rather than intrusion-related deposits. Together with the reliable mineralization ages from robust isotope methods, cooling ages of gold-hosting granites demonstrate that these granites provided P-T conditions close to the ductile-brittle transition and fluid trap-sites for mesozonal orogenic gold mineralization. This enabled the effective ingress of fluids, leading to intensive brecciation, alteration, and formation of both vein-type and disseminated gold ores at ca. 120 Ma.

G12
Assessment of Gold Systems in Brazil: Major Deposits, Camps, and Future Exploration Potential

Lydia M. Lobato[1], Marco A. Costa[2], Steffen G. Hagemann[3], Rosaline C. Figueiredo e Silva[1]
1. Universidade Federal de Minas Gerais, Belo Horizonte, MG, Brazil, 2. Exploration Outcomes, Nova Lima, MG, Brazil, 3. The University of Western Australia Study Research About UWA News Community, Perth, WA, Australia

Gold has had a significant historical role in Brazil since discoveries of deposits by so-called *bandeirantes,* initially in Minas Gerais (MG) and Bahia, then Goiás and Mato Grosso states. The famous, world-class Morro Velho (Quadrilátero Ferrífero (QF) Province, MG) was the largest underground mine in Brazil, starting operation in 1834 and ending in 2003. Figure 1 provides an overview location of the main gold deposits, mine camps, and provinces, some discussed below.

Brazil is number thirtheen among sixteen leading gold producers worldwide, having four world-class gold deposits (>3.2 Moz) in production (Moz): Morro do Ouro (8.9), Cuiabá (6.3), Crixás (4.6), and Fazenda Brasileiro (3.2). Two additional, Morro Velho (11.9 Moz) and Igarapé-Bahia (3.12 Moz), are presently closed but still have resources at depth. Today's resources for the Salobo IOCG deposit (1.94 Moz), operating since 2012 in conjunction with other IOCG deposits (Cristalino, Sequeirinho-Sossego, Alemão), all in the Carajás Province (Amazon region), should make these an additional example of Brazil's world-class gold system. Potential, future world-class deposits presently in operation are Jacobina (2.20 Moz) and Chapada (1.64 Moz).

Brazilian gold systems are dominated by orogenic gold deposits, formed mostly in Neoarchean and Paleoproterozoic greenstone belts, similar to Canada and Australia. A few orogenic gold deposits formed in Neoproterozoic metasedimentary belts including Brazil's largest gold deposit (Morro do Ouro). From the three major cratons, São Francisco, Amazonas, and Rio de la Plata, most gold deposits are associated with the former and its neighboring orogenic belts. The São Francisco and Rio de la Plata cratons are surrounded by Neoproterozoic (900-550 Ma) orogenic belts, whereas the western and northwestern Amazon Craton are surrounded by late Mesoproterozoic belts (~1020 Ma).

Brazil's most important gold systems today are as follows (Fig. 1):

1- Au-only orogenic gold in Archean granite-greenstone belts stands out in the QF (Cuiabá deposit) and Crixás regions; the latter also has Paleoproterozoic-aged belts.

2- Au-rich Archean and Paleoproterozoic IOCGs (various deposits in Carajás).

3- Paleoproterozoic Itapicuru greenstone belt, Bahia (Fazenda Brasileiro deposit).

4- Sedimentary rocks overlying greenstone belt, Bahia; quartz-pebble conglomerate, Jacobina.

5- Neoproterozoic metasedimentary belts with large, low-grade (<1 g/t Au), phyllite-hosted orogenic gold (Morro do Ouro).

6- Neoproterozoic rocks containing an earlier-emplaced porphyry system, subsequently overprinted by an orogenic gold system (Chapada deposit, Mara Rosa camp).

Some future exploration potential (Fig. 1):

(I) Paleoproterozoic plutonic-volcanic rocks in the Tapajós and Alta Floresta-Juruena gold provinces (Amazon) with Au-Cu, Au-Ag systems, where numerous *garimpos* and small deposits are presently operating.

(II) Archean orogenic systems in the QF, including Raposos, São Bento, Córrego do Sítio, and smaller older deposits to be reevaluated. Resources increasing in deposits in the western QF sector (e.g., the Turmalina) along the Pitangui greenstone belt.

(III) Paleoproterozoic greenstone belts in Bahia, Goiás, Tocantins, Pará, and Amapá states should be further investigated for their potential.

(IV) Au-rich IOCGs in the Carajás greater, extended region.

Finally, it is noteworthy that Brazil contains supergene, lateritic enrichments of hypogene gold systems in Cenozoic cover rocks and recent sediments in many parts of the country.

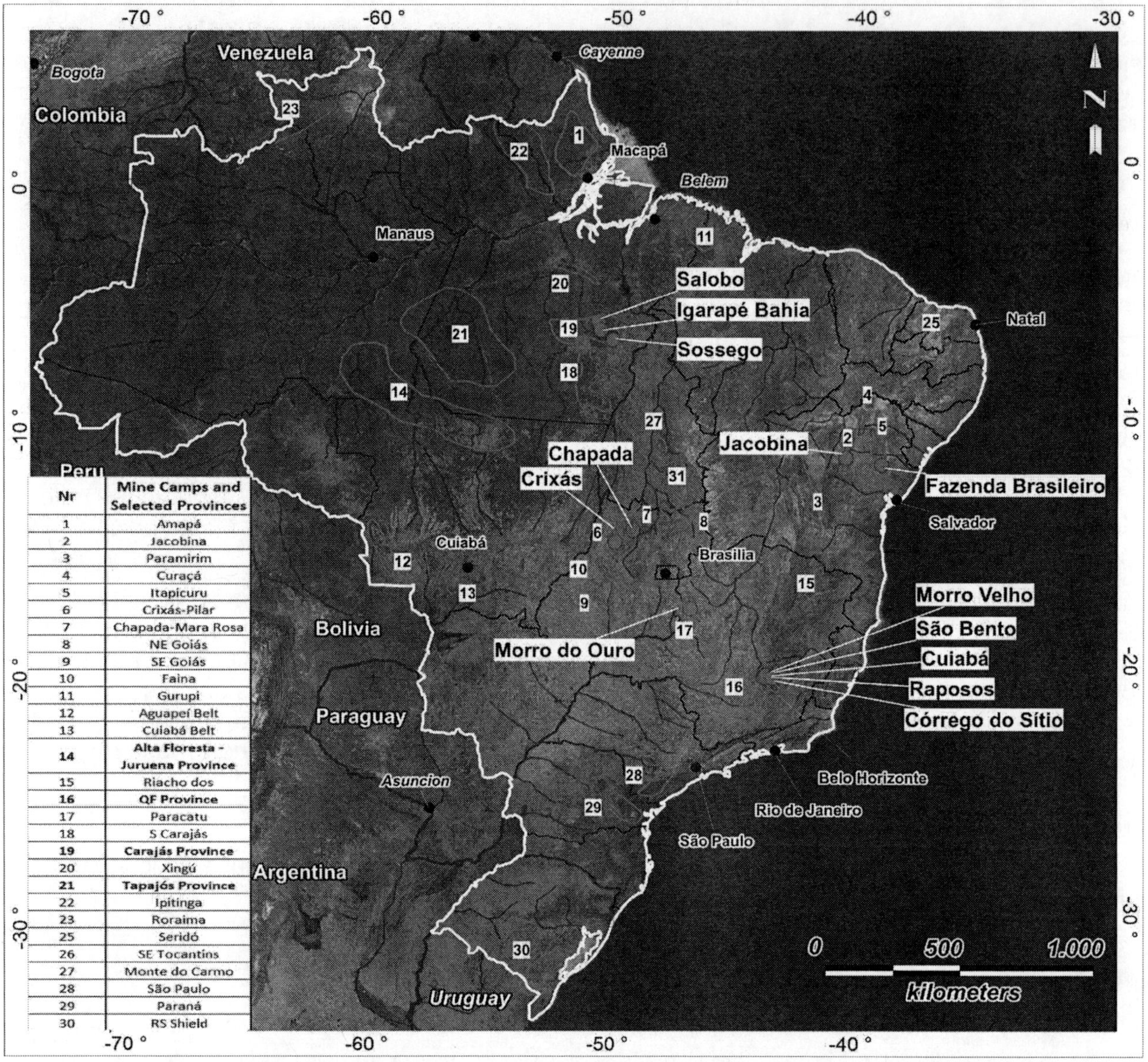

Nr	Mine Camps and Selected Provinces
1	Amapá
2	Jacobina
3	Paramirim
4	Curaçá
5	Itapicuru
6	Crixás-Pilar
7	Chapada-Mara Rosa
8	NE Goiás
9	SE Goiás
10	Faina
11	Gurupi
12	Aguapeí Belt
13	Cuiabá Belt
14	**Alta Floresta - Juruena Province**
15	Riacho dos
16	**QF Province**
17	Paracatu
18	S Carajás
19	**Carajás Province**
20	Xingú
21	**Tapajós Province**
22	Ipitinga
23	Roraima
25	Seridó
26	SE Tocantins
27	Monte do Carmo
28	São Paulo
29	Paraná
30	RS Shield

G13
The Kibali (KCD) Orogenic Gold Deposit: Gold Without Quartz Veins

Doug MacKenzie[1], Andrew Allibone[2], Carlos Vargas[3], Etienne Mwandale[3], Joel Holliday[3]
1. University of Otago, Dunedin, New Zealand, 2. Rodinian(NZ) Ltd, Wanaka, New Zealand, 3. Barrick Gold Corp, Toronto, ON, Canada

The Kibali (KCD) orogenic gold deposit in northeast Democratic Republic of the Congo is one of the largest gold deposits in Africa with a current gold resource of >16 Moz and 2020 production of 808 Koz. The deposit is hosted in the Kibali Greenstone Belt in a package of sedimentary, volcanic, and porphyritic intrusive rocks that were deposited and emplaced between ca. 2629-2625 Ma and then metamorphosed to lower greenschist facies. The metasedimentary rocks were deposited as part of an extensional basinal succession that was subsequently compressed and complexy folded between ca. 2625-2610 Ma. This coincided with generation of late metamorphic gold-bearing fluids that were focused along the basinal margin, forming a 60-km-long regional shear zone known as the KZ Trend.

The deposit is hosted within polymict conglomerates, sandstones, siltstones, black carbonaceous shales, BIF, and chert that interacted with these fluids and were extensively hydrothermally altered and mineralized. A notable feature of the deposit is the lack of fracture-controlled vein mineralization and the dominance of Au-bearing disseminated sulphides which have locally replaced host lithologies.

Mineralized zones are developed within and along the margins of folded Fe-rich BIF, fractured chert rocks, and in carbonaceous shears that developed along the margins of deformed black shales. Economic zones within the deposit are termed the 3000, 5000, 9000,11000, and 12000 lodes and are centred on a series of discrete, locally contiguous folded zones (Fig. 1). The lodes are controlled by and plunge parallel to fold axes of km-scale, NE-plunging, tight to isoclinal folds. In general, lodes hosted within the hinges of fractured cherts are smaller and more rod-like than the more voluminous BIF-hosted lodes which are more curviplanar where they extend around fold hinges and for short distances along limbs.

Rather than a large fracture-feeder system where mineralizing fluids have overwhelmed the host rocks and dominated the chemistry of the deposit, gold-bearing fluids at Kibali permeated along pervasive axial planar foliation and subparallel microshears, and the resultant mineralization and alteration assemblages are strongly controlled by the host lithologies. Sericite-ankerite-pyrite has replaced metamorphic chlorite-muscovite-calcite ± epidote in the metaclastic rocks, sericite-pyrrhotite-pyrite has replaced metamorphic sulphides in graphitic chert and carbonaceous rocks, and siderite-magnetite-pyrite has replaced metamorphic magnetite and Fe-carbonate in BIF. Gold occurs predominantly as microparticulates in the cores of disseminated pyrite grains and pyrite in mm-scale sulphide veinlets. Some gold-bearing sulphides have been remobilised along the margins of post-mineral, intermediate to mafic dykes where they cut across pre-existing mineralized zones. Arsenopyrite is less common and typically occurs as disseminations and rare sulphide veinlets within a distinctive polymict conglomerate horizon in structurally lower parts of the sequence. Mineralized rocks are locally overprinted by a later crenulation cleavage and subsequent generations of more brittle, cataclastic shears and faults that are barren.

Figure 1. A) Geological map with location of the Kibali mine and the Kibali Greenstone belt in NE, DRC. B) Geological plan map of the KCD open pit and location of the cross section. C) Representative cross section through the KCD deposit showing mineralized lodes.

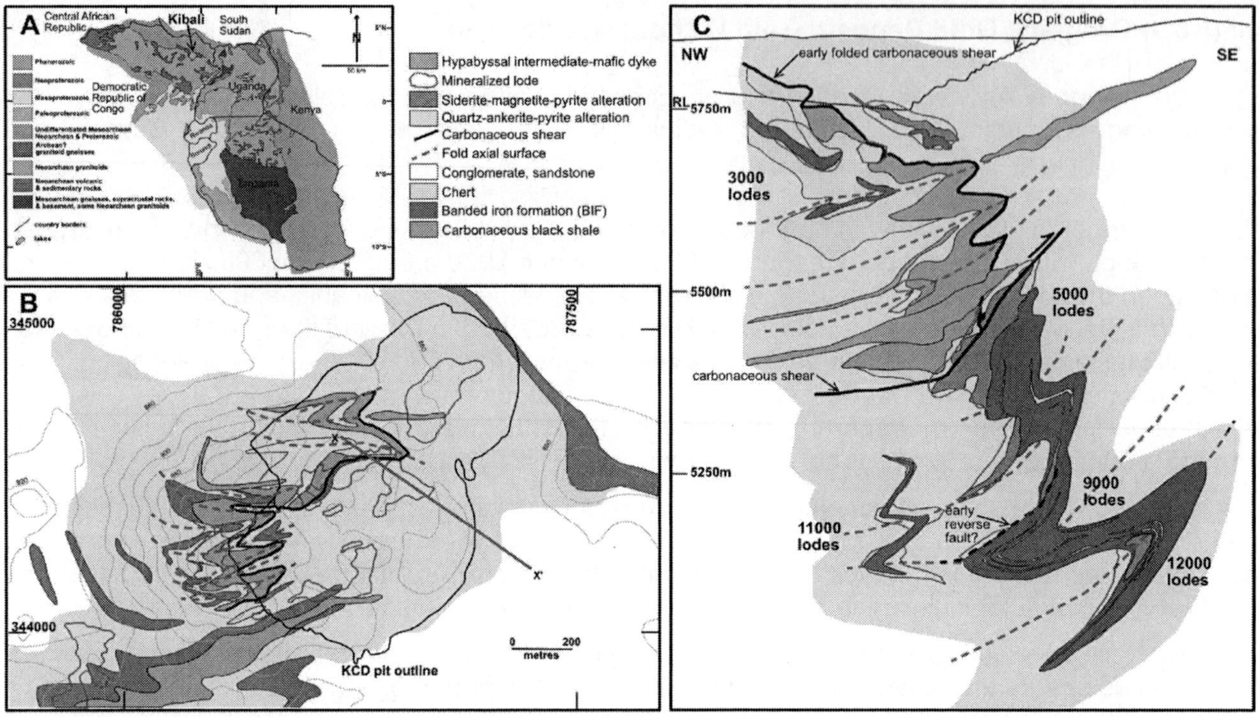

G14
Obuasi: The World's Largest Precambrian Gold Deposit, Ghana, West Africa

Michael J. Nugus[1], Andrew Allibone[4], Nicholas H.S. Oliver[2, 3], Carlos Vargas[4], Richard Jongens[5], Vaughan A. Chamberlain[2]
1. AngloGold Ashanti, Newtown, South Africa, 2. HCOVGlobal, Alligator Creek, QLD, Australia, 3. Economic Geology Research Unit, James Cook University, Townsville, QLD, Australia, 4. Barrick Gold Corporation, Toronto, ON, Canada, 5. Anatoki Geoscience Ltd., Dunedin, New Zealand

Obuasi, with a total Mineral Resource plus past production of approximately 70 Moz Au, is the largest Precambrian gold deposit and one of the very largest in the world. The orebody is dominated by a roughly equal mixture of paragenetically complex gold-bearing quartz veins, and altered (carbonate-muscovite-arsenopyrite-dominant), moderately to strongly deformed, variably graphitic phyllites and meta-greywackes. The steeply dipping lodes extend over an 8-km strike length and to depths of >2.5 km. Obuasi is hosted by ~2135 Ma siliciclastic rocks of the Eburnean Kumasi Basin, which were obliquely shortened along an inverted boundary with the older Eoeburnean Ashanti Belt to the east. Greenschist facies metamorphism (~350°C and 100-200 MPa) and related deformation with strong shear and veining components were coeval with mineralization and related alteration at ~2095 Ma. For this part of the deposit's 100-year research history, we combined relatively recent major regional geological framework analysis, underground and historical mapping, new cross sections, and early and revised 3D models with the drill-core database, and added a layer of interpreted litho-geochemistry and vein density from a portable XRF-gathered dataset, to expand our understanding of the gold endowment, distribution, and controls.

Any discussion of the origin of the Obuasi deposit needs to reflect not only on the giant size of Obuasi itself, but also the large endowment of the wider host fault system over its 400-km strike length (>100 Moz Au). No other regional-scale structure in West Africa, and very few others globally, have comparable gold endowments. Several aspects of the Obuasi deposit indicate that the accumulation of so much gold can be considered a product of the protracted history of deformation, inversion, dewatering, and metamorphism, with much fluid flow focussed regionally near an original major tectono-stratigraphic boundary. The mineralizing fluids were derived primarily from deeper, As-rich metasedimentary sources by basinal fluid expulsion and metamorphic devolatilization triggered by inversion and shortening, followed by transpression (Figure). Continued fluid injection and deformation during and after the metamorphic peak produced changes in gold fineness, sulfide assemblages, repeated dissolution (stylolites), and re-precipitation of mineralized veins, and a change from early deformed shear-related, sulfide-rich lodes to later quartz-rich lodes that plunge down or across the axes of younger transpressional folds. Channelized fluid flow due to reactivation of basin-edge transfer structures (inferred from mapping and geophysics), and/or irregularly distributed gold source rocks, may explain the variation in gold endowment along the former basin boundary.

The deposit, and its understanding, have significant implications for the orogenic gold model. As has been proposed by several authors in the last decade, any assessment of gold deposits that include substantial, poly-deformed metasedimentary rocks needs to consider early tectono-stratigraphic evolution (in particular basin and sub-basin edges and basement-influenced transfer structures) and the potential roles of basinal and early metamorphic fluids–Obuasi being the prime example.

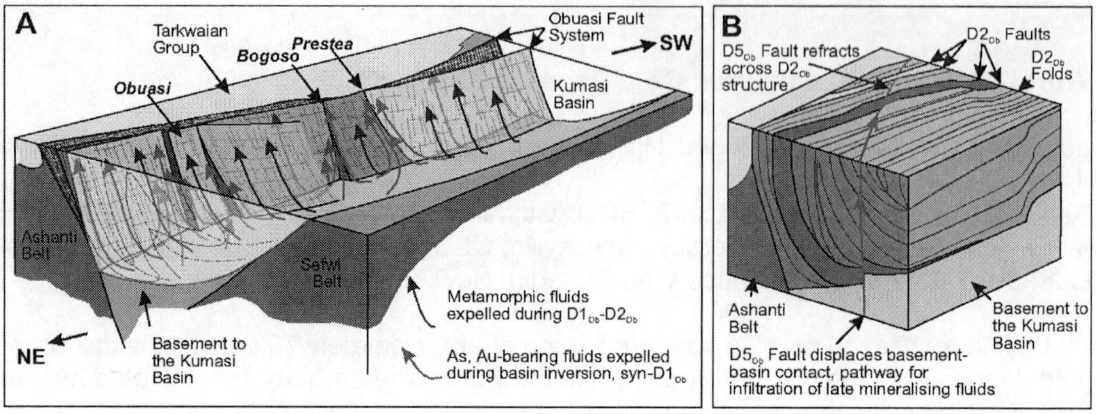
A
Tarkwaian Group
Obuasi Fault System
SW
Bogoso
Prestea
Obuasi
Kumasi Basin
Ashanti Belt
Sefwi Belt
NE
Basement to the Kumasi Basin
Metamorphic fluids expelled during D1$_{Ob}$-D2$_{Ob}$
As, Au-bearing fluids expelled during basin inversion, syn-D1$_{Ob}$
B
D2$_{Ob}$ Faults
D5$_{Ob}$ Fault refracts across D2$_{Ob}$ structure
D2$_{Ob}$ Folds
Ashanti Belt
Basement to the Kumasi Basin
D5$_{Ob}$ Fault displaces basement-basin contact, pathway for infiltration of late mineralising fluids

Figure 10

G15
The Fosterville Gold Deposit, its Geology and Place Amongst the World's Best Orogenic Goldfields

Wessley B. Edgar
Kirkland Lake Gold, East Bendigo, VIC, Australia

The transformation of the Fosterville gold mine from a burning Phoenix into a majestic Swan has been the most significant gold story in the past 100 years of mining in Victoria, Australia.

Located just 20 km east of the world-class Bendigo 22 Moz goldfield, and within a wider Central Victorian province of historical production over 80 Moz, the Fosterville mines were economically rated poorly as recently as 2014 when gold production was slightly over 100 koz pa and Mineral Reserves just 182 koz. In that year, Victoria's total gold production was only 225 koz from just four operating mines, with geological research and exploration investment at extremely low levels.

By the start of 2020, however, an exploration pegging rush had seen all available ground surrounding Fosterville taken, and exploration expenditures increased more than 10 times! The change in mineralization at this single mine has wildly driven international corporate and investor interest, forever changing the perceptions of Victorian gold potential! What geologically changed in less than 5 years of exploration?

Shallow intersections of visible gold were sporadic and isolated and seen as an interesting geological phenomenon rather than as a precursor to recent deep discoveries. This was largely influenced by the initial belief that Fosterville was a sulphide-only gold system, even proposed as a closer analogue to Carlin than Bendigo or Ballarat (now disproven). Sparse visible gold occurrences continued until 2015, when coarse gold-in-vein drill intercepts increased in regularity, particularly where reverse faults intersected an anticline footwall to the Fosterville Fault. Before this, most gold won from arsenopyrite- and pyrite-altered wallrock disseminations were associated with thrust faults and a vein-stockworked syncline. Amazingly, a refractory orebody with grades of 4-5 g/t Au gained a well-defined shear vein orogenic system containing disseminated coarse gold, commonly less than 3mm, at multi-ounce-per-tonne grades and with low nugget effect!

When drill intersections including 1429 g/t Au across 15.15 m were announced in 2017, the newly named Swan orebody was born, and Fosterville transformed into one of the world's most profitable gold mines worth multi-billions. Annual production in 2020 increased to 640,467 oz at a phenomenal grade of 33.9 g/t Au and saw an expanded Mineral Reserve for the Swan zone of 1.25 Moz at 30.6 g/t Au.

The emergence of ultra-high grades in a structural style of multi-phase quartz-carbonate veins was a challenging change for mine geologists where not even very small quartz volumes of ore can go unmined, and small distances have markedly different vein textures, volumes, and orientations affecting gold grades. Mining and ongoing internal geoscience studies are revealing a range of local and internal structural controls and contrasts to other deposits in Central Victoria, especially geochemically. Swan barely covers 500-m strike x 300-m height, and yet generates optimism for all explorers that perseverance in a marginal sulfide gold system at depths about 1 km from surface can ultimately lead to one of the highest-grade gold deposits in the world, and a tier-one gold asset, as it did for Kirkland Lake Gold at Fosterville.

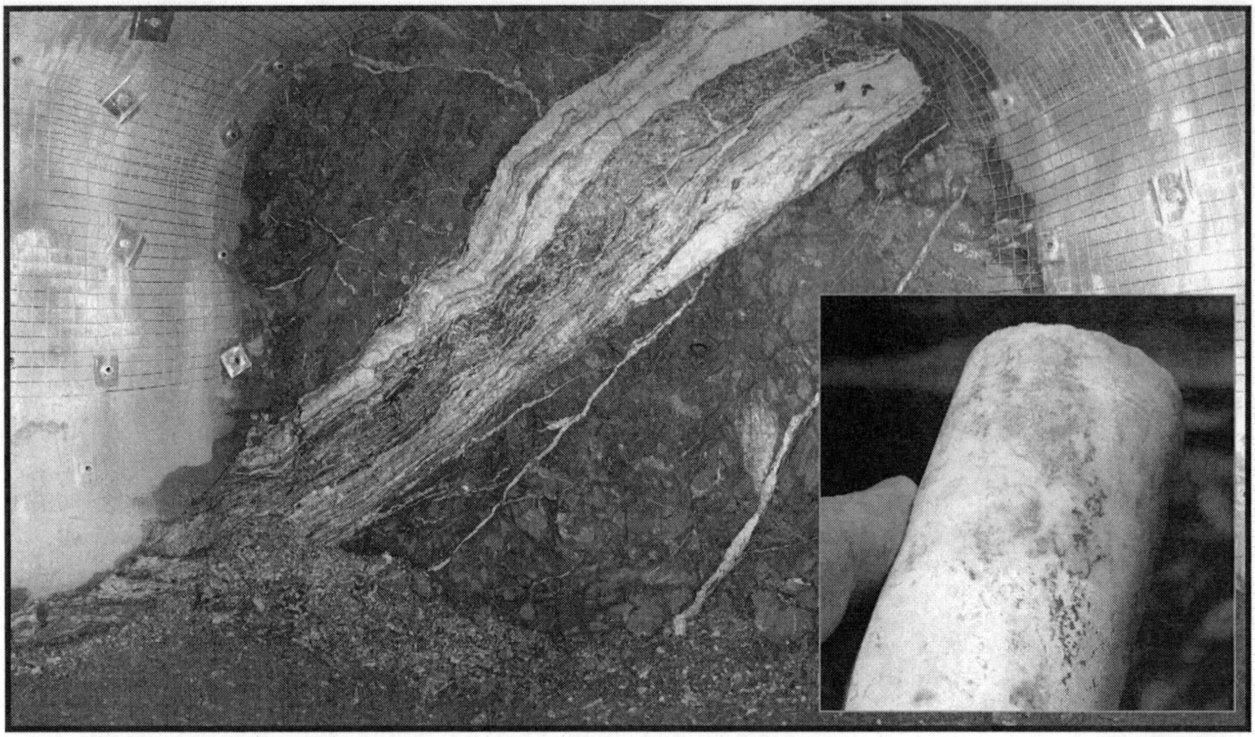

G16
Gold 2020 - Current Understanding and Ongoing Questions

Richard J. Goldfarb[1], Stuart Simmons[2]
1. China University of Geosciences Beijing, Beijing, China, 2. Hot Solutions, Auckland, New Zealand

Since the *SEG 2000 Gold Conference* and *Reviews Volume*, our understanding of the genetic models for major global gold producer types has not advanced significantly, whereas understanding for some lesser-understood deposit types (e.g., Carlin-type, IOCG) continues to be just as poor today. There have been major changes in the past few decades regarding what we mine today, particularly as the gold price has led us to economically favorable development of many more giant sub-gram ore systems and large refractory deposits. Furthermore, improved analytical techniques have led to vast amounts of published geochemical data on all gold deposit types, some helpful and some definitely not, commonly in lieu of necessary, more detailed geological field study of a deposit or district.

Models for shallow intrusion-related (e.g., porphyry, skarn, epithermal) and deeper metamorphic-related (e.g., orogenic) gold are well-accepted. The mainly Cenozoic intrusion-related deposits are products of oceanic and continental arc development in the circum-Pacific and Tethyan belts. The Neoarchean, Paleoproterozoic, and <700 Ma orogenic gold deposits are widely related to metamorphism in Precambrian greenstone belts and younger orogens. Although these widespread ore types formed throughout Earth history, preservation processes control their ultimate present-day spatial-temporal distributions. Many issues critical to explorationists continue to require address, such as defining extent/edges of hydrothermal systems (particularly to aid in vectoring to ores), controls on high-grade ore zones, controls on element abundances (particularly as demand increases for by-product critical elements), and issues of provinciality that may identify more prospective gold belts. When considering a mineral systems approach, we know much about conduits and traps for these major deposit types, but fluid/metal source continues as an issue of some debate. Importantly, are there oxidized intrusions that release multiple pulses of magmatic-hydrothermal aqueous-carbonic fluids to form series of fault-fill and extensional vein systems along the length of a deep crustal fault system? Despite evidence for metamorphic fluid involvement, almost every major orogenic gold deposit has been related somewhere in the recent literature to magmatic processes. So, can fluid exsolution from an oxidized magma lead to formation of orogenic gold deposits? This has important regional targeting implications regarding whether to favor areas with structural traps near greenschist zones or near a particular plutonic suite.

Unlike other gold deposit models, the Carlin model is a local model for the well-studied deposits in Nevada. It provides thorough descriptive data to explore for such 40 Ma replacement-style ores in the Basin and Range, but because it has not been developed from widespread global examples, a genetic model is still not clear. Magmatic, meteoric, and metamorphic fluids have all been suggested as part of the ore-forming process, but the ores are significantly different from those defined as high-sulfidation epithermal, low-sulfidation epithermal, and epizonal orogenic, respectively. Deposits elsewhere in the world defined as Carlin types, such as in China and southeast Asia, are not the same as what is present in Nevada, and it would be a mistake to try to define a global Carlin gold model that includes these examples.

P1
Global exploration trends

Sinead Kaufman
Rio Tinto Exploration, Salt Lake City, UT, USA

Historically, industry success has been tied to spend, albeit with the gratification of discovery lagging investment by several years. From 1994-2008 there appeared to be a direct relationship between spend and discovery success, but in the period following the Global Financial Crisis when the industry ramped up spending again, there were fewer discoveries. The diminishing performance and the increasing degree of difficulty of new discoveries may be attributed to industry consolidation, access challenges, decreasing grades, exploring at depth and under-cover, as well as the changing geopolitical environment. Everything "easy" to find has probably already been found. What is not well understood is the lag in the performance. Many of the discoveries in the early-mid 2000s came from work done in the prior decade. Based on this and the peak in funding levels seen in 2012, the industry should now be seeing a solid pipeline of early stage projects starting to emerge. However, this trend seems elusive with increased exploration spend not translating into a steady stream of quality discoveries. The critical question is how will we address these challenges? For mature exploration terrains the challenges are not only the over-explored ground but also post mineral cover and over-staked land. Even today, with all the tools at our fingertips, we struggle to explore through even a few meters of post mineral cover. This will require changes in the future with out-of-the box thinking.

We need to better understand and manage the risks of working in complex and underexplored terrains. Grassroots opportunities are now overwhelmingly in jurisdictions with high geopolitical risk and this may have played a role in the progressively decreasing exploration budgets in this space. Some companies willing to take the risk in these jurisdictions would benefit from the early entry and access to underexplored opportunities. Others will potentially pay the price of their mineral rights being taken away or get stuck trying to navigate through inadequate mining laws and corrupt practices.

While the majors have maintained dominance in global exploration spend over the last several years, juniors have been a close second with almost 90% of their spend dedicated to greenfield targets. Mergers and acquisitions, however, have changed the exploration landscape. In recent years, the industry saw diminishing competition and participation from mid-tiers while Chinese companies have entered the market with fresh new strategies for resource acquisition and exploration.

We need to challenge norms and the status quo, embrace new technology and advanced data analytics. The combination of highly motivated and skilled exploration geologists with access to excellent data and analytical tools creates explorers with a hunger for success. Future geoscientists will have to navigate through a complex, data saturated environment, with targets and patterns often invisible to the human eye, where machine learning and data analytics play an increasingly important role in target generation and resource evaluation.

The mineral resource sector is facing more pressure to deliver better ESG performance and with exploration being in the frontlines, it is slated to play a significant role in this space. This requires a strong focus on safety and environment to manage the unique challenges in maintaining tenure as well as genuine and transparent community engagement.

ST.102
In Situ LA-ICP-MS Analysis of Fluid Inclusions in Barite and Calcite from Hutti Gold Deposit, India

Kunda Badhe[1, 2, 3], Pei Ni[2], Hari S. Pandalai[3], Junyi Pan[2]
1. Rajiv Gandhi University, Rono hills, Doimukh, Itanagar, AR, India, 2. Nanjing University, Nanjing, China, 3. Indian Institute of Technology Bombay, Mumbai, MH, India

Hutti gold deposit is a classic orogenic greenstone-hosted gold deposit in India. As in most orogenic gold deposits, gold is localized in Hutti along quartz reefs and sulfidized wall rock.The early quartz veins and the quartz reefs that host mineralization have predominantly low-salinity CO_2-H_2O-NaCl fluids with minor CH_4. Late-stage calcite veins are common in the Hutti deposit. In addition, post-ore late-stage barite veins have been reported.

Microanalysis of individual fluid inclusions by laser ablation inductively coupled plasma mass spectrometry (LA-ICP-MS) is a powerful tool for directly determining fluid chemistry in fossil hydrothermal systems. In the present study, we provide the geochemistry of fluid inclusions in barite and calcite that were co-precipitated in veins using the new protocol (LA-ICPMS).

The vein of barite consists of excellent large translucent to transparent crystals of barite with prisms often up to 8 cm in length. In other cases, smaller crystals ranging from a few millimeters to a few centimeters are present.

The post-ore stage barite-calcite veins deposited from fluids of moderate salinity (10-16 wt % NaCl equiv.) and at temperatures close to 145-150°C are also reported in Nevin and Pandalai (2010) and Badhe and Pandalai (2015).

The absolute elemental concentrations are calculated based on the applied internal standard. For example, Na concentration derived from calculated salinity from microthermometric observation.

Fig.1 illustrates the time-resolved LA-ICPMS signals of a liquid-rich fluid inclusion in barite and calcite. Most elements occurring synchronously with the narrow Na peak suggests their presence in the aqueous solution, whereas delayed peaks are attributed to solid minerals. Halogen elements such as Cl and Br were detected in the fluid inclusions and were quantitatively determined.

All fluid inclusions contain NaCl, KCl, and $CaCl_2$. $BrCl_2$ occurs as major dissolved salts with trace amounts of alkaline metals such as Li, Mg, Rb, Sr, Cs, B, and Ba. The ratio of Ca/K and Na/K average values are higher in fluid inclusions in barite than calcite. Mg/K and Ba/K values in barite and calcite are less than 1. Li and B concentrations in fluids are higher in calcite than barite. The dataset is more limited for Fe and Cu; they were below their limits of detection. The halogens Cl and Br have high concentrations.

Fluid inclusion studies on the quartz reefs in the underground exposures and quartz veins in surface exposures in the Hutti schist belt have revealed that the fluids related to early (D1) quartz veins, quartz veins associated with the shear zone, and a few, later, (D2-related) quartz veins are all associated with fluids belonging to the CO_2-H_2O-NaCl-(CH_4) system (Pandalai et al., 2003), while late-stage post-ore cavity-filling barite-calcite veins with fluid inclusions belong to the H_2O-NaCl system (Badhe and Pandalai, 2015).

The high molar Br/Cl ratios and the chemistry of fluids need to be explained in the context of the tectonic setting of an evolving island-arc orogen. The exact origin of these high-saline waters is unclear yet.

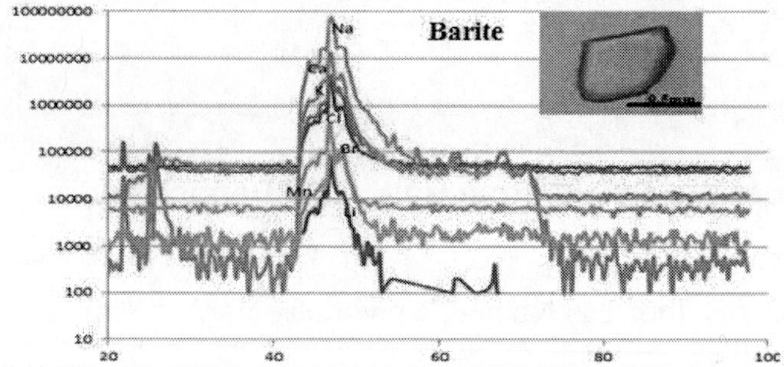

Barite
0.5mm
Na
Ca
K
Cl
Br
Mn
Li

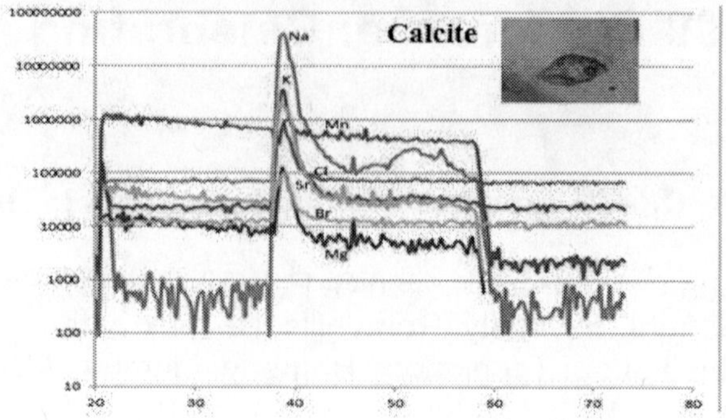

Calcite
Na
K
Mn
Cl
Sr
Br
Mg

ST.103
Physico-Chemical Conditions of Pahardiha- Rungikocha Gold Deposit, North Singhbhum Mobile Belt, Eastern India

Anmol Barla[1], Sahendra Singh[1], Rajarshi Chakravarti[1, 2], Ramesh C. Behera[1]
1. INDIAN INSTITUTE OF TECHNOLOGY (INDIAN SCHOOL OF MINES), DHANBAD, INDIA, DHANBAD, JH, India, 2. INDIAN INSTITUTE OF TECHNOLOGY, ROORKEE, ROORKEE, UK, India

The Pahardiha-Rungikocha gold deposit is located south of the Dalma volcano-sedimentary belt within the North Singhbhum Mobile Belt (NSMB), in the eastern part of India. Gold occurs in the quartz ± carbonate veins within the sheared quartz-chlorite schists. Field observations coupled with previous studies show 3 phases of deformation (D1, D2, D3) within the Pahardiha-Rungikocha gold deposit. The D1 and D2 deformation phases are progressive metamorphism. The late deformational phase, D3, is responsible for the gold mineralization. The total gold ore reserve reported from the deposit is 0.487 million tonnes with an average grade of 3.43 ppm. Gold within the deposit occurs as native particles or specks of visible gold along the cracks/margins of euhedral pyrite. The gold-bearing pyrite is coeval with a generation of chromite-cored magnetite where euhedral magnetite rims show epitaxial overgrowth over the relict chromite cores. In this work, we use chromite-cored magnetite chemistry and bulk rock chemistry to determine the physico-chemical conditions of gold mineralization. The final phase of deformation is responsible for the formation of the chromite-cored magnetite rims, which is coeval with the gold mineralizing event. These magnetite rims formed as a result of the metasomatic process brought about by the interaction of hydrothermal fluids with the wall rock. The chromite-cored magnetite trace element signatures resemble the high-temperature Ca-Fe alteration and high-temperature K-Fe alteration facies reported from various IOCG and IOA type deposits. The host rock chemistry is a consequence of a mixed provenance of felsic and mafic/ultramafic source rocks of the mineralized quartz-chlorite-schists. These fluids were relatively rich in mafic components such as Ni, Mg, Cr, and V, as evident in the chromite-cored magnetite chemistry. A comparative study of bulk rock chemistry of the quartz-chlorite schists and chromite-cored magnetite chemistry reveals that the chemical composition of this gold mineralizing metasomatic fluid was controlled by the ambient wall rock chemistry. The final stage of the deformation event allowed metasomatic Au-enriched fluids and precipitation in quartz ± carbonate veins.

Fig: A) BSE image of chromite-cored magnetite and X-ray elemental map shows the Chromite-rich zone at the core. B) Comparison of trace elements of average magnetite and bulk rock chemistry.

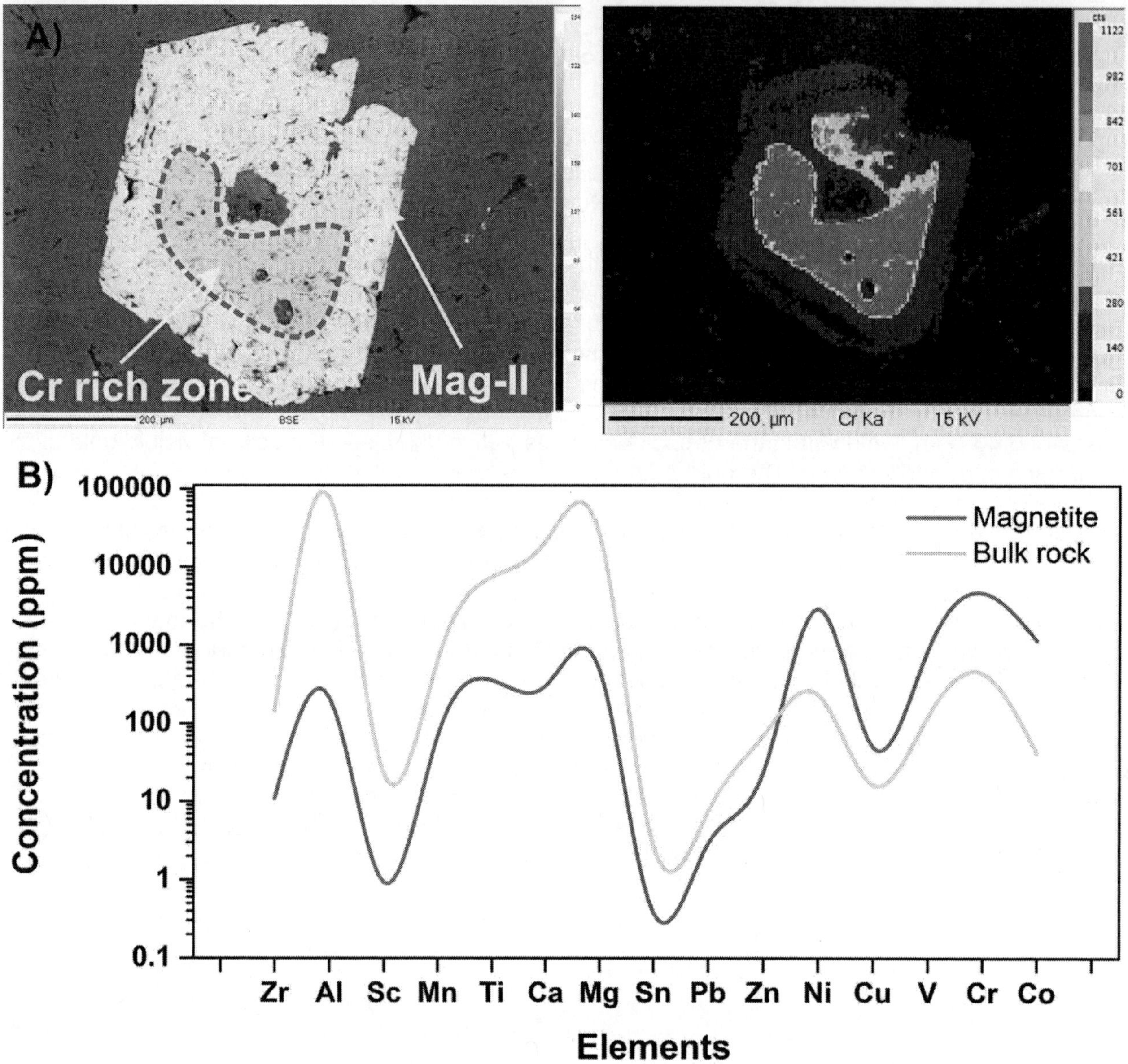

ST.104
New Eyes on Old Rocks: A New Structural Model Reinvents a World-class Gold District – Black Hills, South Dakota, USA

Robert Bergmann[1, 2], Brian Lentz[1, 2], Stephen T. Allard[3]
1. F3 Gold, Minneapolis, MN, USA, 2. Big Rock Exploration, Minneapolis, MN, USA, 3. Winona State University, Winona, MN, USA

The Black Hills region of South Dakota is home to one of the world's richest gold mines, the Homestake Mine. Over 137 years, Homestake produced over 40 million ounces of gold from deformed Proterozoic rocks. Homestake is historically interpreted as a Banded Iron Formation (BIF) hosted, synformal, stratabound deposit with remobilized gold mineralization locally controlled by F3 fold and shear structures. This host-rock driven model guided exploration elsewhere in the Black Hills for decades, resulting in minor successes.

The geology and structure of the greater Black Hills region has been primarily characterized by lithostratigraphy and local, brittle fault structures. However, recent publications suggest a model dominated by transpressional, brittle-ductile shearing along the Dakotan Tectonic Zone (DTZ), a long-lived, deep-seated, through-going regional NW-striking shear zone, strongest along the eastern margin of the Black Hills, but now recognized across the entire uplift.

Critically, the eastern high-strain zone of the DTZ lies along strike with the Homestake gold mine and numerous DTZ shear structures (F3a and F3b), which have been found elsewhere in the hills to be coincident with gold mineralization. Furthermore, previous workers suggest the highest ore grades within Homestake are near these shear structures. This recognition of deep-rooted shearing (DTZ) casts new light on Homestake mineralization controls and suggests that a metamorphic shear zone, lode gold model may be fitting for the Black Hills.

Extensive field work over the last decade tests this new model and encourages exploration for overlooked gold deposits. This work includes over 6,000 surface samples combined with detailed 1:8,000-scale geological mapping, petrography, SEM-EDX, and mineralogical analysis. These data provide compelling evidence correlating DTZ structures across the Black Hills to high-grade gold at surface; they are now being considered the controlling driver of gold mineralization in the region. This suggests the Black Hills region is more similar to other world-class shear-hosted orogenic gold regions and reopens Black Hills to exploration for additional large gold deposits.

ST.106
Geoscience Integration - A Prospect for Gold Exploration Beyond 2021 in Ghana, West African Craton

Benjamin Boadi[1, 2], P V S Raju[1], David D. Wemegah[3]
1. National Geophysical Research Institute, Hyderabad, AP, India, 2. Academy of Scientific and Innovative Research (AcSIR), AcSIR Headquarters CSIR-HRDC Campus, Ghaziabad, UP, India, 3. Kwame Nkrumah University of Science and Technology, Kumasi, Ghana

Orogenic gold deposits preserved in Precambrian greenstone belts such as Superior Province (Canada), the Dharwar Craton (India), and the West African craton (Ghana) contribute a share of 10-200 Moz of the world's gold resource. South-western Ghana is regarded as one of the most prolific regions for gold exploration, with gold making up 90% of the total mineral exports of the country. Currently, Ghana is the leading producer of gold (4.8 Moz) on the African continent, ahead of South Africa (4.2 Moz), with a volume reserve of 8.74 metric tonnes from 2015 to 2020. The geological structure complexity hosting this gold boom is linked to the Eburnean orogeny (ca. 2.1-1.9 Ga). The gold mineralisation on regional and local scale is controlled by major deformational structures (D1-D6), with imprints extending from shallower (~5km) to deeper (>15 km) depth on the cratonic scale. Deeper structures, such as faults and shear zones, acted as conduits for auriferous mineralised fluids with high gold concentrations to advect upward to form gold deposits in second- and third-order structures or fractured rock units in basins and belts. Traditional methods (stream sediments analyses, soil sampling, conventional geological mapping, drilling) of mineral exploration have dominated exploration in Ghana for decades, even though some of them are less successful in areas where bedrock geology is obscured by tropical vegetation, laterite, and thick saprolite soils. Adopting modern geophysical mapping (e.g., gravity, magnetic, seismic tomography, and magnetotelluric) will suffice for a thorough assessment of deeper subsurface structural architecture. To meet gold demand and production beyond 2021, a continuous search for new potential gold targets is required. The advanced development in GIS allows for a more reliable and cost-effective method of combining geoscience data sets to map new potential zones. The use of GIS-based geoscience integration to predict possible mineralized zones in Ghana is still in its early stages. However, in the areas where it has been implemented, the approach has been successful. In this presentation, the Ahafo Gold District of the Sefwi greenstone belt, southwest Ghana, was investigated to delineate potential zones for lode gold mineralisation. GIS-based weighted overlay, fuzzy overlay, and weight-of-evidence techniques were used to integrate evidential layers deduced from digital elevation, geological, and geophysical datasets. The approaches were successful in delineating potential zones for lode gold mineralization that can serve as a guide for further exploration in the study area. Adopting and implementing cost-effective GIS geoscience integration in Ghana will be critical in mapping prospective zones for exploration purposes, allowing basins and belts to fully exploit their mineral potential. The advent of efficient and cost-effective exploration, continuous mining production, and government support will not only increase gold production levels but will also make it a long-term sustainable endeavour.

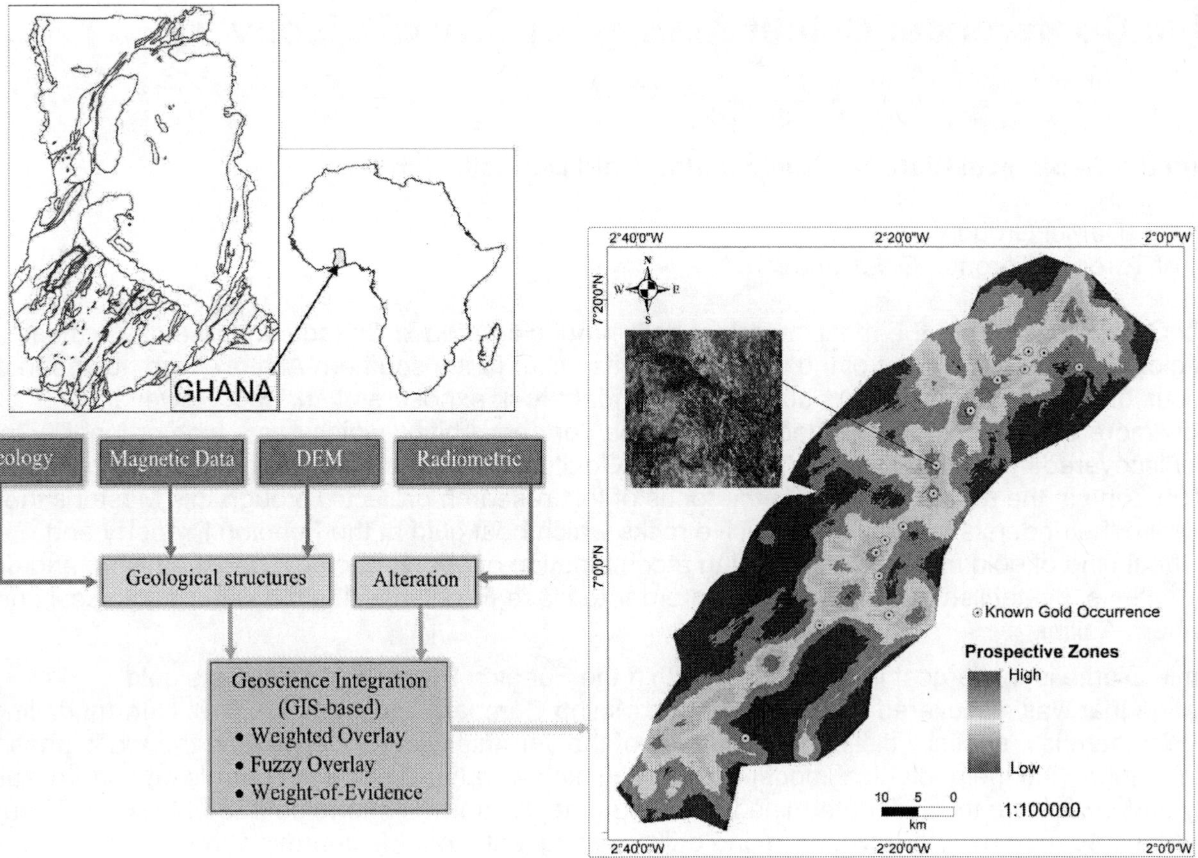
GHANA
Geology
Magnetic Data
DEM
Radiometric
Geological structures
Alteration
Geoscience Integration
(GIS-based)
• Weighted Overlay
• Fuzzy Overlay
• Weight-of-Evidence
2°40'0"W
2°20'0"W
2°0'0"W
7°20'0"N
7°0'0"N
N
W
E
S
Known Gold Occurrence
Prospective Zones
High
Low
10 5 0
km
1:100000

ST.107
Unraveling the Geological History of the Fenelon Gold Deposit, Québec

Joy M. Carter, Daniel Gregory
University of Toronto, Toronto, ON, Canada

The Abitibi Greenstone Belt is the most important gold-producing area in Canada and one of the most important globally. Presently, most of the known deposits occur in the southern Abitibi which has been the primary focus of exploration endeavors due to the abundant rock exposure at surface. However, increasingly more effort is being expended exploring the northern Abitibi, which has a high probability to contain undiscovered mineral resources. The Fenelon Gold deposit, situated along the Sunday Lake Deformation Zone in the northern Abitibi, is the focus of this research project. Through this Master's thesis, I aim to improve the understanding of the intrusive rocks which host gold at the Fenelon Property and constrain the timing of gold mineralization. Using a combination of geochronology, geochemistry, and drill core observations, I establish the local geology in order to place Fenelon within the wider geological context of the northern Abitibi.

The Jeremie Diorite is the largest intrusive body within the Fenelon Property, and it hosts gold mineralization that was discovered by the Wallbridge Mining Company Ltd. in early 2019. Current drilling suggests that there is a spatially distinct mafic phase of the Jeremie Diorite. Delineating the mafic phase is important for improving the geological model and determining whether it has any controls on mineralization. During my undergraduate thesis, I determined the age of the Jeremie Diorite to be 2697 Ma using Thermal Ionization Mass Spectrometry (TIMS). Diamond drill core suggests that the Jeremie Diorite was emplaced into a sedimentary basin composed of greywacke and argillite. A large gabbro sill and related dyke swarm crosscut both lithologies. Age dating of zircons by Laser Ablation Inductively Coupled Plasma Mass Spectrometry (LA-ICP-MS) confirmed the paragenetic relationship between the Jeremie Diorite and the sediments, and the age of the gabbro was determined using TIMS. Detrital zircons extracted during dating reveal the provenance of the sediments and help reconstruct the geological history and crustal evolution of the region. Gold mineralization post-dates the emplacement of the Jeremie Diorite as the gold is hosted in crosscutting smoky quartz veins. Molybdenite found within a gold-bearing vein was analyzed using Re-Os dating techniques to determine the absolute age of the gold mineralization at the Fenelon deposit.

The results of this thesis will significantly improve the geological understanding of the Fenelon deposit, and allow for comparison with established deposits including the neighbouring Detour Lake Gold Mine. There has been very limited research conducted along the Sunday Lake Deformation Zone outside of Detour Lake, and this study contains the first age dates of the rocks and mineralization at the Fenelon Property, setting the geological framework for all future research. Its success will create a strong precedent that will encourage future exploration in the northern Abitibi, generating continued growth for the Canadian mineral resource sector.

ST.109
Developing An Atlas Of Gold Compositions For British Columbia: A New Tool For The Exploration Community

Robert J. Chapman[1], Rory Murphy[1], Mortensen James[2], Britt Bluemel[3], David Banks[1]
1. University of Leeds, Leeds, United Kingdom, 2. University of British Columbia, Vancouver, BC, Canada, 3. Goldspot, Vancouver, BC, Canada

Detrital gold particles can greatly aid mineral exploration campaigns because they are broadly dispersed and create large footprints around lode gold deposits. In areas of complex geology, such as central BC, there may be gold particles from several different potential sources in surface sediments. Consequently, despite exhibiting the chemical and physical durability essential for an indicator mineral, gold particles have traditionally found limited application in this regard. This project has elevated particulate gold to the status of a source-specific indicator using a new approach based on compositional signatures. Natural gold is almost invariably an alloy, and gold particles commonly contain inclusions of other minerals that may be representative of the source. This project, underwritten by Geoscience BC, has established generic compositional signatures of gold from different styles of mineralization and has developed a methodology for interpretation of new data relevant to all scales of exploration activity. Populations of gold particles from a single locality may be classified by their microchemical signature; i.e., the alloy composition (Au, Ag, Cu, Hg, Pd) correlated on a particle-by-particle basis with the mineral inclusions revealed in polished section. This microchemical signature is directly related to the deposit style, but it is only of use to the exploration community if a sufficiently large compositional template is freely available. This project is developing a new, publicly available reference database of detrital gold particle microchemistry to encourage and aid exploration in BC. The provincial database describes 13,999 gold particles from 327 localities covering all major auriferous regions. Alloy compositions were generated by electron probe microanalysis and mineral species present as inclusions were identified using the scanning electron microscope. In addition, trace element compositions of 922 particles have been collected using laser-ablation inductively coupled plasma mass spectrometry. Gold formed in porphyry, epithermal, skarn, and orogenic mineralization is represented, and diagnostic compositional characteristics for each have been identified. Hitherto, new datasets have been interrogated to yield geological interpretations regarding gold provenance by domain experts who can draw upon extensive prior experience. The aim of the project is to empower individual researchers and explorationists to immediately benefit from analysis of gold particles collected during their own routine reconnaissance. Using Exploratory Data Analysis, we have developed a framework and methodology to facilitate classification of new gold data using the new gold Atlas. The training dataset for this approach was curated by researchers with expertise in both the metallogeny of BC and the microchemical signatures of gold formed in different deposit styles. Therefore, use of the classification algorithm will replicate data set interrogation as if performed by experienced researchers. A key deliverable of this project will be the detailed description and workflow of the algorithm, such as Linear Discriminant Analysis, used to classify new data. In this way, the project will generate a new tool immediately available to exploration in BC, but with the potential to expand efficiently to both the Canadian and Alaskan Cordillera. Subsequent workers will engage with the project outcomes according to the workflow in the figure below.

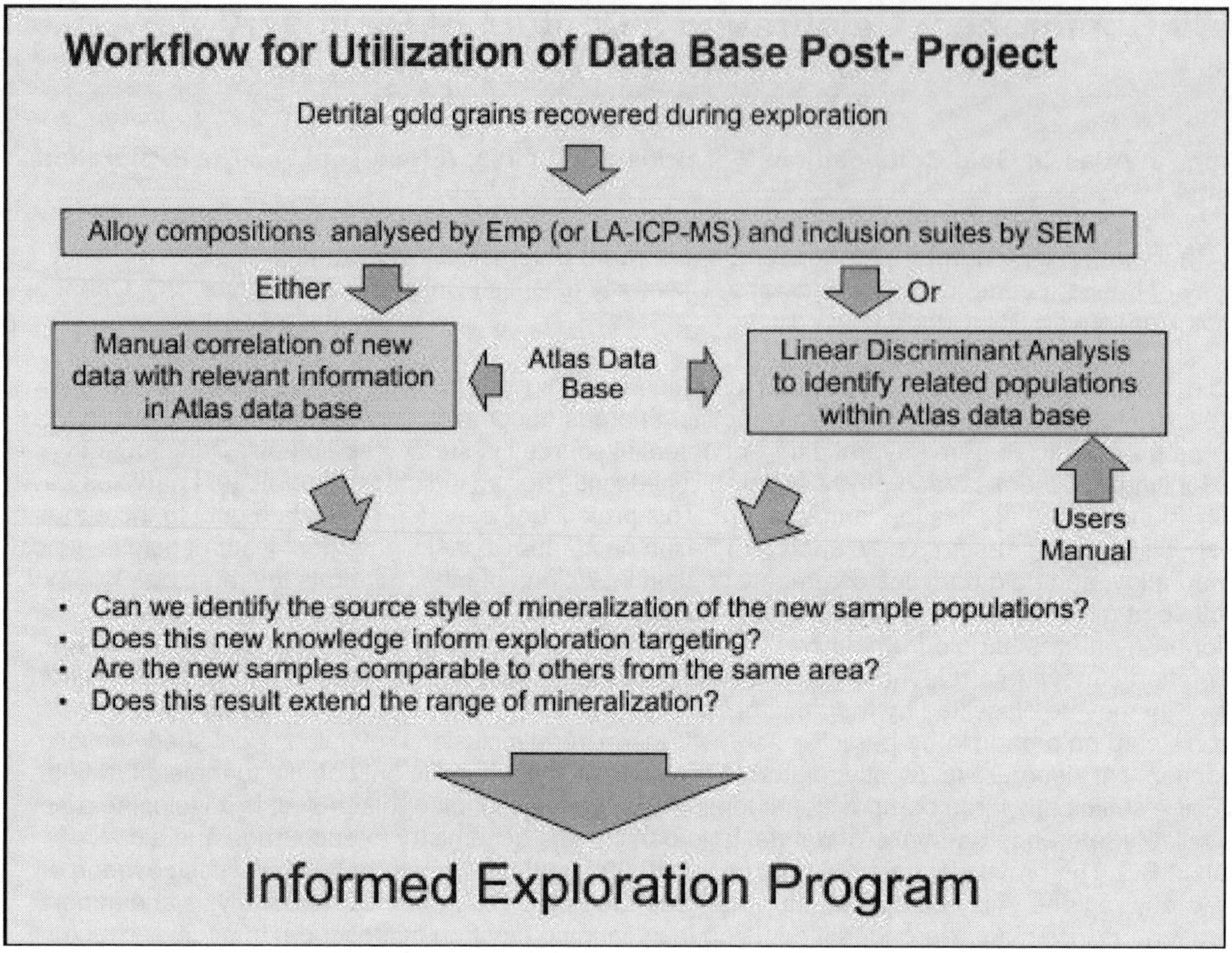
Workflow for Utilization of Data Base Post- Project
Detrital gold grains recovered during exploration
Alloy compositions analysed by Emp (or LA-ICP-MS) and inclusion suites by SEM
Either
Or
Manual correlation of new data with relevant information in Atlas data base
Atlas Data Base
Linear Discriminant Analysis to identify related populations within Atlas data base
Users Manual
• Can we identify the source style of mineralization of the new sample populations?
• Does this new knowledge inform exploration targeting?
• Are the new samples comparable to others from the same area?
• Does this result extend the range of mineralization?
Informed Exploration Program

ST.110
Compositional and Textural Features of Natural Gold and Implications for Gold Particle Studies

Robert J. Chapman[1], David Banks[1], Michael Styles[2], Richard Walshaw[1], Sandra Piazolo[1], Daniel Morgan[1], Matthew Grimshaw[3], Carl Spence-Jones[1], Thomas Matthews[4], Olga Borovinskaya[5]
1. University of Leeds, Leeds, United Kingdom, 2. British Geological Survey, Keyworth, United Kingdom, 3. srk, Cardiff, United Kingdom, 4. Imperial College, London, United Kingdom, 5. Tofwerk, Thun, Switzerland

Gold compositional studies have offered a refinement to the traditional prospector's methodology of "panning back to source." Recently, the scope of research on natural gold has expanded both in geographical terms and to include paragenetic, crystallographic, and isotopic studies. Researchers new to the field may be unaware of the complexity and diversity of heterogeneous alloy textures and inclusion mineralogy and their implications for gold particle studies.

This study is based on our own data derived from individual inspection and analysis of circa 35,000 gold particles from 505 placer and in-situ localities worldwide, coupled with published information describing a further 15,000 gold particles. Our aims are threefold. First, we have characterized textural heterogeneity according to generic compositional and/or crystallographic criteria. Second, we have identified the environment of formation of these features according to hypogene settings (both syn- and post-gold-deposition) and supergene environments. Third, we show how this new understanding should influence future approaches to both sample collection and data interpretation.

Compositional and textural characteristics of gold are easily observed in polished sections using the SEM, and EBSD-EDS systems permit correlation of crystallographic and chemical features. Gold-rich rims (Fig. 1A) are well documented, and various workers have noted both Ag-rich and Au-rich films or tracks, sometimes adjacent (Fig. 1B). Complex compositional variation within individual particles may be sympathetic to crystallographic domains (Fig. 1C), or not (Fig. 1D). Inclusions of other minerals are generally related to hypogene vein mineralogy (Fig. 1E). The application of Time of Flight mass spectrometry with laser ablation to polished sections of gold particles has revealed previously unreported heterogeneity at the sub-micron scale (Fig. 1F, G), which may either indicate tiny inclusions or sporadic concentrations of different elements ("clusters") within the gold alloy.

Characteristics such as mineral inclusions, clusters, and the composition of the primary gold alloy are functions of the initial environment of precipitation and are the only useful features for linking composition to primary ore genesis. Enrichment of pre-existing alloy with Ag may take place either in the late stages of the mineralization process or during subsequent residence in the hypogene setting. Gold rims form in the surficial environment largely by silver depletion, but Au-rich tracks are generated in a two-stage process involving Ag removal along grain boundaries during surface weathering by followed by addition of pure gold.

The present study has quantified the degree and nature of heterogeneity both within and between gold particles. Future studies should be based on populations of gold particles sufficiently large to minimise the potential for unrepresentative sampling. Understanding of the genesis of different characteristics of natural gold permits informed targeting of those relevant to studies of either the hypogene and supergene environments. Finally, new appreciation of the degree of sub-micron heterogeneity has implications for trace element characterization through analysis of a small proportion of the particle because the resulting signature may be unrepresentative.

Figure 1: Examples of compositional and crystallographic characteristics of gold particles. (A-E) SEM BSE images. (F, G) ToF lA-ICP-MS images

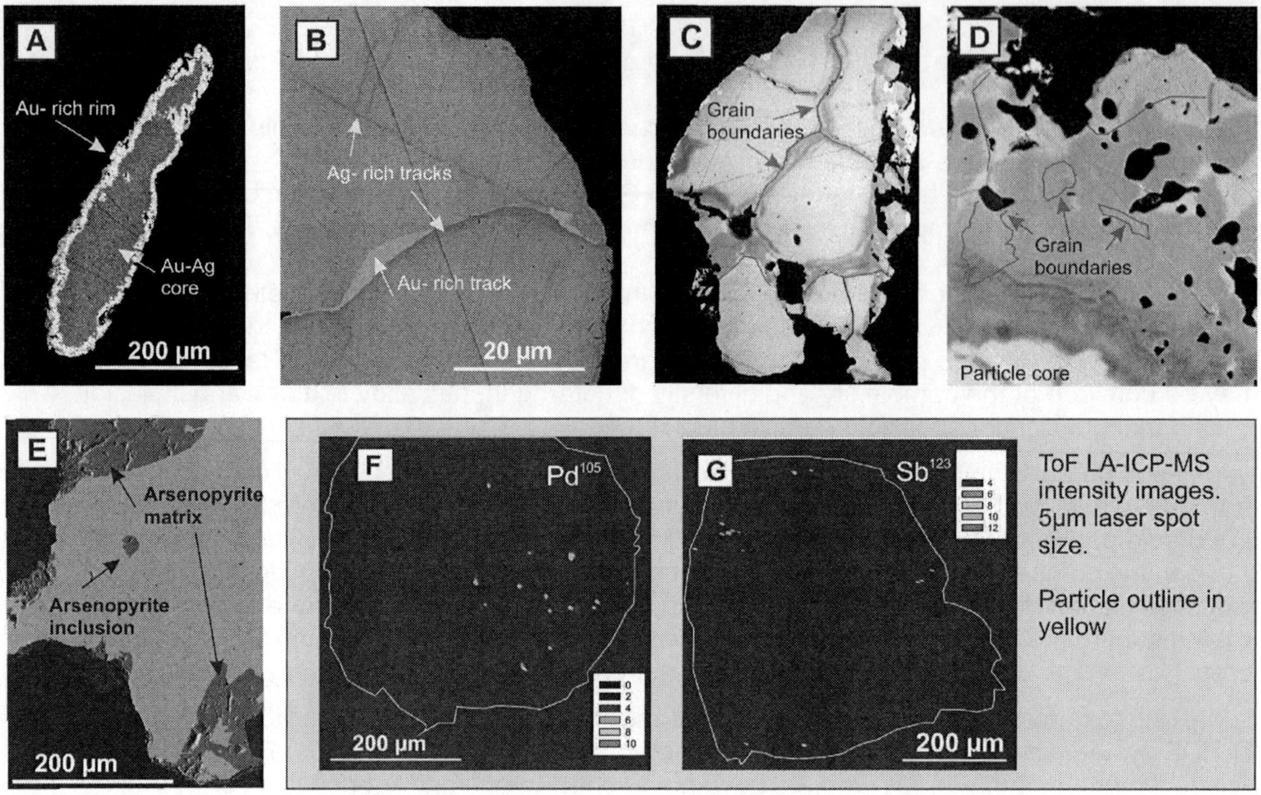

ST.114

Ediacaran-Cambrian Fluid Flow Imprint on Archean Gold Deposits: U-Pb SHRIMP Evidence from the Quadrilátero Ferrífero Metallogenetic Province, Brazil

Tatiana G. Dias[1], Rosaline C. Figueiredo e Silva[1], Lydia M. Lobato[1], Steffen G. Hagemann[2]
1. Universidade Federal de Minas Gerais, Belo Horizonte, MG, Brazil, 2. Centre for Exploration Targeting, School of Earth Sciences, University of Western Australia, Perth, WA, Australia

The constraints on the timing of orogenic lode-gold mineralization in the Quadrilátero Ferrífero (QF), an important metallogenetic province located in southeastern Brazil, have been hotly debated in the literature. The main gold deposits in QF, their tectonic evolution spanning a long history from Archean to Proterozoic orogenic cycles, are hosted in the Rio das Velhas Greenstone Belt (RVGB), including the world-class Cuiabá and Morro Velho deposits. Gold mineralization is classically interpreted to be related to the late stages of Archean deformation and metamorphism, at ca. 2.7 Ga. However, geochronological data obtained from hydrothermal monazite suggests an imprint of the late Ediacaran-Cambrian Brasiliano orogeny. *In situ* U-Pb SHRIMP dating was conducted on hydrothermal monazite from three orogenic gold deposits: Cachorro Bravo (CB), Carvoaria (CV), and Pilar (PL). These deposits form a 15-km-long gold camp in the eastern portion of the QF, are controlled by a NE-trending strike-slip shear zone, and are characterized by different mineralization styles The CB deposit is dominated by replacement-style mineralization, with abundant arsenopyrite, and constrained by mafic dykes. In contrast, the CV deposit is characterized by vein-style mineralization, defined by auriferous quartz lodes with "free" gold and low content of sulfide minerals, and subordinate sulfosalts. The PL deposit occurs mainly as stratabound mineralized bodies, hosted by an iron formation and comprising bands of massive sulfides and quartz-carbonate veins, or in silicified sulfide-schists. The U-Pb on monazite SHRIMP II (B) analyses (Sensitive High Resolution Ion MicroProbe) were conducted at Curtin University, Western Australia. Hydrothermal monazite in mineralized metapelites from CV yielded a discordia in the U-Pb diagram with intercepts at 2514 ± 22 and 555 ± 19 Ma. Three younger, age-equivalent crystals are concordant and yield a U-Pb concordia age of 540 ± 18 Ma, identical within uncertainty to the lower intercept age. Monazite from the CB deposit, inter-grown with arsenopyrite, yielded a robust U-Pb concordia age of 535 ± 11 Ma. Monazite from the silicified sulfide-schist from the Pilar deposit is depleted in U, precluding the calculation of robust U-Pb ages. On the other hand, the Th contents are relatively high, and disclosed a mean $^{208}Pb/^{232}Th$ age of 508 ± 6.4 Ma, coherent with the results obtained from the other deposits. The results demonstrate the growth of hydrothermal monazite both in the Archean, during ore formation, and during Cambrian hydrothermal alteration. The 2.5 Ga upper intercept age could be linked to the opening of the Paleoproterozoic Minas platform basin imposed on the Archean paleocontinent and associated isotopic resetting of monazite, an event also marked by the intrusion of the Lavras mafic dike swarm dated (U-Pb SHRIMP) at 2.55 Ga. These new ages add to the recently growing body of data which suggest Archean and Cambrian ages for xenotime and monazite in gold and iron deposits throughout the QF. The results also reveal the possible influence of hydrothermal fluid flow through the QF mineral systems during the Cambrian, possibly due to gravitational collapse of the Araçuaí orogenic system.

ST.115
Lifting the Cloak of Invisibility: Gold in Pyrite from the Olympic Dam Deposit, South Australia

Kathy Ehrig[1], Cristiana L. Ciobanu[2], Max R. Verdugo-Ihl[2], Marija Dmitrijeva[2], Nigel J. Cook[3], Ashley Slattery[4], Vadim S. Kamenetsky[5]
1. BHP Olympic Dam, Adelaide, SA, Australia, 2. School of Chemical Engineering and Advanced Materials, The University of Adelaide, Adelaide, SA, Australia, 3. School of Civil, Environmental and Mining Engineering, The University of Adelaide, Adelaide, SA, Australia, 4. Adelaide Microscopy, The University of Adelaide, Adelaide, SA, Australia, 5. School of Natural Sciences, Hobart, TAS, Australia

"Invisible gold" in pyrite refers to gold either present within the sulfide lattice or as discrete nanoparticle (<100-nm-diameter) inclusions (NPs), making it undetectable by conventional optical and scanning electron microscopy. Investigation of "invisible gold" in chalcopyrite-pyrite ores from the Olympic Dam Cu-U-Au-Ag deposit, South Australia, confirms the presence of Au in some arsenic-bearing pyrites at concentrations measurable by laser ablation inductively coupled-plasma mass spectrometry (LA-ICP-MS). Arsenic-bearing pyrite in the studied sample shows As-Co-Ni-oscillatory zoning patterns with variable complexity, suggesting grain re-crystallization during replacement by chalcopyrite (Fig. 1a). LA-ICP-MS data obtained from 164 pyrite grains plot below the Au and As solubility limit (Fig. 1b), empirically defined from studies of epithermal and Carlin-type deposits.

Several As-rich pyrite grains were analyzed using Scanning Transmission Electron Microscopy (STEM) and EDX-STEM analysis of foils obtained by Focused Ion Beam methods. Micron-scale, oscillatory zoning patterns observed on back-scattered electron (BSE) images and LA-ICP-MS element maps extend down to the nanoscale. Decoupling between trace elements is common, for example, Ni depletion wherever As and Co are enhanced, with nucleation of discrete Co-As-bearing NPs (cobaltite/safflorite?; Fig. 1c).

Importantly, Au-bearing NPs are identified in all cases, in intimate association with other (sulfo)tellurides (Fig. 1d-f). In addition, abundant cassiterite and rare chalcopyrite NPs are also identified. Some of the largest Bi-Ag-telluride NPs contain electrum as tiny pore-attached NPs within the larger telluride (Fig. 1d, e). Nanometer-sized electrum NPs were also identified in association with chalcopyrite. Silver-Au-telluride NPs form mono- or bi-component NPs (Fig. 1f, g). These NPs occur along trails displaying As-Co-enrichment, or formation of nm-wide lamellae of Bi-Pb-sulfotellurides marking pyrite twin boundaries (Fig. 1h). One wider lamella was identified from the layer stacking as a strongly disordered member of the aleksite series. Coarser tellurobismuthite (Bi_2Te_3), a few µm wide, is associated with altaite (PbTe) at pyrite-chalcopyrite boundaries.

Pyrite displays kink- and screw-dislocations associated with trace element remobilization or NPs nucleation. These defects can be associated with either "marcasitization" or loss of Fe (formation of pyrrhotite) within nanoscale domains affected by fluid percolation and pyrite recrystallisation. Twin planes in pyrite enriched in heavy elements (Bi-Pb-Te) represent zones of weakness and assist element exchange between host mineral and percolating fluids during coupled dissolution reprecipitation reactions (CDRR), analogous to those known for hematite from Olympic Dam. Nanoscale textures in pyrite allow for interpretation of Au-NPs, as Au released from solid solution in pyrite during CDRR associated with marginal chalcopyrite replacement.

Nanoscale analysis lifts the cloak of invisibility for Au in pyrite at Olympic Dam. These results show that confirmation of whether gold occurs as NPs or in solid solution based solely on position above or below the solubility limit of Au in pyrite on a plot of Au vs. As (Fig. 1b) is impossible without corroborative studies at the nanoscale.

Fig. 1. (a) BSE image showing boundary between oscillatory zoned pyrite grains. (b) Plot of Au vs As. Yellow circles: grains analyzed by STEM. (c, e) EDX maps of pyrite and electrum. (d,f,g) Selected NPs. (h) Bi-Pb-Te-enriched twin plane in pyrite.

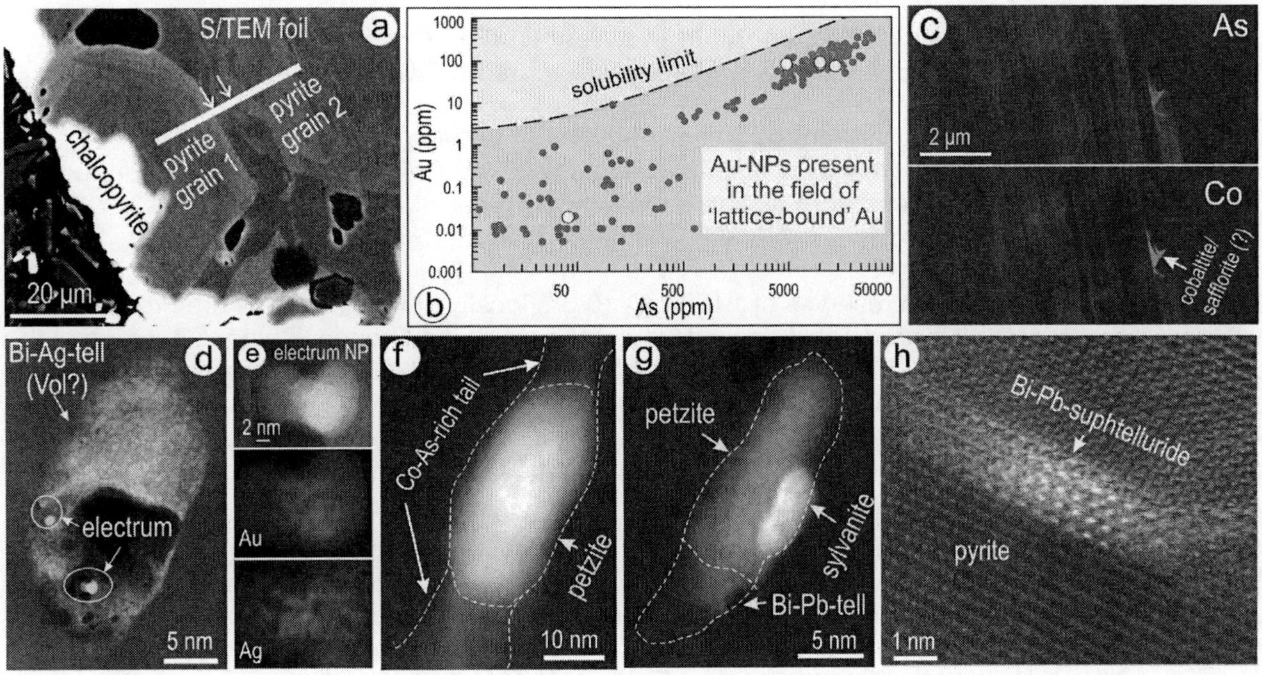

ST.116
Genesis of the World-Class Dashui Gold Deposit in the West Qinling Orogen, China: Constraints from Petrography, Whole-rock Geochemistry, and Mineral Inclusion Composition

Liang Fan[1], Jian-Wei Li[1, 2], Jia Chang[1], Shao-Rui Zhao[1], Zhe Ren[2], Ji-Xiang Sui[3], Shi-Guang Du[1]
1. School of Earth Resources, China University of Geosciences, Wuhan, China, 2. State Key Laboratory of Geological Processes and Mineral Resources, China University of Geosciences, Wuhan, China, 3. Yifu Museum of China University of Geosciences, Wuhan, China

The Dashui gold deposit, with proven reserves of >100 t @ 10 g/t (locally having bonanza ore grades of 1000 g/t), is one of the largest high-grade gold deposits in the West Qinling Orogen, China. Gold ores of this deposit are mainly composed of very fine-grained hematite, microcrystalline quartz, and carbonate. Rare sulfides are common as are ore minerals predominant in other gold deposits over the West Qinling orogen. The distinct mineral assemblages at Dashui has led to a prevailing supergene origin for its formation. Here, we present new geological and mineralogical evidence to suggest a primary origin.

Orebodies are of dendritic, lenticular, or cystic shapes and are mainly hosted in the Lower Triassic limestone and dolomitic limestone, with minor ores in granodioritic dykes that intruded the Triassic sedimentary rocks at 215.36 ± 0.87 Ma. Gold mineralization is closely associated with extensively silicified carbonate breccias that form breccia pipes in high-grade ore zones. The breccias are generally angular to sub-angular, range in size from several millimeters to centimeters, and are frequently cemented by multiple generations of hydrothermal iron-rich silica and calcite. Fragments of hydrothermal silica and calcite are not uncommon and hydrothermally cemented, suggesting multiple brecciation events. Based on field relation and petrographic evidence, gold mineralization and ore-related hydrothermal alteration can be divided into three stages: microcrystalline quartz-hematite stage (I), calcite-hematite stage (II), and calcite stage (III), with gold mostly being deposited in stage I. This main ore-stage is characterized by replacement of carbonate by microcrystalline silica (30–80 μm) and hematite (2–3 μm), typically marked by irregular carbonate residues surrounded by euhedral microcrystalline quartz. Integrated FE-SEM and XRD analysis has revealed fine-grained hematite, kaolinite, apatite, monazite, native gold, and some unknown minerals within euhedral microcrystalline quartz. Stage II features hematite-interlayered calcite, with microcrystalline quartz variably present in the growth zones of calcite. The latest stage is marked by coarse-grained calcite veins surrounding or cutting the orebodies.

Bulk chemical analysis of gold ores reveals that SiO_2 (0.33–87.44 wt %) is positively correlated with Au (R = 0.47; R refers to correlation coefficients) but negatively correlated with CaO (R = -0.99). The results are consistent with petrographic observations, further confirming that gold mineralization is genetically associated with silicification, hematitization, and decarbonization of the host carbonate. Minor, fine-grained pyrite is locally developed in stage I microcrystalline quartz and yields LA–ICP–MS Au signal intensities significantly higher than the background, which have positive correlations to elevated S, As, Hg, Tl, and Ag signals. This observation indicates that minor gold was deposited in As-rich pyrite. Taken together, we suggest that the Dashui gold deposit formed from acidic, Si- and Fe-rich, and oxidized fluids that replaced carbonate to form microcrystalline quartz, hematite, native gold, and locally minor As-rich pyrite. As such, Dashui can be best classified as a Carlin-like gold deposit or simply an epithermal gold deposit.

ST.117
Micron-scale Trace Element Zoning in Carlin-type and Carlin-like Deposits from Nevada and the Yukon Based on Relative Sensitivity Factor Calibration of NanoSIMS Data

Anne A. Fulton[1], Elizabeth Holley[1], Christie Jilly-Rehak[2], Dante Huff[1]
1. Colorado School of Mines, Golden, CO, USA, 2. Stanford University, Stanford, CA, USA

Carlin-type gold deposits (CTGDs), primarily concentrated in north-central Nevada, are the source of some of Earth's largest concentrations of gold. They are characterized by "invisible" (sub-micron) gold hosted within arsenian pyrite overgrowths on precursor barren pyrite grains. The element zoning histories and gold deportment of Carlin ore are not fully understood, partially due to the fine scale of the overgrowths, which can range from nanometers to tens of microns in thickness.

Here we present results from new applications of nanoscale secondary ionizing mass spectrometry (NanoSIMS) to ore-stage sulfide minerals from deposits in Nevada and the Yukon Nadaleen trend. NanoSIMS differs from conventional methods in its ability to detect and quantify chemical variations on the sub-micron scale. We analyzed sulfides on the NanoSIMS using a combination of high-spatial-resolution image analyses and high-precision spot analyses and depth profiles. Due to a lack of sulfide trace element standards for NanoSIMS, concentrations of trace elements were calibrated using their relative sensitivity factors (RSFs). To quantify the RSFs, trace element-rich areas in sulfide grains were measured with both electron probe microanalysis and NanoSIMS to serve as single-point standards. In this study, RSFs for Cu, As, Ag, and Au were determined to be 0.57, 0.03, 4.20, and 0.94, respectively.

High-grade samples from Carlin-type or Carlin-like deposits throughout Nevada (including Turquoise Ridge, Deep Star, and Marigold Red Dot) and Osiris and Conrad in the Yukon Nadaleen trend reveal complex, multi-stage growth patterns in ore pyrite grains with significant variations between deposits. All classic CTGDs from Nevada and the Yukon exhibit the highest concentrations of gold either within the innermost rim zone, adjacent to the precursor pyrite core, or within the outermost rim zone along the edge of the grains. The Carlin-like Red Dot deposit samples show textural and chemical zoning characteristics similar to both magmatically-sourced distal disseminated deposits and classic CTGDs. Red Dot pyrite grains have the same precursor core and ore-stage arsenian pyrite rim pattern as classic CTGDs but contain complex Au and trace element zoning throughout the rims and have higher Ag:Au ratios than typical CTGDs. NanoSIMS $\delta^{34}S$ spot analyses of Red Dot Au-rich rims fall between 3.92 and 7.99‰, which is within the range of $\delta^{34}S$ values for magmatic pyrite in the region.

These micron-scale zoning patterns in both trace elements and sulfur isotopes can reveal clues into changing fluid compositions during hydrothermal ore deposit formation. NanoSIMS methods allow us to see previously impossible levels of detail in ore minerals and can provide better understandings of fluid sources and metal deportment in CTGDs and other deposit types.

ST.119
Orogenic Gold Bonanza Linked to Recycling of Oceanic Exhalative Systems

Marcelo Godefroy Rodriguez[1, 2], Steffen Hagemann[1], Marco Fiorentini[1]
1. The University of Western Australia, Perth, WA, Australia, 2. IGO Ltd, Perth, WA, Australia

As their name implies, orogenic gold deposits are formed in convergent tectonic margins, whereby deformation, metamorphism, fluid migration, and magmatism shape the architecture of the crust. The interaction between crust, mantle, and hydrosphere has hampered our understanding of the fluid and metal sources involved in the genesis of this deposits. However, Mass Independence Fractionation (MIF) and trace elements in pyrite can fingerprint the signature of hydrothermal fluids in Archean deposits. We focus on the $\Delta^{33}S$ and trace element signature of pyrite of the Golden Mile deposit and compare it with the current multiple S isotope data generated from orogenic gold deposits in the Yilgarn Craton of Western Australia. Overwhelmingly, most orogenic gold deposits display small $\Delta^{33}S$ values (~0.2‰), which are consistent with mixing of a minute portion of sedimentary sources ($\Delta^{33}S$ ~ up to 8.0‰) with a dominant contribution of S from the mantle ($\Delta^{33}S$ ~ 0‰). However, pyrite from the high-grade Oroya style (>120 g/t Au and Te), documented in the Golden Mile deposit, display negative sulfur isotope signatures ($\Delta^{33}S$ ~ -0.9‰) and anomalous Au and Te concentrations (up to ~37 and ~182 ppm). Since Au is primarily transported by S complexes (e.g., S_3^- and HS^-) and associated with Te in a number of auriferous ores worldwide, we conclude the three elements were ultimately derived from the same fluid source. We reconcile the signature recorded by $\Delta^{33}S$ values and the Au and Te enrichment in Oroya pyrite with a specific episode of mineralization, whereby fluids and metals were sourced from a VMS system and volatilized via interaction with the metasomatized mantle in a subduction zone.

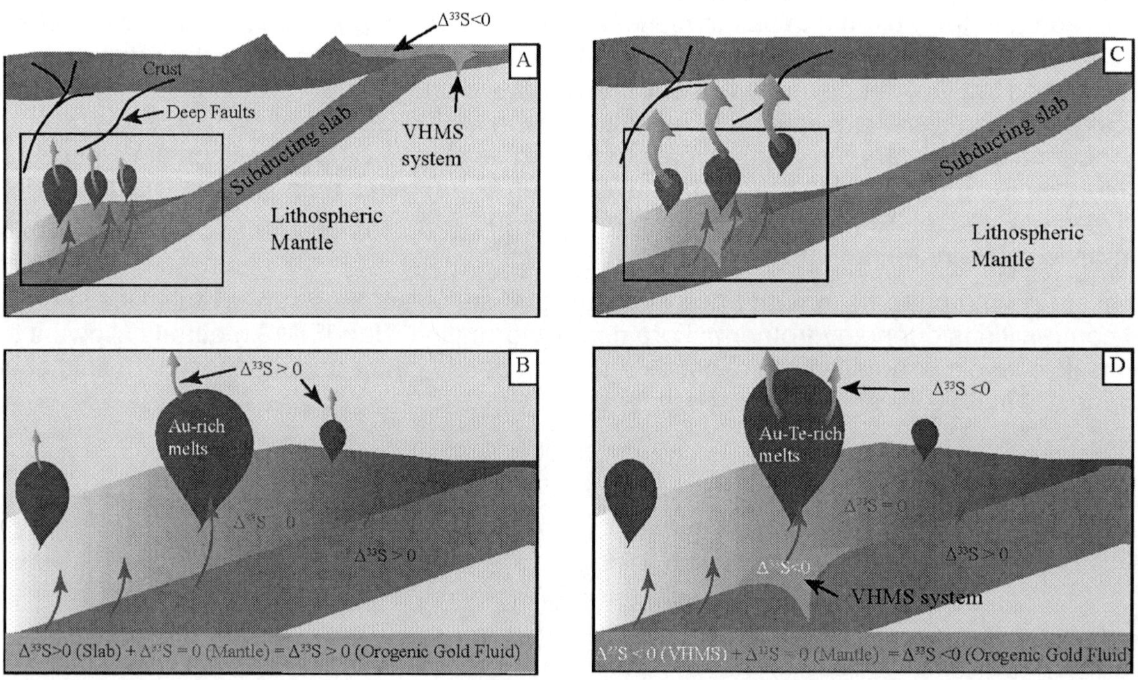

ST.120
Sulfide, Oxide, Sulfate and Stable (S and C) Isotope Footprints of Au in the Kundana Camp, Western Australia

Marcelo Godefroy Rodriguez[1, 2, 3], Adam Bath[3], John Walshe[3]
1. The University of Western Australia, Perth, WA, Australia, 2. IGO Ltd, Perth, WA, Australia, 3. CSIRO, Perth, WA, Australia

The Kundana Au camp totals ~103 t of Au and is mostly controlled by NW-trending shears, emplaced along lithological contacts (e.g., the K2 shear zone along shale and basalt). High-grade mineralization (>30 ppm Au) is spatially related to steeply west-dipping laminated veins and shallow dipping veins within the Pegasus deposit. Discrete mineralization in veinlets (<1 ppm Au) is located 1 km NE of Pegasus, in the Papa Bear prospect. Petrographic observations across the Pegasus deposit and Papa Bear prospect indicate that scheelite, pyrrhotite, arsenopyrite, and variable amounts of sphalerite were precipitated with native Au in veins and altered wall rocks. The above mineral association is consistent with strong correlation coefficients (i.e., >0.5) between Au and Ag, Te, W, As, Pb, Zn, and S as revealed by bulk geochemical analyses. In contrast, CO_2 yields low correlation coefficients (i.e., <0.3) with all the aforementioned elements. The $\delta^{13}C$ values of calcite do not reveal a precise pattern with regards to Au-mineralized intervals. However, there are broad differences between the $\delta^{13}C$ data recorded in the Pegasus deposit and the Papa Bear prospect with mean values of -5.0 (2.3‰ SD) and -0.8‰ (1.9‰ SD), respectively. Systematic SEM-EDS reveal that pyrrhotite ± arsenopyrite domains are dominant in mineralized sites at Pegasus and Papa Bear and transition to pyrite-pyrrhotite, pyrite ± anhydrite domains, whereby Au records background contents of 10 ppb. The $\delta^{34}S$ values of pyrrhotite and pyrite range from -2.7 to 9.0‰ (1.3‰ SD) and from -10.0 to 10.8‰ (1.9‰ SD), respectively. The negative $\delta^{34}S$ values in pyrite and its association with anhydrite (with $\delta^{34}S$ values from 7.2 to 10.2‰, 0.8 SD) suggest that pyrite was replaced by anhydrite under oxidizing. Extreme positive $\delta^{34}S$ values in pyrite, of up to 10.8‰, are inherited from SO_2 that underwent disproportionation to sulfide (H_2S) by reacting with water. In contrast, pyrrhotite in pyrrhotite ± arsenopyrite assemblages displays dominantly positive $\delta^{34}S$ in samples with up to 30 ppm Au, suggesting reduced fluid conditions during mineralization. The $\Delta^{33}S$ values in pyrrhotite and pyrite yield values from -0.2 to 1.4‰ (0.3‰ SD) and from -0.3 to 1.3‰ (0.3‰ SD), with samples with $\Delta^{33}S$ values of up 1.4‰ in pyrrhotite which is spatially coincident with graphitic shales adjacent to laminated veins hosted by the K2 shear zone. The combined mineralogical, geochemical, and multiple S and C isotope data imply that auriferous fluids were initially oxidized as evidenced by pyrite ± anhydrite assemblages and, focusing in shear zones and reacting with graphitic shales, underwent significant reduction. The latter interaction also resulted in S remobilization as evidence by MIF-S recorded in pyrrhotite within auriferous veins.

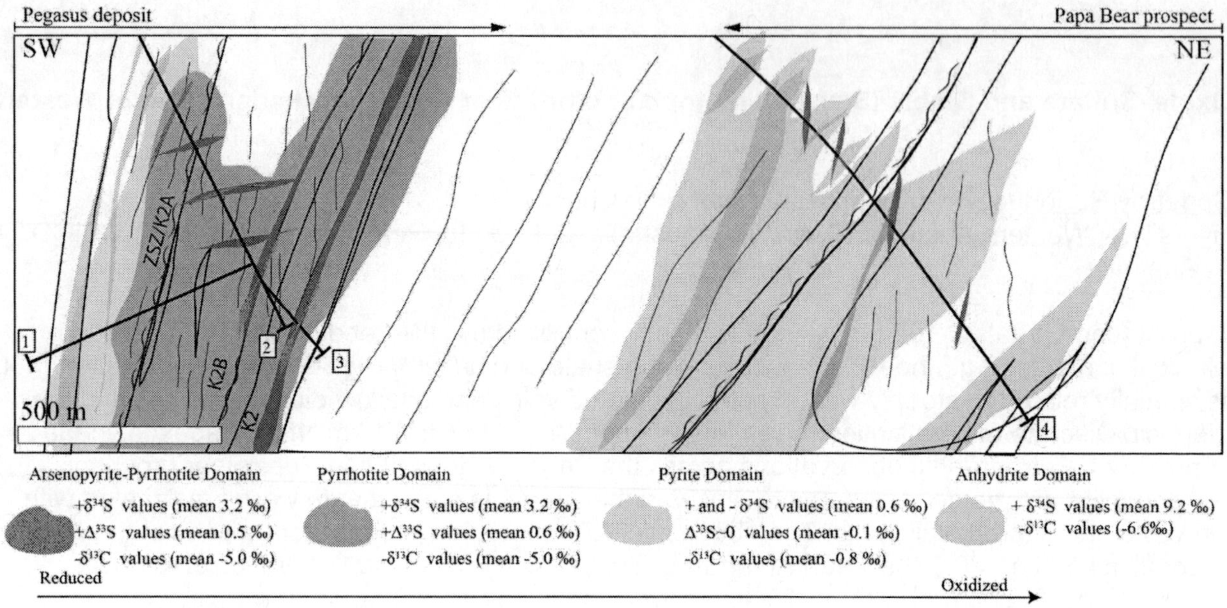

Pegasus deposit
Papa Bear prospect
SW
NE
ZSZ/K2A
K2B
K2
1
2
3
4
500 m
Arsenopyrite-Pyrrhotite Domain
+δ34S values (mean 3.2 ‰)
+Δ33S values (mean 0.5 ‰)
-δ13C values (mean -5.0 ‰)
Pyrrhotite Domain
+δ34S values (mean 3.2 ‰)
+Δ33S values (mean 0.6 ‰)
-δ13C values (mean -5.0 ‰)
Pyrite Domain
+ and - δ34S values (mean 0.6 ‰)
Δ33S~0 values (mean -0.1 ‰)
-δ13C values (mean -0.8 ‰)
Anhydrite Domain
+ δ34S values (mean 9.2 ‰)
-δ13C values (-6.6‰)
Reduced
Oxidized

ST.122
Geochemistry, Mineralogy, and Reconnaissance Cathodoluminescence Study of Gold Vein Samples, Eastern Interior Alaska

Garth E. Graham[1], Ryan D. Taylor[2], Heather A. Lowers[2]
1. US Geological Survey, Golden, CO, USA, 2. US Geological Survey, Denver, CO, USA

A series of gold occurrences, including the active Pogo mine, are located along a broadly east-southeast-trending region in eastern interior Alaska that stretches from the Shaw Creek fault in the west to Black Mountain in the east. Auriferous veins in the occurrences can be hosted in either metamorphic or mid-Cretaceous plutonic rocks. Vein thickness ranges from meter-scale (e.g., Liese veins at Pogo) to cm-scale sheeted veins. The total contained sulfide in veins is generally several percent and ore minerals are commonly very fine grained (<10-100s of microns).

Geochemical analyses of select vein samples from surface and drill cores indicate that a few prospects hosted partially or entirely in ortho- and paragneiss have strong correlations among Au, Bi, and Te. These prospects include the Pogo mine (Bi commonly exceeding ~100 to >1000 ppm; Te mostly >0.5 to 200 ppm; Au ranging to 100+ ppm) and the Boundary and Gray Lead prospects near/at Black Mountain. Despite significant arsenic in these samples, there is no strong As:Au correlation. Other prospects hosted in mid-Cretaceous granitoids include prospects just north of Gray Lead at Black Mountain. These prospects generally have no detectable Te and lower Bi (<100 ppm). In the limited sample suite, no correlation between Au and other elements exists, except in the West Pogo prospect, where enriched Bi and Te, and Sb-Au correlations are reported.

We selected mineralized vein samples from the greater Pogo system, Gray Lead (geochemically similar to Pogo), and Blue Lead (Sb-As-rich prospect) to examine mineralogy and quartz textures using cathodoluminescence (CL). Scanning electron microscope-energy dispersive spectroscopy analysis confirms the presence of Bi-bearing phases in the Pogo and Gray Lead samples. Pogo exhibited Bi-Pb-Sb sulfosalts, Bi-tellurides, chalcopyrite, pyrite, galena. Hill 4021 (on the Pogo property) exhibited arsenopyrite, loellingite, Bi-tellurides, electrum and complex intergrowths of Au- and Bi-bearing minerals. Gray Lead exhibited arsenopyrite, Bi-Te- and Bi-Sb sulfosalts, and Pb-Sb-Ag-Bi sulfosalt, and Blue Lead displayed arsenopyrite, electrum, stibnite, and unknown Sb-Pb-Ag-sulfosalt. Gold occurs as electrum, alone or intergrown with Bi-bearing minerals associated with quartz grain boundaries and cracks, or, rarely, as inclusions in clusters of euhedral arsenopyrite that may be rimmed by Bi-Te sulfides (including at Blue Lead).

We examined CL spectral intensities at the ~ 2.7-2.8 eV position, previously interpreted to reflect variations in Ti concentration. Despite complexity, intensities tend to vary from higher intensities in the interiors of quartz crystals and decreasing outward. Arsenopyrite, Bi-Te-bearing minerals, and gold occur within microveinlets of lowest intensity quartz that clearly crosscuts the older quartz, as well as along grain boundaries. Stibnite is paragenetically youngest at Blue Lead, filling voids and partially corroded earlier quartz generations. Preliminary microprobe analyses of quartz with different spectral intensities hint at the sulfide-related quartz having lower Ti concentrations, consistent with lower temperatures of formation during late quartz (and Au) deposition. Our preliminary observations in CL suggest that, despite variations in ore mineralogy empirically related to host-rock type, similar relative timing and processes were responsible for gold deposition within these widely spaced prospects and perhaps others in the region.

ST.124

New Findings on the Gold Deportment, Quartz Textures, and Fluid Characteristics in the Red Dot Sulfide Zone at Marigold Mine, Nevada

Dante E. Huff[1], Matthew T. Fithian[2], Anne A. Fulton[1], Elizabeth A. Holley[1]
1. Colorado School of Mines, Golden, CO, USA, 2. SSR Mining Inc., Vancouver, BC, Canada

The Battle Mountain mining district hosts a variety of magmatic-hydrothermal ore deposits, as well as sedimentary rock-hosted Au deposits that lack a conclusive relationship to magmatism. Mining of the sedimentary-rock hosted deposits has been primarily limited to oxidized zones, so there is little information available about the hypogene ore characteristics and mineralizing processes. The Red Dot deposit is a sulfide ore zone recently intersected by deep drilling at Marigold Mine.

At Red Dot, early quartz and sulfide veinlet textures resemble D and E veins from porphyry systems. Early-stage pyrite is enriched in Hg and Ni but contains less As than ore-stage pyrite and little to no Ag or Au. Ore-stage quartz lacks fluid inclusions, but produces a yellow cathodoluminescence response typical of low-temperature hydrothermal environments. Ore-stage pyrite occurs as arsenian pyrite rims enriched in Au, As, Ag, Cu, and Sb that overgrow pre-ore pyrite; these rims have $\delta^{34}S$ stable isotope values consistent with porphyry pyrite elsewhere in the district. Alteration styles include decarbonatization, carbonation, sericitization, argillization, silicification, sulfide precipitation, and oxidation. The absence of zonation in the alteration styles suggests Red Dot is distal to any causative porphyry stock. District aeromagnetic data indicate an absence of any buried igneous bodies beneath Marigold, providing additional evidence that Red Dot is distal to any magmatic-hydrothermal fluid source.

Red Dot is ideally situated along a complex structural network capable of focusing fluids from laterally distant sources into the host rocks repeatedly through time. Cretaceous magmatic-hydrothermal fluids related to the emplacement of the Trenton Canyon Stock or local quartz monzonite dikes at Marigold Mine may have caused the early-stage alteration and sulfide mineralization at Red Dot. The early-stage alteration and sulfide mineralization could also be attributed to the emplacement of felsic dikes at the Lone Tree and Trenton Canyon Au deposits during the Eocene. Apatite and zircon (U-Th)/He data support an Eocene formation age for ore-stage quartz and pyrite.

The alteration styles, ore-stage sulfide textures, geochemical signature, base metal and Au mineralization, and distal nature of Red Dot most closely resembles the Cove deposit, 25 km to the south. At Cove, overprinting polymetallic and Carlin-like Au mineralization are genetically related to Eocene magmatic-hydrothermal fluids. We suggest that Red Dot represents a similar expression of magmatic-hydrothermal mineralization, but at a greater distance from the causative intrusion and in different host rocks that underwent ideal preparation for interaction with ore fluids.

ST.126
Geospatial Analysis Delineates Lode Gold Prospectivity in Alaska

Susan M. Karl[1], Douglas C. Kreiner[1], George N. Case[1], Keith A. Labay[1], Nora B. Shew[1], Matthew Granitto[2], Bronwen Wang[1], Eric D. Anderson[2]
1. US Geological Survey, Anchorage, AK, USA, 2. US Geological Survey, Denver, CO, USA

Alaska produces more than 50,000 ounces of gold annually from placer deposits across the state. Many of the placer districts lack known lode sources of proportional size, suggesting potential for undiscovered lode gold. Comprehensive data-driven analyses were conducted in a geographic information system (GIS) framework to assess prospectivity for lode gold in Alaska. The analyses used available geospatial lithologic and geochemical data sets, known mineral sites, and geophysical data to build models for recognizing different types of gold systems. The models are applied to physiographic units defined by stream drainage basins that are approximately 100 km^2 in area. The GIS-based analyses successfully delineate areas known to contain gold systems and predict prospectivity in areas of Alaska where no gold deposits are known. Broad-based analyses for undivided lode gold systems have application for land-use decisions. GIS-based analyses can also be tailored to individual types of gold systems and may help exploration campaigns in areas where high sample densities for diverse data types are available.

Analyses were designed to assess the potential for lode gold systems in Alaska in general, as well as to distinguish the distribution of different types of gold systems, including orogenic, reduced-intrusion-related, epithermal, and gold-bearing porphyry. Multiple parameters that discriminate these types of gold mineralization were applied; geochemistry of rocks and stream sediments is the most heavily weighted parameter. Trace elements and pathfinder elements reflect fluid compositions in different systems and are diagnostic for distinguishing favorable geologic environments. Substantial overlap in geologic environments leads to overlap in parameters and difficulty in distinguishing types of gold mineralization. Owing to the limitations of reconnaissance scale geologic mapping and the uneven distribution of geochemical data available for Alaska, the four types of gold mineralization listed above are difficult to distinguish. To mitigate overlap, we combined gold systems that form in similar geologic environments; one ore-system model is reduced intrusion-related and orogenic systems, and a second model is oxidized epithermal and porphyry systems. This approach achieved better discrimination of gold systems and increased the precision of the distribution of types of gold mineralization.

Prospectivity analysis for lode gold potential in Alaska highlights areas where exposure is good as well as areas under surficial cover. Our results confirm high prospectivity for known gold occurrences and nearby areas, indicating potential for expansion of areas that contain gold deposits. Exploration in these areas may improve the balance between the volume of placer gold produced in a mining district and the relatively low volume of identified lode resources that contributed to the placer deposits, if the lode sources are preserved. Our analyses also indicate high gold prospectivity in a few areas in the state that are not known to contain lode gold deposits. These areas warrant further evaluation, new geologic mapping, systematic geochemical sampling, and application of modern technology for various types of sample analysis. Our results can help focus future investigations in areas where data are lacking, bedrock geology is incompletely understood, and prospectivity can be better defined and constrained with additional data.

ST.127
Zone Refining in an Epithermal Setting: A Gold Liberation Mechanism at the Klaza Deposit, Yukon

Well-Shen Lee[1], Daniel J. Kontak[1], Duane Petts[2], Simon Jackson[2]
1. Laurentian University, Sudbury, ON, Canada, 2. Geological Survey of Canada, Ottawa, ON, Canada

Gold mineralization in hydrothermal systems has long been considered to be a primary process; however, recent advances in analytical methods have highlighted the importance of secondary processes. The zone refining model proposed here discusses the role of secondary Au enrichment via its liberation from primary sulfides by hydrothermal fluids. The model is supported by a case study on the Klaza intermediate sulfidation epithermal deposit (Yukon) using evidence from empirical approaches: drill core logging, optical microscopy, SEM-EDS, CL, LA-ICP-MS element mapping of sulfides (pyrite, arsenopyrite, and sphalerite), and data-driven approaches (unsupervised machine learning workflows derived from the LA data). Integrated petrography and SEM-EDS studies reveal electrum as secondary inclusions in pyrite and arsenopyrite, whereas grains lacking such inclusions display primary As-zonation and banding. LA-ICP-MS element maps reveal that (1) primary zones in arsenopyrite have elevated Au-Sb (Au = 46 to 400 ppm) and (2) elevated Cu-Pb-Ag-Sb-Bi-Ba along fractures in all mapped sulfide grains. These data, combined with the data-driven approach (feature maps constructed with a combination of K-means clustering and principal component analysis), supports a model whereby coupled-dissolution-precipitation (CDP) processes destroyed the primary zonation in pyrite and arsenopyrite which released lattice-bound impurities which then re-precipitated as high-fineness electrum and Bi-Pb-Ag sulfides. This secondary process, which produces relatively pure secondary pyrite and arsenopyrite and upgrades metals (e.g., Au-Ag-Bi-Pb-Cu), constitutes zone refining and is considered an important and relevant mechanism for locally producing gold enrichment.

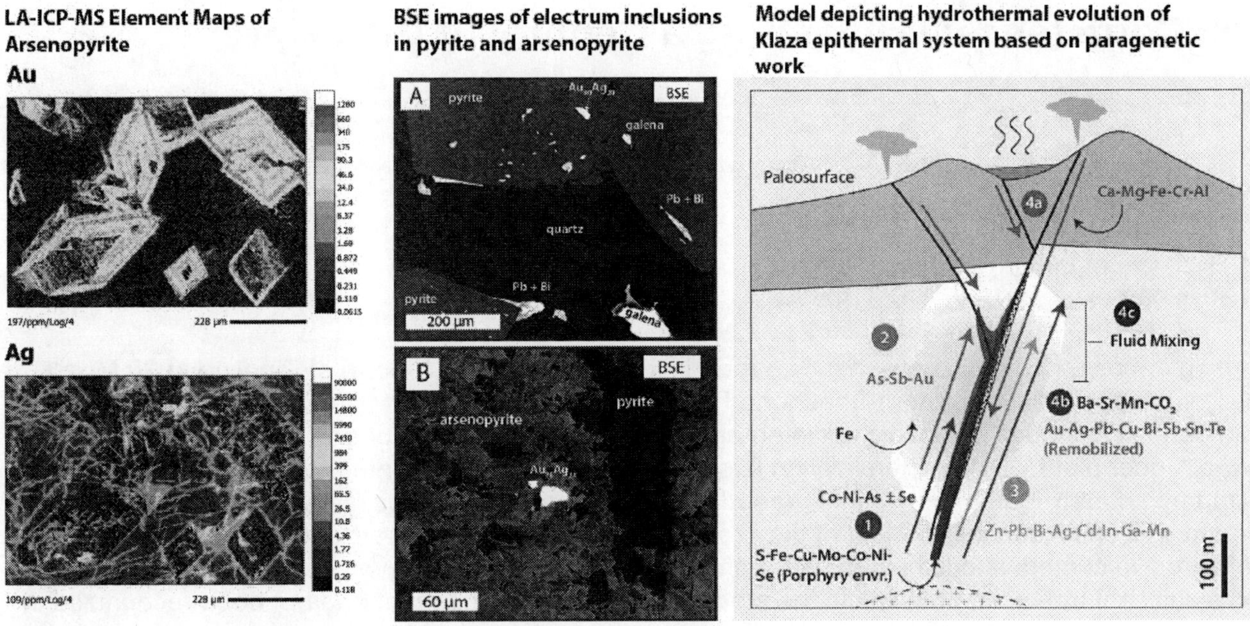

Figure 1

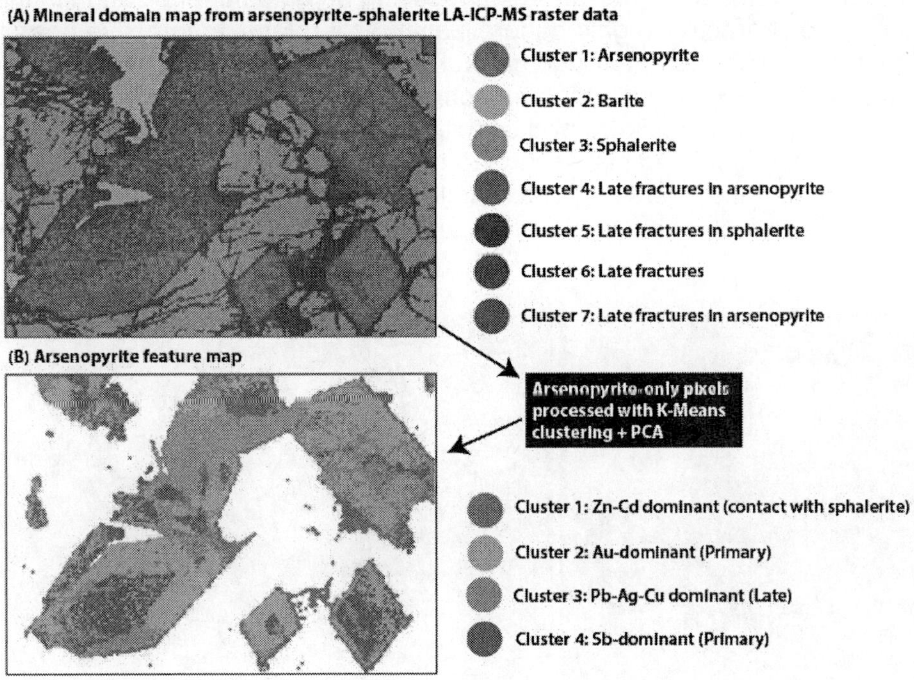

Figure 2

ST.129

The Robertson Deposit: Eocene Intrusion Related Gold Deposit in the Northern Shoshone Range, Nevada

Neal Mankins, John Muntean, Matthieu Harlaux
University of Nevada Reno, Reno, NV, USA

The Robertson deposit, formerly known as “Tenabo,” is located about 7 km North of Pipeline, a 20 Moz+ Carlin-type gold deposit in the northern Shoshone Range, Nevada. Robertson is an Eocene intrusion-related gold deposit hosted in paleozoic sedimentary lithologies, part of the Roberts Mountain allochthon known as the “upper plate.” Gold mineralization is spatially and temporally related to an intrusive complex characterized by an early diorite, polyphase granodiorite, and associated dikes, sills, and plugs that vary in composition from dioritic to rhyolitic. Most of the previously defined gold mineralization occurred near the contacts of intrusive centers and is hosted by hornfels in the overlying lithologies. Ongoing exploration has discovered significant free gold, both within the stock and in distal settings that had not been recognized in the past.

This study is the first in-depth examination of Robertson focusing on gold mineralization in these settings. The purpose of this study is to characterize gold mineralization as it relates to intrusive phases, alteration styles, and structural controls. This is done through careful examination of host lithologies using applied geochronology, petrography, and SEM work. Ongoing petrographic investigations have identified hornfelsing, potassic alteration, endo/exoskarn, and sericitic alteration related to a multiphase ilmenite-bearing granodiorite stock. Native gold has been found to have a strong association with Bi-Te-Se-S, spatially adjacent to or on the surface of arsenopyrite and loellingite grains. Understanding intrusion-related systems such as Robertson is critical for evaluating potential fluid sources that led to prolific mineralization in this district.

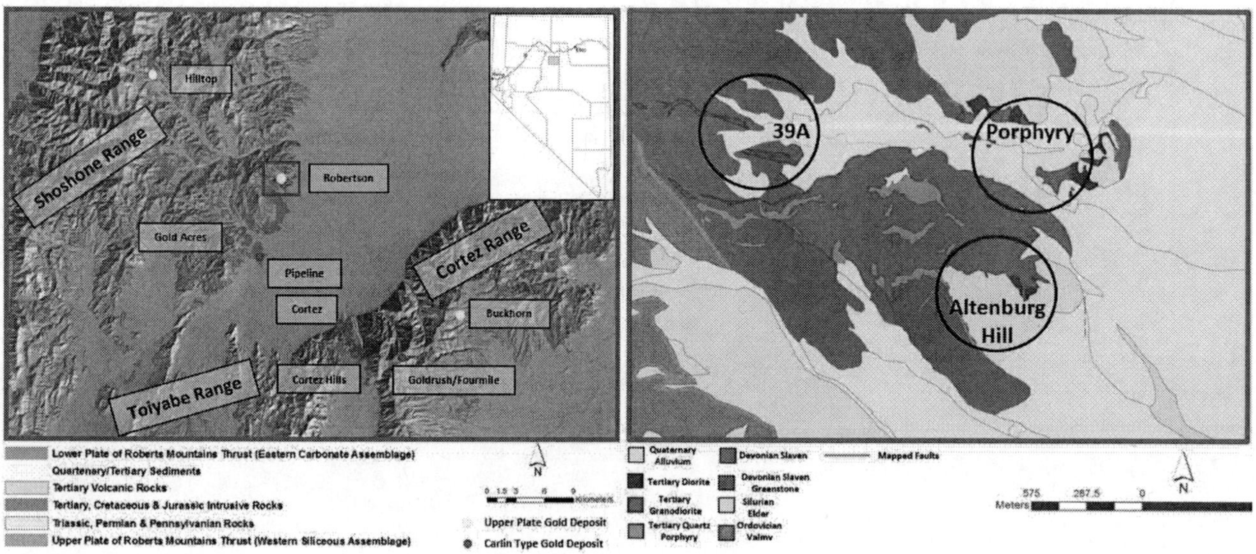

ST.133
Ore Paragenesis of Taka Barr Vein, Sangilo Epithermal Deposit, Baguio Mineral District, Philippines

Chelsi Grace S. Mendoza[1], Jillian Aira Gabo-Ratio[1], Acer Jian Figueroa[1], Pearlyn Manalo[2], Jessamin Belle Demegillo[1], Karl Jabagat[1], Marco Alfredo Barrientos[1], Elisha Jane Maglalang[1], Kotaro Yonezu[3], Ryohei Takahashi[2]

1. University of the Philippines - Diliman, Quezon City, Philippines, 2. Akita University, Akita City, Japan, 3. Kyushu University, Fukuoka City, Japan

The Sangilo Epithermal Deposit is a gold deposit adjacent to the Acupan Epithermal Deposit in the Baguio Mineral District, Philippines. The latter is a well-studied gold deposit hosted by the Virac Granodiorite and Balatoc Diatreme. Despite being operational since 1927, the former lacks studies on its mineralization and is hosted by the Lucbuban Gabbro.

The Taka Barr vein is the most productive and the main vein mined in the Sangilo Epithermal Deposit. It has an average width of 1 meter and an average orientation of due East, dipping 62°S. It is hosted by the Early Miocene intrusive units of the Central Cordillera Diorite Complex. The study investigated a section of the vein in mine level 2, which is at approximately 1200 masl. Three stages of mineralization were identified according to the textures and ore-gangue minerals present.

Stage I of the Taka Barr vein is adjacent to the walls. It exhibits brecciated textures with the hornblende quartz diorite host rocks as clasts and calcite as matrix. Cockade textures were observed around selected clasts near the host rock. Pyrite and chalcopyrite are found in the clast and rarely in the matrix. This stage is barren of precious metals.

Stage II, the gold-bearing mineralization stage, is distinguished by alternating gray bands of collo-crustiform texture. The sulphide minerals comprise almost 1% of the whole rock. Pyrite and chalcopyrite dominate as the major sulfide minerals present. Other sulfide minerals identified were sphalerite and galena. Gold minerals such as electrum, with 93-97 wt % Au, appear as isolated grains. Au-Ag tellurides, on the other hand, occur when sulphide minerals are present. Early-formed minerals include pyrite and galena, followed by late-stage precipitation of sphalerite and chalcopyrite.

Stage III is characterized by comb and drusy calcite crystals. Pyrite, chalcopyrite, and sphalerite were observed in this stage. No precious minerals were identified. Late-stage carbonate deposition of rhodochrosite and bladed calcite was also observed in this stage.

It can be inferred from the textures that the deposit first experienced brecciation before vein precipitation occurred. The collo-crustiform and bladed textures also point to boiling as a possible mechanism for ore deposition.

ST.138
Role of the Sheeted Veins in the Understanding of the Bonikro Intrusion-Related Gold System, Fettékro Greenstone Belt, Côte d'Ivoire

Zié Ouattara[1], Yacouba Coulibaly[2], Marie-christine Boiron[3]
1. Université de Man, Man, Ivory Coast, 2. Université Felix HOUPHOUET-BOIGNY, Abidjan, Ivory Coast, 3. Université de Lorraine, Nancy, France

The Fettèkro greenstone belt is one of the most productive gold belts in Côte d'Ivoire. In its southern part, the intensive exploration has revealed the gold deposits of Dougbafla, Agbahou, Bonikro, and Hiré, thus forming the Oumé-Hiré gold district.

The Bonikro gold deposit presents two main birimian units: the mafic volcanic made of basaltic to andesitic composition in the East and the westerly volcano-sedimentary composed of shales, siltstone, and pyroclastic lavas (basaltic to dacitic). These two units are metamorphosed to the greenschist facies and affected by the north-south Bonikro Shear Zone (BSZ). This major structural event governing the interface between these two units also controls the setting of the Bonikro main intrusive and associated lithologies: granodiorite, pegmatite, acidic lava, and aplite.

Gold is primarily found within the granodiorite and secondarily in the zones where the intrusive and the BSZ are also present. The mineralization, associated with strong hydrothermal activity, is marked by sericitisation, silicification, chloritisation, albitisation, and the presence of three veins generations—sheeted, planar, and transversal.

Among these veins, the sheeted show particular characteristics: (i) they are syngenetic, (ii) they display a low fluid-rock ratio, and (iii) they are scheelite-bearing veins.

The sheeted veins are earlier, thick (1-cm), sub-parallel sets of quartz and feldspar veins. They comprise milky quartz (70 to 80%), albite (5-10%), scheelite (up to 15%), and pyrite (up to 5%). Most of the deposit's visible gold is located in the sheeted veins.

Seventy-two fluids inclusions have been studied in the sheeted (34) and the planar (38) veins by microthermometry at the Université de Lorraine, Nancy, in France. Ten of them have been investigated by the raman spectroscopy method.

The primary fluids associated with the sheeted veins are monophase (carbonic), whereas the planar veins contain both monophase and biphase inclusions. The carbonic fluids are dominated by CO_2-rich CH_4-N_2. The aquo-carbonic contains more N_2 than CH_4, presenting a global composition of H_2O-rich CO_2-N_2 ± NaCl ± CH_4. The homogenization temperatures support that gold has been deposited in these veins from 200° to 290°C in primary fluids. In the secondary fluids, the temperature reaches 300°C and 125 to 250 Mpa.

Elsewhere in the Birimian of West Africa, similar fluids and compositions have been observed. The fluid immiscibility has been invoked to explain the gold deposition process in other deposits; it may have occurred in Bonikro.

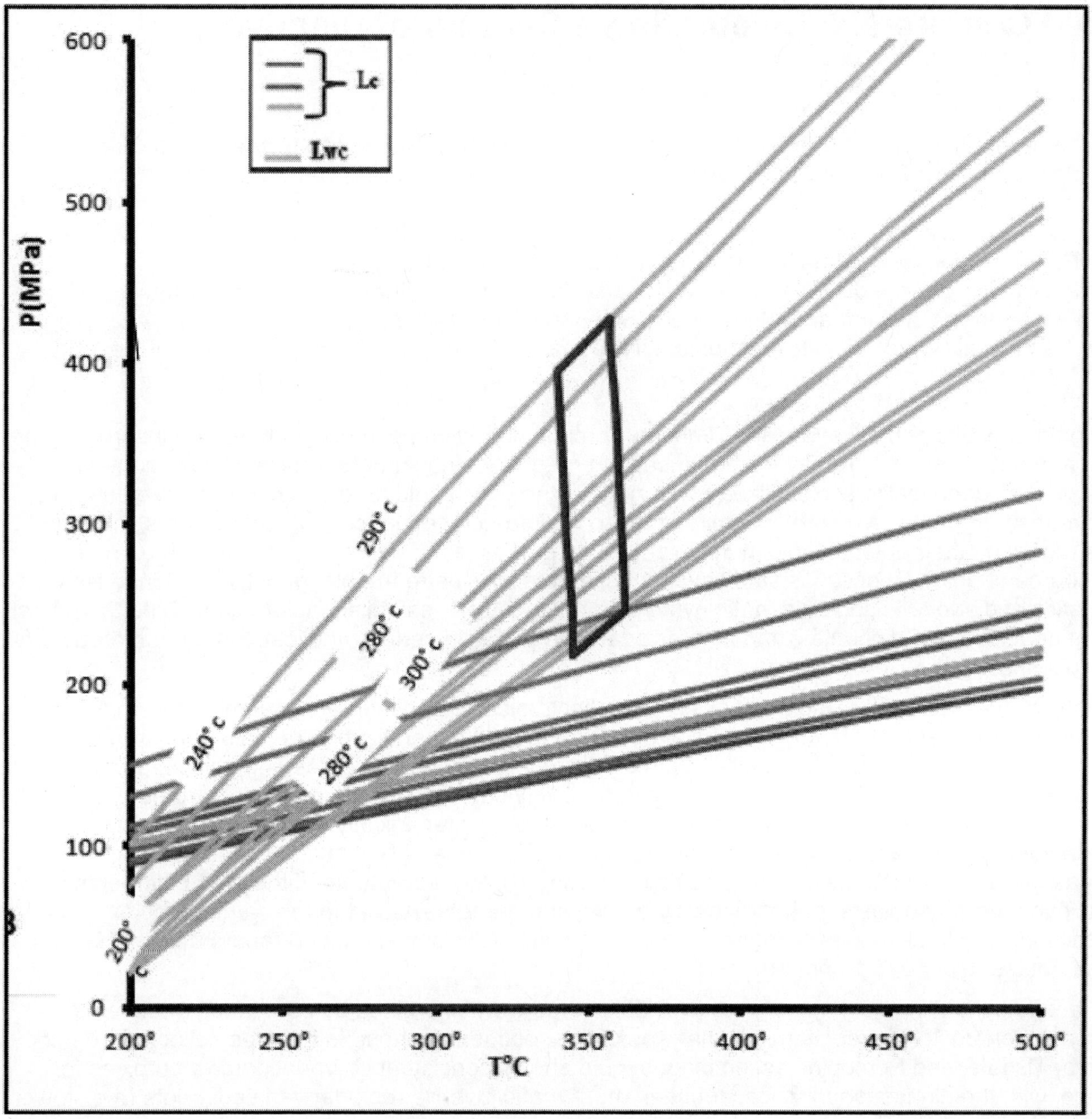

Lc
Lwc
600
500
400
300
200
100
0
P(MPa)
200°
250°
300°
350°
400°
450°
500°
T°C
290° c
280° c
300° c
240° c
280° c
200° c

ST.139

Gold in Sedimentary Pyrite and its Implications for Ore-forming Processes in the World-class Timmins-Porcupine Au Camp and its Surroundings, Abitibi Greenstone Belt, Canada

Jean-Luc Pilote[1], Simon E. Jackson[2], Patrick Mercier-Langevin[1], Benoît Dubé[1], Christopher J. Lawley[2], Duane C. Petts[2], Zhaoping Yang[2], Richard A. Stern[3], Ed van Hees[4], Dave Rhys[5]
1. Geological Survey of Canada, Québec, QC, Canada, 2. Geological Survey of Canada, Ottawa, ON, Canada, 3. Department of Earth and Atmospheric Sciences, University of Alberta, Edmonton, AB, Canada, 4. Ontario Geological Survey, South Porcupine, ON, Canada, 5. Panterra Geoservices Inc., Surrey, BC, Canada

Orogenic gold deposits of the world-class Timmins-Porcupine gold camp in the Archean Abitibi greenstone belt are commonly found in proximity to sequences of fine-grained sedimentary rocks. These contain diagenetic and epigenetically recrystallized pyrite nodules, in which gold tenors can vary from below ~10 ppb to more than 10 ppm. Given the proximity of these nodule-rich sequences to gold systems, important questions regarding their potential role in the ore-forming processes have to be addressed (e.g., one potential source of gold). Hence, this study examines in detail the petrographic, geochemical, and isotopic characteristics of diagenetic and epigenetic pyrite with the aim to: 1) establish factors controlling the primary enrichment of gold in pyrite nodules; and 2) interrogate the possible genetic implications for auriferous ore-forming processes.

Investigation of over 120 rock samples from seven distinct volcanic-sedimentary assemblages, including samples within, proximal to (<100 m), and distal (>100 m) to gold mineralization reveals many regional textural commonalities within pyrite nodules, such as polycrystalline nodule cores and acicular rims. In situ multi-element analyses (n = 454) by laser ablation inductively coupled plasma mass spectrometry (LA-ICP-MS) show that gold content is largely refractory and decreases systematically towards the acicular rim to near barren values. The trace elements Ni, As, Ag, Sn, Sb, Te, Pb, and Bi in pyrite co-vary positively with Au, whereas Mo and Tl generally behave antithetically with Au. Measurements of the same elements in epigenetic pyrite porphyroblasts, euhedral overgrowths, and granoblasts within the same samples show clear distinctions in chemical assemblages and compositions with, for instance, Tl tenors being >100 times greater in nodules (typically >1 ppm Tl).

Possible factors controlling gold abundance in nodule cores were examined by comparing geochemical proxies from whole-rock lithogeochemical analyses of the sedimentary host. In nodules collected from the Kidd-Munro, Tisdale, and Porcupine assemblages, gold shows consistent co-variations with proxies for provenance, with the highest abundances found in more mafic/juvenile rock-derived sediments (e.g., lower La/Sc, higher V/Al and Co/Th).

In situ sulfur isotope ratio ($^{34}S/^{32}S$, $^{33}S/^{32}S$) measurements on pyrite nodules show significant variations in composition ($\delta^{34}S$ = -18.3 to +8.7‰; $\Delta^{33}S$ = -2.1 to +4.2‰), with negative correlations between $\delta^{34}S$ and $\Delta^{33}S$ and, overall, suggest that the nodules formed through processes involving both mass-dependent and mass-independent sulfur isotope fractionation. Metamorphic pyrite unrelated to ore revealed mainly positive $\delta^{34}S$ and $\Delta^{33}S$ values ranging from -1.0 to +11.3‰ and -1.8 to +4.6‰, respectively, whereas ore-associated metamorphic pyrite in three gold deposits (Bradshaw, Bell Creek, and Pamour) show a restricted range in $\delta^{34}S$ values (+0.3 to +3.8‰) with slightly positive mass-independent fractionation signatures ($\Delta^{33}S$ = -0.3 to +0.6‰). These preliminary results suggest that at least some of the sulfur in the ore-forming fluids was atmospheric and was likely sequestered in sedimentary (± volcanic) rocks that now reside at depth. By liberation processes during metamorphism, and given that sulfur is an important transport ligand for gold, it is possible that some gold was also derived from diagenetic pyrite and incorporated into the ore-forming hydrothermal system.

ST.141

Interpretation of Pyrite LA-ICP-MS Maps from the Colosseum Au Mine, Southern California

Nelson Román[1], Daniel D. Gregory[1], Simon E. Jackson[2], Jean-Luc Pilote[3]
1. Department of Earth Sciences, University of Toronto, Toronto, ON, Canada, 2. Geological Survey of Canada, Ottawa, ON, Canada, 3. Geological Survey of Canada, Quebec, QC, Canada

Pyrite LA-ICP-MS mapping has become a common technique for the study of pyrite in ore deposits, since pyrite texture and trace element composition provide unique insights into the evolution of these systems. However, despite the advantages of the use of LA-ICP-MS maps, their interpretation is still challenging, especially when comparing pyrite generations from the same deposit that have similar textures or intricate paragenesis. In this study, we aim to develop a statistically robust approach to identifying chemical zones associated with different pyrite generations, such that, combined with detailed petrographic observations, they allow confident comparison of pyrite from different parts of a deposit, even when pyrite textures are similar and crosscutting relationships are absent.

For this purpose, we applied unsupervised clustering for the examination of pyrite LA-ICP-MS maps from the Colosseum Au mine, southern California. Gold mineralization here is closely associated with pyrite, providing an ideal framework for the study of the relationship of pyrite features with important mineralization events. We utilized five pyrite LA-ICP-MS maps, representative of the two main pyrite generations in the Colosseum mine, which are (1) early pyrite, characterized by coarse, euhedral crystals (>500 μm), frequently fractured, and (2) late pyrite, characterized by aggregates of finer-grained euhedral crystals (<150 μm).

The clustering techniques used were K-means, Clustering Large Applications (CLARA), Hierarchical clustering, Hierarchical K-means, and Gaussian Mixture Models (GMM). Results were validated using clustered maps and biplots of compositional data. The best clustering results were achieved by different methods in each map. Hierarchical clustering (Ward), CLARA, and GMM yielded the best results. In each map, four to five clusters were identified, which correctly differentiated between early pyrite, fractures in early pyrite, and late pyrite, matching petrographic observation.

Clustered maps reveal that early pyrite is characterized by compositional zoning, which occurs mostly as concentric zones. Conversely, late pyrite does not show intra-grain compositional zoning. Compositional transition from early to late pyrite is represented by a decrease of Co, Ni, and Te contents, coupled with an enrichment of Cu, Ag, Sb, Pb, and Tl contents from early to late pyrite. Late pyrite has consistently higher Au concentrations than early pyrite, with median Au concentrations of 1.1 and 0.06 ppm, respectively, emphasizing the association of late pyrite with economic mineralization. However, among the zones identified in early pyrite, a group shows higher As and Au concentrations, with medians higher than 2,000 and 0.7 ppm, respectively. Moreover, Au in fractures in early pyrite reach abundances over 100 ppm. This is consistent with the occurrence of anomalous, economic areas in the deposit not necessarily associated with late pyrite. The changes between early and late pyrite could be related to a shift in the physicochemical conditions during pyrite deposition, which also led to efficient Au mineralization. This aspect will be investigated in subsequent studies.

These results show that clustering techniques applied to pyrite compositional maps facilitate chemical zone identification in pyrite grains, allowing direct comparison between pyrite generations and the extraction of relevant information in a petrographic and paragenetic context.

ST.142

Hydrothermal Alteration and Gold Mineralization at the Gruyere Gold Deposit (Yamarna Terrane, Yilgarn Craton, Western Australia)

Ravi Schreefel[1], Steffen G. Hagemann[1], Chris M. Fisher[1], Anthony I. Kemp[1, 2]
1. Centre for Exploration Targeting, School of Earth Sciences, The University of Western Australia, Perth, WA, Australia, 2. School of Earth Sciences, The University of Western Australia, Perth, WA, Australia

The Yamarna terrane is situated at the north-eastern margin of the Yilgarn craton (Western Australia) and is part of the Eastern Goldfields Superterrane (EGST). The largely unexposed terrane encompasses the Yamarna greenstone belt (~2737-2666 Ma) and Dorothy Hills greenstone belt (DHGB; ~2840-2810 Ma). The Gruyere gold deposit (135.54 Mt at 1.31 g/t for 5.73 Moz Au; Gold Road Resources Limited, ASX Announcement 15-02-2021), mined since 2019, is the eastern-most gold mine of the EGST and is situated within the DHGB at an inflection in the regional-scale Dorothy Hills shear zone. Only limited geological investigations were conducted on the evolution of the Gruyere gold deposit. Therefore, it is unclear if gold mineralization was part of an early (orogenic and/or intrusion-related?) mineralization and deformation event at ~2810 Ma, or whether mineralization was part of the younger mineralization event at ~2665-2620 Ma that gave rise to widespread gold mineralization in the EGST.

Gold mineralization in the Gruyere gold deposit is hosted by the Gruyere granitic intrusion (GYI; 2834 ± 3 Ma; U-Pb on zircon using LA-ICP-MS). The steeply east-dipping GYI intrudes tholeiitic, low-Th meta-basalts to the east and intermediate to mafic volcanoclastic rocks to the west. Numerous steeply ENE-dipping, post-mineralization, mafic dikes (<5 m width) intrude the DHGB sequence.

Hydrothermal alteration associated with gold mineralization of the GYI can be subdivided into three alteration zones. The distal alteration zone (<1.0 g/t Au) is pervasive with a steeply ENE-dipping foliation, and consists of a dark-green, biotite-chlorite ± albite ± sericite ± carbonate ± pyrite assemblage. The intermediate alteration zone (0.5-3.0 g/t Au) has a distinct feldspar-porphyritic texture and consists of a beige albite-biotite-chlorite ± sericite ± carbonate ± pyrite ± pyrrhotite ± arsenopyrite ± gold assemblage. Tabular quartz-veins with slip-fibre textures, which are striking subparallel to the earlier biotite-foliation, are characteristic for the intermediate zone. These quartz ± albite veins have distinct chlorite-biotite selvages with pyrite-pyrrhotite ± arsenopyrite ± galena ± chalcopyrite ± gold ± electrum ± telluride assemblages. The proximal alteration zone (>2.0 g/t Au; <5 m width) shows complete obliteration of the igneous textures, and consists of a yellow-beige sericite-albite ± biotite ± chlorite ± pyrite ± arsenopyrite ± pyrrhotite ± gold ± electrum ± telluride assemblage. Massive, irregular quartz-veins with granitic wall-rock slivers crosscut the tabular quartz-veins and are characteristic for the proximal zone. Sulphide-gold assemblages of the proximal zone are predominantly located within altered granitic slivers and are in equilibrium with biotite-chlorite assemblages within the irregular quartz-veins. Accessory minerals such as rutile, ilmenite, apatite, monazite, xenotime, and zircon are in many places observed at the vein selvages, and share common grain boundaries with pyrite, arsenopyrite, and gold. These equilibrium assemblages suggest a hydrothermal origin of the accessory minerals.

Further geochemical work is ongoing to constrain the precise gold mineralization age. The age of the Gruyere intrusion will then be compared to the age of gold mineralization in order to constrain a robust genetic model that can be used for exploration targeting of concealed gold mineralization in the Yamarna terrane.

ST.144

Multiple Sulfur Isotope and Trace Element Analysis of Pyrite in the Kanowna Belle Deposit, Western Australia: Insights into Fluid Oxidation and Gold Precipitation Processes

Dennis Sugiono[1], Crystal K. LaFlamme[2], Nicolas Thébaud[1], Laure Martin[3], Dany Savard[4], Marco Fiorentini[1]
1. CET-University of Western Australia, CRAWLEY, WA, Australia, 2. Université Laval, Quebec City, QC, Canada, 3. CMCA-University of Western Australia, Crawley, WA, Australia, 4. Laboratoire des Matériaux Terrestres-UQAC, Chicoutimi, QC, Canada

In orogenic gold deposits, the processes by which gold is precipitated from hydrothermal fluids to form Au-rich veins remain controversial. Textural observations and geochemical analyses on orogenic gold deposits have come up with different interpretations on how gold is precipitated, such as fluid mixing, fluid-rock interaction, wall-rock carbonation, and/or phase separation (boiling). The Kanowna Belle deposit hosts gold mineralization in quartz-pyrite-gold veins hosted in volcano-sedimentary and granitic rocks of the Kalgoorlie Terrane, Australia. Pyrite hosted in the gold-rich veins demonstrates evidence of three phases of growth zoning (Pyrite core, mantle, and rim). Using in-situ multiple sulfur isotopes and trace element geochemistry coupled with vein paragenesis, we monitor different phases of hydrothermal fluid evolution, including the critical phase when gold precipitation occurs. Results show that the pyrite core is Au-poor (Au ≤149 ppm), with moderate As (≤2.5 wt %) and Te content (≤416 ppm) and elevated Ni contents (≤4,022 ppm). Pyrite mantle is Au-rich (≤2,251 ppm) with high As (≤4.5 wt %) and Te (≤829 ppm) content and lower Ni (≤1,111 ppm) content. The late pyrite-rim are Au- (≤264 ppm) and As- (<1.4 wt %) poor, with low Te (≤229 ppm) and Ni (≤2113 ppm) contents. The sulfur isotope composition of the pyrite comprises heavier and more constrained values in the pyrite core ($\delta^{34}S$ = -3.3 to +4.2‰), followed by a lighter range with wider variation in the pyrite mantle ($\delta^{34}S$ = -8.4 to +0.1‰), and a heavier range in the pyrite rim ($\delta^{34}S$ = -6.0 to +2.8‰). Importantly, each pyrite zone preserves consistent $\Delta^{33}S$ values of $\Delta^{33}S$ = +0.3 ± 0.2‰, indicating a single sulfur reservoir was retained during the whole pyrite precipitation process. Therefore, the changes of trace element contents and $\delta^{34}S$ content within the pyrite growth zones should be related to fluctuations of fluid oxidation and pressure conditions. Gold-rich pyrite is precipitated during fluctuations of SO_4^{2-}/H_2S conditions associated with phase separation induced by a rapid drop of fluid pressure during the "fault-valve" process. Our findings demonstrated that the "fault-valve" is an important process in precipitating gold from a single Au-rich fluid reservoir and is an integral process for the formation of Orogenic gold deposits.

ST.145

Geological and Structural Characterization of the Odyssey Project, Malartic Mining Camp, Abitibi, QC: Age and Structural Controls on Gold Emplacement

Eliott Théas[1], Stéphane De Souza[1], Stéphane Perrouty[2], Ehouman N'dah[3]
1. Université du Québec à Montréal, Montréal, QC, Canada, 2. Laurentian University, Sudbury, ON, Canada, 3. Mine Canadian Malartic, Malartic, QC, Canada

Felsic-intermediate intrusive rocks are present in most orogenic and intrusion-related gold systems, and commonly act as local structural or/and chemical traps for mineralizing fluids. This study aims at characterizing the gold-mineralized fracture network of a complex ore system and its host intrusive rocks in order to determine the nature of the main controls on mineralization.

Discovered in 2014, auriferous zones of the Odyssey project are located 2 km east of the Canadian Malartic open pit gold mine and could represent a significant underground extension of this world-class deposit (~16 Moz Au; reserves and past-production). Mineralized zones are developed within the Larder Lake-Cadillac fault zone, a major crustal break that marks the contact between the Abitibi and Pontiac subprovinces (Fig. 1A). Twenty oriented drill cores from the North, South, and Internal zones of the Odyssey project have been logged and sampled for mineralogy, lithogeochemistry, and U-Pb geochronology.

The North and South zones (1.56 Moz indicated and inferred gold resources) are developed along the faulted contacts of a sub-alkaline to alkaline porphyritic quartz monzodiorite intrusion that was emplaced between the metamorphosed ultramafic rocks of the Piché Group and sedimentary rocks of the Pontiac Group. These zones are ~10 m thick, dip >60° toward the south, and are composed of discontinuous quartz ± calcite ± pyrite ± gold veins and disseminated pyrite ± gold in the host intrusion. Westward, the North zone can be mapped almost continuously into the Sladen fault, one of the main metallotects of the Canadian Malartic system.

The Internal zones are more complex and appear to be spatially-associated with dikes crosscutting the main intrusion, regardless of their composition. The dikes are sub-alkaline and range from mafic to felsic, with equigranular, porphyritic, and aplitic textures. The Internal zones consist of a complex assemblage of sericite-quartz-pyrite-gold shear zones and various types of veins, the most common ones being quartz ± calcite ± pyrite ± gold veins and silica flooding zones, which are associated with carbonate ± pyrite ± chlorite ± hematite alteration and a locally developed N-S- to NNW-SSE-trending mineralized fracture cleavage. Ore shoots are steeply plunging and appear to be developed in the vicinity of the dikes. The auriferous veins locally contain hydrothermal titanite, paragenetically associated with gold and yielding a U-Pb age of ~2661 Ma (ID-TIMS), almost identical to the 2664 Ma Re-Os molybdenite age for mineralization, determined in previous studies at the Canadian Malartic deposit.

This study suggests that, independently of genetic considerations, the Internal zones are spatially associated with dikes that played a key role in the emplacement of mineralization. The North zone of the Odyssey project and those of the Canadian Malartic deposit are in structural continuity, and dating of hydrothermal titanite suggests that mineralization was contemporaneous, increasing the overall extent of the Canadian Malartic ore system to more than 5 km along-strike.

Figure 1: A) Simplified geological map of the Canadian Malartic mine area and location of the Odyssey project. LLCfz: Larder Lake – Cadillac fault zone. B) Section across the Odyssey project.

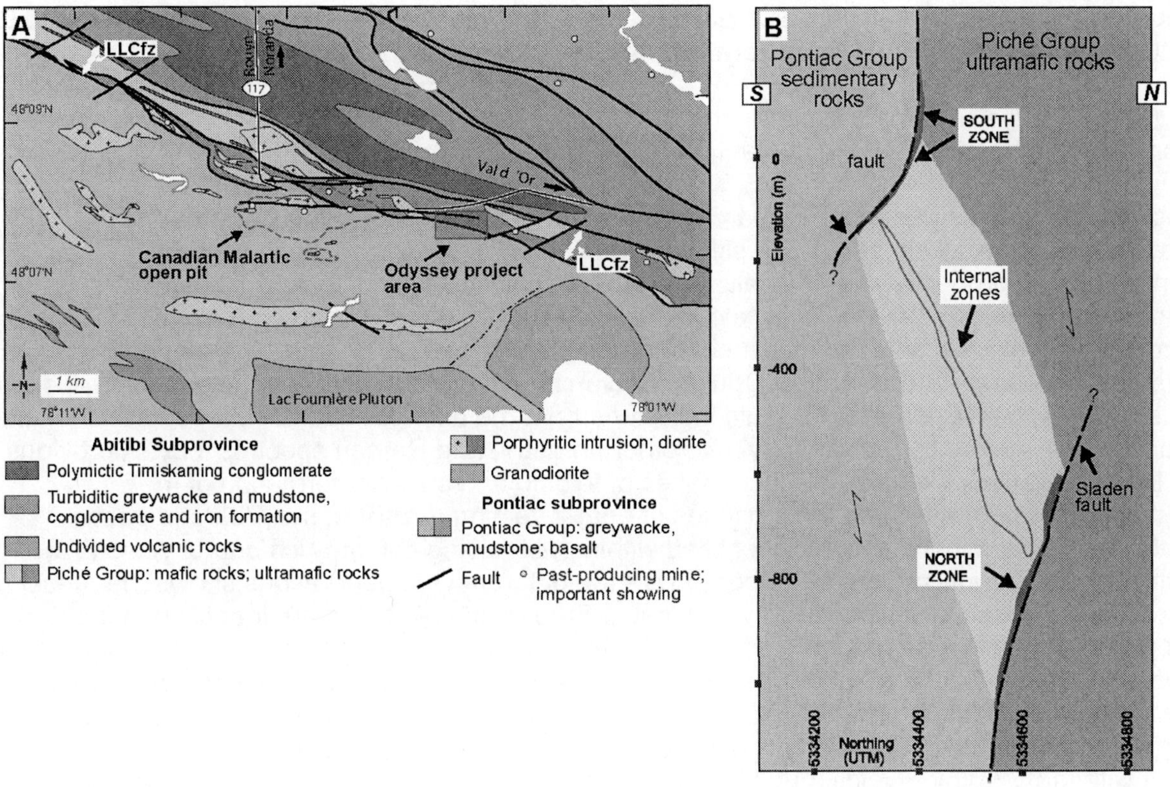

ST.147

A Fluid Inclusion Study of the Auriferous Quartz Veins in the Mesoproterozoic Karagwe-Ankole Belt (Byumba, Rwanda): Petrography, Microthermometry, and Raman Spectroscopy

Sander Wouters[1], Stijn Dewaele[2], Philippe Muchez[1]
1. KU Leuven, Leuven, Belgium, 2. Ugent, Gent, Belgium

Despite a resurgence of mineral deposit research in the area, gold deposits in the Central African Mesoproterozoic Karagwe-Ankole belt (KAB) still have many unanswered questions concerning their genesis. Central African orogenic gold mineralization has been linked to fold-and-thrust belt formation accompanied by shear movement. However, a genetic link between the auriferous quartz veins and early-Neoproterozoic post compressional magmatism (G4 granites) is still under debate. The aim of this research is to characterize the quartz vein mineralizing fluids at Byumba (Rwanda) in order to help resolve this long lasting discussion. In order to achieve this, fluid inclusions from the different quartz vein generations at the Byumba deposit were studied by petrography, microthermometry, and Raman spectroscopy. A previous study at the Byumba deposit identified distinct phases of folding and shear deformation of the metasediments, associated with three main phases of quartz veining (pre-[V1], syn[V2], and post-folding[V3]). Primary gold mineralization is identified within post-folding chlorite-rich quartz veins (V3), primarily in the form of sub-micron gold. Petrography of the different vein generations shows the presence of many leaked/decrepitated inclusions in all vein types. Three main fluid types are identified in the Byumba quartz veins (see Figure 1). Type I are primary H_2O-CO_2-X fluid inclusions which consist of an aqueous and a liquid carbonic phase (Lw-Lc) and other gasses (X). Type II primary CO_2-H_2O-X fluid inclusions are dominantly carbonic liquid with a rim of H_2O (Lc-Lw). Type III are aqueous inclusions (Lw-V) found inside secondary trails and have not been studied. Fluid inclusions inside a V1 quartz vein show many resemblances with the V3 quartz veins, but contain more secondary trails. Syn-folding V2 quartz vein inclusions were not suited for further microthermometric analysis. They are too small to be measured or have been leaked/decrepitated. Auriferous V3 quartz veins show large cloudy clusters of primary inclusions. Microthermometric data of primary fluid inclusions of Type I and II is similar in the pre- and post-folding quartz veins. Melting temperatures of CO_2 (Tm_{CO2}) are on average between -61.0 and -59.0°C, with a higher average for Type I inclusions. These temperatures indicate the presence of other gasses inside the carbonic phase, verified by subsequent Raman spectroscopy, which shows significant amounts of N_2 and CH_4. Homogenization of the carbonic phase (Th_{CO2}) is variable within a single assemblage, mostly between 3.5 and 9.5°C. Within Type I fluid inclusions it was often not possible to see the formation of frozen CO_2 because it was completely used up in clathrate formation. Clathrate melting temperatures (Tm_{CLATH}) are mostly between 10.5 and 11.0°C. The primary fluids present in the post-folding auriferous quartz generation have a H_2O-CO_2-CH_4-N_2 composition, which is often associated with a metamorphic origin. However, in a next step, a more detailed analysis of the major and trace element composition of the inclusions will be performed to determine the possible influence/mixing with other likely fluid types, such as magmatic hydrothermal fluids.

Figure 1: Overview of the different fluid types inside an auriferous V3 quartz vein at Byumba.

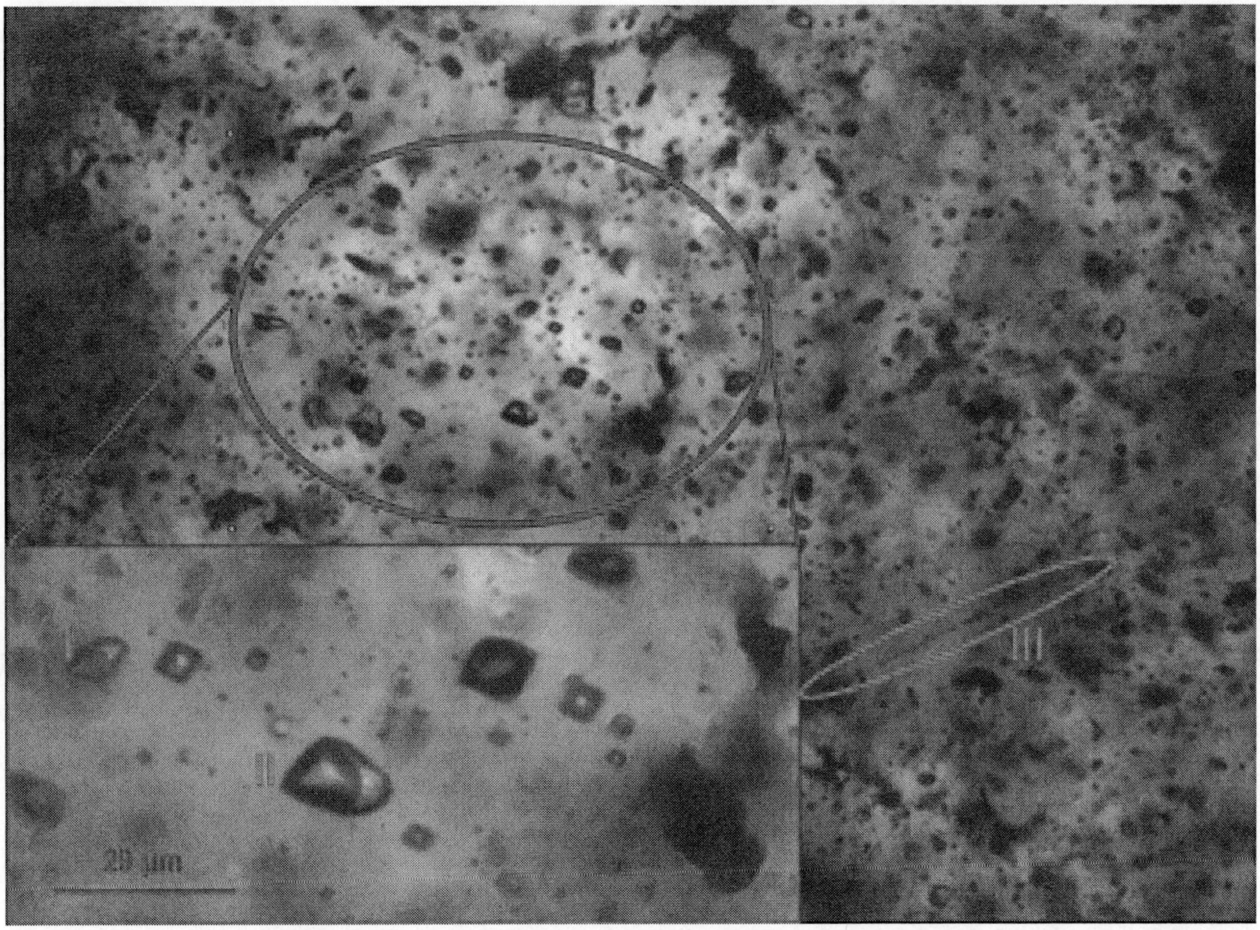

ST.148
Fe and Cu Isotope Compositions of Sulfides in Several Vein Gold Deposits, China

Jiuhua Xu, Hui Zhang
University of Science and Technology Beijing, Beijing, China

Fe and Cu isotope data are useful in studying ore-forming process, including magma-fluid exolution, evolution, and source of ore-matierals, but papers on vein gold deposits were rare as of yet. Pyrite and chalcopyrite samples of vein gold deposits from the Xiaoqinling (XQL) in west Henan, the Dongping (DP) of north Hebei, and southern Altay were selected to analyze Fe and Cu isotope compositions in this study.

The pyrite samples were all collected from an earlier stage (pyrite-quartz stage) of studied deposits. The results show that the δ^{56}Fe values of pyrite in the XQL have a relatively narrow range varying from -0.026 to 0.409‰ (7 samples) for the Wenyu-Dongchuang gold deposit, and from 0.298 to 0.558 ‰ (3 samples) for the Dongtongyu deposit. The δ^{56}Fe values of pyrite from DP range from -0.010 to 0.255‰ (4 samples) with little variation, whereas those in the Sarekoubu (SR) gold deposit of southern Altay change remarkably from -0.498 to 0.759‰ (6 samples). The δ^{56}Fe values of chalcopyrite from SR deposit are from 0.393 to 0.417‰, and those from the Wulasigou (WL) deposit in southern Altay are from -0.175 to -0.122 ‰ (6 samples).

The δ^{56}Fe values of pyrite from vein gold deposits are mostly positive. These results are somewhat different from the δ^{56}Fe values of pyrite in the Jiaodong Peninsula of east NNC (Zhu and Jiang, 2018), especially for the pyrite in quartz veins (Linglong deposit), which has a lighter δ^{56}Fe range from -0.78 to 0.79‰ (16 samples). Combined with sulfide isotope compositon, it is suggested that the δ^{34}S of pyrite is somehow related to δ^{56}Fe. Different gold deposits have different areas in δ^{56}Fe vs δ^{34}S diagram (Fig.1).

The δ^{65}Cu values of chalcopyrite in later stages (polymetallic sulfides and dark grey quartz stage) from the Sarekoubu gold deposit have a narrow range, which are from -0.280 to 0.922‰ (5 samples) with error ranges of 0.02-0.06‰ (2SD). The δ^{65}Cu values of chalcopyrite in late overprint gold-bearing quartz veins have a narrow δ^{65}Cu range, from -0.14 to -0.02‰ for the adjacent Tiemurte Pb-Zn-Cu deposit and from 0.149 to 0.169‰ for the Wulasigou Cu-Zn-Pb deposit. These results show that gold mineralization in southern Altay is different from that of VMS/SEDEX which have a wide range of δ^{65}Cu values (Fig.2).

This work is funded by NSFC (41672070).

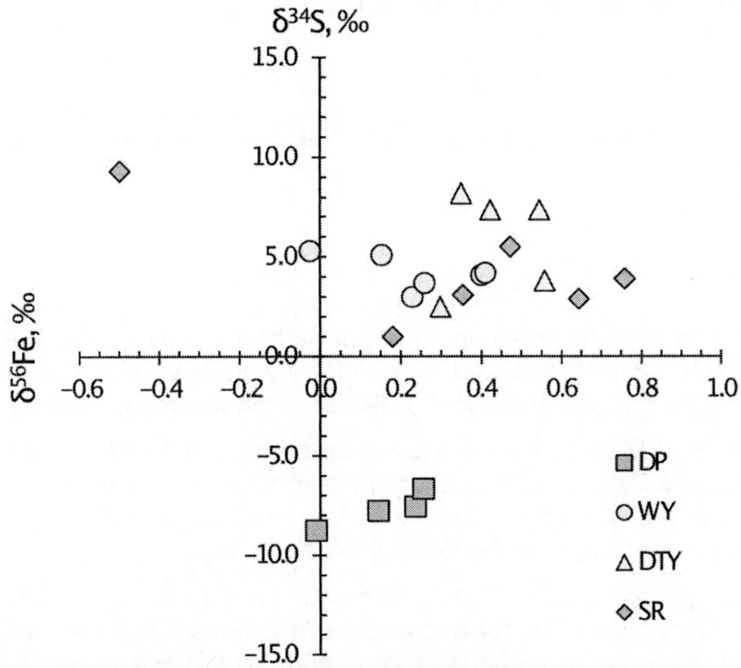

Fig.1 δ⁵⁶Fe vs δ³⁴S diagram of several vein gold deposits

DP-Dongping gold deposit in north Hebei province, China; WY-Wenyu-Dongchuang gold deposit in Xiaoqinling, west Henan, China; DTY-Dongtongyu gold deposit in Xiaoqinling; SR-Sarekoubu gold deposit in southern Altay, China

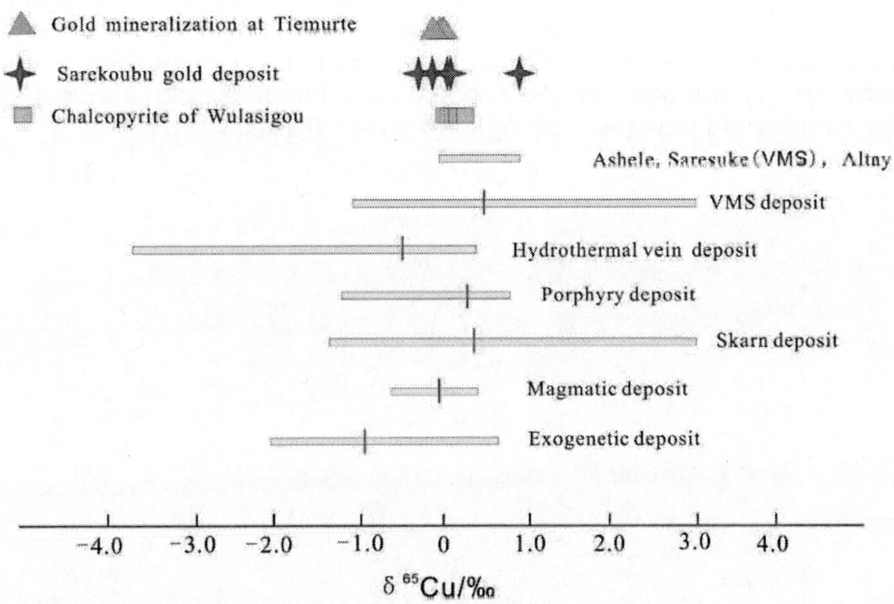

Fig.2 Comparison of Cu isotope composition of chalcopyrite in gold mineralization at southern Altay with some other ore deposits

(Modified after Wang and Zhu., 2010; Data of Altay are our unpublished work)

ST.149

The 12.5 Moz Au Sergeevsko-Klyuchevskoe IRGS Cluster (Transbaikalia, Russia) and Yanshanian Gold Metallogeny

Alexander Yakubchuk, Konstantin Lobanov, Sergei Shmatov
Orsu Metals Corp, White Rock, BC, Canada

The Sergeevsko-Klyuchevskoe cluster of gold deposits, containing about 390 t Au and grading 1.3-1.5 g/t Au, is the tenth largest hardrock gold endowment in Russia and is one of the world's largest intrusion-related gold deposits. The cluster is part of the 1,400 t Au Mogocha district (Transbaikalia, Russia).

The Sergeevsko-Klyuchevskoe stockwork (3 x 0.7 km) extends west-east, being structurally controlled by the dextral extensional duplex (Fig. 1A). Disseminated sulfides and quartz-tourmaline veins within an alteration envelope are found in Permian and Late Jurassic intrusive rocks. The latter consists of seven magmatic phases, some of which are genetically related to Au-(Mo) mineralization.

The Au mineralization of Transbaikalia associates with Mo-(Cu). Traditionally, the metallogenic belts are interpreted to be controlled by the Mongol-Okhotsk suture, formed at the end of the Middle Jurassic (~162 Ma) after collision between the southeastern edge of the Siberian craton and the Argun and Bureya terranes of the Central Asian orogenic supercollage, generally coinciding with the area of Late Jurassic igneous rocks. However, gold-bearing deposits of different types (intrusion-related, epithermal, and porphyry), but close in age (159-143 Ma), are grouped along transverse mineral trends striking across the suture (Fig. 1B). This indicates post-collisional timing of mineralization.

These Transbaikalian metallogenic belts occur on the northern periphery of the Yanshanian metallogenic province, occupying an area of 4,500 x 3,300 km, mostly in eastern China (Fig. 1B). The individual belts of this province are controlled by magmatic arcs and giant faults, stretching for thousands of kilometers parallel to the Pacific coast. The displacement along the faults reaches hundreds of kilometers. They are kinematically related to the northward collision between the Yangtze craton up to the southern periphery of the Siberian craton, part of overall northward translation along the western Pacific margin.

Figure 1: (A) Geology of the Sergeevsko-Klyuchevskoe gold cluster and (B) its position in the Yanshanian areal gold province of eastern Asia. Note control of giant sinistral strike-slip faults.

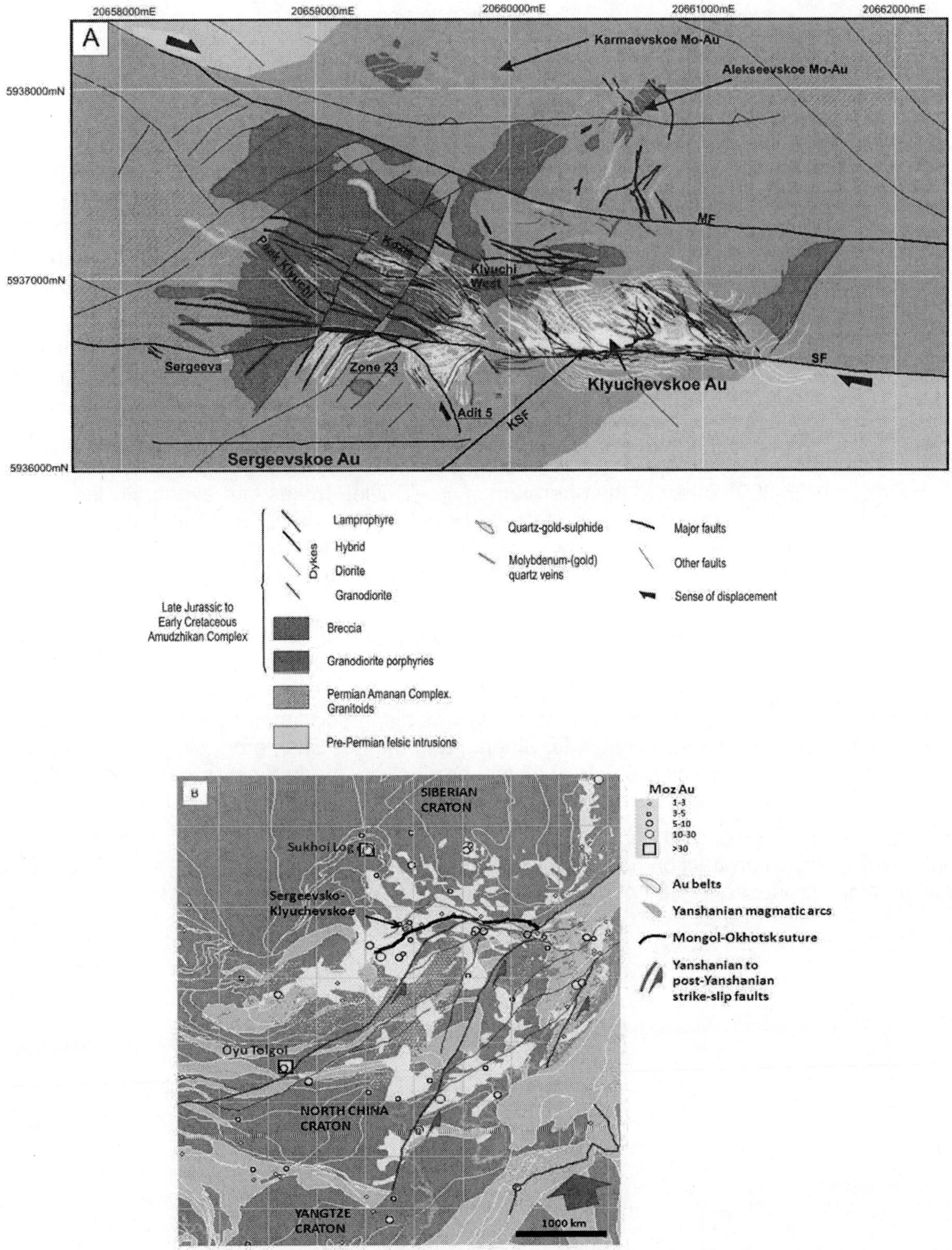

ST.150
1991-2020: Russian Gold Production

Alexander Yakubchuk
Orsu Metals Corp, White Rock, BC, Canada

During 30 years after the collapse of the USSR, Russia has produced more than 6,000 t (194 million ounces) of gold, about 10% of the annual world production. By 2020, the total Russian gold production increased 2.3 times to 1991 level (Fig. 1), whereas the world production went up ~1.5 times. In 1991-2020, Russian gold was extracted from more than 100 hardrock deposits and many placers (about 4,500 placers historically). Eastern Siberia and the Far East are undisputed regional leaders in Russian world-class hardrock gold mining.

Excluding gold extracted from scrap, placers provided 79.2% of 143.1 t of Russian gold in 1991, and in 2020, their share was 24% from the total of 331.1 t Au. In absolute terms, this was 30% less than in 1991 (113.4 vs. 79.5 t), meaning that gold from placers is still a significant contributor to the overall production. On the other hand, gold production from hardrock deposits has grown 8.5 times from 29.7 t (in 1991) to 251.6 t (in 2020).

The fall in gold prices in 1998-2001 affected the operation of the "Soviet" mines and eventually led to the closure of some of them. However, the production of hardrock gold in Russia began to increase in 1997 after modernization at the Olimpiada mine in Krasnoyarsk and Kubaka in Magadan. In 2002, the production of hardrock gold exceeded alluvial gold, and in 2017, Russian gold mining exceeded the all-Soviet peak of 1989.

Taking the average annual production of 6 t (193 thousand ounces) of gold as the minimum criterion for a world-class mine, in 1991, only Kuranakh deposits met this criterion, contributing 46% to the extraction of the hardrock gold (excluding by-product gold) or 30.5% of the total hardrock production, including by-product gold. The rest of the hardrock gold mining in Russia was dispersed among the numerous small underground mines and a few open pits in the Urals, Siberia, and the Far East, producing from 200 to 1,500 kg of gold per year. Some have been operating in such low-productivity mode for many decades and even more than 150 years. In 1991, a world-class gold mining industry in Russia simply did not exist.

From 1991 to 2020, more than 6 t of gold during at least one year was produced at only 18 hardrock deposits. At only 9 mines, such production has been maintained consistently or for 5-10 years. In 2020, the 15 largest Russian mines produced 129.8 t of gold, or 55.8% of the hardrock gold production.

Fig. 1: Gold production from hardrock deposits (in tons) and distribution of placers in Russia in 2020.

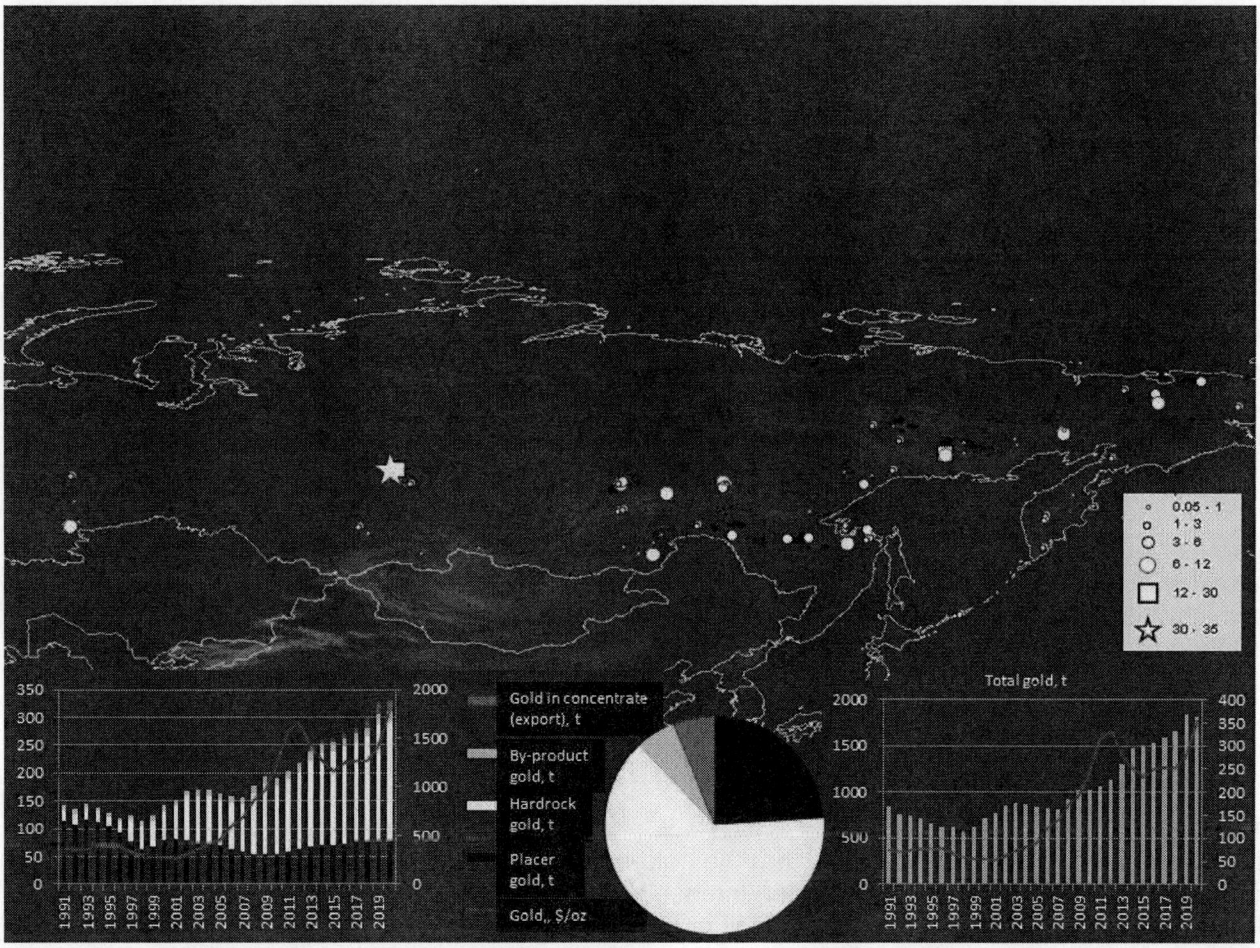

ST.151

The Genesis of the Telluride-Rich Epithermal Gold Deposits: A Perspective of Thermodynamic Modeling

Degao Zhai[1], Anthony Williams-Jones[2], Jiajun Liu[1]
1. China University of Geosciences Beijing, Beijing, China, 2. McGill University, Montreal, QC, Canada

Telluride minerals are regularly minor in most epithermal gold deposits, and only relatively few ore deposits are dominated by telluride minerals, for example Emperor, Cripple Creek, and Golden Sunlight. Those telluride-rich deposits, however, commonly are characterized by very high-grade ores in which both gold and tellurium are economic metals, and especially tellurium has been selected as one of the critical metals that are important for our modern technologies. However, how Te was incorporated into the epithermal system and the critical factors dominating both Au and Te precipitation from the epithermal environment are still unclear. In this contribution, we select the Sandaowanzi Au-Ag-Te deposit in NE China to address those issues.

The Sandaowanzi gold deposit has a total reserve of ≥28 tonnes Au and an average grade of 15 g/t. The deposit has attracted broad attention for two reasons: (1) it is the first example of a gold deposit in NE China, in which >95% of the gold is present as coarse-grained tellurides; and (2) it contains bonanza Au- and Ag-telluride ores, with grades of Au, Ag, and Te up to 35,000, 25,000 and 30,000 g/t, respectively. Gold-bearing veins are hosted in 20 NW-trending, syntaxial, layered quartz veins and tension gashes within almost coeval trachyandesites and andesitic breccias. Alteration halos are developed around the quartz veins and tension gashes, which consist of an inner silicic-pyritic zone, two intermediate zones containing quartz-illite-sericite and quartz-adularia, and an outer zone containing a quartz-calcite-kaolinite-chlorite assemblage.

Fluid inclusion microthermometric data show that the gold-silver telluride ores deposited from an epithermal system emplaced at a depth of <1 km that boiled intensely and cooled from a temperature in excess of 300°C to a temperature less than 200°C at a pressure that approached hydrostatic. Although fluid boiling is commonly invoked to explain the genesis of adularia-sericite (low-sulfidation) epithermal deposits, this mechanism cannot explain the genesis of the Sandaowanzi Au-Ag-Te deposit. Instead, based on physicochemical modeling results, we show convincingly that the deposit owes its origin to an H_2Te vapor, which was derived from the coeval alkaline magma and condensed in the dominantly meteoric ore fluid, and sulfidation of the host rocks contributed to the deposition by destabilizing the gold and silver bisulfide species.

This study emphasizes the importance of fluid-rock interaction (sulfidation) and the mixing of low-density H_2Te-bearing magmatic fluids with meteoric waters in producing large Au- and Ag-telluride deposits.

L2
Large-Scale Multidimensional Mineralization Processes: A Bottom-Up Approach to Predictive Exploration for Tier 1 Deposits

Andrew Wurst
Barrick Gold Corporation, Montreal, QC, Canada

Not all places on the planet are created equal when it comes to metal endowment and the processes of ore deposit formation.

A large portion of metal endowment is often contained in a few key districts or subprovinces, including one or more so-called Tier 1 deposits. These are "company-making" mines that are very large, long lived, and low cost. Discovery of new Tier 1 deposits is exceedingly rare, and only a handful of concealed discoveries have been made in the last 20 years. To meet global metal needs for the future, additional Tier 1 discoveries will need to be made in the coming decades, and a scaled predictive approach is required to increase our chances of discovery under covered terranes.

Local- to district-scale mineralizing processes leading to these well-endowed areas are generally well studied at the shallow crustal scale but are poorly understood at the continental and deeper crustal scale. Mineralization is often studied after discovery and development, generating a new deposit model, which in turn drives further exploration. The model-driven approach can be successful but leads to search spaces being exhausted relatively quickly in districts and the discovery of a swarm of smaller, less economic deposits or "bycatch."

Notably, this model and paradigm driven-approach is unsuccessful at predicting the size and/or grade of a deposit. Significantly, models do not predict the occurrence in time and space of Tier 1 deposits.

The key features contributing to Tier 1 deposits include enhanced components of plumbing (long-lived translithospheric structures), fertility (source of fluids and metals), and long-lived and wide-reaching geodynamic activity (global plate reorganization events). These features are interdependent and corelated in both time and space. While the understanding of the importance of plumbing, fertility and geodynamics, as well as preservation and exposure for the older deposits, is not new, we have been limited by the inherent challenges of studying the architecture and processes of the mantle and subcontinental lithospheric mantle.

Recent insights have changed our understanding of larger-scale temporal and spatial geologic processes. This includes improved dating to constrain epochs of ore deposit formation and various geophysical and geochemical techniques and proxies to understand deep architecture. These insights invariably raise further questions, and new tectonic thinking may be required to reconcile features such as the long-lived continental lineaments that locate the position of Tier 1 deposits.

Understanding the earth's deep architecture and its impact on locating Tier 1 deposits requires a "bottom-up approach" focused on the interplay between deep crustal architecture and shallower crustal mineralization processes in four dimensions. This fundamental shift in exploration strategy is crucial to predicting the occurrence of Tier 1 deposits undercover and will boost exploration success across a range of commodities far into the future.

L3
The Earth Model: Using Geologic Data and Global Processes to Enable Predictive Exploration

Graeme Nicoll, Andrew Davies
Halliburton, Milton Park, United Kingdom

Exploration success is related to our ability to accurately predict the nature of the subsurface. In hydrocarbon exploration, it is necessary to determine if effective reservoirs, seals, and source rocks are present and if the timing of hydrocarbon expulsion happened after trap formation. To assist in this task, we have developed a data constrained scalable 4D model of the Earth that integrates an understanding of different geologic processes to allow informed prediction. Although tailored to the hydrocarbon industry, this model has many applications for mineral exploration.

The first process to integrate is an understanding of global plate tectonics. For this, we use a geodynamic model that provides insights into the evolution of tectonic settings and features, such as collision or subduction zones, that may control depositional processes and petroleum systems or influence the timing of trap formation. The geodynamic model also allows data to be reconstructed into its true paleogeographic position, providing greater insights than is possible from only a modern worldview. As many mineral deposits are intrinsically linked to geodynamic processes, this tool can also be used to high-grade areas to explore for new mineral provinces and deposits under cover.

Before disparate geologic data can be reconstructed, they must first be integrated into a common chronostratigraphic framework. For this, we use a global sequence stratigraphic model covering the past 800 m.y., generated through the interpretation of many thousands of stratigraphic sections from across the globe. Sequence stratigraphy provides insights into the processes controlling sedimentary geometries, architecture, facies, and lithologies, allowing these factors to be predicted away from data. Stratigraphic data interpreted using this framework yields hundreds of thousands of data points on sedimentary systems through time.

From temporally attributed and paleogeographically restored data, it becomes possible to create global gross depositional environment maps for key stratigraphic time intervals. These in turn can be used to construct global paleodigital elevation models (PDEMs) that are bolstered using statistical information on modern landscapes and their relationship to plate tectonic settings. PDEMs can be used as inputs for paleoclimate models, the outputs of which provide additional insights into the factors controlling sediment supply and enabling the likely distribution of key petroleum system elements (e.g., source rocks, reservoirs) to be understood and even predicted away from data control.

These combined insights allow predictive global maps of sedimentary facies and their composition, including potential reservoir quality and source rock effectiveness to be created. Using a set of plate-scale structure-contour maps derived from the integrated subsurface data allows the effects of burial on petroleum system elements effectiveness to be assessed, facilitating basin screening and play risking. This same framework can also be used to identify sediment-hosted mineral deposits.

By combining time-attributed data and knowledge of different geologic processes, the 4D model of the Earth provides insights from reservoir to global scale, allowing informed decision-making. Emerging digital solutions promise the ability to significantly increase the accuracy of such models and the efficiency with which they are built.

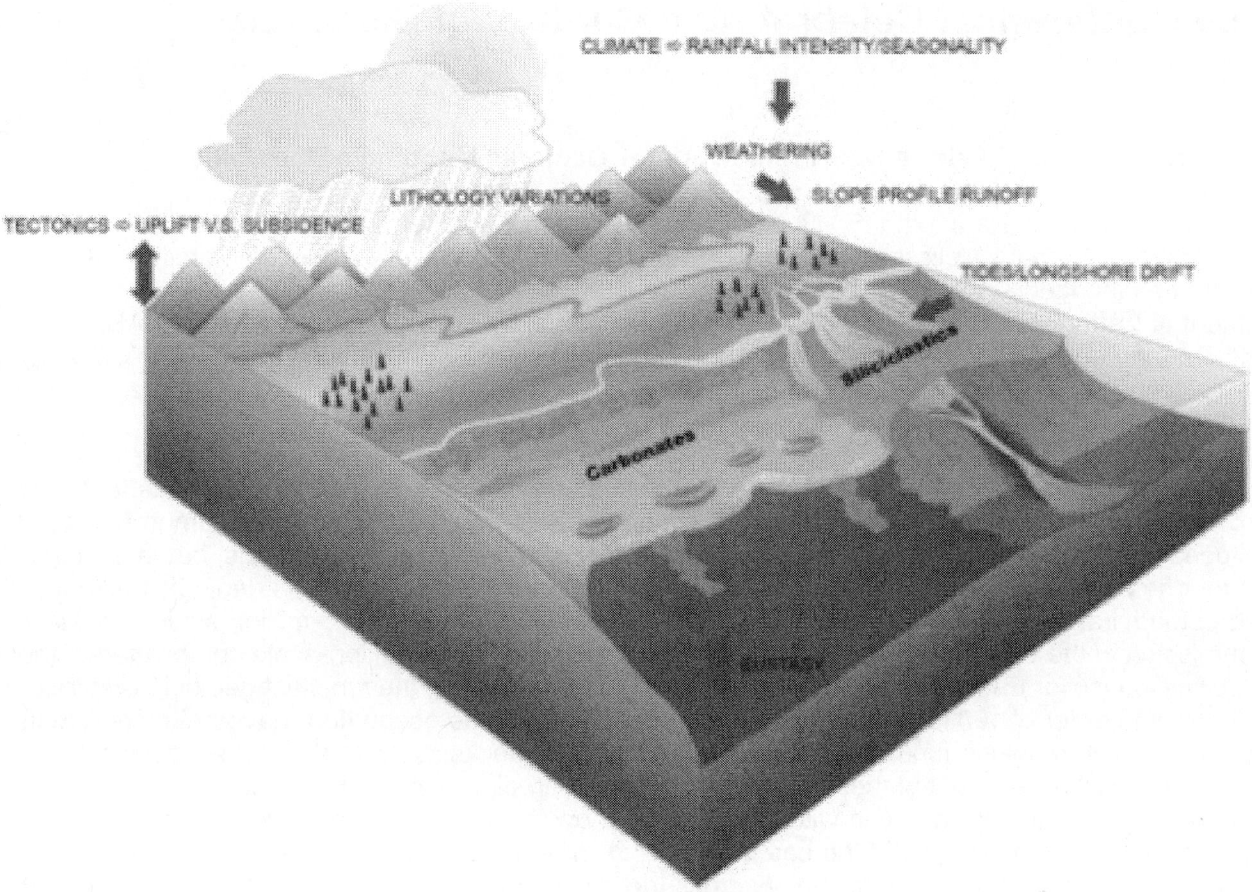
CLIMATE ⇨ RAINFALL INTENSITY/SEASONALITY
WEATHERING
SLOPE PROFILE RUNOFF
LITHOLOGY VARIATIONS
TECTONICS ⇨ UPLIFT V.S. SUBSIDENCE
TIDES/LONGSHORE DRIFT
Siliciclastics
Carbonates
EUSTASY

L4
1:1 Million Map of the Lau Basin: A New Framework for Geologic Mapping of the Seafloor

Margaret Stewart[1], Mark D. Hannington[2, 3], Justin Emberley[2], Alan T. Baxter[2], Anna Krätschell[3], Sven Petersen[3], Philipp A. Brandl[3], Melissa Anderson[3, 4], Patrick Mercier-Langevin[5], Rebecca Mensing[2], Kaitlyn Breker[2], Marc Fassbender[2]
1. Department of Earth and Evnironmental Sciences, Mount Royal University, Calgary, Calgary, AB, Canada, 2. Department of Earth and Environmental Sciences, University of Ottawa, Ottawa, ON, Canada, 3. GEOMAR Helmholtz Institute for Ocean Research Kiel, Kiel, Germany, 4. University of Toronto, Toronto, ON, Canada, 5. Geological Survey of Canada, Quebec, QC, Canada

Geologic maps are critical to our understanding of Earth evolution and processes, including plate tectonics, crustal growth, basin evolution, and the formation of mineral and energy resources. The geomorphology of the global oceans has been determined from satellite altimetry and derived gravity models, but less than 5% of the seafloor has been explored in detail, leaving the vast majority unmapped and its geologic makeup poorly understood. Large-scale geologic maps have so far been limited to the 1:35-million-scale compilation of the Commission of the Geological Map of the World (2014) and the 1:17-million-scale compilation of the Circum-Pacific Council for Energy and Resources (2013). At these scales, the smallest geologic features depicted are on the order of 100 km. More detailed geologic maps of the ocean floor have been created in areas where local high-resolution multibeam bathymetry, acoustic backscatter, and direct seafloor observations exist. However, most of these maps are of limited areal extent and rarely correlate lithostratigraphic units regionally or basin-wide. To illustrate how this can be achieved for a larger area, we have prepared the first geologic map of the Lau Basin at 1:1 million in a single sheet and correlated assemblage- and formation-level geologic units across the entire basin.

The Lau Basin is characterized by some of the fastest growing crust on Earth, associated with high heat flow and prolific magmatic and hydrothermal activity, and it is the type locality for intraoceanic back-arc basin formation. It has been the target of more than 50 research cruises since 1970, most collecting multibeam bathymetry and half collecting whole-rock samples, making it one of the best-studied regions of the deep ocean. The available geophysical and sampling data provide the necessary criteria for recognition of geologic formations and are the basis of an internally consistent geologil legend. The map covers more than 1,000,000 km^2 of the arc-back-arc system, subdivided into nine assemblage types: active arc (6% by area), back-arc rifts and spreading centers (19%), transitional arc-back-arc crust (13%), relict arc crust (40%), forearc crust (9%), relict back-arc crust (8%), and undivided arc-back-arc assemblages (<5%), plus oceanic assemblages, intraplate volcanoes, and carbonate platforms. The distribution of these assemblages is consistent with a recent microplate breakout north of a major crustal-scale fault (Peggy Ridge) and accelerated basin opening accommodated by seven different spreading centers. These observations provide important clues to the geologic evolution and makeup of arc and back-arc crust and processes of crustal growth, including the emergence of melt and fluid pathways for magmatic-hydrothermal mineralizing systems.

L5

A New Model for the Early Crustal Architecture of the Archean Abitibi Greenstone Belt, Canada: Implications for Gold Mineralization

Kate Rubingh, G Tuba, N St-Jean, Mostafa Naghizadeh, Eric Roots, Graham Hill, Rajesh Vayavur, B Lafrance, RL Sherlock
Mineral Exploration Research Centre, Harquail School of Earth Sciences, Laurentian University, Sudbury, ON, Canada

The Abitibi greenstone belt (AGB) is host to some of the largest economic mineral deposits of the Archean Superior Province, which are predominantly associated with two major deformation zones, the Porcupine Destor Deformation Zone, and the Larder Lake Cadillac Deformation Zone (LLCDZ). The LLCDZ extends laterally across the AGB for 250 kilometres, exhibits a variable local geometry, defined by segments that vary in both dip and strike orientation, and is defined by a 10- to 300-m-wide alteration and deformation zone, at the contact between older mafic and ultramafic rocks and younger clastic sedimentary rocks. There is a spatial association with world-class gold deposits that lie predominantly in the footwall to this fault, at the intersection with cross faults. Deposits include the Kerr Addison mine, which historically produced over 11 Moz of gold. Gold mineralization is typically associated with fuchsite-altered ultramafic rocks and albite-altered mafic volcanic rocks and to a lesser extent with alkalic intrusions.

To develop an overall understanding of gold endowment for the AGB and its association with major deformation zones the LLCDZ was examined in terms of structure, alteration, and mineralization in the Larder Lake area, combining field mapping, drill hole structural analysis, and new geophysical data. The LLCDZ juxtaposes younger Timiskaming clastic sedimentary rocks (2674-2670 Ma) against older ultramafic and mafic volcanic rocks of the Larder Lake group (ca. 2710-2704 Ma). Previous interpretations for the evolution of this deformation zone proposed i) a crustal -cale extensional fault that formed during lithospheric-scale extension, due to crustal thinning that was coincident with alkaline magmatism (ca. 2680–2670 Ma) and synorogenic deposition of the Timiskaming sediments and gold mineralization, ii) a reactivated thrust fault that formed initially as a result of segment linkage during accretion-related deformation, or iii) a late transcurrent fault system.

New geophysical data for the LLCDZ defines a vertical depth extent of 27 km and variable alteration zone from seismic data, with a corresponding high conductivity in the structural hanging wall. Magnetotelluric data also identifies linear zones of high conductivity that correlate with the plunge and trend of ore deposits, located at the intersection of the LLCDZ and NE-trending cross faults. Furthermore, field observations and drill core provide evidence of a primary gradational nature for the LLCDZ, preserved in a low-strain zone, which has redefined the contact between the Timiskaming sedimentary rocks and the Larder Lake Group volcanic rocks as a strongly modified unconformity.

Contrary to previous interpretations for this deformation zone, this new model presents evidence that the LLCDZ did not originate as a lithospheric-scale extensional fault, reactivated thrust fault, or transcurrent fault but instead formed as a primary stratigraphic contact that was later reactivated. This primary contact acted as a plane of anisotropy that localized the subsequent development of the fault and migration of gold-bearing hydrothermal fluids. These fluids were focused along an inherited basin architecture, as suggested by the apparent control on gold mineralization, by a network of high-angle NE cross faults.

L6

The Formation of Iron Oxide Copper-Gold (IOCG) and Iron Oxide-Apatite (IOA) Deposits: New Insights from Field Studies and Lab Experiments

Adam C. Simon[1], Maria A. Rodriguez-Mustafa[1], Martin Reich[2], Fernando Barra[2], Irene del Real[2], Jose Tomas Orvalle Ortégá[2], John F. Thompson[3], Tristan Childress[1], Nikita L. La Cruz[1], Artur P. Deditius[4], Laura D. Bilenker[5], Ilya Bindeman[6], Daniel R. Blakemore[1], Robert Holder[1]
1. Department of Earth and Environmental Sciences, University of Michigan, Ann Arbor, MI, USA, 2. Department of Geology and Andean Geothermal Center of Excellence (CEGA), FCFM, Universidad de Chile, Santiago, Chile, 3. PetraScience Consultants, Vancouver, BC, Canada, 4. School of Engineering and Information Technology, Murdoch University, Murdoch, WA, Australia, 5. Department of Earth, Ocean, & Atmospheric Sciences, University of British Columbia, Vancouver, BC, Canada, 6. Department of Earth Sciences, University of Oregon, Eugene, OR, USA

Iron oxide copper-gold (IOCG) and Kiruna-type iron oxide-apatite (IOA) deposits are commonly spatially and temporally associated with one another and with coeval magmatism. In many districts, field observations reveal that IOCG mineralization transitions with depth to S-Cu-Au-poor IOA mineralization and/or that IOA mineralization transitions laterally to IOCG mineralization. Several genetic models have been proposed to explain the evolution of IOCG deposits and IOA deposits as separate systems, but no model has successfully explained the geochemical signature of the deposits or the hypothesized genetic continuum between them. Here, we use trace element concentrations in magnetite and pyrite, Fe and O stable isotope abundances of magnetite and hematite, H isotopes of magnetite-hosted fluid inclusions and actinolite, and S in pyrite from IOCG and IOA deposits in Chile and Peru to develop a new genetic model that explains IOCG and IOA deposits as a continuum produced by a combination of igneous and magmatic-hydrothermal processes. The $\delta^{18}O$, $\Delta^{17}O$, $\delta^{56}Fe$, δD, and $\delta^{34}S$ data unequivocally fingerprint a silicate magma as the source of the ore fluid in IOCG and IOA deposits. Magnetite-hosted silicate melt inclusions and aqueous fluid inclusions, along with magnetite trace element compostions, reveal crystallization from silicate melt and magmatic-hydrothermal fluid. The trace element abundances of magnetite and pyrite, and Mg-in-magnetite and Fe# actinolite thermometry, reveal a systematic continuum between IOA and IOCG deposits that is consistent with formation from a cooling magmatic-hydrothermal fluid. The data cannot be explained by existing models. We developed a new genetic model that explains the geochemical data and field observations: 1) magnetite crystallizes as a near-liquidus phase from an intermediate to mafic silicate melt; 2) magnetite microlites serve as nucleation sites for fluid bubbles and promote volatile saturation of the melt; 3) the volatile phase coalesces and encapsulates magnetite microlites to form a magnetite-fluid suspension; 4) the volatile phase scavenges Fe, Cu, Au, S, Cl, P, Si, and other fluid-mobile elements from the melt; 5) the suspension ascends from the host magma during regional extension; 6) as the suspension ascends, originally igneous magnetite microlites grow larger by sourcing Fe and O from the cooling magmatic-hydrothermal fluid, and apatite and gangue minerals (e.g., actinolite) precipitate from the ore fluid; 7) in deep-seated crustal faults, magnetite crystals are deposited to form a IOA deposit due to decompression of the magnetite-fluid suspension; 8) the further ascending fluid transports Fe, Cu, Au, and S to shallower levels or lateral distal zones of the system where hematite, magnetite, and sulfides precipitate to form IOCG deposits. To test this model in the laboratory, we conducted experiments at appropriate PTX conditions to simulate the evolution of intermediate silicate melt during decompression. The experimental results demonstrate that magmatic-hydrothermal fluid bubbles nucleate on magnetite and form magnetite-bubble pairs that bouyantly ascend through the magma during decompression, accumulating in an upper layer that grows during reequilibration. Regional tectonic stress changes promote ascent of the magnetite-fluid suspension along preexisting faults, followed by precipitation of IOA mineralization and then IOCG mineralization as the fluid cools.

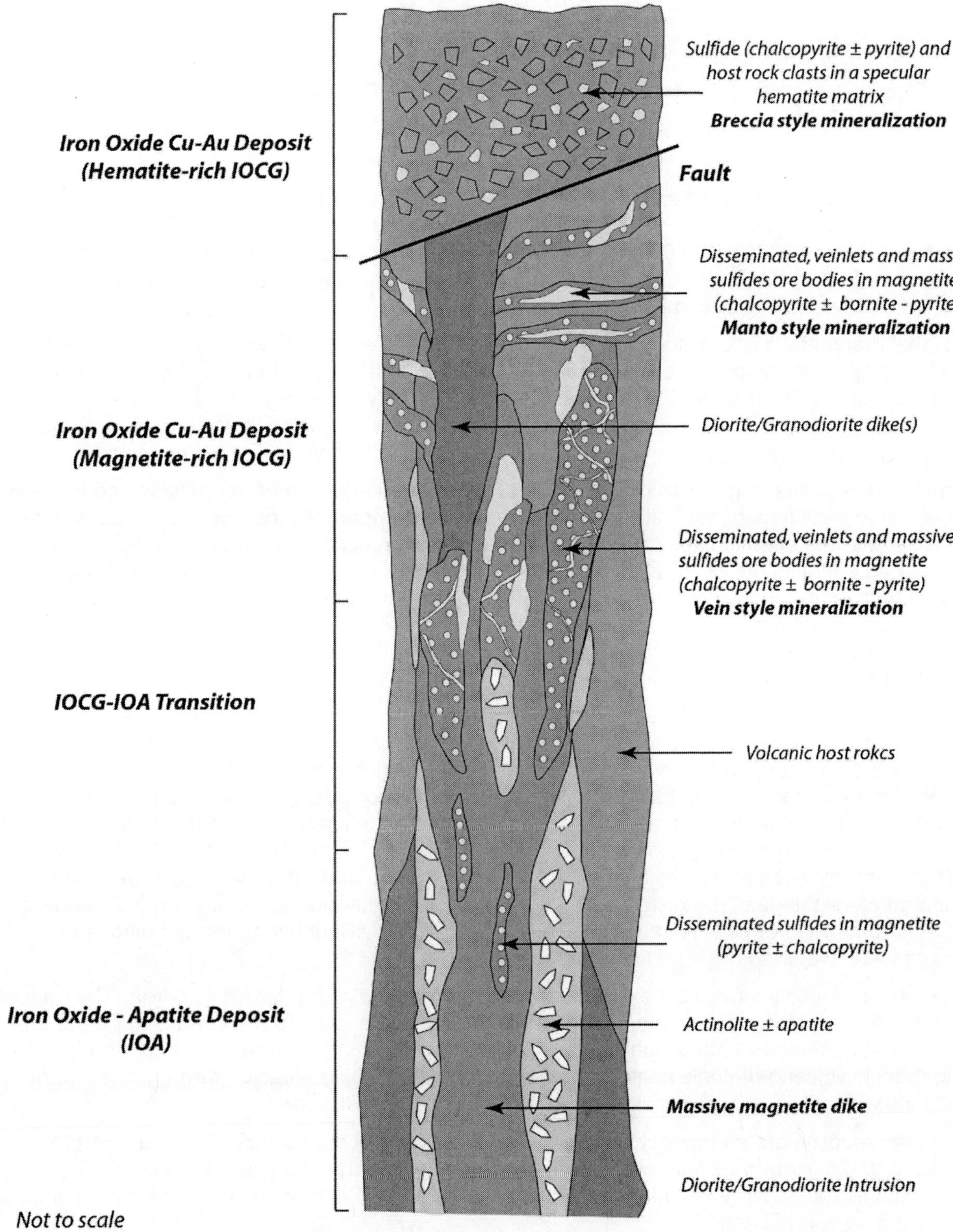

Iron Oxide Cu-Au Deposit (Hematite-rich IOCG)
Iron Oxide Cu-Au Deposit (Magnetite-rich IOCG)
IOCG-IOA Transition
Iron Oxide - Apatite Deposit (IOA)
Not to scale
Sulfide (chalcopyrite ± pyrite) and host rock clasts in a specular hematite matrix
Breccia style mineralization
Fault
Disseminated, veinlets and massive sulfides ore bodies in magnetite (chalcopyrite ± bornite - pyrite)
Manto style mineralization
Diorite/Granodiorite dike(s)
Disseminated, veinlets and massive sulfides ore bodies in magnetite (chalcopyrite ± bornite - pyrite)
Vein style mineralization
Volcanic host rokcs
Disseminated sulfides in magnetite (pyrite ± chalcopyrite)
Actinolite ± apatite
Massive magnetite dike
Diorite/Granodiorite Intrusion

L7
The Lindgren Classification and Postcollisional Metallogeny: A Case Study from the Western Tethyan Metallogenic Belt

Sabina Strmic Palinkas
Norges Arktiske Universitet, Tromsø, Norway

The Tethyan Metallogenic Belt is a 12,000-km-long orogenic belt that extends from the Alps in Central Europe to Indochina (Fig. 1). The belt hosts numerous metallogenic provinces formed in different stages of the Wilson cycle from Permo-Triassic to Recent times.

The postcollisional magmatism and associated ore deposits are well exposed along the Balkan Peninsula where, during the Oligocene-Pliocene period (31.0–1.8 Ma), calc-alkaline to shoshonitic magmatism affected the Vardar Zone (VZ) and the Serbo-Macedonian Massif (SMM). The volcanic to subvolcanic magmatism in an extensional tectonic setting was accompanied by widespread hydrothermal activity that produced numerous hypothermal, pyrometasomatic, mesothermal, and epithermal ore deposits (Fig. 1).

The hypothermal ore deposits (e.g., Buchim, Borov Dol, Kadica, Ilovica) are mostly spatially and temporary related to subvolcanic intrusions of the latitic and latitic-andesitic composition that intruded Precambrian gneisses, mica-schists and amphibolites of SMM. The mineralization occurs in forms of stockworks and disseminations along contacts of the intrusions and their host rocks. Chalcopyrite, bornite, and cubanite are the main ore minerals, associated with variable amounts of pyrite, magnetite, hematite, and native Au. The available fluid inclusion data accompanied by various geothermometers have revealed that the mineralization was formed in a temperature range between 380° and 680°C, at depths from 6 to 12 km, and from highly saline fluids (25-75 wt % NaCl equiv).

The pyrometasomatic ore deposits include proximal and distal skarn deposits. The proximal skarn deposits (e.g., Damjan), formed in a direct interaction of the Oligocene magmatism and carbonate-rich flysch lithologies of VZ, are characterized by the magnetite- and hematite-rich mineralogy, associated with abundant retrograde skarn minerals (epidote, amphiboles, chlorites, quartz, and carbonates) and minor amounts of pyrite and pyrrhotite. In contrast, distal skarns formed by infiltration of magmatic fluids through Paleozoic marbles of SMM (e.g., Sasa, Toranica) and Upper Triassic limestones of VZ (e.g., Trepca), host Pb-Zn-Ag-sulfide mineralization that overprints prograde (anhydrous), and retrograde (hydrous) Ca-Fe-Mg-Mn-silicate mineral assemblages. The distal skarns were formed at temperatures between 350° and 475°C, depths of 1-4 km, from moderate salinity aqueous fluids (4-12 wt % NaCl equiv) under conditions of progressive increase in water activity and sulfur fugacity.

The Carlin-like Allchar deposit is a typical example of mesothermal mineralization in the area. The deposit is predominately hosted in calcareous sedimentary rocks at intersections of high-angle faults in permeable stratigraphy of VZ. The mineralization is spatially and temporally associated with a Pliocene high-K calc-alkaline to shoshonitic volcano-plutonic center. The Au-As-Sb-Tl-rich mineralization formed at temperatures >200°C from moderate salinity aqueous-carbonic fluids (3-21 wt % NaCl equiv).

The epithermal ore deposits are common. In general, these deposits occur as Pb-Zn-sulfide-quartz ± carbonate veins (e.g., Zletovo) or as Cu-Au dissemination (e.g., Plavica) in shallow portions of postcollisional volcanic rocks. They were formed at temperatures <300°C from low- to moderate-saline aqueous fluids (<15 wt % NaCl equiv).

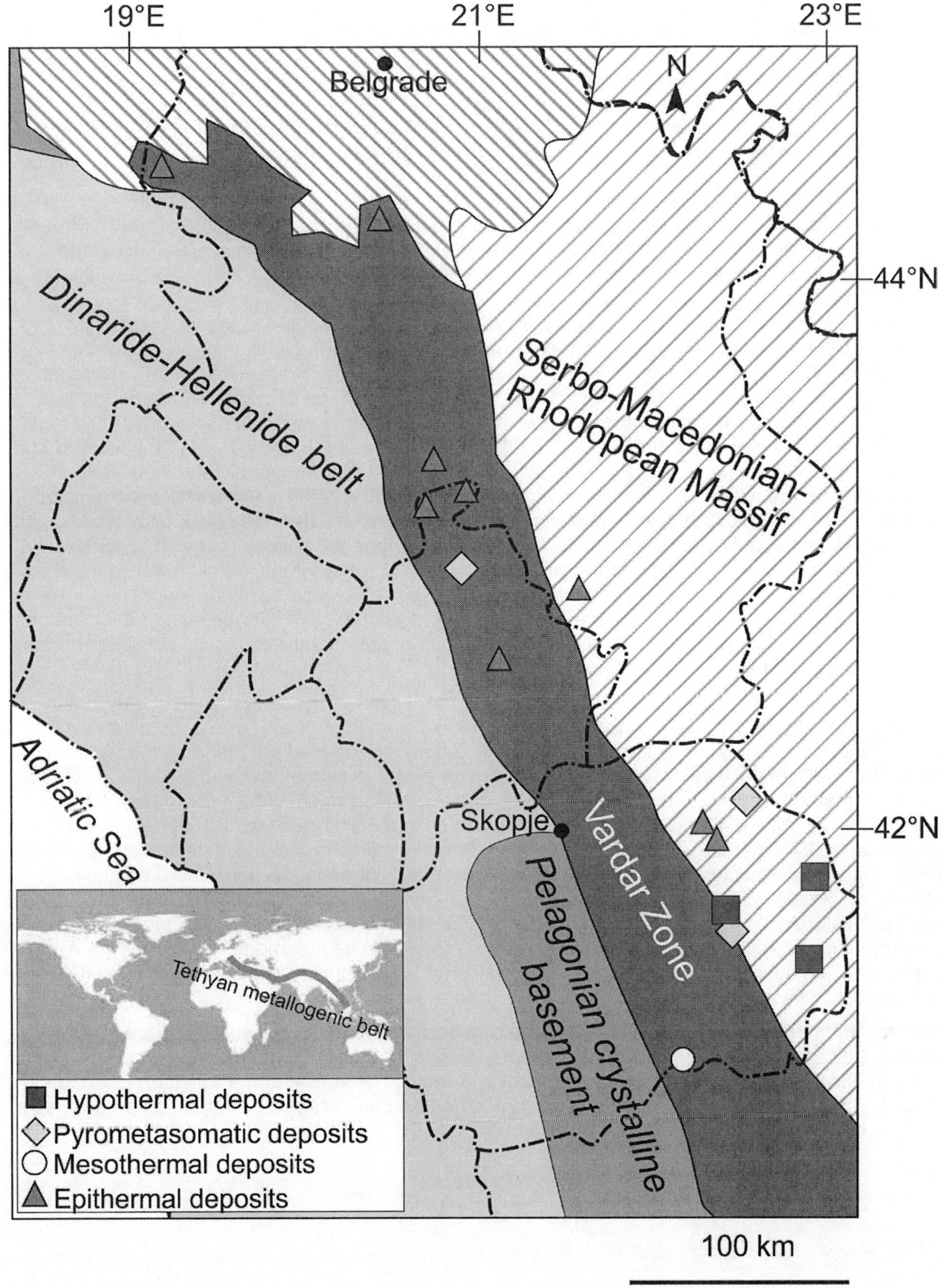
19°E
21°E
23°E
Belgrade
N
44°N
42°N
Dinaride-Hellenide belt
Serbo-Macedonian-Rhodopean Massif
Adriatic Sea
Skopje
Vardar Zone
Pelagonian crystalline basement
Tethyan metallogenic belt
Hypothermal deposits
Pyrometasomatic deposits
Mesothermal deposits
Epithermal deposits
100 km

L8

Redox Controls on Eocene Metallogeny in the Great Basin, U.S.A.: Bridging Porphyry Cu-(Mo-Au) and Reduced Intrusion-Related Au, with Implications for Carlin-Type Deposits

Curtis L. Johnson[1], Michael W. Ressel[2], Philipp P. Ruprecht[3]
1. Center for Research in Economic Geology, University of Nevada, Reno, NV, USA, 2. Nevada Bureau of Mines and Geology, University of Nevada, Reno, NV, USA, 3. Department of Geological Sciences and Engineering, University of Nevada, Reno, NV, USA

Late Eocene (~42-34 Ma) ore deposits of the Great Basin, U.S.A. have contributed the vast majority of this region's precious and base etals, with the majority of Au production occurring in Nevada and most Cu, Ag, Pb, Zn derived from Utah. Eocene Au deposits in Nevada are almost entirely disseminated in calcareous sedimentary rocks, and here we distinguish four styles: 1) gold skarns, 2) distal disseminated deposits, 3) Carlin-type gold deposits, and 4) shallow-formed, paleosurface Carlin-type deposits. We propose the four styles of SHDG are part of a continuum that spans high-temperature and relatively deeply emplaced to low-temperature and shallow-formed, all during a brief period from ~42-34 Ma, temporally and spatially coincident with Eocene arc magmatism but predating large-magnitude Cenozoic upper-crustal extension. The ubiquitous Au-rich nature of deposits in Nevada is in stark contrast to Utah, which contains world-class porphyry systems dominated by Cu, most notably the 38-39 Ma Bingham Canyon porphyry Cu-Au-Mo system. This progressive change in Eocene deposits from Au-rich in the west to Cu-rich in the east mimics changes in the thickness and composition of the middle and upper crust. Nevada is dominated by a >10-km thickness of reduced, carbonaceous units in the Neoproterozoic through Paleozoic slope, and basin setting in north-central Nevada, while Utah contains a thinner sequence of more oxidized, locally evaporite-bearing, coarse clastic, and carbonate shelf sequences.

Here we present a detailed study of 21 regional mineralization-related intrusions across the Great Basin (Eocene to Oligocene) that shows, through analyses of Ce speciation (Ce^{3+}/Ce^{4+}) and determination of magmatic oxidation state from igneous zircon, initially oxidized mantle-derived arc magmas (~fayalite-magnetite+quartz, FMQ +1-3) were systematically reduced to ~FMQ -1 as they evolved in Nevada, whereas magmas in Utah retained, and in some cases increased, their oxidation state (~FMQ +2-3). Nevada intrusions exhibit a wide range of 0-100% Ce^{4+}/Ce_{Total}, while Utah magmas are dominantly >70%, with substantially more 100% Ce^{4+}/Ce_{Total} measured in zircon. In conjunction, igneous apatite in magmas from Utah is shown to be more enriched in S (dominantly 0.1-1 wt % SO_3) than in Nevada, where apatite is dominantly <0.1 wt % SO_3—likely a function of the elevated magmatic oxidation state in Utah and coupled increase in S solubility due to the transition from S^{2-} to S^{6+} dominance at ~FMQ +0.5. We conclude the assimilation of reduced, carbonaceous sedimentary units in Nevada is a fundamental control on the Au-rich metallogeny of Nevada due to the modification of magmatic redox conditions to optimal levels (~FMQ) for the highest melt solubility and extraction efficiency of Au and reduced S into fluids.

We propose continental arcs emplaced into thick slope and basinal sedimentary sequences are an effective means to form reduced, Au-rich, Cu-poor intrusion-related deposits. Due to the abundance of reduced S in fluids and the buffering capacity of wall rock, these geologic settings are ideal for the transport of Au to low temperatures typical of Carlin-type Au deposits and should be targeted in regional exploration programs for new Carlin-type districts.

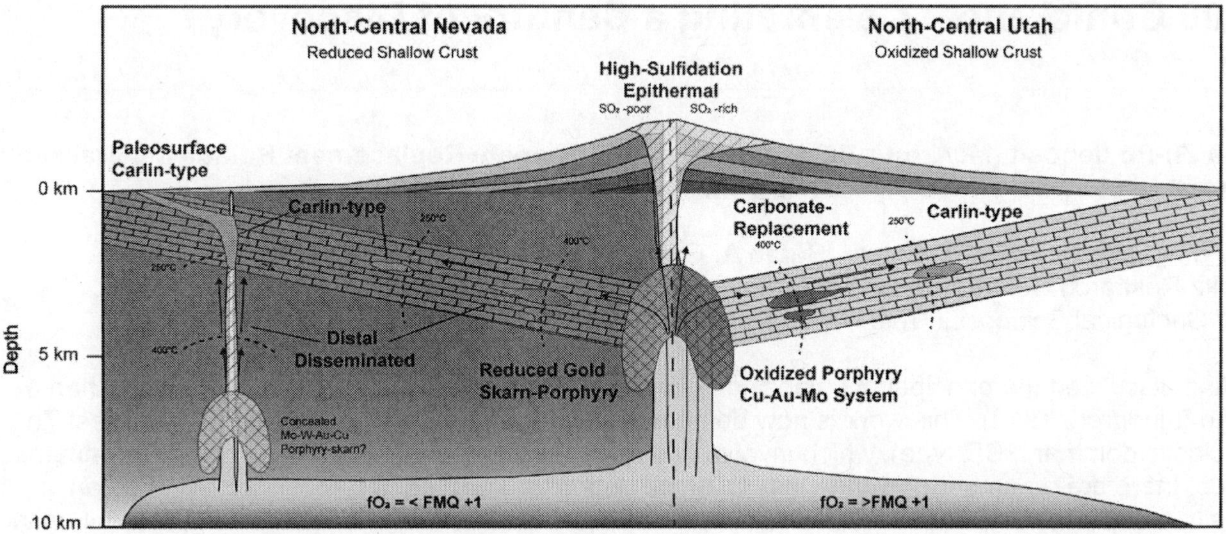
North-Central Nevada
Reduced Shallow Crust
North-Central Utah
Oxidized Shallow Crust
High-Sulfidation
Epithermal
SO₂ -poor
SO₂ -rich
Paleosurface
Carlin-type
0 km
Carlin-type
250°C
400°C
250°C
Carbonate-
Replacement
400°C
250°C
Carlin-type
Depth
5 km
400°C
Distal
Disseminated
Reduced Gold
Skarn-Porphyry
Oxidized Porphyry
Cu-Au-Mo System
Concealed
Mo-W-Au-Cu
Porphyry-skarn?
fO₂ = < FMQ +1
fO₂ = >FMQ +1
10 km

L9

The Teena Zn-Pb deposit (McArthur Basin, Australia): Carbonate Replacement Sulfide Mineralization During Burial Diagenesis

Joseph M. Magnall[1], Nicholas Hayward[2], Sarah A. Gleeson[3]
1. Helmholtz Research Centre, Potsdam, Germany, 2. Teck Australia Pty Ltd, Perth, WA, Australia, 3. Institute of Geological Sciences, Freie Universität, Berlin, Germany

Lindgren first discussed the principal of balancing volumes in mineral replacement reactions more than a century ago (Lindgren, 1912). This work is now being revisited in recent genetic models of the largest Zn deposits (clastic dominant; CD-type), which involve host-rock replacement. In the Proterozoic Carpentaria Zn Province, there are a number of world-class CD-type deposits. The Teena deposit is hosted by an exceptionally well preserved subbasin in the Carpentaria Province and in the same stratigraphic unit as the nearby McArthur River (HYC) deposit. A combination of data sets (geophysical, structural, geochemical, sedimentological, and petrographic) have been combined to produce a model for subbasin development that addresses key aspects of the structural and hydrothermal evolution of the Teena mineral system. Petrographic data show that sulfide mineralization postdated the formation of fine-grained diagenetic pyrite (py1) and dolomite nodules and resulted from lateral subsurface fluid flow at least several hundred meters below the most reducing sedimentary facies. To evaluate the deposit-scale role of mineral replacement reactions, a large whole-rock lithogeochemistry data set (n = 2,705) was developed in the Teena subbasin. The unmineralized protolith (dolomitic siltstone) and mineralized end members have been defined and evaluated in a chemical mass balance analysis. The results clearly show the addition of hydrothermal sulfides (and associated trace elements) and corresponding loss of dolomite. Although key assumptions relating to lithofacies variability should be carefully considered, the extent to which molar volumes are balanced in a dolomite replacement reaction indicates that this was the primary mineralization pathway in the Teena subbasin. As metals were likely transported in an oxidizing fluid, disequilibrium with the host rock and in situ acid generation potential would have been important factors for reaction permeability and preferential replacement of carbonate units. High-grade mineralization formed at the stratigraphic redox boundary between an oxidized footwall sequence and a host-rock facies that contained the necessary components (dolomite, organic matter, reduced sulfur) for effective subseafloor replacement.

L10
Unravelling Complex Alteration by Dating Hydrothermal Titanite: Implications for the Use of Geochemical Vectoring Tools in the Propylitic halos of Porphyry-Type Ore Systems

Lisa Hart-Madigan
LODE, Department of Earth Sciences, Natural History Museum, London, United Kingdom

New geochemical exploration techniques that aim to detect orebodies at depth are currently being developed. In this endeavor, minerals in the propylitic halo of porphyry deposits have been of particular interest. Across the porphyry alteration halo, systematic changes have been identified in the chemical compositions of propylitic minerals, such as chlorite, which form the basis of tools that can vector towards mineralization. These tools are particularly useful in districts where orebodies may be blind and only the propylitic facies is visible at surface. However, porphyry systems form in active tectono-magmatic settings, so primary alteration can be overprinted by later events, particularly in older porphyry camps such as the Paleozoic Oyu Tolgoi Cu-Au deposits of southern Mongolia. For the application of mineral vectoring methods to be successful in such districts, an explorer must be confident that the mineral assemblage under investigation is related to the target porphyry event and that this alteration event can be isolated from preceding or overprinting events. Here, we present LA-ICP-MS U-Pb dates on in situ propylitic titanite from the Oyu Tolgoi district that identify three principal phases of hydrothermal activity that coincide with known magmatic events (Fig. 1): Devonian porphyry mineralization (~372 Ma); the intrusion of granodiorite plutons and andesite dikes in the Carboniferous (~350-310 Ma); and the emplacement of the Permian Khanbogd Granite alkaline batholith (~290-272 Ma). The titanite age-constrained alteration assemblages were then used to develop a mineral chemical classification scheme for other samples that allows chlorite in Devonian porphyry-related assemblages to be distinguished from the two later alteration events. By isolating the porphyry-related chlorite, it has been possible to demonstrate the validity of chlorite vectoring as applied to the known deposits and to generate new exploration targets in the district.

Fig. 1. Diagram comparing A) the hydrothermal titanite U-Pb ages from this study, to B) probability density plots of compiled igneous ages from intrusive units in the OT district.

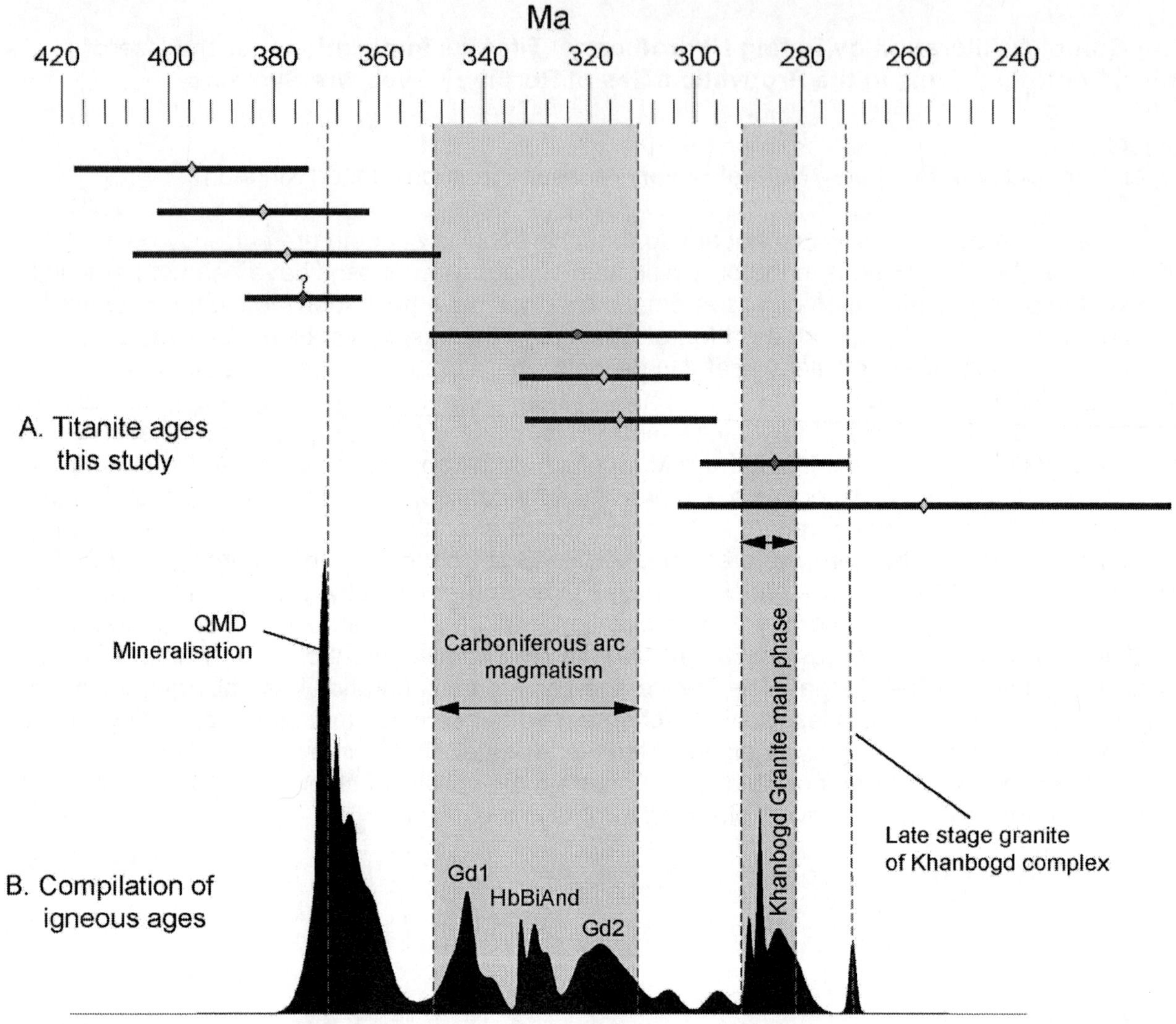

BC.1
100 Years Later: Reviving Lindgren's Legacy in the Tintic District

Graham Boyd[1], Harold P. Hanneman[1], Shawn Vandekerkhove[1], Catherine Fitzgerald[2], Ben M. Frieman[1]
1. Ivanhoe Electric Inc, Vancouver, BC, Canada, 2. Apollo Gold and Silver Corp, Vancouver, BC, Canada

The Tintic Mining District is one of the premier polymetallic mining camps in Utah, alongside Park City and Bingham Canyon, operating continuously from 1871 to 1983 with total historical production from the Main Tintic District estimated at 2.18 Moz Au, 209 Moz Ag, 116 kt Cu, 589 kt Pb, and 63 kt Zn. Past production is wholly derived from carbonate replacement deposits (CRDs) and fissure veins, which cut across the Paleozoic quartzite and carbonate stratigraphy, and late Eocene-Oligocene igneous rocks of the Deep Creek-Tintic Belt. Given the Tintic District's early prominence, the USGS conducted geological investigations that led to the seminal 1919 USGS publication on the district by Lindgren and Loughlin. This touchstone work established the groundwork for modern, empirical models for our understanding of the formation of hydrothermal deposits in the Tintic District. Critical observations made by Lindgren and his contemporaries on the structure and metal zonation of the ore zones helped to vector towards extensions and/or undiscovered ore and enabled later mineral discoveries in other parts of the district (e.g., the subeconomic Southwest Tintic Porphyry deposit; 559 Mt at 0.28% Cu).

Historical data and land consolidation efforts in the Main Tintic District over the past ~40 years have recently enabled district-scale evaluation, the application of modern exploration paradigms, and cutting-edge technologies. This has facilitated a system-scale approach to exploration efforts in Tintic to further establish the geometry of the hydrothermal system responsible for mineralization and to elucidate the potential for undiscovered CRD/porphyry targets. Integrated data sets include detailed surface and underground mapping with petrographic, fluid inclusion, and geochemical sampling over 15 km^2, in addition to soil geochemistry and SWIR analysis. These data sets identified robust multifaceted anomalies indicative of porphyry-style alteration and mineralization in the Silver City area that was originally posited by Lindgren ~100 years prior. Geophysical data sets include 3-D inversions of a 72-km^2 highly detailed, deep-penetrating Typhoon™ IP survey and a 2,850-km^2 airborne magnetic survey. This proprietary IP survey yielded high-quality resistivity and chargeability data to over 1.5-km depth, resulting in an unparalleled ability to compare the geometry of the historical workings and alteration around known orebodies with deeper and less constrained portions of the system. Through integration of the geophysical data, geochemical results, and digitization of ~625 km of historical underground workings and ore-stopes, a 3-D model of the district was constructed. This provides an unprecedented level of detail that guides exploration towards underexplored domains where CRD mineralization is likely and to prospective zones where porphyry-style mineralization like that observed at Bingham Canyon may occur.

The Tintic Mining District has a legacy of quality data acquisition and analysis accompanied by successful, novel exploration efforts that resulted in further discoveries since Lindgren's time. By continuing to dig deeper into the fundamental observations made by Lindgren and his contemporaries and through the application of robust scientific first principles that he championed, while leveraging modern systems-scale understanding and exploration tools, it is likely that further discoveries in the Tintic District will occur in the next 100 years.

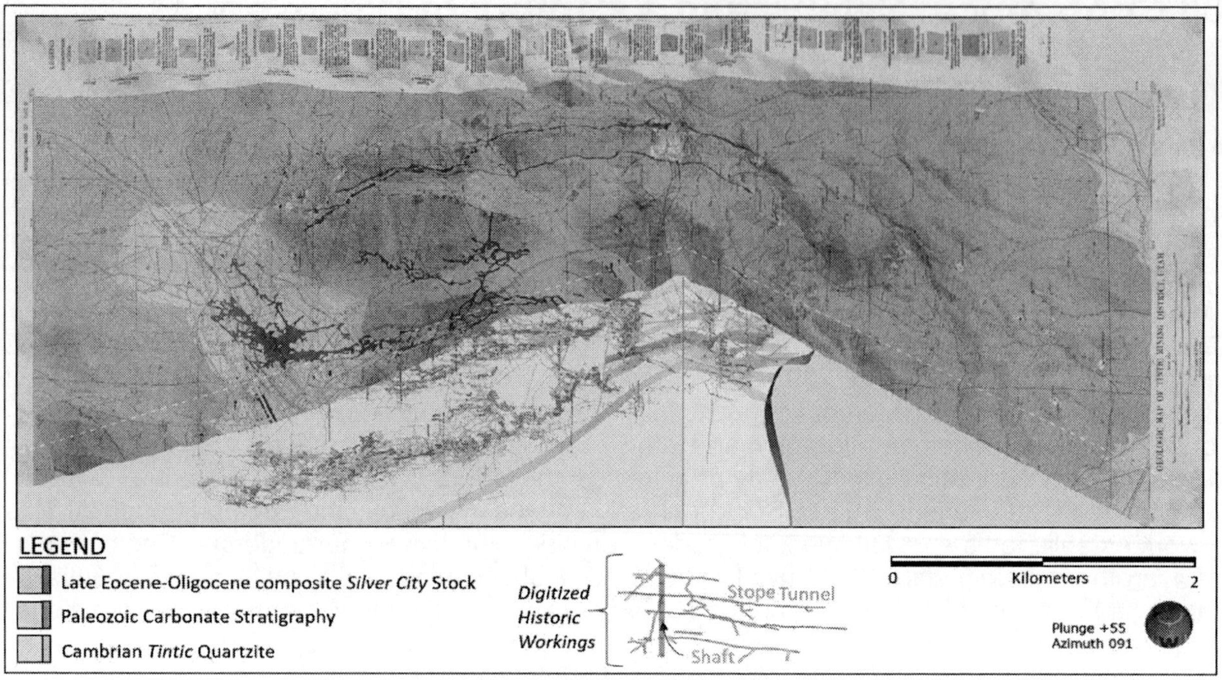
LEGEND
Late Eocene-Oligocene composite Silver City Stock
Paleozoic Carbonate Stratigraphy
Cambrian Tintic Quartzite
Digitized Historic Workings
Stope Tunnel
Shaft
0
Kilometers
2
Plunge +55
Azimuth 091

ST.153
Technology Metals and Metamorphism: Linking Geologic Environments, Processes, Tools, and Techniques

Anna Bidgood, Murray Hitzman
iCRAG, University College Dublin, Dublin, Ireland

Technology metals are used in products such as computers, batteries, and magnets critical to the green technology needed to help decarbonise our societies. As global demand for these metals increases, new thinking is required to improve our understanding of where and how deposits containing these metals can form. The study of metamorphic geology covers a range of geologic environments and processes, many of which are integral to the study of deposits host to technology metals. Titanium and graphite are examples of technology metals hosted in or sourced from metamorphic rocks by processes such as prograde metamorphism.

Metamorphic rocks grow new minerals in response to new physical or chemical conditions such as heating, burial, deformatio, and fluid infiltration. As such, metamorphic rocks provide a record of changing crustal conditions in different geologic environments through time. The scope of the discipline of metamorphic geology is therefore extremely broad, overlapping with other disciplines in terms of lithologies, geological environments, tools, and techniques. A key example of this overlap is hydrothermal metamorphism, where growth of new minerals occurs in response to introduction of fluids, dissolved species, and heat, influencing metamorphic reaction progress at a range of conditions. For example, the contentious genesis of the Broken Hill deposit in New South Wales, Australia, provides a window into the ways that sedimentary, exhalative, metamorphic, and/or hydrothermal processes may have been important factors in concentrating or diluting metals.

When it comes to technology metals, the majority of our current data sets focus on ore minerals host to high concentrations of targeted elements. To better understand these ore systems, and perhaps imagine new undiscovered types of deposits, we need a broader understanding of the natural processes that enrich and deplete these metals. Data required includes measuring subeconomic concentrations in minerals from a range of geological environments, as well as developing a much fuller understanding of the effect that different processes have on elements concentrations. Just as we are understanding that the fertility of copper porphyries may reside in the history and composition of portions of the mantle and crust associated with and individual subduction zone, we should also consider the environment of deep crustal melting that may lead to derivation of Sn- and W-rich melts but may also leave behind other elements in the partially melted residuum that are concentrated by other physiochemical means. More research defining the life cycle of elements, including the technology elements in rock masses subjected to prograde metamorphism and melting is required to help us perceive currently poorly understood means of elemental concentration.

ST.154

The Temporal Evolution of the Candelaria IOCG System, Chile: Insights from U-Pb LA-MC-ICP-MS of Apatite, Magnetite, Titanite, and Ar-Ar of Actinolite

Daniel R. Blakemore[1], Maria A. Rodriguez-Mustafa,[1], Adam C. Simon[1], Robert Holder[1], Irene del Real[2], Martin Reich[2], Fernando Barra[2], John F. Thompson[4], Willis E. Hames[3]
1. University of Michigan, Ann Arbor, MI, USA, 2. Universidad de Chile, Santiago, Chile, 3. Auburn University, Auburn, AL, USA, 4. PetraScience Consultants, Vancouver, BC, Canada

Iron oxide copper-gold (IOCG) and iron oxide-apatite (IOA) deposits are important sources of copper and iron, respectively, and also host a diverse array of metals critical to modern technology and green energy. Many studies have interpreted the commonly recognized spatial and temporal relationship between IOCG and IOA deposits to indicate that the two deposits types are members of a single mineralizing system. Here, we focus on samples from the Candelaria IOCG deposit, Chile, that reveal a clear transition from magnetite-actinolite-dominated mineralization at depth, to chalcopyrite-dominated mineralization at shallow levels of the system. Textural evidence has been interpreted to suggest that the system evolved upward with time, where early magnetite-actinolite was subsequently overprinted by chalcopyrite-dominant mineralization. This transition makes Candelaria an ideal deposit to study to better understand the hypothesized temporal relationship between IOA and IOCG deposits.

We analyzed samples from a 1,000-meter drill core from the Candelaria deposit using laser ablation-multicollector-inductively coupled plasma-mass spectrometry (LA-MC-ICP-MS) to measure U and Pb isotope abundances of apatite, magnetite, and titanite. The data yield dates of 115.4 ± 0.7 Ma for apatite, 126.3 ± 4.9 Ma for magnetite, and 114.9 ± 1.1 Ma for titanite (2se uncertainties) when calculated by mineral without taking depth into consideration. Age dates for each mineral type were also separately calculated by sample depth in order to determine if there is a difference in age throughout the deposit (Fig. 1). No resolvable difference in age by mineral type was observed when calculated by depth. When dates are calculated together by mineral type without separating by depth, dates for apatite and titanite are consistent with published Re-Os dates for molybdenite of 114.2 ± 0.6 and 115.2 ± 0.6 Ma, which were interpreted to constrain the timing of sulfide mineralization. An Ar-Ar date of 121.2 ± 0.6 Ma for actinolite from one sample falls within the uncertainty of the combined magnetite date. When calculated by mineral type without taking depth into consideration, if the magnetite and actinolite dates represent crystallization ages, the data are consistent with two separate hydrothermal events at Candelaria: the first at 126-121 Ma (actinolite, magnetite) and a second at 115 Ma (apatite, titanite, and main sulfide mineralization). These ages are consistent with published ages of ~122 and ~115 Ma (U-Pb zircon) for the two major early phases of the Copiapo batholith. Because magnetite is not a conventional geochronometer and no external matrix-matched standard yet exists, interpretation of this date is cautioned. However, with appreciable uranium in many of these samples, magnetite U-Pb LA-MC-ICP-MS has exciting potential for geochronology, especially with respect to Fe-rich deposits such as IOCGs and IOAs. The dates calculated in this study will contribute to the overall understanding of the temporal evolution of the Candelaria mine and of IOCGs more broadly.

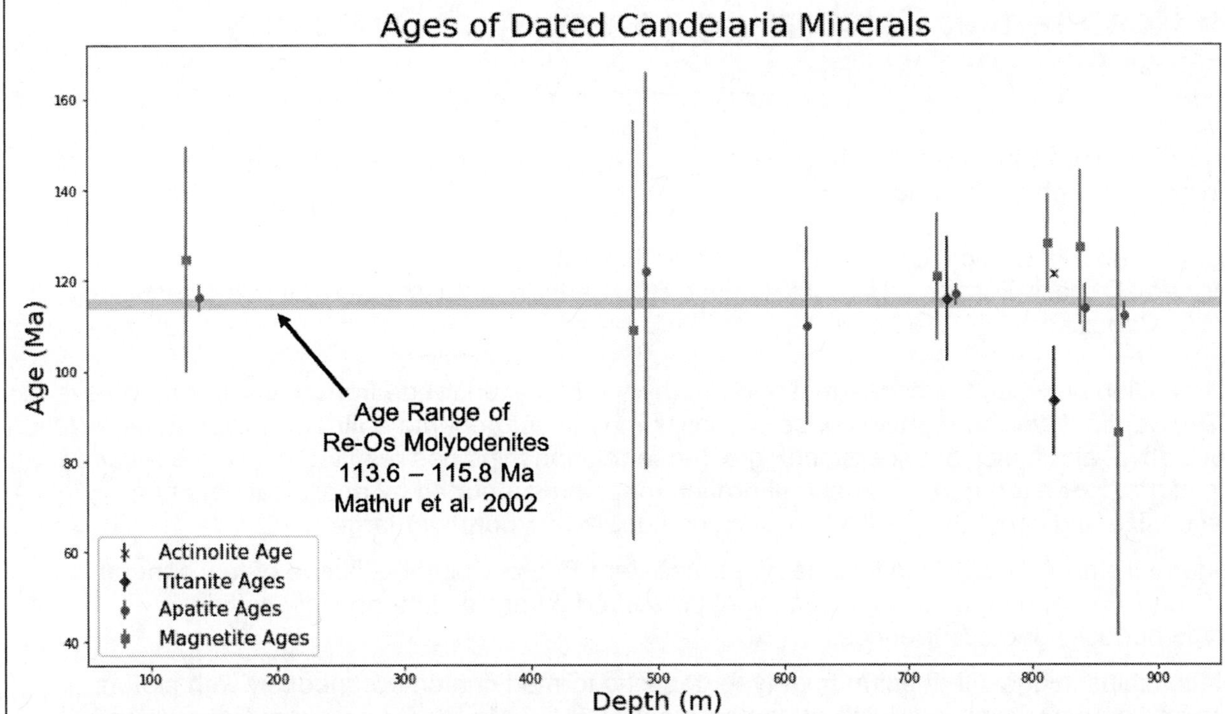

Figure 1. U – Pb ages of apatite, titanite, magnetite and an Ar – Ar date of actinolite, calculated by depth from the Candelaria mine plotted with 2se uncertainties.

ST.155

Vertical Variability from Epithermal Gold to Porphyry Copper Mineralization: Toodoggone District, Northern British Columbia, Canada

Farhad Bouzari[1], Thomas Bissig[2], Craig Hart[1], Hildebrando Leal-Mejía[1]
1. MDRU - Mineral Deposit Research Unit, Vancouver, BC, Canada, 2. Bissig Geoscience Consulting, Vancouver, BC, Canada

The vertical transition between the epithermal and porphyry realms can range from a few 100 m to several kilometers. The depth at which porphyry copper deposits may occur beneath epithermal-type mineralization is highly speculative, and a lack of understanding of those relationships has resulted in failures related to deep-drilling efforts. However, geological and alteration mapping supported by analytical data can effectively recognize and constrain the likely occurrence depth of a porphyry target.

The Toodoggone district in the Stikine terrane of northeastern British Columbia is one of few mineral districts in BC and worldwide that hosts several well-preserved Jurassic high- and low-sulfidation epithermal-type deposits and occurrences.

New age relationships suggest that epithermal-type deposits formed contemporaneously with pluton emplacement and porphyry-type alteration at depth. The 193.8±0.8 Ma Re-Os age from Baker mine overlaps with nearby 193.4±2.6 Ma K-feldspar-altered granodiorite. Similar relationships are evident for epithermal-type mineralization at Alunite Ridge and the 189.6±2.1 Ma K-silicate-altered granodiorite, which hosts porphyry copper mineralization at the Sofia.

Mineral assemblages indicate that some of the epithermal-type veins that were previously classified as low-sulfidation have alteration features indicative of shallow-level porphyry systems. Quartz-sericite-pyrite alteration lacking kaolinite occurs at Baker mine and Black Gossan. White micas at these localities are moderately crystalline (>1.75 index) and K-rich on the basis of SWIR data. Occurrences of pyrophyllite with quartz-sericite alteration at the Brenda indicate shallow porphyry rather than alteration related to epithermal environment.

The occurrence of banded quartz-magnetite±pyrite±chalcopyrite veins in drill holes at the Baker mine indicates a transition toward porphyry alteration 100-150 m below the surface. Such K-silicate alteration occurs in several valley cuts into the deeper-exposed plutons in the district such as at Black Gossan, suggesting that porphyry-level alteration is exposed.

Fluid inclusion analysis indicates near-surface exposures of high-temperature (up to 556°C) and saline (up to 50 wt % NaCl equiv) fluids associated with quartz veins at the Baker mine that are characteristic of porphyry-type alteration at depth. More importantly, these fluids occur only 100-200 m below the surface where lower-temperature (<358°C) and dilute (<2.6 wt % NaCl) fluids caused pervasive sericite alteration and gold-bearing quartz veins. Similar low- to moderate-temperature fluids were also recognized at Black Gossan and Alunite Ridge.

The transition to a porphyry center is further supported by high concentrations of Cu, Mo, W, and Sn, which are typically enriched in the core of porphyry system, relative to Sb, As, Ag, Li and Tl, which are enriched in shallow levels. A normalized element ratio, the MDRU Porphyry Index, is developed (Fig. 1). This index indicates that Sofia and Baker mines have a distinct porphyry character, and prospects such as Shasta, Cliff Creek, Black Gossan, and Alunite Ridge have notable porphyry proximity.

Results show that a range of mineralization types are exposed within a ~1-km vertical profile in the Toodoggone district that formed episodically over from 196 to 186 Ma. The vertical framework proposed herein provides vectoring considerations for the exploration of porphyry-type copper deposits in areas previously explored for only epithermal mineralization.

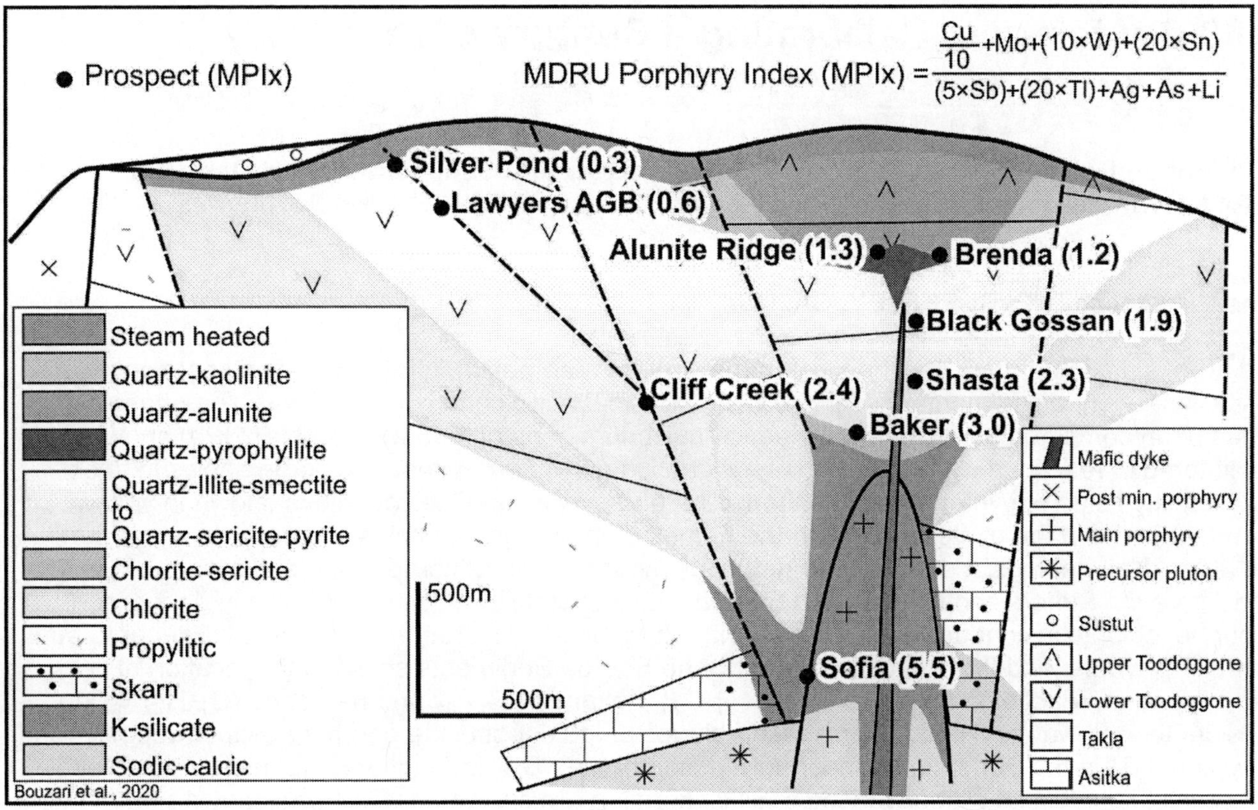

Figure 1: Generalized cross section of the Toodoggone district showing vertical setting of mineral deposits and prospects relative to the typical porphyry alteration zoning (based on Sillitoe, 2010).

ST.156

Redox-Sulfidation Relationships of Porphyry Copper, IOCG and Sediment-Hosted Deposits to Source-Rock Hydrocarbons and Brines: The Tops of Deep Sediment-Hosted Porphyries

George H. Brimhall
Clementine Exploration LLC, BERKELEY, CA, USA

Presented here is a comparative geochemical framework for magmatic-hydrothermal and sediment-hosted ore deposits in relation to sedimentary source rocks of basinal brines and hydrocarbons. The phase diagram has been computed using thermodynamic equilibrium constants (300°C and 50 MPa) showing mineral-solution-gas reactions within CO_2/CH_4-H_2S gas fugacity coordinates. The figure shows in thick black lines stability fields of pyrite, magnetite, hematite, and pyrrhotite. Chalcopyrite is shown in yellow surrounded by bornite in the pyrite field and in the hematite and magnetite fields chalcopyrite is surrounded by bornite alone. Key geological domains are depicted in red in the four corners of the diagram: I: Granitic Rocks and Alteration Buffers show the nested Fe end member stability fields of actinolite in dark green, biotite in purple, chlorite in light green, and muscovite. II: Magmatic Hydrothermal fluids, III: Evaporite and Hydrothermal Anhydrite, and IV: Hydrocarbon and Brine Source Sediments showing the location of diagenetic maturation reactions of kerogen (C_{128} $H_{68}O_7$), hexane $C_6H_{14\ aq}$ and n-decane ($C_{10}H_{22\ gas}$) to methane, siderite, and water. Using mineral assemblages as guideposts the porphyry copper deposit trend is shown where "BT" is the Butte high-temperature assemblage, "BC" is Bingham, and high-grade Butte veins with covellite-chalcocite with advanced argillic alteration are shown as "BV." Chalcopyrite-bearing Orbicular Actinolite alteration "CO" completes the array of porphyry systems from magma to host sediments. The Olympic Dam IOCG deposit is "OD" and Ernest Henry is "EH," which lacks bornite. Although the geotectonic framework is different, porphyry and IOCG deposits show a similar path. Anhydrite in porphries, however, contrasts sharply with fluorite in the IOCGs, which reflects extremely high HF_{aq} concentrations. In contrast, a family of reactions involving Hg and As plot with shallower negative slopes connecting the spectrum of basinal petroleum/brine reduced H_2S and metamorphic fluids with magmatic fluids as buffer-controlled mixing lines revealing varied source of brines, petroleum, and metals in hot spring Hg, Au, and fluorspar Pb-Zn. Orogenic Au and reduced intrusion-related Cu lack barite (OG-IRC). Sediment-hosted deposits occur along these buffered reaction lines in accord with the amount of amagmatic sediment-derived species, carbonaceous hydrocarbons, and brines relative to magmatic sources. These reaction lines bridge inorganic and organic sources of ore-forming components and operate when fluid potential gradients support advection. The Hg and As mineral equilibria bridging the sedimentary source rocks to porphyry trends are the missing link that ensures that stratiform ores fall on a spectrum from syngenetic to magmatic processes, thus explaining enigmatic features like high-K fluids and rutile. CO_2, CH_4, and H_2S are sufficiently high that the patterns described are substantiated by Raman spectroscopy of fluid inclusions. Chalcopyrite-bearing orbicular actinolite alteration along the porphyry trend at both Bingham and the Clementine Prospect with soil Hg-Ag-Zn anomalies and peripheral barite and arsenopyrite at depth is reason to rethink the nature of the tops of deep porphyries confined by regional fold-and-thrust anticlines. Furthermore, passive tectonic margin host-rock stratigraphy ensures that deep sediment-hosted porphyry deposits afford multiple opportunities for Leadville Colorado-style high-grade replacement mineralization along multiple Sloss-sequence stratigraphy unconformities.

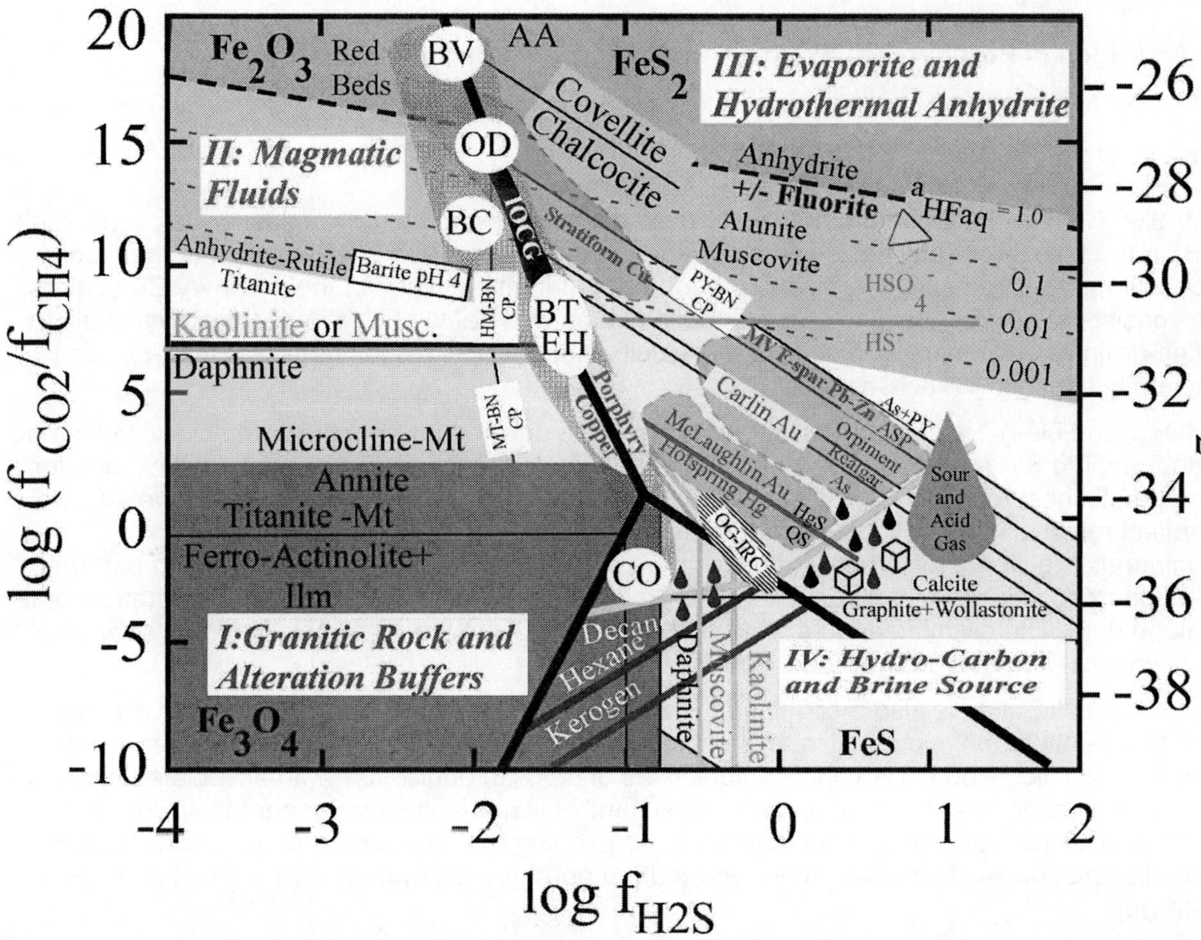
Fe2O3
Red Beds
BV
AA
FeS2
III: Evaporite and Hydrothermal Anhydrite
II: Magmatic Fluids
OD
Covellite
Chalcocite
Anhydrite +/- Fluorite
aHFaq = 1.0
BC
IOCG
Stratiform Cu
Alunite
Muscovite
Anhydrite-Rutile
Titanite
Barite pH 4
HM-BN CP
PY-BN CP
HSO4
0.1
BT
EH
Kaolinite or Musc.
0.01
HS-
Daphnite
0.001
MV F-spar Pb-Zn
Carlin Au
MT-BN CP
Porphyry Copper
As+PY
ASP
Orpiment
Realgar
As
Microcline-Mt
McLaughlin Au
Hotspring Hg
HgS
QS
Sour and Acid Gas
Annite
Titanite -Mt
OG-IRC
Ferro-Actinolite+ Ilm
CO
Calcite
Graphite+Wollastonite
I:Granitic Rock and Alteration Buffers
Decan
Hexane
Kerogen
Daphnite
Muscovite
Kaolinite
IV: Hydro-Carbon and Brine Source
Fe3O4
FeS
log (fCO2/fCH4)
log fH2S
log fO2
20
15
10
5
0
-5
-10
-4
-3
-2
-1
0
1
2
-26
-28
-30
-32
-34
-36
-38

ST.157
Sodic-Calcic Alteration in Porphyry Cu Systems: Definitions, Footprints, and Origins

Kevin Byrne
Teck Resources Limited, Vancouver, BC, Canada

The Highland Valley Copper (HVC) porphyry Cu (±Mo) district is hosted in the Late Triassic Guichon Creek batholith (GCB) in the Mesozoic Quesnel island arc of the Canadian Cordillera. Fracture-controlled sodic-calcic alteration is important because it forms a large footprint (34 km^2) outside of the porphyry Cu centers. This alteration consists of epidote ± actinolite ± tourmaline veins with halos of K-feldspar-destructive albite (1–20 X_{An}) ± fine-grained white mica ± epidote. Sodic-calcic-altered rocks exhibit depletion in K, Fe, and Cu and are generally enriched in Na, Ca, and H_2O.

Calculated $\delta^{18}O_{fluid}$ and δD_{fluid} values of mineral pairs in isotopic equilibrium from the sodic-calcic veins and alteration range from 4 to 8‰ and -20 to -9‰, respectively, which contrasts with the whole-rock values for least-altered magmatic host rocks ($\delta^{18}O$ = 6.4 to 9.4‰ and δD = -99 to -75‰). The whole-rock values are suggested to reflect residual magma values after D loss by magma degassing, while the range of hydrothermal minerals requires a mixed-fluid origin with a contribution of magmatic water and an external water source. Supporting evidence for a seawater-derived fluid entrained in the porphyry Cu systems comes from boron isotope data. The calculated tourmaline $\delta^{11}B_{fluid}$ values from the sodic-calcic domains reach 18.3‰, which is consistent with a seawater-derived fluid source.

Sodic-calcic and sodic alteration is also recognized in several Laramide-age porphyry Cu systems in the SW US. Where Na-Ca alteration is present in these systems, it is best developed at relatively deep levels of porphyry systems. The calculated δD isotopic compositions of fluids in equilibrium with associated alteration minerals span a broad range (–46 to –1‰) and are consistently less negative than magmatic fluids. The isotopic compositions of minerals and the geologic alteration zoning patterns have led other researchers to conclude that sodic and sodic-calcic facies in and around the porphyry Cu systems formed by incursion of saline external fluids.

The studies at HVC and in the SW US imply that sodic-calcic facies can form in different geologic environments from a variety of external fluids. At HVC the presence of Na-Ca alteration in the GCB demonstrates that seawater-derived fluids can infiltrate batholiths and porphyry systems at deep levels (4–5 km) in the crust and highlights that complex paleohydrology and nonmagmatic fluid sources result in atypical porphyry Cu alteration zones. Key factors for generating widespread sodic-calcic assemblages, K-depleted rocks, and local leaching of Fe and Cu are the presence of permeability and conditions such that the external saline fluid is heated on ingress to the porphyry system.

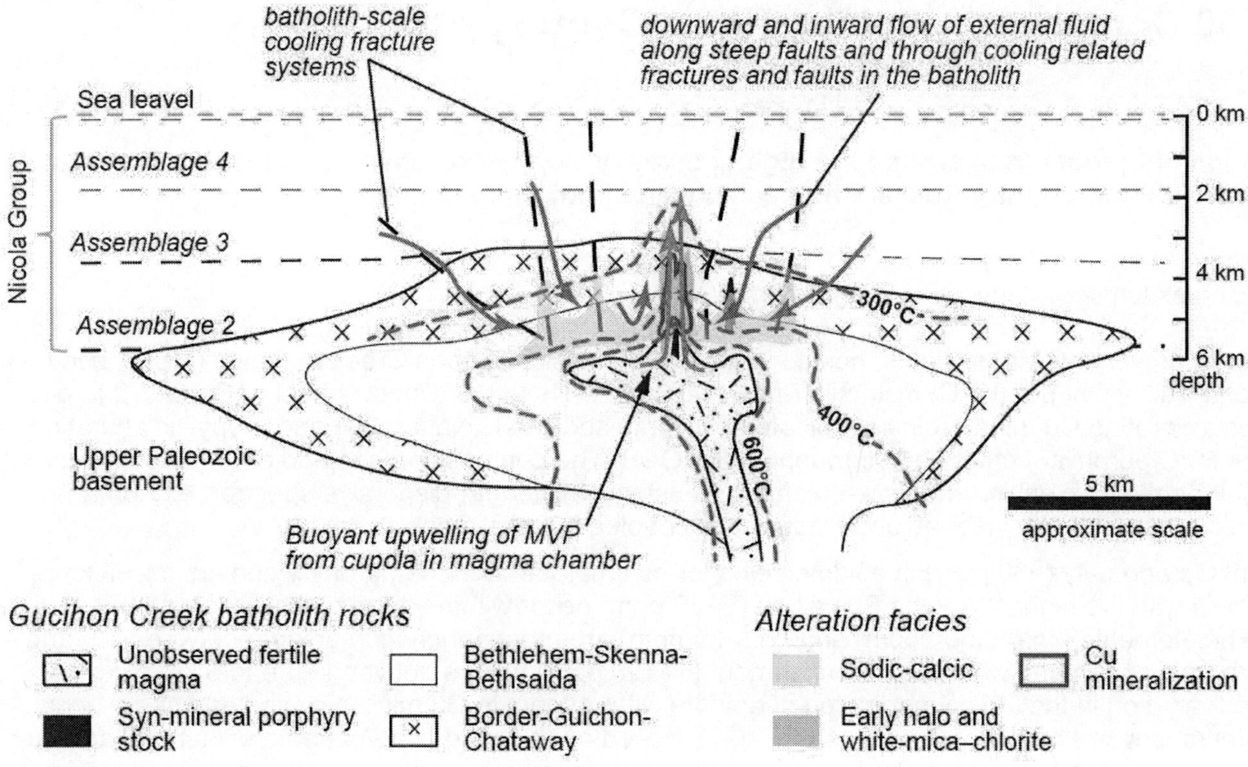

batholith-scale cooling fracture systems
downward and inward flow of external fluid along steep faults and through cooling related fractures and faults in the batholith
Sea leavel
Nicola Group
Assemblage 4
Assemblage 3
Assemblage 2
0 km
2 km
4 km
6 km
depth
300°C
400°C
600°C
Upper Paleozoic basement
5 km
approximate scale
Buoyant upwelling of MVP from cupola in magma chamber
Gucihon Creek batholith rocks
Unobserved fertile magma
Syn-mineral porphyry stock
Bethlehem-Skenna-Bethsaida
Border-Guichon-Chataway
Alteration facies
Sodic-calcic
Early halo and white-mica–chlorite
Cu mineralization

ST.158

Composition of Epidote from around the Highland Valley Copper Porphyry Deposits and the nearby Nicola Batholith and Country Rocks: Prospectivity and Footprint Insights

Kevin Byrne
Teck Resources Limited, Vancouver, BC, Canada

Four major porphyry Cu-(Mo) systems, hosted in the Late Triassic Guichon Creek batholith (GCB), occur in the Highland Valley Copper (HVC) district in British Columbia. The HVC district is host to Canada's largest and longest operating Cu mine. Veins of epidote occur with sodic-calcic alteration and propylitic alteration around the HVC porphyry centers and throughout the GCB. The composition of epidote from these veins and from other geologic environments were analyzed using wavelength-dispersive spectrometry on an electron microprobe and by laser ablation-inductively coupled plasma-mass spectrometry.

A coherent Ge anomaly (>30 ppm) in epidote veins forms a halo around the porphyry centers for ~5 km. Epidote veins with the highest As and Sb values (~>20 ppm) occur within ~4 km of the porphyry Cu centers. The elements Mn (>6,000 ppm) and Zn (>25 ppm) are most enriched in epidote between 1.5 and 4 km of the porphyry centers, whereas Pb enrichment (~>120 ppm) occurs between ~4 and 6 km away. Systematic variation in the Mn concentration of epidote with respect to Cu mineralization has been documented at several porphyry Cu deposits (Fig. 1). Based on this study and patterns evident in the other porphyry systems, the Mn concentration in epidote increase inwards through propylitic alteration to a maximum at the transition from propylitic to higher-temperature assemblages and then decreases sharply inward to Cu-mineralized domains

Epidote veins from the fertile HVC porphyry district can be discriminated from the epidote occurring in nonprospective geologic environments by a combination of compositional features: large positive Eu anomaly, low sum HREE concentration, moderate negative REEN slope, and principally by localized analyses of >30 ppm As and Ge and >10 ppm Sb. Considerable range in trace element concentrations and complicated major and minor element zonation patterns, however, occur on an intrasample basis in the HVC sample suite. This variability necessitates careful characterization and a high number of analyses and presents challenges in determining representative sample strategies for exploration.

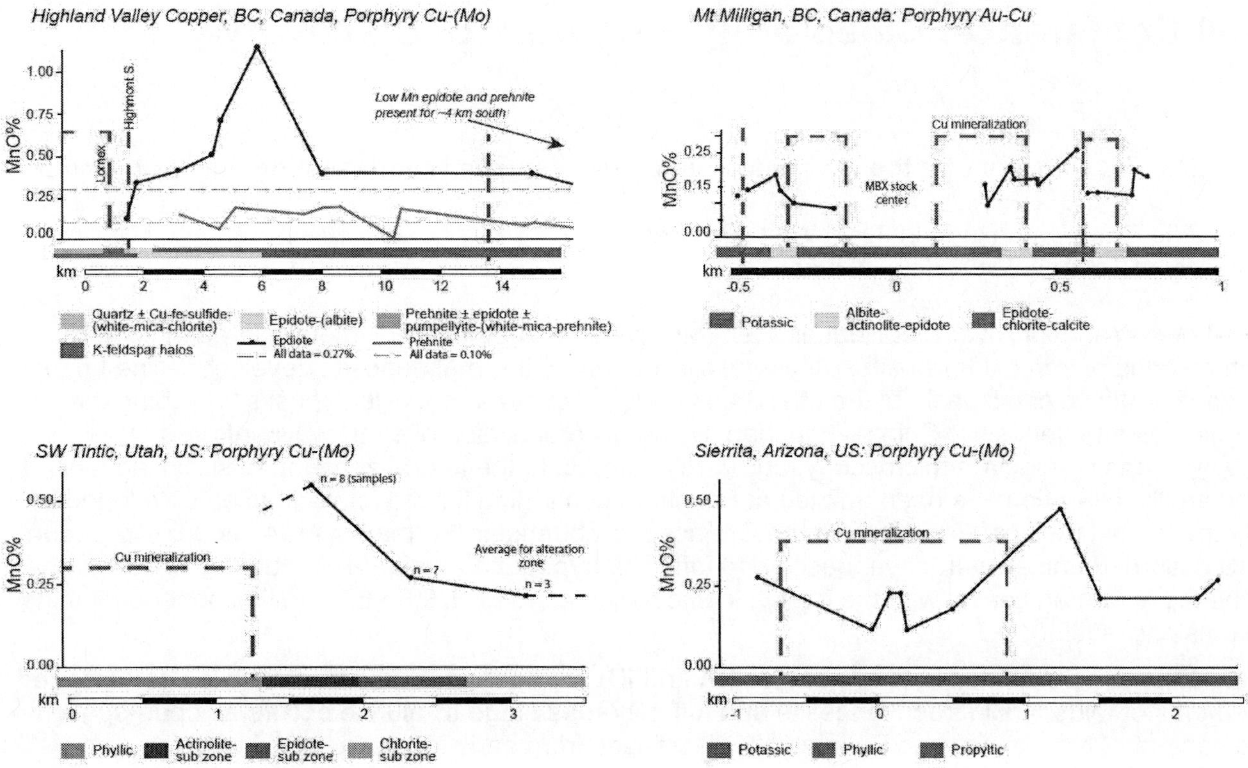
Highland Valley Copper, BC, Canada, Porphyry Cu-(Mo)
MnO%
Lornex
Highmont S.
Low Mn epidote and prehnite present for ~4 km south
km
Quartz ± Cu-fe-sulfide-(white-mica-chlorite)
K-feldspar halos
Epidote-(albite)
Epidote
All data = 0.27%
Prehnite ± epidote ± pumpellyite-(white-mica-prehnite)
Prehnite
All data = 0.10%
Mt Milligan, BC, Canada: Porphyry Au-Cu
MnO%
Cu mineralization
MBX stock center
km
Potassic
Albite-actinolite-epidote
Epidote-chlorite-calcite
SW Tintic, Utah, US: Porphyry Cu-(Mo)
MnO%
n = 8 (samples)
Cu mineralization
n = 7
Average for alteration zone
n = 3
km
Phyllic
Actinolite-sub zone
Epidote-sub zone
Chlorite-sub zone
Sierrita, Arizona, US: Porphyry Cu-(Mo)
MnO%
Cu mineralization
km
Potassic
Phyllic
Propyltic

ST.159
Characteristics and Genesis of Lithocaps in Middle-Lower Yangtze River Metallogenic Belt, China

Taofa Zhou, Jing Chen, Xuanxuan Li, Noel White, Yu Fan
Hefei University of Techonology, Hefei, China

The Middle-Lower Yangtze River Metallogenic Belt (MLYB) is an important polymetallic area in eastern China, with a series of iron oxide-apatite, porphyry, skarn, and epithermal deposits developed. The Luzong Mesozoic continental volcanic basin of the MLYB hosted significant nonmetallic deposits, including the second biggest alunite deposit of China—Fanshan. By detailed analyses of short-wave infrared spectroscopy, scanning electron microscopy, and X-ray diffraction, the lateral zonation of silicic, advanced argillic, and argillic alteration has been defined at Fanshan, confirming it is a typical lithocap developed in MLYB. The main host rocks of Fanshan are the Cretaceous Zhuanqiao Formation (134.1 ± 1.6 Ma; zircon LA-ICP-MS dating), dominantly trachyandesite and tuff. The hypogene alunite Ar-Ar dating shows that the Fashan lithocap is formed coeval with the host volcanic rocks, at 131.2 ± 6.6 Ma, with lithospheric thinning and an extensional setting.

Hypogene alunite can be divided in to two stages (IA and IB) at Fanshan. Stage IA alunite is characterized as replacements of feldspars in trachyandesite and tuff, whereas stage IB alunite occurs as open-space infills. The content of Na, Ca, Sr, and Ba in alunite decreases from early to late stage by electronic microprobe and LA-ICP-MS analyses, which is governed by the decrease of temperature of fluid. A discrimination diagram of Ca + Sr + Ba-Na/(Na + K) content of alunite has been developed to indicate the temperature (depth) of lithocap based on these data.

Results of fluid inclusion study and the O-D isotopes of alunite indicate that the fluid forming Fanshan lithocap is a mixture of magmatic hydrothermal fluid and meteoric water. Fluid reacted with the permeable volcanic rocks, and the lateral outflow helps to develop such an extensive silicic and advanced argillic alteration at Fanshan. Fluid evolved from acidic and oxidized at the early stage to a neutral environment with low temperature. Fluid is not conducive to the dissolution of metal elements during the late stage, resulting in precipitation of metals, which indicate a potential prospect at the bottom of the Fashan lithocap. A series of intermediate-sulfidation copper-silver vein deposits have been discovered within the Fashan district, indicating that the Fashan and Luzong basin is a fertilized system for forming epithermal deposits.

The detailed case study of Fanshan lithocap shows a slightly different pattern from other lithocaps with calc-alkaline volcanic rocks developed in the arc setting. Fanshan is predominantly hosted by alkaline volcanic rocks, and the low potassium fugacity prohibited the formation of pyrophyllite. High-sulfidation epithermal and porphyry deposits are not discovered yet to be related to Fashan, but intermediate-sulfidation epithermal Cu-Ag vein deposits are closely related to Fashan. Other small lithocaps developed in the Luzong basin have also been studied in this research, and they all illustrate the similar patterns. The intracontinental lithocaps seem to have different characteristics from the classic model, and more research should be focused on this topic for further exploration strategy in alkaline rock-hosted lithocaps.

ST.161

The Role of Volatiles in the Metal Budget of Magmatic Sulfide Deposits: Insights from the Lower Crustal Valmaggia Ultramafic Pipe, Ivrea Zone, Italy

Maria Cherdantseva, Marco Fiorentini
Centre of Exploration Targeting, School of Earth Sciences, The University of Western Australia, Perth, WA, Australia

A large body of literature has argued for a long time that "dry" magmas can produce the wide range of mineralised bodies observed in nature. The pieces of evidence for the presence of volatile-bearing phases have been largely relegated to the occurrence of secondary processes related to hydration and carbonation during alteration and metamorphism. However, the development of analytical and experimental techniques has recently allowed us to document the near-ubiquitous presence of high-temperature volatile-rich phases (e.g., mica, amphibole, and carbonate) associated with magmatic sulfides, as well as the preservation of melt-vapor and fluid inclusions in high-temperature phases, together with the widespread occurrence of pegmatoids in mineralized mafic-ultramafic intrusions. Moreover, scientists have recently documented the spatio-temporal relationship between magmatic sulfides and platinum group minerals together with mineral phases enriched in CO_2, H_2O, Cl, F, and P (e.g., carbonate, apatite, amphibole, biotite). This common association has questioned the role of volatiles in the genesis of magmatic ore systems, more specifically in the formation of magmatic sulfide and reef-type platinum group element (PGE) deposits.

This study was dedicated to the close investigation of the interaction of high-temperature volatile-rich phases with base metal sulfides in the lower crustal Valmaggia ultramafic pipe, Italy. The analysis of polished thin sections using the Maia Mapper helped to establish the textural and morphological relationship of coexisting metal- and volatile-rich phases (Fig. 1). This work was complemented by the thorough examination of fluid and melt inclusions in high-temperature minerals, such as olivine and plagioclase. Using the Australian Synchrotron XFM technique, the focus was laid on the spatial distribution of nanonuggets of PGE along the margins of sulfide globules with carbonate phases. Furthermore, to unveil the source of carbon we conducted in situ carbon isotope measurements of the carbonates associated with the sulfide globules using the IMS1280 ion probe. Results are used to test the hypothesis that the observed magmatic calcite embracing sulfide globules does not only represent the frozen product of a CO_2 supercritical liquid that physically entrained dense sulfide liquid upwards across the lithosphere, but also evidence of a fractionation process that partitioned precious metals between sulfide liquid and exsolving volatiles. In this scenario, it is proposed that the exsolution of a CO_2-bearing fluid in the middle-lower crust may promote metal fractionation and priming of ascending hydrothermal fluids in precious elements including Au and PGE.

Fig. 1. Maia Mapper image of the peridotite from the Valmaggia ultramafic pipes representing the intimate spatial association of sulfides with calcite. Cal – calcite, Pn – pentlandite, Po – pyrrhotite.

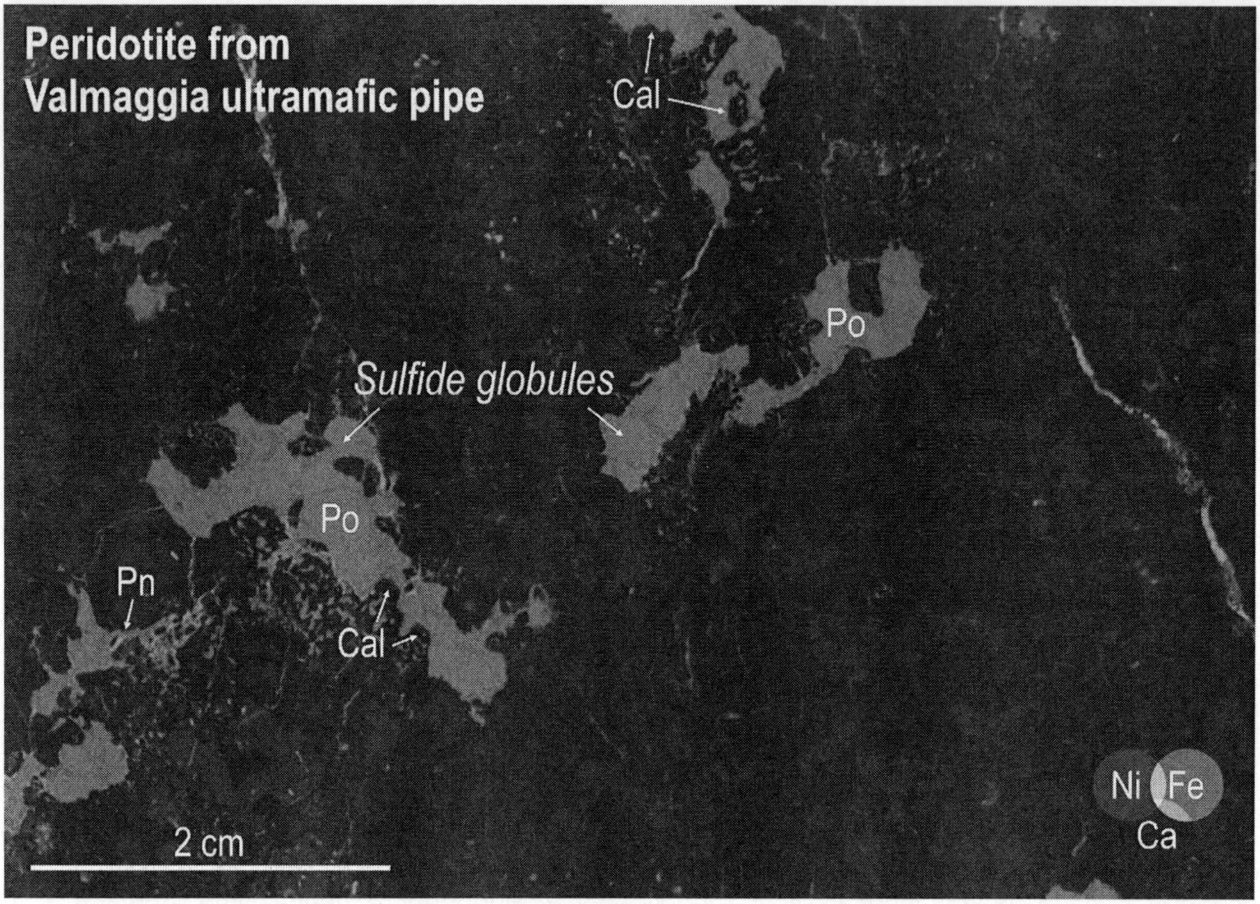

ST.162
Exploration-Stage Geometallurgy: A Value-Added Study of the Gold Hill Trend, Black Hills, South Dakota, USA

Thomas Chudy[2], Brian Lentz[1, 3], Robert Bergmann[1, 3], Eric Nowariak[3]
1. F3 Gold, Minneapolis, MN, USA, 2. PMC Ltd., Maple Ridge, BC, Canada, 3. Big Rock Exploration, Minneapolis, MN, USA

The Gold Hill Trend is a newly (re)discovered gold district in the southwestern portion of the Precambrian core of the Black Hills, South Dakota. The 20-km^2 district is characterized by pervasive tourmaline(schorl-dravite)-graphite alteration and a distinctive Au-Bi+/-Te signature to mineralization hosted in sulfide-poor, shear-parallel veins crosscutting Paleoproterozoic metalamprophyre/alkalic metatuffs and pelitic schists, with gold grades up to 42.6 g/t in grab samples. To better understand the mode of occurrence of the precious metals, their variability and implications for ore processing routes, exploration-level metal deportment studies were completed. For this purpose, composites (>5 kg per composite) representing three different areas within the district were prepared from multiple individual samples and submitted for exploration-level gold deportment studies. The applied methodology includes crushing and milling to approximately 250 μm followed by sizing and gravity upgrading of three size fractions. Representative portions of the fraction heads and all processing products were assayed to obtain a mass-balanced distribution of the gold. In addition, polished block sections were prepared from all products and submitted for systematic scanning electron microscopy scans targeting the overall mineral content including gold-bearing particles. Statistically significant data (based on up to 1,100 gold grain observations per composite) was collected on the gold grain composition, gold grain size distribution, gold association with other minerals, and their liberation characteristics at the applied grind size. The results indicate that the gold mineralization in the southern portion of the district is evenly distributed within the size fractions and occurs predominantly as free native gold (>80% Au). Major gold associations include Bi- and Te-bearing mineral species as well as quartz. In the northern and central areas, the native gold shows a bimodal distribution across the size fractions with the gold reporting dominantly to the coarsest and finest fractions indicating a difference in liberation characteristics, which can be attributed to an overall finer gold grain size and a stronger association with hard silicates such as quartz (in addition to the primary association with Bi-Te-bearing minerals). These results indicate that ore of this composition is amenable to gravity concentration (albeit at a finer grind) followed by CIL (carbon-in-leach) extraction. Therefore, delineating zones with a high proportion of gravity recoverable gold will constitute a priority for further development work as it will improve the overall project economics, especially at the beginning of an operation. Studies of gold-bismuth mineralization from similar deposits attributes the close spatial and mineralogical associations of bismuth and gold observed in this study to reworking or remobilization of preexisting gold-sulfide-rich deposits during shallow intrusion of granitoids. The nearby Harney Peak Granite and related satellite intrusions are identified as a possible source for mineralizing fluids, further supported by granite related metasomatic tourmalinites similar to the host rocks described in this study. The presented gold deportment data represents a value-added study that provides detailed and statistically relevant information about the mineralized system and considers potential processing routes, which initiates a decision-making process for future development work informed by project economics, which go beyond simple grade and tonnage.

ST.164
Ore Paragenesis, Fluid Inclusion and Sulfur Isotope Systematics of the Balatoc Breccia and Vein Orebodies, Acupan Epithermal Deposit, Philippines

Acer Jian Figueroa[1], Jillian Aira Gabo-Ratio[1], Pearlyn Manalo[2], Ryohei Takahashi[2], Hinako Sato[2], Aljess Ramos[3]
1. University of the Philippines Diliman, Quezon City, Philippines, 2. Akita University, Akita, Japan, 3. Benguet Corporation, Itogon, Philippines

The Acupan epithermal deposit is characterized by NE-trending epithermal veins transecting the Balatoc Diatreme and the Virac Granodiorite. Mineralization hosted by the Balatoc Diatreme occurs as veins and breccias. The veins transform to stockwork at the eastern portion, while the breccias, called GW orebodies, occur along the margins of the diatreme. This study characterizes the mineralization of the Balatoc epithermal veins and breccia-hosted GW orebodies using ore-gangue mineralogy, fluid inclusion characteristics, and sulfur isotope systematics.

The breccia and vein mineralization within the Balatoc Diatreme has five stages, distinguished by the dominant gangue mineral: gray quartz (Stage I), white quartz (Stage II), clear quartz (Stage III), calcite (Stage IV), and gypsum (Stage V). Stage I is the main gold mineralization stage. It is characterized by gray quartz with pyrite + marcasite + arsenopyrite + electrum + sphalerite ± chalcopyrite. Stage II is typified by white quartz associated with pyrite + electrum + chalcopyrite. Stage III is composed of clear quartz hosting pyrite + stibnite + chalcopyrite + galena + sphalerite + electrum (± pyrite + marcasite + arsenopyrite). Stages IV and V are associated with calcite and gypsum, respectively, hosting trace amounts of pyrite and sphalerite.

Microthermometry of the fluid inclusions hosted by Stage I and Stage II vein quartz revealed homogenization temperatures (Th) ranging from 220°-240°C and 260°-320°C, respectively. Calculated salinity values range from 1.7-8.7 wt % NaCl eqiv for Stage I and 2.1-13 wt % NaCl equiv for Stage II. The fluid inclusions hosted in Stage III quartz breccia cement have homogenization temperature ranging from 290°-300°C and salinity ranging from 1.05-6.01 wt % NaCl equiv. A boiling trend is recognized for the stages associated with the epithermal veins (Stages I and II), while an isothermal mixing trend is prominent for Stage III mineralization in the GW breccia orebodies.

Measured $\delta^{34}S$ values of pyrite and sphalerite from Stage I to Stage IV veins and breccias narrowly range from 0.8 to 1.5‰, which suggests that H_2S is the dominant sulfur species. On the other hand, the $\delta^{34}S$ values measured from the pyrite of Stage V breccia orebody range from -1.6 to -1.5‰, indicating the partitioning of the heavier isotopes to the gypsum sulfate. This depicts a shift from reducing to oxidizing conditions in the latter part of the mineralization.

ST.166
A Simple Recipe to Unravel Veladero Mine Mineralization Controls, Argentina

Diego M. Guido[1], Leandro Sastre[3], Gerardo Páez[1], Conrado Permuy Vidal[1], Matías Galina[2], Luciano López[1], Sebastián Juarez[3], Sebastián Jovic[1]
1. INREMI (CONICET & UNLP), La Plata, Argentina, 2. Secretaría de Minería de la Nación, Buenos Aires, Argentina, 3. Barrick Gold, San Juan, Argentina

The Veladero mine, a 50/50 joint venture operation between Barrick Gold and Shandong Gold, is in the Andean Cordillera in Argentina. It is a large open-pit, heap-leach Au-Ag mine in San Juan province that started production in 2005 and has recovered approximately 8.2 million ounces of gold and 16.6 Moz of silver from approximately 319 million tonnes (Mt) of ore averaging 1.09 g/t Au and 14.9 g/t Ag as of December 31, 2017.

The deposit is situated at the El Indio Gold Belt, a 120- by 25-km north-trending corridor composed of a Permian basement and characterized by Miocene volcanic and intrusive rocks with associated epithermal deposits.

Veladero is a high-sulphidation epithermal deposit hosted by volcanic breccias, tuffs, and volcaniclastic sediments. Hydrothermal alteration shows a typical silicified core grading outward into advanced argillic, argillic, and propylitic alteration haloes.

In order to expand the brownfield exploration of the epithermal deposit, a new series of studies were completed in the 3.5-kilometer-long NNW-striking mineralized zone that comprise three orebodies: Amable (mined out) in the south, Cuatro Esquinas in the centre, and Filo Federico in the north.

The studies comprise open-pit mapping and logging of 189,585 meters of drill cores and RC cuttings to make detailed geologic, structural, alteration, and mineralization descriptions that were compiled along 70 east-west cross sections, 3 km long and with a 50-meter separation. In addition to these geologic cross sections, a systematic study of the hydrothermal alterations was carried out along the same east-west sections, but with a 200-meter separation. The alteration was characterized with portable TerraSpec with measurements every 10 meters and confirmed by X-ray diffraction and petrographic studies.

The Veladero mine general NNW-striking structural corridor of the epithermal mineralization is reflecting the influence of combined stratigraphic and structural controls, with strong influence of a premineralization intermediate-composition volcanic depocenter that was affected by pre- to synmineralization faulting, intrusions, and two overlapped maar-diatreme complexes (Type I and II phreatomagmatic breccias) that increase permeability for different high-sulphidation events, represented by two main silica flooding pulses.

Circulation of hydrothermal fluids (alteration and silica injections) in Federico is favored by the lithological control of the volcaniclastic sequence (mainly ignimbrites and phreatomagmatic breccias) present in the NNW-striking depocenter (Federico mineralization style). In Amable, alteration and mineralization areas are aligned with N-S and NE directions, evidencing greater structural control (Amable mineralization style).

Au and Ag concentrate in NNW-striking subhorizonal level inclined 15° east, with 300 to 400 meters of vertical development (3,900 to 4,400 m altitude). WNW and N-S fault intersections seem to control the source of the fluids, inferred to the South (Amable) and East (Zorro fault) of Veladero mine, as evidenced by high-temperature minerals: dickite, pyrophyllite, and topaz. This favors the presence of high-grade feeders vectoring towards those sectors.

The geologic model, together with data interpretation, provides ideas to improve exploration and recommendations to increase understanding of the Veladero mine geology and mineralization.

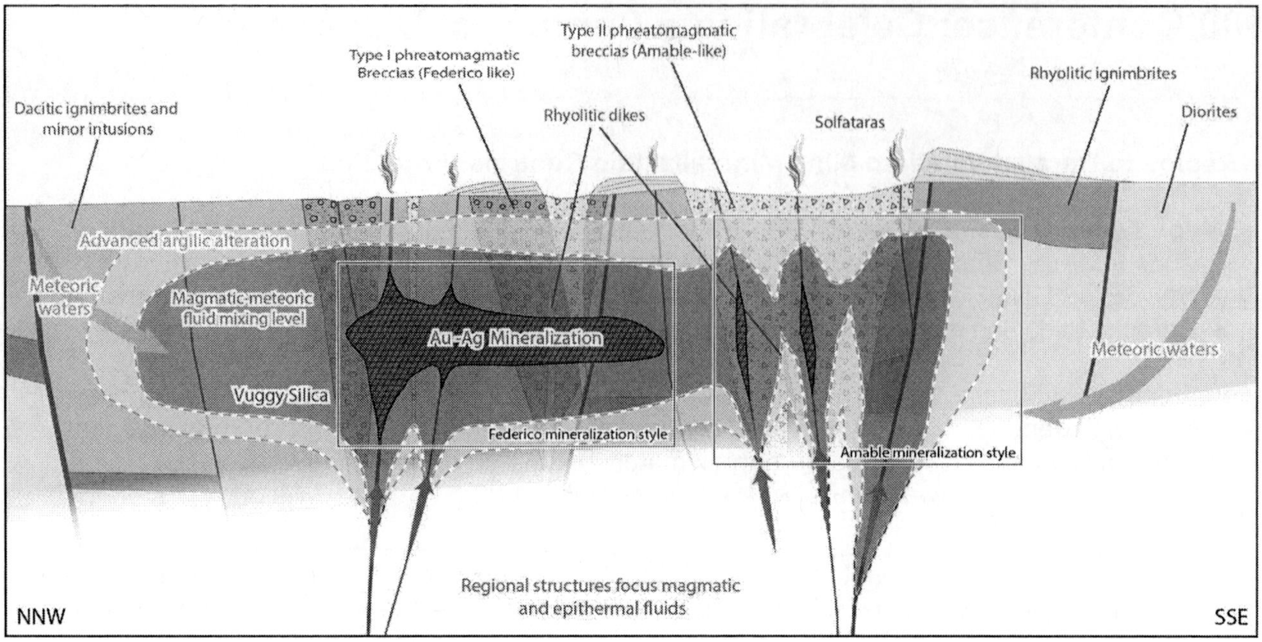

Type II phreatomagmatic breccias (Amable-like)
Type I phreatomagmatic Breccias (Federico like)
Rhyolitic ignimbrites
Dacitic ignimbrites and minor intusions
Rhyolitic dikes
Solfataras
Diorites
Advanced argilic alteration
Meteoric waters
Magmatic-meteoric fluid mixing level
Au-Ag Mineralization
Vuggy Silica
Federico mineralization style
Amable mineralization style
Meteoric waters
Regional structures focus magmatic and epithermal fluids
NNW
SSE

ST.167

Ore-Forming Processes in Critical Mineral Deposits: Can We Link Hydrothermal Experiments to Exploration in the Field?

Alexander P. Gysi[1, 2]
1. New Mexico Bureau of Geology and Mineral Resources, New Mexico Tech, Socorro, NM, USA, 2. Department of Earth and Environmental Science, New Mexico Tech, Socorro, NM, USA

Rare earth elements (REEs) mineral deposits are commonly found in carbonatite and (per)alkaline igneous systems (e.g., Bayan Obo in China and Strange Lake in Canada) where REE transport, partitioning, and deposition occurs via a complex interplay between magmatic and hydrothermal processes. Field observations yield important clues about the mineralogy and geochemistry of these deposits, but our understanding of the relationships between ore-forming processes and alteration zones is still quite rudimentary in comparison to porphyry deposits. The hydrothermal fluids are characterized by an enrichment in alkalis and are commonly more F rich and S poor, with Cl^-, F^-, CO_3^{2-}, and/or SO_4^{2-} as the dominant ligands controlling REE mobility. Fluid-rock interaction associated to the formation of REE deposits results in the observed exotic alteration styles such as fenitization (alkali metasomatism) and Ca-F metasomatism. The deposition of REEs is also controlled by hydrothermal carbonate, fluoride, and phosphate minerals (e.g., monazite-(Ce) and bastnäsite-(Ce)), which can form in a wide temperature range. Many questions remain, however, unresolved, which makes it difficult to interpret geochemical data sets of these ore deposits: What fluid types control the transport of REEs in different deposits? Are these fluids acidic/alkaline? Why do we find deposits enriched in either light/heavy REEs? How can hydrothermal fluids mobilize, fractionate, and further concentrate the REEs?

The complex behavior of the REEs can be captured via well-constrained hydrothermal laboratory experiments. For this purpose, a new Ore Deposits and Critical Minerals Research Laboratory was established at the New Mexico Bureau of Geology and Mineral Resources. Solubility data generated from our experimental laboratory are implemented in the MINES thermodynamic database (https://geoinfo.nmt.edu/mines-tdb/) and permit predicting fluid-rock interaction scenarios at depth as a function of pressure, temperature, and fluid compositions. These modeling predictions can be linked to field observations at various scales, particularly vein microtextures, alteration styles, and rock/mineral geochemistry. Batch-type experiments indicate that the solubility of synthetic monazite (La to Gd) and xenotime (Y, Tb to Lu) is higher than previously described between 100° and 250°C. Simulations of xenotime solid solutions in a quartz-muscovite vein reproduce the observed chondrite-normalized REE signatures of natural xenotime-(Y). Comparison of the simulations with data compiled from the Pea Ridge IOA-REE deposit in Missouri and from the Bear Lodge carbonatite in Wyoming indicates that both temperature and fluid salinity play a significant role in modifying the REE signatures of xenotime. At high temperature (>250°–300°C), the simulated xenotime solid solutions display a depletion in Er, Tm, Yb, and Lu similar to the Bear Lodge deposit, whereas at lower temperature, xenotime signatures are comparable to the Pea Ridge deposit and are enriched in the heavy REEs. Hydrothermal mixing experiments permit further synthesizing REE-doped calcite, fluorite, and apatite crystals between 100° and 300°C, providing new insight into REE mineral-fluid partitioning. Using these novel approaches, I will show how REE signatures in minerals can now be linked to temperature-dependent alteration styles and bulk-rock geochemistry in carbonatites and alkaline igneous systems with applications to exploration vectoring in critical mineral deposits.

ST.168

New Horizons in Magmatic Cu-Ni-PGE Sulfide Exploration: Expanding the Search Space to Alkaline Systems

David A. Holwell[1], Daryl E. Blanks[1], Marco L. Fiorentini[2]
1. University of Leicester, Leicester, United Kingdom, 2. University of Western Australia, Perth, WA, Australia

Magmatic Ni-Cu-PGE sulfide deposits are classically hosted in mafic-ultramafic complexes formed from magmas that are the product of moderate to high degrees (10-25%) of mantle melting. Such ores typically have Ni/Cu ratios around 1-20, with deposits formed from the most MgO-rich magmas generated from the highest degrees of melting, komatiites, being the most Ni rich. However, there is an increasing recognition of magmatic sulfide occurrences in alkaline ultramafic systems, generated from lower (<10%) mantle melting. A key consequence of the lower degree of partial melting in such systems is that the melts formed are more volatile rich, as components such as H_2O and CO_2 are incompatible. Thus, the lower the partial melt, the more volatile-rich the magma. This also impacts on the metal budget. Impartial or incongruent melting of mantle sulfide will liberate elements such as Cu, Au, and Te preferentially over Ni, Co, and some of the PGEs, which will remain in any residual mantle sulfide.

Consequently, magmatic sulfides in alkaline systems are typically hosted by ultramafic rocks that contain hydrous and volatile phases like amphibole, phlogopite, apatite, and carbonate. There is a continuum in composition from carbonatite-phoscorite, through shonkinites and alkalic ultramafics to thoeliites. The volatile content decreases through this continuum, and the metal ratios change, with initially high Cu/Ni, Au/PGE, and Te/PGE, which reduce to a more conventional Ni-Cu-PGE assemblage in volatile-poor tholeiitic melts. Thus this "alkalic DNA" is characterised by sulfides that are Cu dominant over Ni and have elevated Au concentrations and abundant telluride minerals as part of the assemblage. Notably, sulfides are often intimately associated with carbonate that suggest CO_2 played a critical role in buoyant transport from the mantle into the crust.

The alkaline magmatic sulfide systems can be observed throughout the lithosphere, from lower crustal pipes and intrusions (e.g., in the Ivrea Zone, Italy) where sulfide and carbonate are hosted by amphibole-rich ultramafic rocks. In mid-crustal conduits and alkaline-mafic-carbonatite complexes (e.g., Mordor Complex, Australia), sulfides are associate with carbonate as large blebs. Notably, upper crustal expression of these systems are not, however, ultramafic-mafic intrusions, but evolved alkaline intrusions that can be host to Cu-Au-Te-(Pd)-rich porphyry deposits and Au-Te epithermal systems.

Whilst the upper crustal porphyry-epithermal deposits can be giants (e.g., Cripple Creek), the question remains as to whether such systems can generate significant magmatic sulfide orebodies in the mid to lower crust. We propose that the Cu-Au-Te-rich orebodies of the Curaça Valley, Brazil; the Okiep district, South Africa; and the Mbesa prospect, Tanzania are all mid-crustal expressions of the accumulation of magmatic sulfides from such systems and form a continuum towards classic, volatile-poor, Ni-Cu-PGE deposits. As such, exploration targeting in regions of alkaline mafic-ultramafic and carbonatitic magmatism has the potential to broaden and open up new search spaces for more Cu-Au-Te-rich magmatic sulfide deposits in settings that do not necessarily conform to the conventional model for Ni-Cu-PGE deposits.

ST.169

New Constraints on Petrogenesis of the Bajo de la Alumbrera Cu-Au Porphyry System and Other Porphyries of the Farallon Negro Volcanic District (NW Argentina)

Madeleine Ince[1], Steffen Hagemann[1], Marco Fiorentini[1], Tony Kemp[1], Nora Rubinstein[2], Eduardo Zappettini[3]
1. Centre For Exploration Targeting, University of Western Australia, Crawley, WA, Australia, 2. Instituto de Geociencias Básicas, Aplicadas y Ambientales de Buenos Aires (Universidad de Buenos Aires–CONICET), Buenos Aires, Argentina, 3. Servicio Geológico Minero Argentino (SEGEMAR), Buenos Aires, Argentina

The Farallon Negro Volcanic District is a well-endowed area in the Sierras Pampeanas of Argentina. It includes key Cu-Au porphyry and low-sulfidation epithermal deposits including Bajo de la Alumbrera, Agua Rica, and Farallon Negro. This study investigates the petrogenesis of the Bajo de la Alumbrera porphyry Cu-Au deposit as well as mineralised and unmineralised satellite porphyry stocks including El Durazno, San Lucas, and Las Pampitas. The aim is to understand the petrogenesis of the porphyry systems of the Farallon Negro Volcanic District, using whole-rock geochemistry (TE, REE) and zircon isotope analyses (U-Pb, $\delta^{18}O$, Lu-Hf, TE). Ten least altered porphyry samples were analysed: four from Bajo de la Alumbrera, three from Las Pampitas, two from San Lucas, and one from El Durazno.

The samples were dated by SHIRMP U-Pb analysis. The El Durazno porphyry was emplaced at 8.0 ± 0.3 Ma, making it the oldest intrusion. The Las Pampitas intrusions were emplaced between 7.3 ± 0.1 and 7.03 ± 0.2 Ma, and the San Lucas porphyries are a similar age at 7.3 ± 0.2 and 7.2 ± 0.2 Ma. The Bajo de la Alumbrera porphyries are the youngest, forming from 6.9 ± 0.2 to 6.7 ± 0.2 Ma. All intrusions show a similar whole-rock TE and REE pattern, with an enrichment in LILEs relative to HFSEs. There is overall enrichment in REEs, a flattening of the slope in LREEs, and a trough in MREEs, suggesting amphibole fractionation. The Eu/Eu° ratio is between 0.80 and 0.92, indicating no plagioclase fractionation in the source melt. The El Durazno zircons exhibit a mantle-like $\delta^{18}O$ and Lu-Hf signature ($\delta^{18}O$ = 5.7±0.1‰; εHf = -0.5 ± 0.4), whereas Bajo de la Alumbrera, Las Pampitas, and San Lucas have a supracrustal signature ($\delta^{18}O$ >6.5‰, εHf <-2), suggesting assimilation of crustal material. Because of cohesive isotopic values and lack of inheritance and zircon cores, it is likely that assimilation of crustal material did not occur during emplacement or magma fractionation and instead was incorporated into the source melt. It can be inferred that the Bajo de la Alumbrera, Las Pampitas, and San Lucas intrusions are cogenetic. Zircon Ce/Ce* ratios reveal an oxidised source; however, there is no difference in zircon trace element signatures between fertile and infertile intrusions with respect to Cu-Au fertility.

Based on these data and other studies, a petrogenetic model has been proposed for the formation of the porphyry stocks in the Farallon Negro Volcanic District. From 9.5 to 8.0 Ma there is mantle-derived, extrusive magmatism and the emplacement of the El Durazno porphyry. From 7.5 to 6.0 Ma a magma chamber develops, and magmatism is intrusive. This period encompasses the emplacement of the Las Pampitas, Bajo de la Alumbrera, and San Lucas porphyries, as well as Cu-Au mineralisation of the Bajo de la Alumbrera porphyry. The magma source also changes from mantle-derived to a mix of mantle-derived and crustal. Understanding the petrogenesis and changing source compositions of these porphyry stocks is key to unravelling the components necessary for Cu-Au mineralisation in porphyry deposits.

ST.170

Multistage Mineralization Indicated by the Re-Os Geochronology of Low-Level, Highly Radiogenic Sulfides at the Kamioka Pb-Zn Skarn Deposit, Central Japan

Mizuki Ishida[1], Takumi Sakai[1], Junichiro Ohta[2, 3], Moei Yano[3, 1], Yasuhiro Kato[1, 2, 3]
1. Department of Systems Innovation, School of Engineering, The University of Tokyo, Tokyo, Japan, 2. Frontier Research Center for Energy and Resources (FRCER), School of Engineering, The University of Tokyo, Tokyo, Japan, 3. Ocean Resources Research Center for Next Generation (ORCeNG), Chiba Institute of Technology, Chiba, Japan

The Hida metamorphic belt in central Japan is thought to be a remnant of the Asian continent prior to the opening of the Japan Sea around 15 Ma. The Kamioka skarn deposit, located in the southeastern part of the Hida metamorphic belt, is the largest Pb-Zn producer in Japan. Despite the extensive literature on the deposit geology, geochronology of the Kamioka deposit has been a topic of much controversy [1,2,3]. Although the orebodies are spatially associated with the Triassic to Jurassic igneous and metamorphic rocks, reported ages are mainly restricted to Paleocene, which may represent the episodes of overprinting mineralization [1].

Here we investigated the Re-Os isotope composition of sulfide (chalcopyrite, arsenopyrite, galena, sphalerite) from four orebodies in the Kamioka deposit. Data points (n = 28) were highly scattered on the $^{187}Re/^{188}Os$ vs. $^{187}Os/^{188}Os$ diagram and could not define an isochron age. Therefore, we calculated the single model ages for 10 samples, which were low-level, highly radiogenic (i.e., ^{187}Os composes more than half of the total Os [4]). One chalcopyrite separate from the Sasahira orebody yielded a model age of 176 ± 19 Ma, which is close to the ages of the Hida younger granites found near the deposit (209–176 Ma, determined by zircon U-Pb, whole-rock Rb-Sr and hornblende K-Ar [5]). Model age of 136.7 ± 19 Ma for one arsenopyrite separate from the Sanotsu orebody is in good agreement with the whole-rock Sm-Nd age of ore from the different part of the deposit, dated ca. 150 Ma [2]. The model age is also within the range of the K-feldspar K-Ar ages of the Inishi migmatite (56–162 Ma [3]), which is in close contact with the Kamioka orebodies.

In contrast, eight arsenopyrite separates from the Izumidaira fault yielded much younger model ages ranging from 41 to 19 Ma. While no igneous rock of such age is known in the Kamioka area, fission-track thermochronometry of zircon collected along the Atotsugawa fault zone, on top of which the Kamioka deposit is located, suggests the presence of hydrothermal fluid activity along faults until 15 Ma [6].

Based on the evidence above, we consider all of our model ages to be geologically meaningful, indicating that the Kamioka deposit was formed from multistage processes that occurred over millions of years. Details of different mineralization episodes and directions for future studies will be discussed in the presentation.

[1] S. Akiyama (1980) *Mining Geology,* 30(6), 345-362.

[2] Ministry of International Trade and Industry (1998) *A Report on Analysis for General Geological Structures*. 198.

[3] K. Sato and S. Uchiumi (1990) *Mining Geology,* 40(6), 389-396.

[4] H. J. Stein et al. (2000) *Economic Geology,* 95(8), 1657-1672.

[5] Y. Takahashi et al. (2010). *Gondwana Research,* 17(1), 102–115.

[6] S. Sueoka et al. (2019) *Journal of Asian Earth Sciences: X,* 2, 100011.

ST.171
Geologic and Tectonic Setting of Uranium Mineralisation in the Damara Orogen

Thomas L. Jones, Judith Kinnaird, Paul Nex
University of the Witwatersrand, Johannesburg, South Africa

The southern Central Zone of the Damara Orogen, situated in the Erongo region of Namibia, is a well-known uraniferous province. In addition to the Rössing and Husab uranium mines, exploration for primary granite-hosted uranium mineralisation is actively ongoing beneath the desert plains. Understanding the tectonic setting of a mineral system is integral to understanding the processes that focused mineralisation into economic concentrations. This understanding forms a key component of modern approaches to exploration.

Primary uranium mineralisation is hosted by leucogranite sheets, which intruded into the mid-crust of the Damara Orogen during NW-SE collision between the Congo and Kalahari cratons at ~550-500 Ma. At the regional scale, pressure-temperature-time (P-T-t) paths show that uraniferous leucogranites intruded into the Damara Orogen at a relatively late stage, ~20-30 m.y. after the onset of crustal thickening. Their intrusion overlapped with peak metamorphic conditions of ~800°C/4-5 kbar, which was associated with in situ partial melting and the formation of migmatites. While the exact source of this late-stage high-T metamorphism is equivocal, the absence of a clear mantle signature in any Damaran granites, combined with elevated concentrations of radioactive elements in Damaran metasediments, a near-isobaric heating path, and an appropriate timescale for radioactive heating (~20-30 m.y.), suggests that internal crustal heat production may have been significant. Leucogranites have a predominantly S-type to slightly I-type chemistry, consistent with partial melting of metasedimentary to metaigneous crustal sources.

Geologic mapping shows that several domes in the lower Swakop and Khan rivers are cored by (1) overwhelming volumes of Damaran-aged granites and migmatites, which (2) crosscut rafts and/or locally continuous horizons of Damaran metasediments. Granite migration was structurally controlled; at deeper levels in the cores of domes, melt intruded pervasively along foliation planes and as crosscutting dikes, forming hectometre-scale granite-injection complexes. In the overlying metasediments, granite transport is restricted to crosscutting dikes, which intruded either as ductile fractures or as self-propagating hydrofractures. Paradoxically, many dikes exposed along the Swakop River strike consistently NE-SW, seemingly parallel to the sigma-3 principal stretching direction; this contrasts with theoretical models predicting dikes should be oriented normal to sigma-3.

Granite-hosted uranium mineralisation at the deposit scale is closely associated with a redox boundary at the Khan-Rössing Formation contact. However, an explanation for why uraniferous leucogranites are so localised to this contact, when barren leucogranites are regionally widespread throughout the cores of domes, is the most significant outstanding question. Distinct shear zones, which might act to localise the intrusion of uraniferous leucogranites, are not consistently developed. This likely reflects the high-temperature conditions at the time of granite intrusion, which led to widespread rheological weakening of the mid-crust and acted to prevent strain localisation. Similarly, post-Damaran deformation is relatively limited and there is little evidence for major fault zones, which might have caused postdepositional modification of the deposits.

ST.172

Vectors to Porphyry Copper Mineralization in Carbonate Rocks at the Bingham Canyon Mine, Utah, USA

Michael J. Kirschbaum[1], Zhaoshan Chang[1], Paul Agnew[2], Adam Pacey[2], Phillip Nickerson[3], Andrew Sasso[3], Kim Schroeder[3], Kathleen Gundy[3]
1. Colorado School of Mines, Golden, CO, USA, 2. Rio Tinto, Bundoora, VIC, Australia, 3. Rio Tinto, Salt Lake City, UT, USA

The Bingham Canyon porphyry Cu-Mo-Au occurrence (Utah, USA) is a highly productive mineral system with an extensive hydrothermal alteration halo. Such systems likely cause subtle, far-field alteration patterns, which may be useful in vectoring toward the system center. This study presents initial results from our investigation into the effects of this hydrothermal system on two laterally extensive carbonate units, the Jordan and Commercial limestones, by analyzing the trace elements of calcite in veins and in adjacent carbonate wall rocks and will compare the findings with signals from other systems such as the Candelaria Cu-Au deposit, Chile [1].

Bingham Canyon comprises a multiphase sequence of Eocene-aged igneous rocks emplaced within a folded and thrusted Paleozoic carbonate and siliciclastic sequence [2], producing proximal Cu-Mo-Au porphyry-style mineralization and adjacent Cu-Au skarn deposits in carbonate host rocks, mainly in the Jordan and Commercial beds. Outboard from the porphyry-skarn zones (~0.5-2.5 km) there are Pb-Zn-Ag vein and carbonate replacement orebodies, and even more distal (~7-9 km) sediment-hosted Au deposits.

A series of vein and wall-rock samples were collected beginning at the Bingham Canyon open pit and extending 17 km to the W-SW along a transect terminating at the subeconomic Stockton porphyry system. LA-ICP-MS analysis was completed on vein and wall-rock calcite. Preliminary analysis reveals a decreasing trend in Fe, Mg, Ba, and several REEs from the Bingham Mine outboard to a distance of ~3-5 km, while Mn and Zn display an increase to about 2-km distance before dropping off to background levels ~4 km outboard. These initial findings are based on 12 samples, which we will corroborate through further LA-ICP-MS analysis of additional samples along the transect to produce a robust spatial data set.

ST.173
Celebrating Lindgren and Others: Rebooting Mineral Deposit Classifications

David V. Lefebure[1], Samuel F. Cantor[2], Clinton P. Smyth[2]
1. University of Victoria, Victoria, BC, Canada, 2. Minerva Intelligence Inc., Vancouver, BC, Canada

Mineral deposit classifications employ selected properties to identify deposit types and to facilitate concise communication of complex detail. To avoid confusion, and often argument, they need to ensure that the properties they reference are also part of equally well structured classification systems. Knowledge engineering applications of artificial intelligence (AI) provide the tools necessary for these purposes, which materially contribute to Lindgren's quest for a complete mineral deposit classification. Additionally, a 21st-century mineral deposit classification is required to facilitate communication not only between humans but also between humans and computers and between different computer systems.

The authors have developed a prototype mineral deposit classification system that harnesses research documented by Lindgren, Stanton, Von Cotta, and many others to meet these objectives. The Multi-hierarchical Mineral Deposit Classification (MMDC) uses plate-tectonic settings, mineral systems parameters, host earth materials, ore textures, mineralogy, commodities, and other properties to characterise deposit types. As a starting point, the MMDC has been populated by drawing on deposit profiles published by the British Columbia Geological Survey (BCGS).

The MMDC has a graph structure, rather than a tree structure, meaning that a particular deposit type may qualify as a subtype of more than one supertype. This structure can help economic geologists to better understand the distinctions between similar deposit models, such as various types of massive sulphide orebodies (Fig. 1). The ability to use classification graphs becomes extremely important when tackling a complete classification of all metallic, industrial mineral, precious stone, and gemstone deposits.

The Aristotelian Class Editor (ACE) is a free, online taxonomy editor that is used for this project to capture deposit model properties sourced from published mineral deposit models. ACE provides a logically consistent framework for organizing mineral deposit properties and managing the resulting hierarchies of deposit-type classes that are created from combining those properties. Some of these classes will be novel groupings of mineral deposits with matching properties, which may constitute a "new" deposit type. The formally defined ACE files are well suited for use in AI systems engineered to assist in minerals exploration, geometallurgy, and environmental protection.

The MMDC is built on internationally documented terminology and data exchange standards. Used in ACE, these standards enable the coexistence of any number of different deposit models. Mining industry workers and academic researchers can collaborate to select the properties and related attributes they wish to use to classify mineral deposits, name their models, compare them efficiently with other models, and choose the models for their particular purpose. New models can be added and are automatically categorized. MMDC files can be exported in Excel or JSON formats for use with other software to explore the relationships between models, between deposits and models, or between specific deposits.

Fig. 1. A graph structure is used to classify volcanogenic massive sulphide and selected sedimentary-hosted stratiform and stratabound mineral deposit models. It uses a plate tectonic hierarchy and a portion of an earth environment hierarchy to separate these deposit types more effectively than a traditional tree structure.

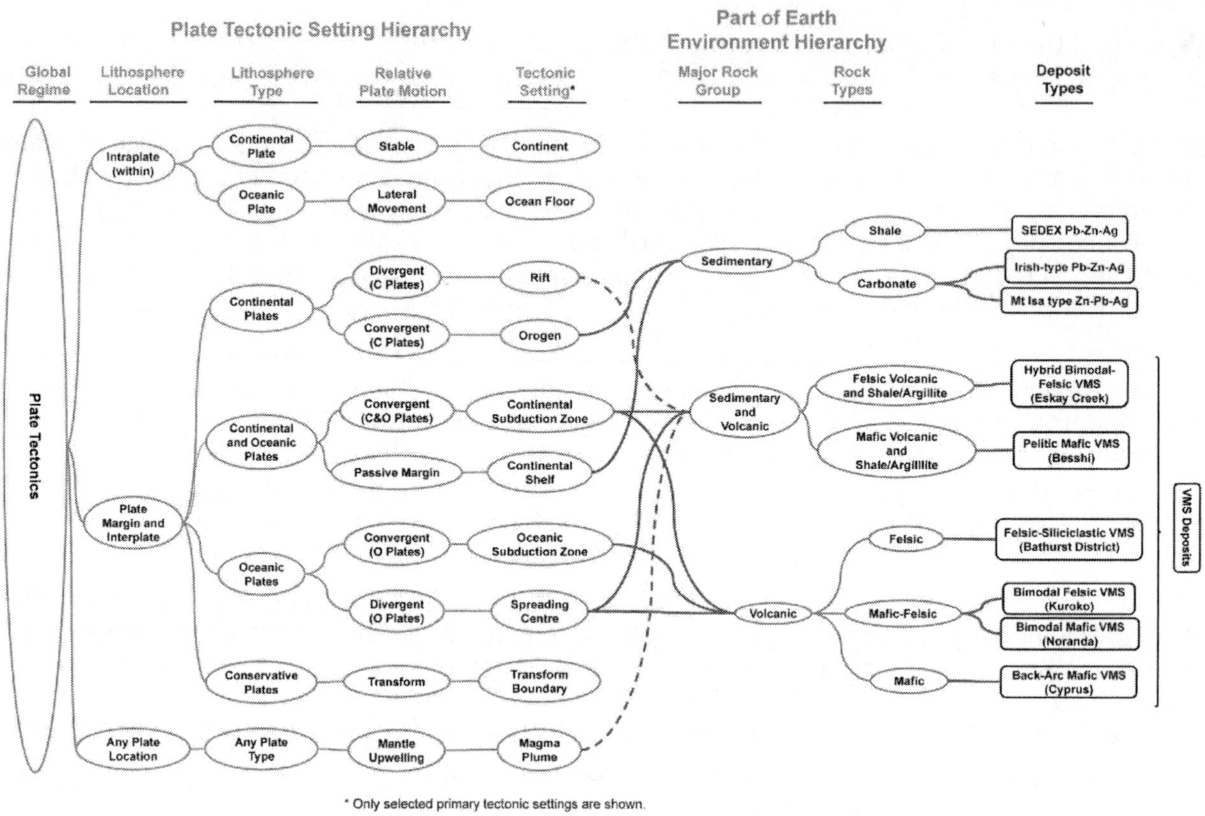

ST.174
Sr and Nd Sources for Epigenetic Cu ± Au and IOCG Deposits in the Kiruna Mining District Based on Rb-Sr and Sm-Nd Isotope Data

Leslie Logan[1], Joel Andersson[1], Fernando Tornos[1, 2], Tobias E. Bauer[1]
1. Luleå University of Technology, Luleå, Sweden, 2. Instituto de Geociencias (CSIC-UCM), Madrid, Spain

The Kiruna Mining District is host to a variety of ore deposit types including the giant Kiirunavaara iron oxide-apatite (IOA) deposit, smaller-scale syngenetic to epigenetic Cu±Au deposits (e.g., Pahtohavare), and iron oxide-copper-gold (IOCG) deposits (e.g., Rakkurijärvi). A recent structural investigation in the Kiruna area has highlighted that a significant brittle overprint is associated to Cu mineralization and can be correlated with a tectonic period ca. 80 m.y. after the Kiirunavaara IOA emplacement (Andersson et al., 2021, in press), bringing into question the time frame for Cu±Au precipitation/remobilization with respect to regional tectonic processes. Utilizing a mineral systems perspective, new whole-rock Rb-Sr and Sm-Nd isotope data from the Pahtohavare and Rakkurijärvi ore deposits in the southern Kiruna Mining District (Fig. 1) are compared to fluid and metal source candidates cropping out directly in the district.

The Pahtohavare Cu±Au mineralization is accompanied by pervasive albite-rich alteration especially in graphitic schist horizons (hosted in the Kiruna Greenstone Group (KGG)). The sample analyzed represents an albitized graphite schist and has an $^{87}Sr/^{86}Sr_{1.88\ Ma}$ value of 0.7205 and an $\varepsilon Nd_{1.88\ Ma}$ value of -5.3. The early Rakkurijärvi IOCG Na-Ca-(Fe) alteration was sampled from a weakly mineralized (chalcopyrite and pyrite, ±Au) outcrop of Kurravaara Conglomerate and has an $^{87}Sr/^{86}Sr_{1.86\ Ma}$ value of 0.7040 and an $\varepsilon Nd_{1.86\ Ma}$ value of -5.3. The data for the deposit samples were calculated using an estimated age of mineralization/alteration. Mixing trends from 1/Sr versus initial $^{87}Sr/^{86}Sr$ values for all isotopically preserved samples indicate distinct patterns for the two ore-altered samples. The Pahtohavare albitized graphite schist shows a Sr mixing trend between the KGG source candidates and a ca. 1.88 Ga granite. The Rakkurijärvi Na-Ca-(Fe)-altered conglomerate shows a mixing trend with a magnetite-gabbro and a KGG gabbroic sill. The different trends suggest the fluid pathways and Sr sources for each deposit were distinct but that the bimodal ca. 1.88 Ga magmatism contributed Sr in the ore-bearing fluid for both deposits, suggesting the ore deposits to have formed simultaneously or after the emplacement of these intrusions. While a later Rb-Sr isotope reset prevents the direct comparison of some felsic ca. 1.88 Ga intrusions in the district, the Sm-Nd system was preserved. Epsilon Nd values show a partial overlap of the ore-altered samples with the felsic igneous intrusions and a complete overlap with previously published εNd values from Perthite Monzonite Suite intrusions regionally. The significance of the different crustal lithologies and the ca. 1.88 Ga intrusive activity as possible Sr-Nd-metal sources is an important finding in this study and has implications on the regional behavior of ore forming systems in northern Norrbotten.

Fig. 1. The southern Kiruna Mining District showing sample locations and lithologies (modifed after Offerberg, 1967).

References

Andersson, J.B.H., Bauer, T.E., and Martinsson, O., 2021, Structural evolution of the central Kiruna area, northern Norrbotten, Sweden: implications on the geological setting generating iron oxide-apatite and epigenetic iron and copper sulfides: Economic Geology, in press.

Offerberg, J., 1967, Berggrundsgeologiska och flygmagnetiska kartbladen, Kiruna NV, NO, SV, SO: 29J.

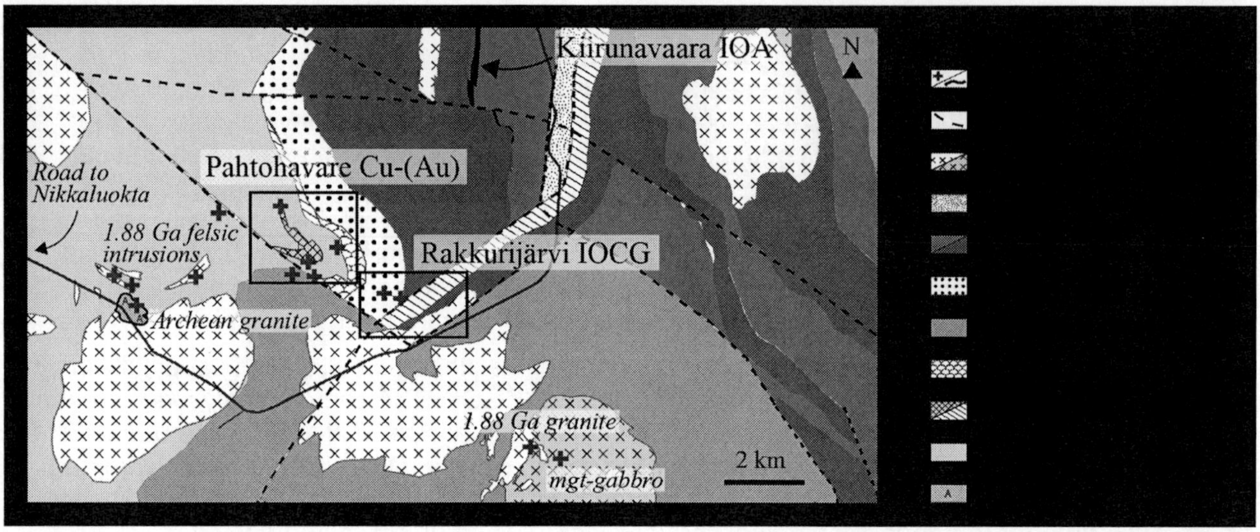

ST.176

Magmatic and Metallogenic Evolution of a Greenstone Belt: The Example of the Chibougamau Area, NE Corner of the Abitibi Subprovince, Canada

Lucie Mathieu[1], Pierre Bedeaux[1], David B. Snyder[2], Saeid Cheraghi[2], Bruno Lafrance[2], Phil Thurston[2]
1. Centre d'études sur les ressources minérales (CERM), Département des Sciences appliquées, Université du Québec à Chicoutimi (UQAC), Chicoutimi, QC, Canada, 2. Harquail School of Earth Science, Mineral Exploration Research Centre, Laurentian University, Sudbury, ON, Canada

Magmatism is a major component of crustal evolution and is paramount to mineralizing processes, as magmas carry metals from the mantle to the upper crust. In addition, magmatic fluids are an important carrier of precious and base metals in many mineralized systems (porphyry deposits, intrusion-related gold systems – IRGS, volcanogenic massive sulphides – VMS). The importance of fluids exsolved from magmas remains a source of much debate for some contexts, such as IRGS and orogenic gold deposits, while the amount of metal transported by magmas, throughout the evolution of greenstone belts, remains to be quantified. Addressing these issues requires (1) regional-scale studies, to evaluate the distribution of magmatic systems with distinct chemical characteristics (including metals and volatiles contents) and to unravel the volume, architecture, and duration of magmatic processes. It also requires (2) deposit-scale investigations to evaluate the processes leading to fluid exsolution and circulation, as well as metal precipitation. To address these issues, we take the example of the Chibougamau area, NE corner of the Neoarchean Abitibi Subprovince (greenstone belt), which is characterized by an abundance of Cu-Au and Au magmatic-hydrothermal mineralizing systems. In the Chibougamau area, metals accumulated during the synvolcanic and syntectonic periods to form porphyry-like and IRGS deposits, respectively. Using geochemical and geophysical methods, the architecture of the crust is addressed, with emphasis on the evolution and duration of magmatic systems. Combining new seismic data (Metal Earth profile) with known stratigraphic, structural, and magmatic records, we propose that the study area originated as normal (i.e., thick) Archean oceanic crust that formed at or before 2.80 Ga and that evolved through terrane imbrication at 2.73-2.70 Ga. Shortening caused rapid burial, devolatilization, and partial melting of hydrated mafic rocks to produce tonalite magmas that may have mixed with mantle-derived melts to produce the tonalite-trondhjemite-diorite (TTD) suite associated with observed Cu-Au magmatic-hydrothermal mineralization. Tonalite-trondhjemite-granodiorite (TTG) suites are most abundant in the Abitibi greenstone belt except in the Chibougamau area, where TTD suites dominate. This and other aspect of the Chibougamau magmatic record (e.g., abundance of sanukitoid plutons) may, in part, explain the fertility of this area for magmatic-hydrothermal mineralizing systems.

Fig. 1. Geologic map of the Abitibi Subprovince showing the distribution of the main lithologies. The map is modified from the Ministère de l'Énergie et des Ressources Naturelles of Québec (MERN), Sigeom data set, and the Ontario Geological Survey (OGS). The projection is UTM NAD83 Zone 17N. Numbers locate the Chibougamau (1) and Chester (2) plutons, as well as the gold-endowed Cadillac-Larder Lake fault zone (3). (From Mathieu et al., 2020, doi: 10.3390/min10030242.)

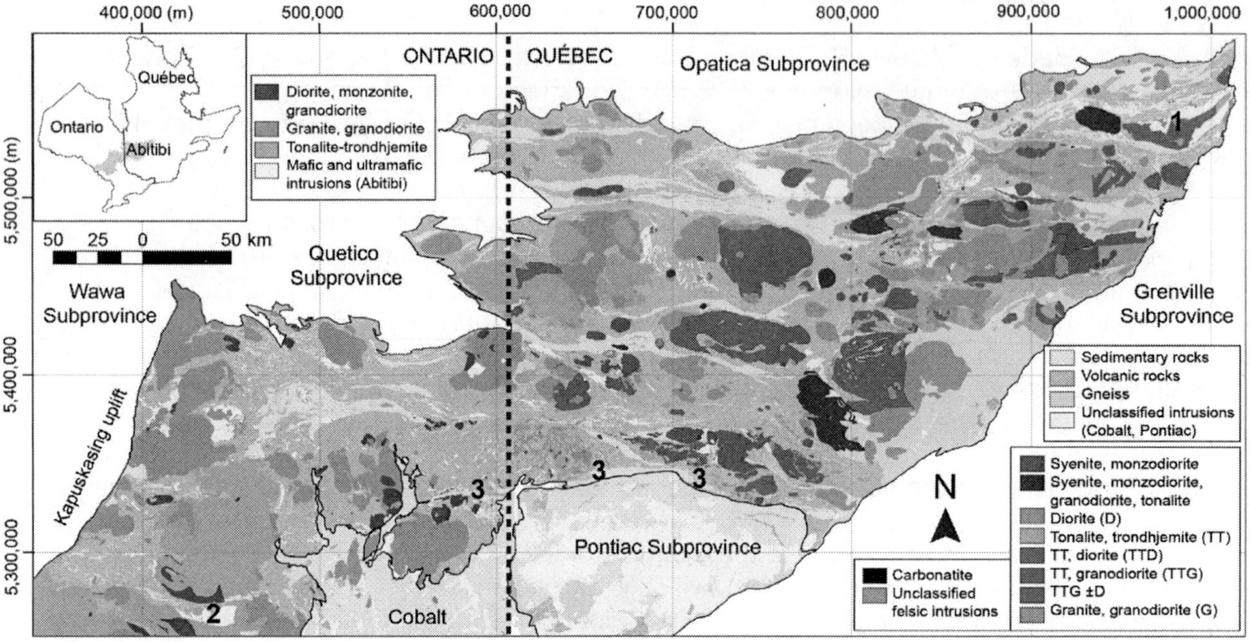

ST.178

Tracing Crustal-Scale Fluid Pathways Under Cover with Magnetotelluric Imaging: Examples from the Central United States

Benjamin S. Murphy[1, 2], Jan Marten Huizenga[3, 4], Paul A. Bedrosian[1]

1. Geology, Geophysics, and Geochemistry Science Center, United States Geological Survey, Denver, CO, USA, 2. Geomagnetism Program, United States Geological Survey, Golden, CO, USA, 3. Economic Geology Research Centre, James Cook University, Townsville, QLD, Australia, 4. Department of Geology, University of Johannesburg, Auckland Park, South Africa

Magnetotelluric (MT) imaging work in Australia has shown a spatial correlation between major iron oxide copper-gold/iron oxide-apatite (IOCG-IOA) deposits and zones of elevated bulk electrical conductivity that transect much of the crust. In the United States, three-dimensional MT imaging similarly reveals subparallel northwest-striking high-conductivity zones at mid- to lower-crustal depths beneath Mesoproterozoic IOCG-IOA deposits in the Missouri Iron Province. Both in Missouri and in Australian districts, these conductive anomalies have generally been interpreted as zones of fluid flow and crustal metasomatism; however, the exact cause of elevated electrical conductivity has previously been unclear, particularly because these anomalies are found within stable Precambrian crust. Here, we interpret these conductors specifically to be the result of graphite precipitation associated with magmatically driven $CO_2 \pm H_2O$ fluid flow through the crustal column. We propose that graphite precipitated during cooling through ~800-500°C along crustal-scale, tectonically controlled fluid pathways either via the oxyexsolution reaction $6Fe_2TiO_4 + CO_2 = 2Fe_3O_4 + 6FeTiO_3 + C$ or via carbon saturation in a C-O-H fluid with oxygen fugacity near the fayalite-magnetite-quartz buffer. In either case, the CO_2 involved in these reactions would have exsolved from mafic to intermediate magmas at mid- to lower-crustal depths and been concentrated along deep-rooted, tectonically controlled pathways; saline magmatic fluids that could drive mineralization are likely derived from intrusions at shallower crustal levels. In this interpretation, graphite is then a marker of magmatic fluid movement along localized crustal-scale conduits that could underlie near-surface magmatic or magmatic-hydrothermal mineralization. Therefore, electrical conductivity anomalies that are associated with such graphite precipitation would demarcate zones of localized, deep-rooted magmatic activity. Although they do not directly mark the locations of mineral deposits, these anomalies do indicate regions that once hosted an appropriate lithotectonic environment for mineral system formation. We apply this model to interpret magnetotelluric images in conjunction with other geophysical and geochemical datasets within the central United States. We highlight three specific regions where a thin (<1 km) veneer of Phanerozoic overburden obscures Precambrian basement rocks that may be prospective for IOCG/IOA mineralization: northeastern Illinois, northwestern Indiana, and northwestern Ohio.

ST.179

Petrography and Ore Microscopy of the Homelode Structural Trend, Black Hills, South Dakota, USA; Understanding Trace Element Zonation Patterns on a District Scale

Eric Nowariak[3, 1], Stephen T. Allard[2, 3], Robert Bergmann[1, 3], Brian Lentz[1, 3]
1. F3 Gold, Minneapolis, MN, USA, 2. Winona State University, Winona, MN, USA, 3. Big Rock Exploration, Minneapolis, MN, USA

The Homelode Trend is a newly discovered, ~6-km-long belt of vein-hosted gold-silver mineralization in the north-central portion of the Black Hills, South Dakota. This study incorporates petrographic analysis and trace element geochemistry across the district and aims to describe district-scale mineralogic and geochemical zonation of mineralization across the district. Mineralization is hosted in multistage quartz-carbonate-sulfide veins within sheared and folded phyllites, schists, and siliceous metagraywackes. Gold-silver mineralization is associated with variable amounts of pyrite, arsenopyrite, boulangerite, sphalerite, chalcopyrite, and other arsenic, antimony, and bismuth sulfides and sulfosalts. Hydrothermal alteration within the district is dominated by sericite dominated "bleached" zones, coarse muscovite in vein selvages, and tourmaline-graphite-rich zones. Many of the veined zones exhibit brecciated fragments of wall rock and pockets of euhedral quartz terminations, indicative of open space filling and a higher-level emplacement within the crust compared to other occurrences and deposits throughout Black Hills. Based on mineralogic, geochemical, and deformational features, the Homelode Trend has been classified as an epizonal orogenic gold deposit. In plan view, the district exhibits an elliptical zoning pattern within trace element geochemical data sets with an arsenic-zinc-copper-rich core transitioning to a lead-antimony-arsenic-silver-rich periphery. Petrographic analysis consistently describes a paragenesis of an arsenic-rich mineralization event followed by base metal- and/or antimony-rich mineralization event. Gold and silver mineralization, occurring as electrum and various silver-bearing sulfides, has been documented within both mineralization styles. The district-scale zonation may be attributed to evolving fluid chemistry from a higher-temperature arsenic-zinc-copper-rich mineralizing center, transitioning to a more distal, lower-temperature lead-antimony-silver signature. Alternatively, this zonation may be attributed to multiple pulses of distinct mineralizing fluids overprinting preexisting mineralization.

ST.180
Lithogeochemical Vectors and Mineral Paragenesis of Hydrothermal REE-Bearing Fluorite Veins and Breccia Deposits in the Gallinas Mountains, New Mexico

Evan Owen[1], Alexander Gysi[1, 2], Virginia McLemore[2]
1. New Mexico Tech Department of Earth and Environmental Sciences, Socorro, NM, USA, 2. New Mexico Bureau of Geology and Mineral Resources, Socorro, NM, USA

The Gallinas Mountains district, located in Lincoln and Torrance Counties, New Mexico, is host to hydrothermal REE-bearing fluorite veins and breccia deposits. The rare earth elements (REEs) are found in bastnäsite-(Ce) ([La,Ce]CO_3F), which is also the primary ore mineral mined in several important carbonatite deposits (e.g., Mountain Pass in California; Bayan Obo in China). Minor production of REEs, fluorite, Cu, Pb, Zn, Ag, and Fe has been recorded in the Gallinas Mountains district between the early 1900s and the 1950s. The REE-bearing fluorite veins and breccias are hosted in Permian sedimentary rocks as well as genetically related trachyte/syenite sills and dikes emplaced between 30 and 28 Ma. Previous studies have described the REE occurrences in the Gallinas Mountains, but the controls of hydrothermal processes on the transport and deposition of REEs in the district remain unclear. In this study, we combine microtextural observations with mineral and bulk-rock chemistry of hydrothermal REE-bearing fluorite veins and breccias to determine the vein types and alteration styles and establish a detailed mineral paragenesis. The goal of this study is to determine lithogeochemical vectors towards REE-rich zones in the district by linking thin-section and deposit-scale observations with mineral and bulk-rock geochemistry. This district is an exceptional natural laboratory for studying the role of hydrothermal processes for transport/deposition of REEs in an alkaline F-rich magmatic-hydrothermal system because very few deposits worldwide have such well-preserved and exposed geology.

Hand samples of hydrothermal veins and breccias containing fluorite ± calcite ± barite ± bastnäsite-(Ce) were collected from outcrops, prospect pits, and mine dumps. Optical microscopy was used to identify minerals and determine the textural features and crosscutting relationships of the different fluorite veins. The veins were classified into i) hematite-fluorite veins, ii) barite-bearing bastnäsite-fluorite veins, and iii) barite-bearing (fluorite)-calcite veins. Nearly all of the barite crystals in the fluorite veins display dissolution textures (skeletonized crystals) with infilling of mostly fluorite and minor calcite, suggesting that barite is part of an earlier paragenetic mineral assemblage. Bastnäsite-(Ce) is commonly found in veins containing barite and occurs either as disseminated crystals in the fluorite veins or together with fluorite infills around large barite crystals. A few of the barren fluorite-calcite veins display an intergrowth with euhedral barite crystals indicating that these could be part of an earlier barite paragenesis. These textural observations suggest a key control of REE mineralization in the Gallinas Mountains district by a coupled dissolution of barite-bearing (fluorite)-calcite veins and precipitation of later bastnäsite-fluorite veins. Geochemical bulk-rock data collected from the New Mexico Bureau of Geology and Mineral Resources database were analyzed using the IMDEX ioGAS™ program to define geochemical signatures of rock types, alteration styles, and vein types. Preliminary data analysis indicates a positive correlation between Ba, F, and total rare earth oxides (TREO). These trends corroborate with the observed vein microtextures, suggesting that the interaction of a hydrothermal fluid with the barite-bearing (fluorite)-calcite veins represents a key process for defining geochemical vectors in the district.

ST.181
Iron-Carbonate Precipitates as Footprint for Hidden Miocene Mineralized Bodies in Jaraña, Puno, Peru

Diana Pajuelo[1], Michael Rowe[2], Kathleen Campbell[2]
1. San Marcos University, Lima, Peru, 2. University of Auckland, Auckland, Auckland, New Zealand

Hot spring deposits are surface manifestations that precipitate around the active vents of continental hot springs within geothermal systems. Sinter (opaline silica) is surface evidence of epithermal systems at depth, and travertine (calcite or aragonite) represents a distal outflow facies of some geothermal system. Globally, the trace element concentrations of precious metals (Au, Ag) and base metals (e.g., Cu, Pb, Zn) in sinter and travertine are related to ore-mineralizing processes at depth.

Peru is characterized by ore deposit minerals exhumed along the Central Andes at high altitudes (>4,000 m.a.s.l), in the same hydrothermal systems as active hot springs. The Jaraña hot springs group is located in the Puno region, southern Peru, located 90 km to the east of the active volcanic arc in the Sillapaca Cordillera. The Jaraña hot springs are structurally influenced by local faults and are hosted within silicified Miocene andesites. The andesitic rocks contain copper sulfide and sulfosalt minerals and may suggest porphyry copper mineralization at depth.

The Jaraña hot springs generate thermal pools and outflow channels of variable depths (~5 cm to 2 m). The thermal waters have a nearly neutral pH (6-6.7) and temperatures ranging from 55° to 74°C. Thermal waters have an alkali-chloride and acid sulfate composition. Moreover, the active hot springs deposits contain minerals such as calcite, halite, and gypsum and chemical enrichments in Ag, Cu, Pb, As, Sn, and Fe. This study focusses on hot spring precipitates below the spring water surface (subaqueous).

The Jaraña subaqueous hot spring deposits are laminated and show markedly different colors (black, orange, white) and luster (metallic, vitreous), clearly suggestive of mineralogical and compositional diversity. X-ray diffraction patterns indicate they comprise a mixture of predominantly calcite and Fe oxides. Petrographic analysis indicates a combination of calcite crystals, polymictic rock fragments, microbial filaments, and diatoms. Cathodoluminescence was used to identify different stages of carbonate deposition, horizons of crystallization, and carbonate fragments. Four-acid digestion results show an inverse relationship between Fe and Ca concentration, consistent with varying proportions of Fe oxide and carbonate material. Furthermore, in situ chemical analysis of individual horizons by laser-ablation ICPMS contributed to identifying variations in silica in vitreous horizons, abundances of Fe, Zn, Pb, and Mo, and the distribution of these elements in the carbonate layers. In addition, the banded horizons contained high levels of As ranging from 174 ppm to 7.12 wt % and correlated to Fe. Samples have variable concentrations of Cu (0.2-268 ppm) and show a positive correlation with base metals such as Zn. However, in samples with high Fe concentrations, Cu is uncorrelated to Fe.

This research documents compositional variability and the concentration of Fe-bearing phases, in distinct laminated horizons of the Peruvian iron carbonate subaqueous precipitates. Trace element and metal enrichments in the iron carbonates may indicate interaction with deeper Cu orebodies, suggesting iron carbonate deposits could be used as a footprint for hidden Miocene or older mineralized bodies at depth.

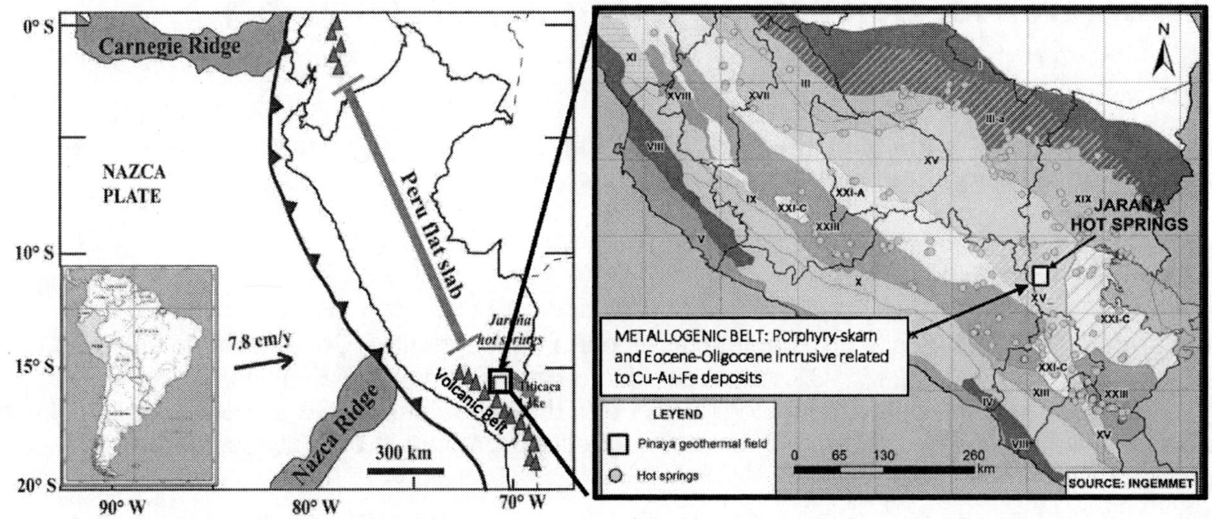

Carnegie Ridge
NAZCA
PLATE
Peru flat slab
7.8 cm/y
Jaraña
hot springs
Titicaca
lake
Volcanic Belt
Nazca Ridge
300 km
0° S
10° S
15° S
20° S
90° W
80° W
70° W
JARAÑA
HOT SPRINGS
METALLOGENIC BELT: Porphyry-skarn and Eocene-Oligocene intrusive related to Cu-Au-Fe deposits
LEYEND
Pinaya geothermal field
Hot springs
0 65 130 260 km
SOURCE: INGEMMET

ST.182
The Timing of Mineralization in the Mina Justa IOCG System, Peru: Insights from In Situ U-Pb MC-ICP-MS Ages for Apatite, Magnetite, and Titanite

Maria A. Rodriguez-Mustafa, Daniel Blakemore, Robert Holder, Adam C. Simon
University of Michigan, Ann Arbor, MI, USA

Iron oxide copper-gold (IOCG) deposits are major sources of Fe (in magnetite), Cu, and Au (in sulfides). They are also usually enriched in Ag, Mo, Co, As, U, and Rare Earth Elements (REE), which are critical metals for developing renewable energy technologies. In Peru, the Mina Justa IOCG deposit consists of two mineralization styles: manto and hydrothermal breccia. Magnetite is present in both styles, but Cu mineralization is concentrated in the latter.

Constraining the timing of Fe and Cu mineralization at Mina Justa is fundamental to understanding the duration and type of processes that generated this mineral deposit and ultimately to test the hypothesis of a genetic link with other deposits in the area. Previous workers used Ar/Ar geochronology to date alteration minerals (actinolite and microcline) from Mina Justa. The samples yielded ages ranging between 157 and 95 Ma, and the authors interpreted magnetite to have formed at around 110 Ma and Cu sulfides at around 100 Ma, suggesting two separate mineralization events in the deposit. However, the relationships between the alteration minerals and the mineralization stages are poorly constrained; hence, the proposed ages do not provide direct information about the timing of mineralization.

We characterized samples from the manto and breccia units at Mina Justa via scanning electron microscopy and identified magnetite, apatite, and titanite grains for U-Pb isotope measurements via in situ multicollector-inductively coupled plasma-mass spectrometry (MC-ICP-MS). This innovative technique allows us to perform spot analyses for detailed geochronological studies. Additionally, its application for dating magnetite opens the door to date mineralization in ore deposits that lack “traditional datable minerals” like molybdenite.

Our preliminary results (Fig. 1) are the first to date ore minerals at Mina Justa and indicate that magnetite, titanite, and apatite in the deposit are older than previously published, inferred ages of mineralization. Further work will focus on directly dating the sulfide minerals to assess the timing of Cu mineralization.

The results of this project will contribute to understanding the timescales of Fe and Cu mineralization at Mina Justa and clarify the time framework in which this deposit formed. This knowledge will help to identify the tectonic setting propitious for the occurrence of IOCG deposits in the Andes and elsewhere, allowing for the refinement of geophysical and geochemical exploration techniques to find new prospects that will optimize the exploration process and maintain the supply of resources for future generations.

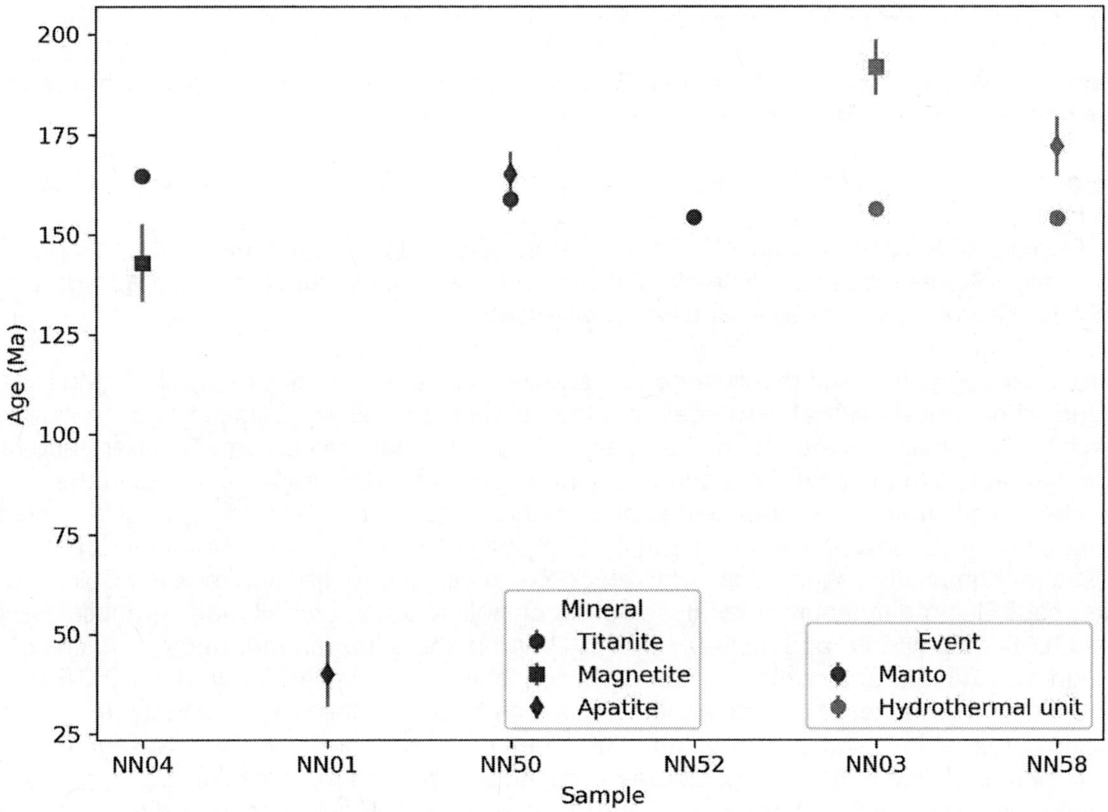

Fig. 1: Compilation of ages obtained for the Mina Justa IOCG deposit in Peru. All but one sample indicate that apatite, magnetite, and titanite at Mina Justa formed before 140 Ma. The different symbols represent the mineral analized and the colors correspond to the mineralization style. Error bars correspond to 2 standard deviations and might not be visible if the uncertainty is less that the size of the marker.

ST.183
Fluorine-Rich Mafic Lower Crust in the Southern Rocky Mountains: The Role of Preenrichment in Generating Fluorine-Rich Silicic Magmas and Porphyry Mo Deposits

Joshua M. Rosera[1, 2], Ryan E. Frazer[3, 2], Ryan D. Mills[2], Kristin H. Jacob[4], Sean P. Gaynor[5, 2], Drew S. Coleman[2], G. L. Farmer[4]
1. United States Geological Survey, Reston, VA, USA, 2. University of North Carolina at Chapel Hill, Chapel Hill, NC, USA, 3. United States Geological Survey, Denver, CO, USA, 4. University of Colorado Boulder, Boulder, CO, USA, 5. University of Geneva, Geneva, Switzerland

Fluorine-rich ferroan leucogranites and rhyolites can be shown to be associated with porphyry Mo ("Climax-type") deposits and other critical mineral resources, but their origins and relationships to continental lithospheric source rocks remain unclear. It has been especially challenging to determine where and how they become enriched in F. We tested if F enrichment is inherited from deep crustal sources in the southern Rocky Mountains by directly measuring amphibole compositions of lower crustal mafic granulite xenoliths from the State Line District in northern Colorado and by analyzing whole-rock elemental abundances and Sr, Nd, and Pb isotope compositions for a suite of Cretaceous to Oligocene igneous rocks in Colorado to evaluate their sources. Electron microprobe analyses of amphibole in lower crustal mafic granulite xenoliths show they contain 0.56–1.38 wt % F and 0.45–0.73 wt % Cl. Amphibole thermobarometry indicates equilibration at high-to ultrahigh-temperature metamorphic conditions (820°-940°C and 0.5-1.0 GPa) prior to entrainment in Devonian kimberlites. Mass balance calculations based on these values indicate that some parts of the mafic lower crust in Colorado are at least 3.5 times more enriched in F than average mafic lower crust. Batch melt modeling indicates low-degree partial melts derived from similar rocks with F-rich amphibole can yield silicic melts with >2,500 ppm F, similar to estimated F melt concentrations for fluorite-bearing leucogranites in the region.

Laramide-aged intrusions (75–38 Ma) predate F-rich magmatism in Colorado and have isotopic compositions consistent with mafic lower crust ± mantle sources, but many of these intrusions contain elevated Sr/Y ratios (>40) that suggest amphibole was a stable phase during magma generation. The F-rich igneous rocks from the Never Summer igneous complex and Colorado Mineral Belt also have isotopic compositions that overlap the lower crustal mafic granulite xenoliths, as well as lower Sr/Y (<40) and high Nb and Y abundances. We suggest that F-rich silicic melts in the southern Rocky Mountains were sourced from mafic lower crust and that fluid-absent breakdown of amphibole in rocks that experienced ultrahigh-temperature metamorphism was a key process in their generation. Experimental data indicate that F-rich amphiboles, such as those observed in the xenoliths, break down at higher T (>100°C greater) than typical lower-F counterparts. As such, these source rocks only melted during periods of unusually high heat flow into the lower crust, such as during an influx of mantle-derived magmas related to rifting or the post-Laramide ignimbrite flare-up in the region. Some rocks that predate the F-rich magmatism could also have been sourced from lower crust, but in this period lower crust melted under conditions in which amphibole was a stable phase or only low-F amphibole was available for breakdown, and therefore those rocks lack distinct F enrichment. These data have direct implications for the genesis of porphyry Mo mineralization, in particular in the Colorado Mineral Belt, because they indicate that F is inherited from crustal sources and that preenrichment of the crust could be a necessary condition for later anatexis and generation of F-rich magmas.

ST.184
Field Analysis of Contrasting Epigenetic Gold and Copper Vein Mineralization, Rouyn-Noranda Mining District, Québec

Marina D. Schofield[1], Bruno Lafrance[1], Harold L. Gibson[1], K. H. Poulsen[2]
1. Laurentian University, Sudbury, ON, Canada, 2. Consultant, Ottawa, ON, Canada

The Rouyn-Noranda mining district of Québec is best known globally for its 20 Cu-Zn (±Au ±Ag) volcanogenic massive sulfide (VMS) deposits including the giant Horne Cu-Au system. The district also hosts 19 gold-quartz-carbonate vein deposits and several subeconomic Cu-Ag occurrences comprised of vein to disseminated quartz, chalcopyrite, pyrite, pyrrhotite ±sphalerite ±galena ±Ag-Bi tellurides. Historically, there was no genetic distinction among the deposit types and all were inferred to be epigenetic, hypozonal, and magmatic-hydrothermal in origin. In the mid 1960s, a syngenetic origin became widely accepted for VMS mineralization worldwide and the gold-quartz-carbonate veins were described as epigenetic. The origins of the vein and disseminated Cu-Ag deposits were much more uncertain and were interpreted to be related to VMS or porphyry-copper deposits. They are similar to the epigenetic Au veins in that involve infilling of brittle fractures, locally contain vuggy quartz and petrogenetically late sulphides, and are overprinted by regional deformation. However, the two vein styles have very different metal tenors and alteration signatures and occur in different structural orientations. The Au veins contain ~3–6 g/t Au and are associated with hematite ± ankerite ± sericite ± chlorite alteration. They uniformly strike NNW, dip moderately towards the E, and are typically spatially associated with lamprophyre dikes that crosscut the Flavrian-Powell Intrusive Complex. In contrast, the Cu-Ag veins contain ~7 wt % Cu, ~98 g/t Ag, and <0.3 g/t Au and are associated with spotted to massive, sericite, chlorite alteration. They are dominantly hosted in the volcanic strata along NE-striking structures that are infilled by felsic dikes and are interpreted as synvolcanic structures that formed during extension. The Au veins locally crosscut the Cu-Ag veins and are folded and overprinted by an E-W regional cleavage that formed during the main NNE-directed deformation event, which affected the ca. 2704–2695 Ma volcanic rocks and younger ca. 2679–2669 Ma Timiskaming Group sedimentary rocks of the Rouyn-Noranda mining district. They are interpreted to have been emplaced early during this deformation event. The association of these Au veins along the margins of narrow lamprophyre dikes suggests the presence of preexisting structures, which were likely reactivated during later deformation and acted as structural traps for later Au-quartz vein mineralization. Similarly, the presence of the Cu-Ag veins along the margins of the NE oriented felsic dikes suggests continued reactivation of synvolcanic structures during progressive extension. The superposition of multiple mineralization events, each a product of different genetic processes that utilize preexisting structures and indicate a fertile geodynamic environment, is of importance to our understanding of the broader metallogeny of the district. With more deposits being discovered undercover, a better understanding of how ore systems evolve and are superimposed in metal-endowed districts is essential to help guide exploration, especially in complex Archean terranes.

ST.185
Characterization of the Footprint of Hydrothermal Ore-Forming Processes in the Cooke's Peak Pb-Zn-Ag-F District, New Mexico

Cody D. Schwenk, Nicole Hurtig, Alexander Gysi, Virgil Lueth
New Mexico Tech, Socorro, NM, USA

The Cooke's Range in Grant and Luna Counties, New Mexico, is host to multiple mining districts and mineral occurrences (i.e., Hadley Pb-Zn-Cu-Ag sulfide district, Cooke's Peak Pb-Zn-Ag-F carbonate replacement district, Fluorite Ridge fluorite vein district, and the Northern Cooke's Peak fluorite vein occurrences). These deposits combined have produced over 8 million pounds of Pb, 6 million pounds of Zn, 70,000 oz. of Ag, and 100,000 tons of fluorite since their discovery in 1876. Fluorite Ridge and the Northern Cook's occurrences have a higher proportion of fluorine to metal content and were primarily mined for fluorite, while the Hadley and Cooke's districts are more enriched in Pb-Zn-Cu-Ag and were mainly mined for these metals. The proximity of these deposits, the spatial association with Eocene magmatism and large-scale faults, as well as the similarities in mineral paragenesis and fluorite morphology suggest that these subdistricts may have originated from the same hydrothermal system. Nonetheless, they have previously been attributed to different geologic events ranging from the intrusion of the Cooke's Peak granodiorite at 38.8 Ma ± 1.1 to the Rio Grande Rift extension at 25-10 Ma. The Central and Hadley districts have been attributed to the development of a hydrothermal system related to intrusion of the Cooke's Peak granodiorite, while the Fluorite Ridge district and Northern Cooke's Peak occurrences have recently been attributed to Rio Grande Rift-type deposits involving fluids sourced from basinal brines with a minor magmatic component. This study will use a combination of field work, fluid inclusion petrography and microthermometry, and fluorite trace element geochemistry to characterize the footprint of hydrothermal processes in each district.

Field observations of sulfide and fluorite veins crosscutting relationships and their association to faulting and dikes will be used to determine the timing of ore formation relative to magmatism and normal faulting in the Cooke's Range. Representative samples of Pb-Zn-Ag-F-mineralized zones from each subdistrict provide the basis for studying the mineral paragenesis of the different vein types. The focus of this study is on the different fluorite generations in these veins and their REE chemistry as well as their fluid inclusion assemblages. LA-ICP-MS will be used to analyze REE elements and other trace elements in fluorite. In the Central Cook's Peak district, three distinct fluorite generations can be distinguished (Fig. 1a), of which the first generation coincided with sulfide mineralization and two-phase fluid inclusions (10-15 vol % vapor to 90 vol % liquid; Fig. 1b); the latter two fluorite generations show fewer inclusions with much smaller vapor proportions of ~3-5 vol %. (Fig. 1c). These observations indicate cooling of hydrothermal fluids resulting in variations of mineral paragenesis along the fluid pathway and likely trace element compositions within fluorite, providing new insights into ore-forming processes. The combination of microscopic analysis with multidistrict-scale observations is a novel approach for this range, as these analytical methods were unavailable when research in the region was first conducted.

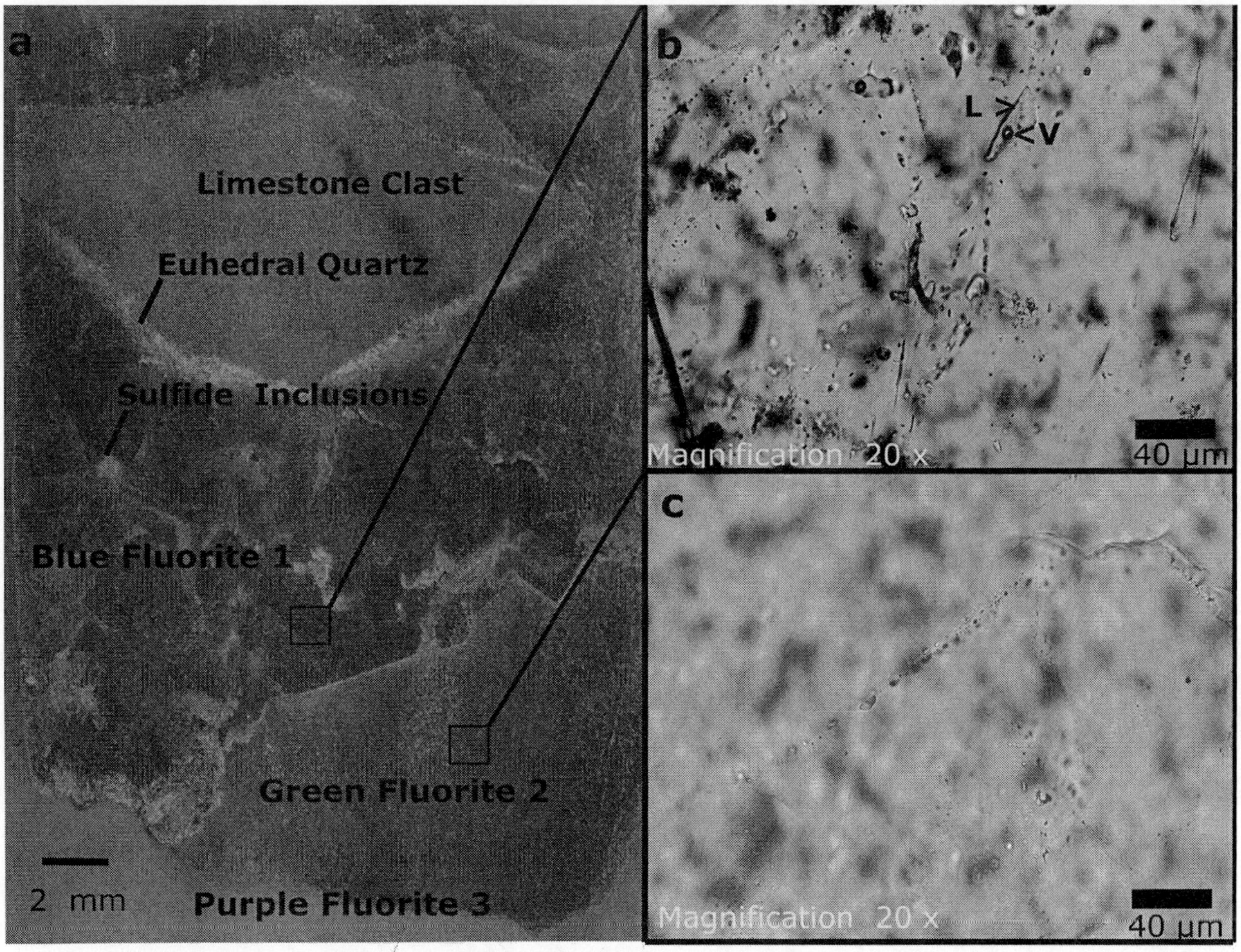

Figure 1: a) Slabbed section showing three fluorite generations b) pseudosecondary 2-phase fluid inclusions (FI) in fluorite 1 and c) secondary 2-phase FI in fluorite 2.

ST.186

Ore-Metasomatic Zonation of the Freedom Area, One of the Ore Centers of the Malmyzh Au-Cu Porphyry Deposit

Vasilii V. Svistunov
Lomonosov Moscow State University, Moscow, Russia

The classic model of ore-metasomatic zonation of porphyry deposits is described by world geologists in many studies. As a rule, the deposit includes the central part of relatively high temperature metasomatites (Na-Ca, K, chlorite-sericite, and sericite zones) with Cu-Au-Mo mineralization. A relatively low temperature zone of propylitic metasomatites with a growing content of Zn, Pb, Ag ± Mn is usually located along the periphery of such a core. The picture of ore-metasomatic zonation in rocks observed on the surface allows us to indirectly assess the erosion level.

The Malmyzh porphyry Au-Cu deposit is located in the Russian Far East in the Khabarovsk Territory, in the middle part of the Amur River. It was opened by international geologists team in 2005-2010. Currently, exploration continues. Mining will start in few years. Resources of Malmyzh deposit are about 5 Mt copper and 300 tons gold.

Regionally, the Malmyzh is based on the western flank of the Sikhote-Alin orogenic belt. The area of the deposit is sea-terrigenic rocks (sandstones, siltstones) that intruded stocks and dikes of intermediate rock (diorite, granodiorite), with some of these stocks linking centers of Au-Cu porphyry mineralization.

"Freedom" prospecting area is one of these centers. Mineralization of this area mainly locates in large body of magmatic-hydrothermal breccias, which framed quartz porphyritic diorite. Orebodies represent stockworks, developed in intensively altered rock. Veins and veinlets have chalcopyrite ± pyrite-quartz composition, with small quantities of molybdenite, galena, sphalerite, bornite, and chalcocite.

Investigation of ore zonation is based on results of exploration and data of chemical analyses of drill core samples.

It has been established that the most high-temperature central part of orebodies in the Freedom area, consisting of quartz-sericite and less common potash metasomatites, characterized the Mo-Au-Cu mineralization. Peripheries of bodies consist of phyllic metasomatites and lesser propylite with increasing content of silver, arsenic, manganese, lead, and zinc.

The analysis revealed two groups of elements with increase positive relationship: first, copper, gold, and molybdenum and second, lead, zinc, and manganese. A higher correlation was observed between copper and gold (correlation coefficient -0.77).

The established relationships between elements are also displayed in their spatial distribution. Investigation of spreading of ore elements in a lateral direction shows geochemical changes from the centers orebodies to their flank: (Cu-Mo)–(Cu-Au)–(Zn + Mn)-Pb. Also, studies of the distribution of elements in a vertical direction shows that molybdenum, relatively of gold-copper mineralization, takes the high-ore level position. Elements of second group correlation (Pb, Zn Mn) consitute the low-ore level. The gold concentrates on the high-ore level, and the ratio of Cu / Au increases with depth. Also, chalcopyrite/pyrite ratio increases from high- to low-ore level.

The character of lateral and vertical zonation allow suggests a high-ore level of erosion in the Freedom area. There is the prospect of further orebodies with high-Cu (± Au) ore on the depth horizons.

The studied features of ore-metasomatic zonation of the Freedom prospecting area correspond with classic models for Au-Cu porphyry deposits in numerous published stuides.

ST.187

Breaking Up the Link Between IOCG Genesis and Granitic Magmatism at 2.6-2.5 Ga in the Carajás Province, NW Brazil

Poliana I. Toledo[1], Carolina P. Moreto[1], Gustavo H. Melo[2], Lena V. Monteiro[3], Fernando M. Matos[4], Juliana A. Carvalho[4], Carlos A. Medeiros Filho[5]
1. Campinas University (UNICAMP), Campinas, SP, Brazil, 2. Ouro Preto University (UFOP), Ouro Preto, MG, Brazil, 3. São Paulo University (USP), São Paulo, SP, Brazil, 4. VALE, Parauapebas, PA, Brazil, 5. Canal Geoquimico, Natal, RN, Brazil

A genetic link between magmatism and IOCG genesis is supported in some ore districts by the temporal association between granites and mineralization and the presence of magmatic fluids, identified by fluid source tracers (e.g., Olympic Dam, Salobo, and Candelaria). At the Carajás Mineral Province, NW Brazil, a set of ca. 2.6-2.5 Ga IOCG deposits (Salobo, Igarapé Bahia, Grota Funda, GT46) were previously interpreted as genetically linked with contemporary granitic intrusions, restrictedly described in the northern part of the province. Despite the timing correlation and evidence of magmatic fluid sources (i.e., stable isotopes and fluid inclusions), the association between IOCG and granite genesis is yet questionable. This is mainly because the ca. 2.6-2.5 Ga granite discordant ages are constrained by metamict, U-rich, and texturally complex zircon grains that show high Pb loss and do not display preserved igneous features. In addition, similar ages on zircons from strongly deformed and hydrothermally altered rocks have been interpreted as resetting ages due to shear activity along the Cinzento Shear Zone, the host structure of IOCG deposits, and targets at the northern part of the Carajás Province. Therefore, we studied the morphologies, textures, and trace element compositions of a set of ca. 2.6-2.6 Ga zircons from variably deformed and metasomatized granitoids, placed along the Cinzento Shear Zone. We included granites from the Salobo (Old Salobo Granite) and GT46 (Igarapé Cinzento granitoids) IOCG deposits as well as regional intrusions, previously interpreted as part of the Igarapé Gelado Intrusive suite. Zircon textures, morphologies, and trace element compositions show complex patterns of multistage fluid-mineral reactions. Relict magmatic zircon domains, with preserved or faint concentric zoning and depleted in REE+Y, HFSE, and nonformula elements (e.g., Ca, Fe), are overprinted and overgrown by textureless, spongy, or convolute zoning and fractured and inclusion-rich domains. These domains are enriched in REE+Y, U, Ca, Ta, and Nb and commonly Zr and Hf depleted. They also show negative Ce/Ce* and Eu/Eu* anomalies, reflecting partition to possible cogenetic hydrothermal minerals (e.g., monazite, allanite, and albite). Diffusion-reaction and dissolution-reprecipitation processes can account for multiple overprinted fluid-zircon reactions, facilitated in more reactive metamict porous grains, and cause, respectively, isotopic disturbances and resetting of the U-Th-Pb isotope system. IOCG hydrothermal fluids, which are alkaline (Na-K-rich) and F-Cl rich, can partially dissolve zircon and mobilize its incompatible elements, inducing zircon re-equilibration and the precipitation of REE+Y-, U-, Ca-, Ta-, and Nb-rich new domains. As these fluids circulated regionally, they have produced a broad hydrothermal footprint in zircon grains from granitic rocks along the Cinzento Shear Zone. Since the geochronological record of 2.6-2.5 Ga is intrinsically related to modified zircon domains, we interpret these registers as hydrothermal event ages in the Northern Carajás Sector, rather than granitogenesis. The magmatic ages of these intrusions must be older and possibly related to the widespread 2.74-2.74 Ga magmatism characterized in the province. Accordingly, we rule out the genetic link between ca. 2.6-2.5 Ga granitic magmatism and IOCG genesis at Carajás, since the temporal IOCG-granitogenesis association is unsupported.

ST.189
Timescales and Processes of Au- and Cu-Fertile Magmatic-Hydrothermal Reservoir Accumulation in Continental Arcs

Daniel Wiemer[1], Steffen G. Hagemann[1], Anthony I. Kemp[1], Nicolas Thébaud[1], Trevor Ireland[2], Carlos Villanes[3]
1. University of Western Australia, Crawley, WA, Australia, 2. Australian National University, Canberra, ACT, Australia, 3. Compania Minera Poderosa S.A., Lima, Peru

Continental arc systems are loci of world-class Cu-Au porphyry and intrusion-related orogenic Au vein deposits. The subduction environment facilitates high magma and fluid fluxes, acting as a heat engine and promoting metal transport. How and when, during the lifetime of an accretionary orogen, Cu and Au fertile magmas are generated and ore-producing fluids are expelled remains elusive.

We present combined structural, geochemical, and zircon trace element, U/Pb and Lu/Hf isotope data from the Carboniferous Eastern Andean Cordillera, northern Perú. Our geochemical data demonstrate that primitive, mantle-derived calc-alkaline diorite magmas are hydrous, oxidized, and enriched in metals (Cu, Au) but lack evidence for crustal assimilation. Field relationships indicate repeated intrusion of these fertile magmas into midcrustal levels. Hornblende-dominated crystallization and fractionation led to successive batch segregation of granodiorite melt upon diorite recharge. Petrographic observations attest to hornblende textural breakdown during magmatic recharge, providing a mechanism for ore-fluid expulsion in the plutonic realm.

Zircon U-Pb dates cover the entire ~370-310 Ma evolution of the Eastern Andean Cordillera (Fig. 1). Age peaks indicate magmatic pulses of 2-5 m.y. duration, followed by 3-4 m.y. gaps that correlate with truncation of zircon internal growth structures. Concomitant to magma recharge, subsets of distinct zircon domains display trace element concentrations indicative of interaction with hydrothermal fluid.

Based on zircon Hf isotope evolution, a slightly unradiogenic source component ($\varepsilon Hf_{(i)}$ = -2) at 370 Ma is inferred (Fig. 1). We interpret this signature as representing a mantle source that was already fertile at the onset of subduction-related melt generation. Source fertilization is ascribed to influx of incompatible elements and metals derived from subducted sediments. Most juvenile $\varepsilon Hf_{(i)}$ zircon values indicate that the source fertility diminished over time, with $\varepsilon Hf_{(i)}$increase to +4 until 342 Ma (Fig. 1). This may be due to (i) consumption of subducting sediments or (ii) progressive contribution of uncontaminated asthenospheric mantle. Both scenarios imply slab retreat and associated compressive stress relaxation, supported by structural data. During ~338-332 Ma emplacement of the upper crustal batholith, no juvenile magma input is detected, suggesting magma discharge from the midcrustal reservoir, corroborating subduction cessation and lithospheric relaxation, promoting magma ascent. The world-class Pataz-Parcoy gold vein system formed directly subsequent to batholith emplacement (Fig. 1).

We argue that compressive stresses in continental arcs inhibit magma discharge, allowing for incremental midcrustal fertile reservoir accumulation. Tapping of such reservoirs is triggered once the buoyancy threshold, as a function of prevailing stresses, is overcome. Episodic magmatic activity during subduction may be explained through either (i) episodic partial melt generation, (ii) intracrustal feedback processes driven by magma volume or melt segregation processes, or (iii) external slab-related tectonic events.

Fig. 1. Time chart of the Eastern Cordillera, northern Perú, highlighting episodic magmatic activity associated with incremental midcrustal, fertile reservoir accumulation, and corresponding mantle source depletion (εHf). Progressive slab retreat and lithospheric relaxation led to discharge manifested in volcanism and batholith emplacement and, in response, the formation of the world-class Pataz-Parcoy orogenic-mesothermal gold vein system.

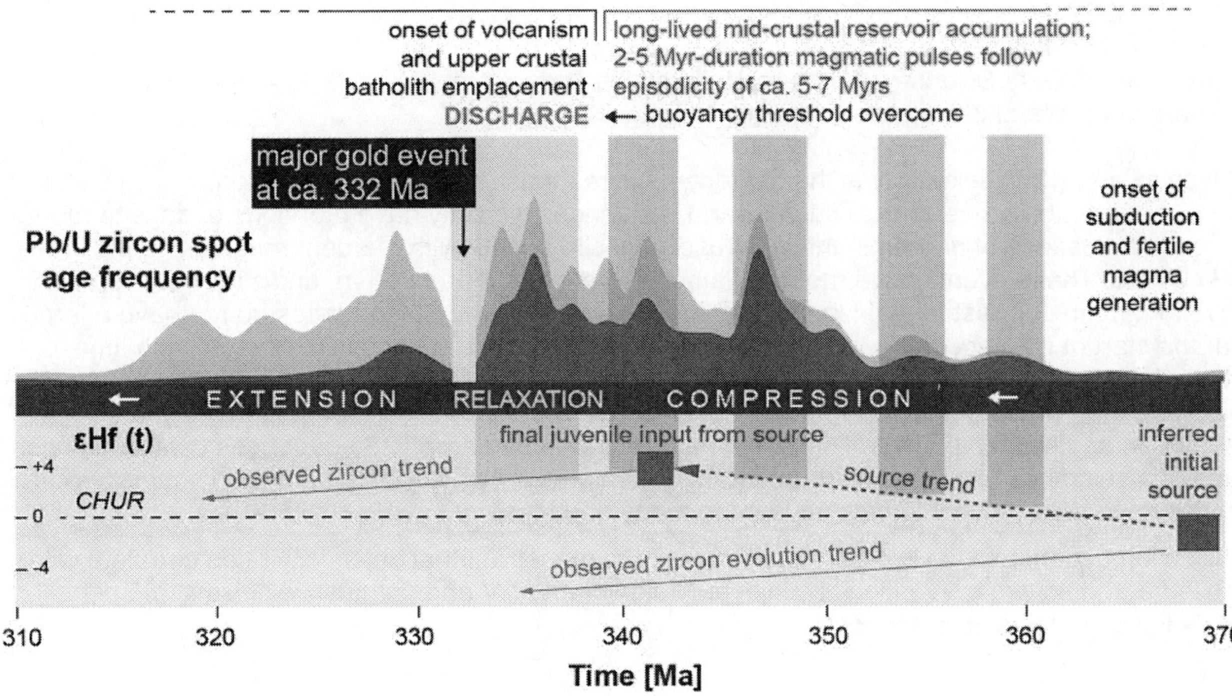

S1

Dynamics of Permeability, Flow, and Ore Deposition in Overpressured, Fault-Controlled Hydrothermal Systems: Constraints from Contemporary, High Fluid Flux Faults

Stephen Cox[1], Yohei Yukutake[2]
1. Research School of Earth Sciences, Australia National University, Canberra, ACT, Australia, 2. Hot Springs Research Institute of Kanagawa Prefecture, Odawara, Japan

Injection-driven swarm (IDS) seismicity is the characteristic response to injection of overpressured fluid into faults within low-permeability host rocks. Such swarm sequences typically have durations of days to months and involve up to thousands of predominantly microseismic slip events with moment magnitudes M_w in the range $-2 < M_w < 2.5$. These events have rupture diameters between 1 and 200 m, and slips between 0.05 mm and several mm. In contrast to mainshock-aftershock sequences, IDS sequences do not have a large rupture near the start of the sequence and do not exhibit Omori Law decay of seismicity rate with time. However, they do exhibit Gutenberg-Richter frequency-magnitude relationships with b-values ranging between 0.85 and 3.5. The largest IDS ruptures typically have $M_w < 4.5$, corresponding to a rupture diameter <1,000 m and slip up to several cm. Both natural and engineered IDS sequences exhibit diffusion-like migration of a seismicity front as a fluid-pressure front migrates away from the injection source. Natural IDS sequences have recurrence intervals of years to many decades.

Deep fluid injection experiments in low-permeability rocks demonstrate that seismicity rates correlate with injection pressure and flow rate. Additionally, injected fluid volume ΔV and cumulative seismic moment release ΣM_0 have a relationship of the form

$$\Sigma M0 = \alpha G \Delta V$$

where G is the shear modulus and α is a geometric factor. The cumulative moment release during natural, injection-driven swarms is used to quantify injected fluid volumes, flow rates, and flow histories associated with fluid release from a deep, intraplate fluid reservoir (Nový Kostel, Czech Republic, 2000–2011) and an upper crustal magmatic-hydrothermal system (Hakone caldera, Japan, 2001-2015). Individual IDS sequences are associated with injection of 104–105 m^3 of fluid over periods of days to weeks at rates of 20-400 Ls^{-1}. Flow histories on decadal timescales constrain fluid production rates in these systems and provide new insights about the architecture and dynamics of flow during fault-valve behaviour in the seismogenic mid- to upper crust.

Insights provided by the dynamics of seismicity style and fluid flow in engineered and natural high fluid flux faults have application to understanding flow dynamics in fault-related hydrothermal ore systems, such as orogenic Au deposits, that form by injection-driven failure in overpressured fluid regimes. The typical dimensions of lode-hosting faults indicate cumulative rupture areas between 1 and 3 km^2 and require ΣM_0 of 10^{14}-10^{15} Nm, cumulative slips less than several cm, and injection of 10^4–10^5 m^3 of fluid during each swarm. The net slip in lode-hosting faults indicates slip accumulation commonly via more than 10^3 swarms. Accordingly, net injected fluid volumes as high as 10^8 m^3 are required to produce a 50 t Au deposit. This corresponds to deposition of several 10s of kg of Au per swarm or extraction of 0.6 ppm Au from the advecting, Au-bearing fluid. Swarm recurrence intervals indicate that such deposits can form in 10^4–10^5 years; however, flow is active during only a small fraction of that time.

S2

Creation and Destruction of Permeability in the Porphyry Cu Environment

Richard M. Tosdal[1], John H. Dilles[2]
1. , Folly Beach, SC, USA, 2. Oregon State University, Corvallis, OR, USA

Porphyry Cu deposits are crustal-scale features requiring deep crustal formation of a hydrous and oxidized magma, magma ascent along extant permeability fabrics to create an upper crustal convecting magma chamber, volatile saturation of the magma chamber, and finally the episodic upward escape of an ore-forming hydrothermal fluid and a phenocryst-rich porphyry magma to produce ores. Three fluid regimes involved are the deep magma ± volatile zone at lithostatic pressure, an overlying zone of transiently ascending magmatic-hydrothermal fluids that breaches ductile rock at temperatures ca. 700° to 400°C, and an upper brittle zone at temperatures <400°C characterized by hydrostatically pressured nonmagmatic and magmatic fluids. Critical steps include the formation of the magma chamber, magmatic vapor exsolution and collection of a hydrothermal fluid in cupola(s), and episodic hydrofracturing of the chamber roof in order to create the permeability that allows hydrothermal fluid to rise along with a phenocryst-bearing magma. The interplay between stress produced by far-field tectonics and stress produced by buoyant magma and magmatic hydrothermal fluid creates the fracture permeability that extends from the cupola through an overlying ductile zone where temperatures exceed ~400°C into an shallower brittle zone where temperatures are less than ~400°C. As a consequence, during each fluid escape and magma intrusion event, the rising hydrothermal fluid ascends, depressurizes, cools, reacts with wall rocks, and precipitates quartz plus sulfide minerals, which progressively seal the permeability fabric. A consistent vein geometry in porphyry Cu deposits is formed by steeply dipping, mutually crosscutting veins. The principal orientation consists of closely spaced sheeted veins whose geometry reflects the far-field stress and likely formed in advance of a rising porphyry magma. Subsidiary veins may be orthogonal to the main vein orientation as radial or concentric veins that reflect magma expansion and extensional strain in the wall rocks as they are stretched by ascent of the buoyant magma. Episodic magmatic-hydrothermal fluid-driven hydrofracturing creates permeability that is destroyed as well as enhanced by vein and wall-rock mineral precipitation or dissolution, and by wall-rock hydrothermal alteration, depending upon fluid and host-rock compositions. The pulsing character of porphyry Cu systems, in part produced by permeability creation and destruction, creates polyphase overprinted intrusive complexes, vein networks, and alteration mineralogy that reflects temporal temperature fluctuations beginning at magma temperatures but continuing to low temperatures. Temperature oscillations locally allow external nonmagmatic fluids to access principally the marginal areas but also in some cases the center of the porphyry Cu ore zone at ca. <400°C between magmatic fluid and porphyry dike emplacement events. Over time, the upper part of the source magma chamber crystallizes downward, accompanied by diminishing magmatic fluid input upward, leading to cooling and isothermal collapse of the porphyry system. Cooling permits the access of external circulating groundwater into the waning magmatic-hydrothermal plume. Magmatic-hydrothermal fluids dominate at temperatures >400°C at pressures transient between lithostatic and superhydrostatic. The external, nonmagmatic saline formation waters or meteoric waters dominate the surrounding and overlying brittle crust at temperatures <400°C at hydrostatic pressures.

S3
Structural Modification of VMS Deposits

Bruno Lafrance, Harold Gibson, M. S. Stewart
Harquail School of Earth Science, Mineral Exploration Research Centre, Laurentian University, Sudbury, ON, Canada

Ancient VMS deposits formed in extensional geodynamic environments and were deformed during later convergent accretionary events. How tectonic structures form in VMS deposits is influenced by primary features. During deformation, strain is typically taken by the sulfide bodies because common sulfides in VMS deposits are weaker than volcanic host rocks. The deposits act as shear zones, undergo hinge thickening and limb attenuation during folding, and are deformed into elongate bodies parallel to regional fold hinges and stretching lineations. A tectonic foliation may form as a sulfide banding in the interior of VMS lenses due to the shearing and flattening of primary textural and compositional heterogeneities, and as a banded silicate-sulfide tectonic foliation along the margins of the VMS lenses due to transposition and shearing of primary silicate (exhalites)-sulfide layers (Fig.1). Cusps, piercement cusps and veins, and durchbewegung structures are other structures that form as a result of the strong competency contrast between the massive sulfide deposits and host volcanic rocks.

Some features in VMS deposits may have both primary and tectonic components. One example is the vertical stacking of VMS lenses, which may be primary and due to rapid burial of lenses by volcanic and sedimentary deposits during long-lived upflow of hydrothermal fluids or tectonic and due to thrusting and isoclinal folding of VMS lenses. A second example is the elongation of VMS lenses that may have a primary component due to the deposition and coalescence of sulfide lenses along linear synvolcanic faults or a tectonic component due to remobilization of sulfides parallel to linear structural features in the host volcanic rocks. Careful mapping of volcanic lithofacies and primary and tectonic structures is needed to assess the nature of structures in VMS deposits. This is hindered by the low temperatures of recrystallization of sulfide minerals, as recrystallization may conceal or destroy deformation fabrics and structures in VMS lenses. Discontinuous and abrupt lithofacies changes in the host volcanic rocks may hamper mapping and the definition of large-scale structures because of the added complexities in correlating rock units. The weak development of tectonic fabrics, which are commonly used as structural datum for correlating generations of structures, and the strong strain partitioning in weaker fault zones, which may undergo multiple reactivation events, also complicate the structural analysis of volcanic terranes. To mitigate these issues, mapping of volcanic rock should be done with structural mapping to delineate repeated stratigraphic panels across early thrust faults and reactivated later faults and to identify regional folds characterized by abrupt changes in strata orientation from limbs to hinge. If well-layered sedimentary rocks are present, the sequence of deformation events that affected the volcanic rocks and their VMS deposits can be determined by first mapping structures in the sedimentary rocks, then correlating these structures with those in volcanic rocks.

Fig. 1. Tectonic foliations in deformed VMS lenses: (a) Ribbon ore with layers of chalcopyrite from the Devonian, Neves Corvo deposit, Portugal. (b) Tectonic foliation defined by deformed and transposed pyritic stringer zone from the Mobrun deposit, Noranda camp, Quebec.

S4

Synbasin Transform Faults and Their Influence on Late-Stage Orogenic Gold Mineralization: Examples from the Mosquito Creek Basin, Pilbara Craton, Western Australia

Anthony A. Morey[1], James Farrell[2]
1. Galt Mining Solutions, West Perth, WA, Australia, 2. Millennium Minerals Ltd, Perth, WA, Australia

The Mosquito Creek Basin (MCB) is located within the southeastern part of the Archean Pilbara Craton of Western Australia and is dominated by siliciclastic sequences that were deposited ca. 2926 Ma and subsequently buried, metamorphosed, and intensely shortened at ca. 2905 Ma. Overprinting relationships indicate that orogenic gold deposits of the MCB are hosted by late-stage structures that are spatially associated with two steeply dipping, broadly east-west shear systems: the Middle Creek and Blue Spec shear zones (Fig. 1).

Although a late-stage mineralization paradigm for the MCB is geologically sound, this study has found that many so-called late-stage faults could be synbasin transform faults that have remained active through the evolution of the basin. The synbasin faults are primarily identified by their variable stratigraphic offsets, ranging from over 2 km within lower stratigraphic sequences to less than 100 m in higher stratigraphic sequences. Although rotational offests along transform faults could still be active, the variable offsets can most readily be explained if they remained active throughout the whole history of the basin. In the northern part of the MCB, abundant slump structures and sedimentary folds are also observed, also showing that synsedimentary deformation was active during basin development. During basin inversion, the synbasin transform faults have remained at a high angle to the dominant MCB trend, possibly enhancing fluid-focusing mechanisms and structural contrasts assisting epigenetic mineralization. Given these outcomes a sequential D_1 to D_n deformation regime, commonly used to help understand mineralization events, cannot fully explain the genesis of gold mineralization.

By using examples from the MCB and other locations within Western Australia, this paper aims to show that although challenging, identifying preexisting structures can lead to new and effective ways to target for gold mineralization from the province to deposit scale.

Fig. 1. Geology map of part of the Mosquito Creek Basin, Pilbara Craton, showing the main gold deposit locations and the long-lived, synbasin transform faults interpreted from this study.

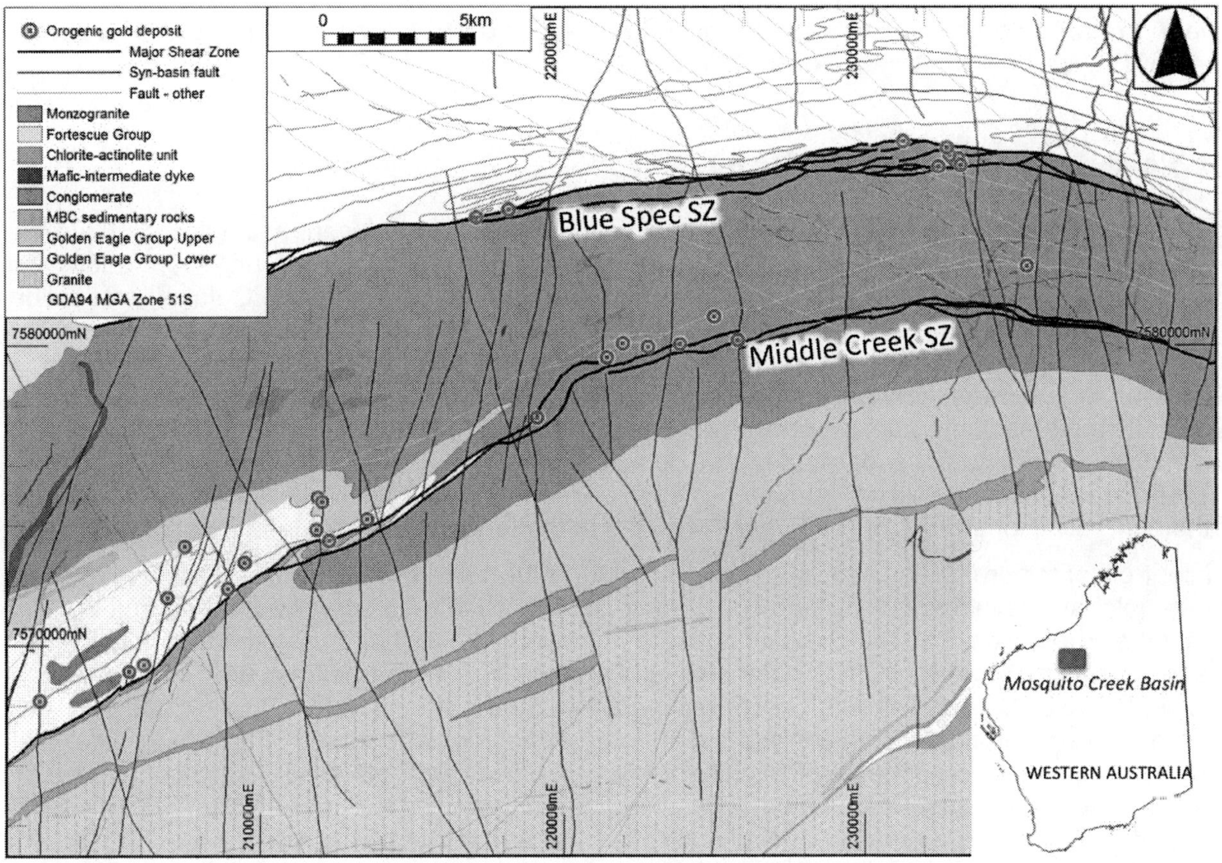

S5

Links Between Local Fluid Sinks (orebodies) and Regional Fluid Flow Paths: Formation of Gold-Quartz Reefs of the Barberton Mines, South Africa

Caitlin Jones, Lucas Pintos Cerda, Alexander Kisters
University of Stellenbosch, Stellenbosch, South Africa

High-grade gold-quartz-carbonate-sulphide orebodies of the Sheba-Fairview Mining Complex (SFC) in the Barberton Greenstone Belt highlight the controls and relationships between localized, high-permeability fluid sinks (economic-grade gold mineralization) and regional, pervasive fluid flow patterns. Gold mineralization (>350 t) in the SFC is hosted by a thick (>5 km), lithologically heterogeneous, well-bedded, low-grade metasedimentary sequence and minor ultramafic rocks (serpentinites and talc-carbonate schists) of the Onverwacht Group, greywacke-shale packages of the Fig Tree Group, and quartzite-conglomerate of the Moodies Group. The metasedimentary rocks are preserved in two complexly refolded, distinctly arcuate regional-scale synclinal structures, separated by the central Sheba Fault. Vein-type quartz-sulphide orebodies (reefs) are developed along the brittle (-ductile) faults of variable orientation and kinematics with the bulk of the mineralization (>85% of orebodies) confined to the hinge region of the late-stage arcuation.

Three main controls of mineralization can be distinguished. (1) Quartz-carbonate stockworks, with commonly free gold, are developed along and in the immediate footwall of the central Sheba Fault in competent Moodies quartzites. The fault crosscuts strata at low angles, and pervasive carbonatization of the siliceous Moodies wall rocks testifies to extensive fluid channelling in the fractured (veined) damage zone of the Sheba Fault. Pods and shoots of economic-grade mineralization (Golden Quarry and Royal Sheba) occur in gentle fault undulations, corresponding to dilational jogs and zones of fluid focussing. (2) Quartz-carbonate-sulphide (± graphite), typically shoot-like orebodies occur along lithological contacts between Fig Tree Group metaturbidites and Onverwacht Group ultramafics in the hanging wall of the Sheba Fault. These orebodies form as a result of strain localization and subsequent fluid focussing during late-stage refolding and flexural slip along lithological contacts with pronounced rheological contrasts. (3) Volumetrically small, but high-grade, sharply bedding-discordant, low-angle thrusts or steep reverse faults in both Moodies and Fig Tree Group rocks are far removed from the central Sheba Fault. The typically highly silicified brittle structures accommodate progressive shortening of the controlling fold structures past the fold lock-up stage.

The localization of orebodies in the hinge zone of the refolded synclines illustrates regional, more pervasive fluid flow paths following regional hydraulic gradients, from the limbs into the hinge zone during refolding of the large-scale folds. Bedding-parallel quartz veinlets, sulphide mineralization, and alteration are ubiquitous and testify to this pervasive fluid flow being mainly accommodated along bedding planes, promoted by bedding slip during flexural-slip folding. This renders crosscutting structures, such as the Sheba Fault, or high-angle thrusts and reverse faults, ideally orientated to tap into this regional-scale fluid flow pattern. Lithological contacts between ultramafic and metasedimentary rocks present slip horizons where strain has been localized as a result of pronounced rheological contrasts. The results also highlight that the controls of gold mineralization in the SFC shares many similarities with Phanerozoic turbidite-hosted gold deposits rather than the commonly greenstone-hosted gold of many Archaean greenstone belts.

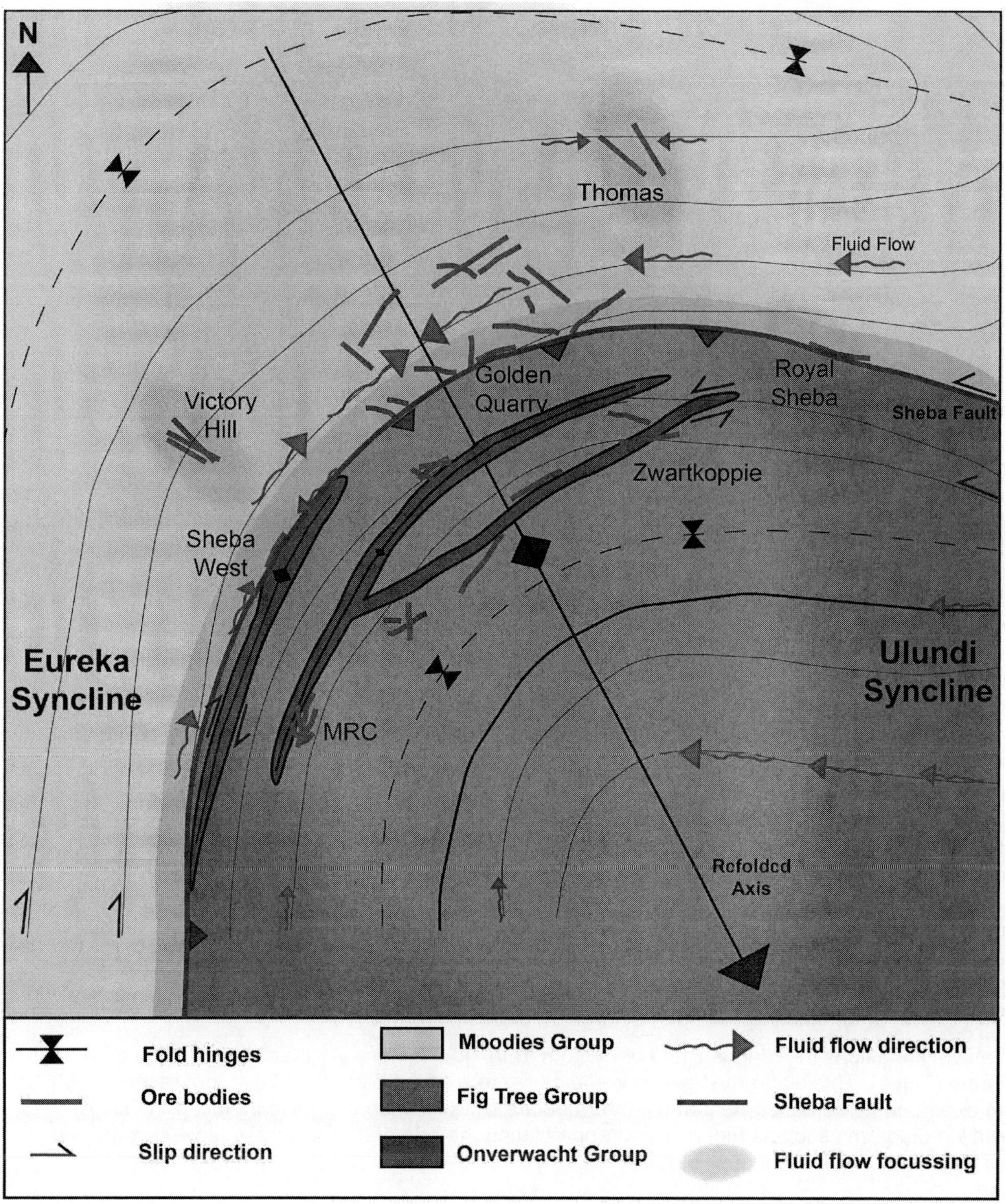
N
Thomas
Fluid Flow
Victory Hill
Golden Quarry
Royal Sheba
Sheba Fault
Zwartkoppie
Sheba West
Eureka Syncline
Ulundi Syncline
MRC
Refolded Axis
Fold hinges
Ore bodies
Slip direction
Moodies Group
Fig Tree Group
Onverwacht Group
Fluid flow direction
Sheba Fault
Fluid flow focussing

S6
Structural Geology and Hydrothermal Ore Deposits

Thomas Blenkinsop
School of Earth and Ocean Sciences, Cardiff University, Cardiff, United Kingdom

Strong structural controls on mineralization are characteristic of hydrothermal ore deposits. These arise because the permeabilities of crustal rocks are too low to allow deposit formation on realistic timescales unless rocks are deformed. A thorough structural investigation involves geometrical, kinematic (displacement and strain), dynamic (stress), and rheological analyses: each step can make important contributions to understanding hydrothermal mineralization. This review advocates a workflow based on these steps and introduces some recent relevant advances in structural geology. The focus is on gold, but most aspects apply to hydrothermal deposits in general.

Deformation zones and networks of deformation zones are the fundamental structures that control mineralization. Geometry relates orebody shape to controlling structures. Networks of deformation zones can be analyzed using topology to evaluate their connectivity and mineralizing potential. Kinematic analysis determines the locations of permeability creation and mineralization. Recent knowledge about shear zone kinematics has advanced significantly by taking into account variable amounts of pure and simple shear. These variations change likely directions of mineralization in predictable ways if the kinematics of the shear zones are known. A fresh appreciation that multiple orientations of deformation zones may form simultaneously and symmetrically about the principal strain axes has commensurate implications for orebody controls. Dynamic analysis is necessary for a mechanical understanding of deformation, fluid flow, and mineralization and for numerical modeling. Several important new tools for dynamic analysis are ready for application in exploration and mining. Rheology is the relationship between deformation (kinematics) and stress (dynamics): rheological contrasts are critical for the localization of many deposits.

Numerous hydrothermal gold deposits, especially the largest, have evidence for multiple mineralizing events that may be separated by tens to hundreds of millions of years. Reactivation of structures is common in these cases. A range of orientations of preexisting structures may be reactivated, given that they are weaker than intact rock.

Hydrothermal gold deposits form in contractional, strike-slip, and extensional tectonic settings. There may be great variations in the spatial scales over which a given stress regime applies, and stresses may change on rapid timescales, so that it is inadvisable to infer local tectonics from deposit-scale patterns and vice versa. It is essential to place mineralizing events within a complete geologic history in order to identify the effects of pre-, syn-, and postmineralizing structures on deposit geometry.

The mechanical conditions for various types of hydrothermal gold mineralization can be depicted on a failure mode diagram (Fig. 1). This diagram shows pore fluid pressure and differential stress, but recent rock mechanics experiments demonstrate that the intermediate stress is also a significant factor in failure. Adding detail to diagrams such as Figure 1 and appreciating the role of the full stress tensor are prime challenges for the future of ore deposit studies.

Fig. 1. Failure mode diagram showing fields of pore fluid pressure and differential stress for various types of hydrothermal gold mineralization.

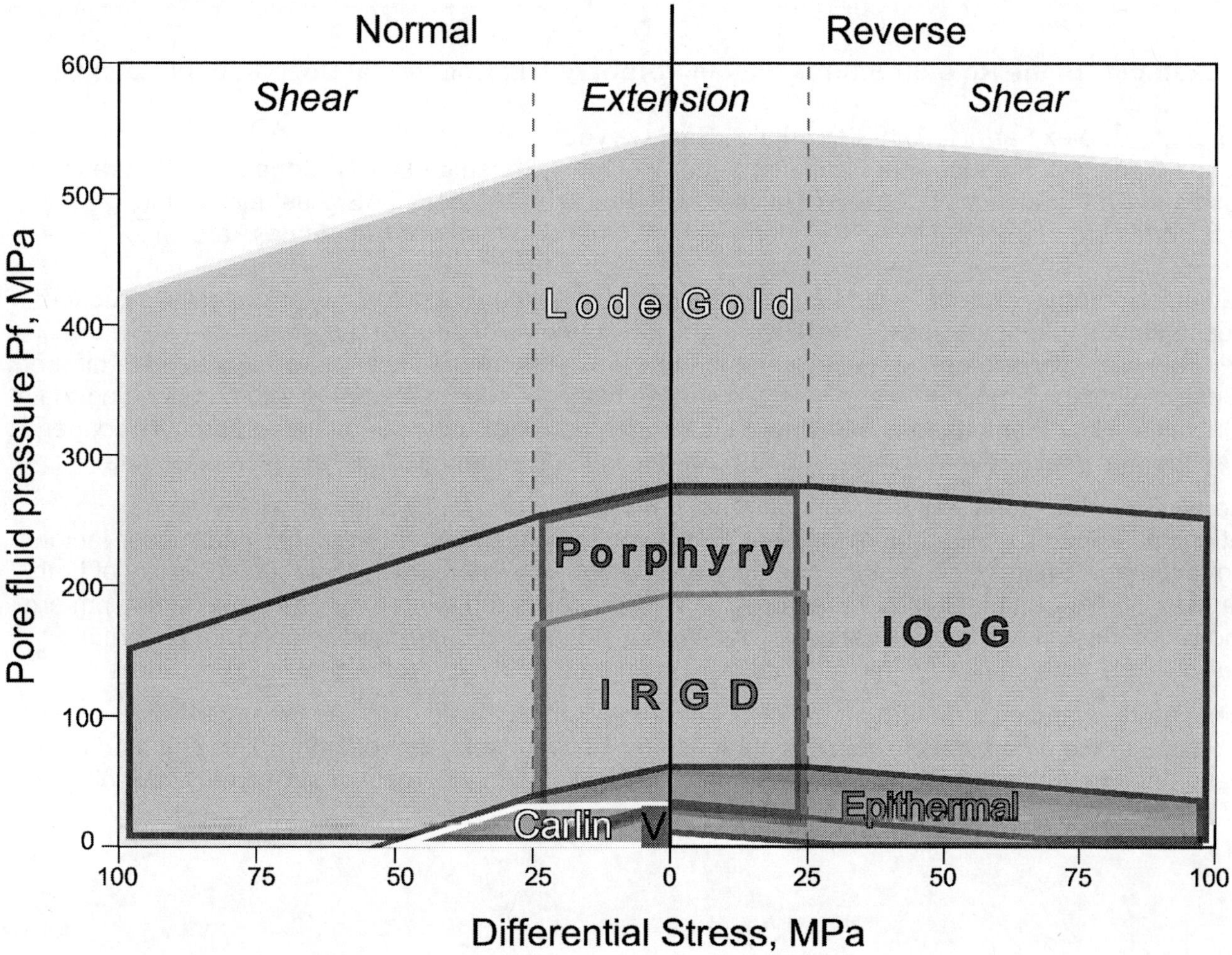

S7
Practical Examples of the Role Structures Play in Porphyry and Epithermal Deposit Exploration

Stephanie Sykora[1], Alex Farrar[2], Rob Sievwright[3], Saúl Galvez[4]
1. Exploration Geologist, Independent Consultant and OreQuest Consultants, BC, Canada, 2. Centre for Ore Deposit and Earth Sciences (CODES), University of Tasmania, Hobart, TAS, Australia, 3. First Quantum Minerals Ltd., London, United Kingdom, 4. First Quantum Minerals Ltd., Lima, Peru

In exploration for porphyry copper (± gold ± molybdenum) and epithermal gold ± silver deposits, structure plays an essential role on many scales. Here we highlight examples of structural patterns that show the variability within porphyry and epithermal systems and how this information is captured and used in different stages of exploration for targeting. In particular, examples highlight surface mapping and core logging that focused on vein and fault orientations and kinematics from a series of early to advanced porphyry copper ± gold and epithermal gold ± silver projects in South America, North America, and Papua New Guinea.

On a regional scale, the identification of fundamental translithospheric faults and opportune stress configurations prior to and at the time of magma (and porphyry) emplacement are critical first-order factors to locating favourable areas of interest for porphyry exploration. On the deposit scale, the interplay of both regional and local stresses manifest as structures. In several cases, an overarching regional stress pattern is clear; however, nested within are local areas of different strikes and dips, which highlight a proximal stress control, likely from shallowly emplaced intrusions and magmatic-hydrothermal fluid pressures.

In each case, understanding structure on the regional or ore-deposit scale, such as the underlying controlling faults or the differences in vein orientations and dips, is useful for vectoring. This can lead to either location of different porphyry dike phases and/or determining the location of structurally controlled ore.

S8

Coincident Fold-Fault-Vein Geometric Patterns at Galore Creek and Sulphurets Cu-Au Districts: Reflections of Cryptic Basin Architecture and Transcrustal Magma Conduits

Gayle Febbo[1], Lori Kennedy[2], JoAnne Nelson[3], Bram I. van Straaten[3]
1. Kingfisher Metals Corp., Terrace, BC, Canada, 2. University of British Columbia, Vancouver, BC, Canada, 3. British Columbia Geological Survey, Victoria, BC, Canada

The Galore Creek and Sulphurets Cu-Au porphyry-epithermal districts are host to combined measured and indicated resources of 9.2 Moz Au with 11.4 Blbs Cu and 95.0 Moz Au with 13.1 Blbs Cu, respectively. The districts are part of a protracted Late Triassic-Early Jurassic Cu-Au metallogenic epoch related to island-arc magmatism in Stikinia, British Columbia. Structural regimes in the study areas include 1) pre-212 Ma extension along N- and E-striking faults at Galore Creek (preporphyry); 2) a 212 (?)-208 Ma fold-and-thrust deformation event expressed as dextral transpression at Galore Creek (pre-/synporphyry; NE-SW-oriented σ_1) and as a significant angular unconformity between Stuhini and Hazelton groups at Sulphurets; 3) latest Triassic-early Jurassic dextral transtension and basin formation at Sulphurets (pre-/synporphyry); and 4) the Skeena fold-and-thrust belt (120-110 Ma) at Sulphurets (postporphyry; early NW-SE-oriented σ_1; late E-W-oriented σ_1).

The latest Triassic Galore Creek deposits cut and emplace to the east of a >2-km-wide, N-striking high--strain zone and folded corridor (~212-208 Ma), which includes a crenulated, prehigh-strain zone, N-striking, 40° W-dipping dextral-normal mylonite. Sheeted and stockwork K-feldspar-rich veins (1-5 vol %) in the district were emplaced along structural corridors of various orientations, including subparallel to a S-dipping normal fault segment in the footwall of a district-bounding, E-striking, S-dipping thrust fault. The >4-km-long Central-West Fork magmatic-hydrothermal trend parallels a N-striking, 45° W-dipping fault array that is defined by normal offset of stratigraphy across hydrothermally healed paleostructures and by spatially coincident, N-trending fold culminations. Both brittle-ductile deformation zone trends and hydrothermal-magmatic healed paleostructures parallel cryptic predeformation and preporphyry N- and E-striking basin architecture at Galore Creek.

The Early Jurassic Sulphurets district includes the Kerr, Sulphurets, Mitchell and Iron Cap (KSM), Snowfield, and Brucejack deposits. It is bounded to the north by the E-striking Johnstone fault and to the east by the N-striking Brucejack fault. Early Jurassic units thicken into the district across both faults, indicating that a coeval basin structural framework provided the E- and N-striking magmatic-hydrothermal pathways for metal deposition. Mid-Cretaceous deformation inverted the Early Jurassic basin, resulting in N- to NW-plunging isoclinal folds of porphyry veins at Kerr deposit, and W-plunging open to isoclinal folds of veins at Mitchell and Snowfield deposits. These fold arrays are parallel to sheeted quartz vein corridors (20-90 vol %) within the Kerr, Mitchell, and Brucejack deposits.

The spatio-geometric correlation of late Triassic and mid-Cretaceous fold patterns with hydrothermally healed paleostructures at Galore Creek and Sulphurets, due in part to basin inversion, can potentially be used to constrain the geometry and strike length of ore-focusing structures. The long-lived nature of these recurring patterns hints at the geometry of transcrustal magma conduits during the formation of porphyry clusters in Stikinia.

S9

Regional to Licence-Scale Structural Controls Derived from Structural Mapping, Geophysical Analysis, and 3D Modeling of New Luika Gold Mine, Lupa Terrane, SW Tanzania

Corné Koegelenberg[1, 3], Jonathan N. Gloyn-Jones[1], Ian J. Basson[1, 3], Paul W. Mbuya[2], Alfred M. Kulwa[2], Shineni Ramadhani[2], Dustan Daud[2]
1. Tect Geological Consulting, Somerset West, South Africa, 2. Shanta Gold, Dar es Salam, Tanzania, 3. Stellenbosch University, Stellenbosch, South Africa

The Lupa Goldfields and encompassing Paleoproterozoic Lupa Terrane in southwestern Tanzania is known for its high prospectivity and is host to numerous artisanal gold mines and several relatively small- to medium-scale gold deposits that range broadly between 150 and 4,000 kt at 1.5–5 g/t Au. Shanta Gold's New Luika Mine is currently the largest open and underground mine within the Lupa Goldfields and comprises at least 10 different deposits over an area of 18 km^2 with a combined resource of 1,105 koz at 2.56 g/t Au (2021). However, despite the conspicuous prospectivity of the region, comparative recent discoveries or extensions to known high-grade (>4 g/t) deposits has been elusive. The Lupa Terrane predominantly comprises a suite of Archaean granitiods (ca. 2.74 Ga) intruded by Palaeoprotrozoic granites and gabbrodiorites (ca. 1.96 Ga) that reflect reworking of the southern margin of the Tanzania Craton during protracted deformation of the Ubendian Orogeny (2.1–1.88 Ga). Relatively late-tectonic auriferous quartz veins account for economic mineralization (ca. 1.88) Ga within narrow, mostly ductile, mylonitic shear zones developed under greenschist metamorphic conditions. From a regional- to deposit-scale, mylonite zones are generally continuous and interconnected, which emphasizes an extensive, structurally controlled fluid conduit system. However, along these mylonite zones and away from known deposits, surface exploration records increasingly variable and often discontinuous along-strike auriferous mineralization, which suggests that new deposits or deposit extensions are likely not exposed on surface but hidden at greater depths.

To advance the understanding on the structural controls on mineralization, this contribution presents a regional- to deposit-scale structural analysis that integrates aeromagnetic, field mapping, drill hole, and soil geochemistry data over New Luika Mine and surrounding areas. Deposit-scale, fault-fill, auriferous vein systems and associated mylonite fabrics are structurally and kinematically described and modeled in 3D, with the aim of establishing the structural controls on fluid flow and associated auriferous mineralization.

Results depict a 1st-order or regional network of wide, incipient to intensely strained, oblique, left-lateral (transpressional-) shear zones (or "shear zone corridors"), which act as mineralized fluid conduits across several tens of kilometres. These shear zone corridors host well-developed, deposit-scale 2nd- to 3rd-order mylonite zones that are mineralized, to between ca. 1 and 3 g/t Au. Increased structural dilation and silicification along 2nd- to 3rd-order mylonite zones within $^{1)}$ 1st-order release bends and/or $^{2)}$ that are unfavourably orientated with respect to the regional stress configuration exhibit increased auriferous grades (ca. 2-5 g/t Au). More locally, a combination of favourable kinematics, variable rock rheology, and/or the reactivation of preexisting contacts or structures cause steep- to shallow-plunging mesoscale release bends on 2nd- to 3rd-order mylonites further accounting for highly mineralized (>6 g/t Au) ore-shoot formation. Exploration following this rationale has significantly added to the resource at New Luika Gold Mine, while also providing a framework for regional exploration in the Lupa Goldfields, thereby reemphasising the value of careful structural analysis in such structurally convoluted terranes.

Figure: Structural architecture and kinematics of the main shear zone corridor encompassing New Luika, Luika South, and Bauhinia Creek Deposits.

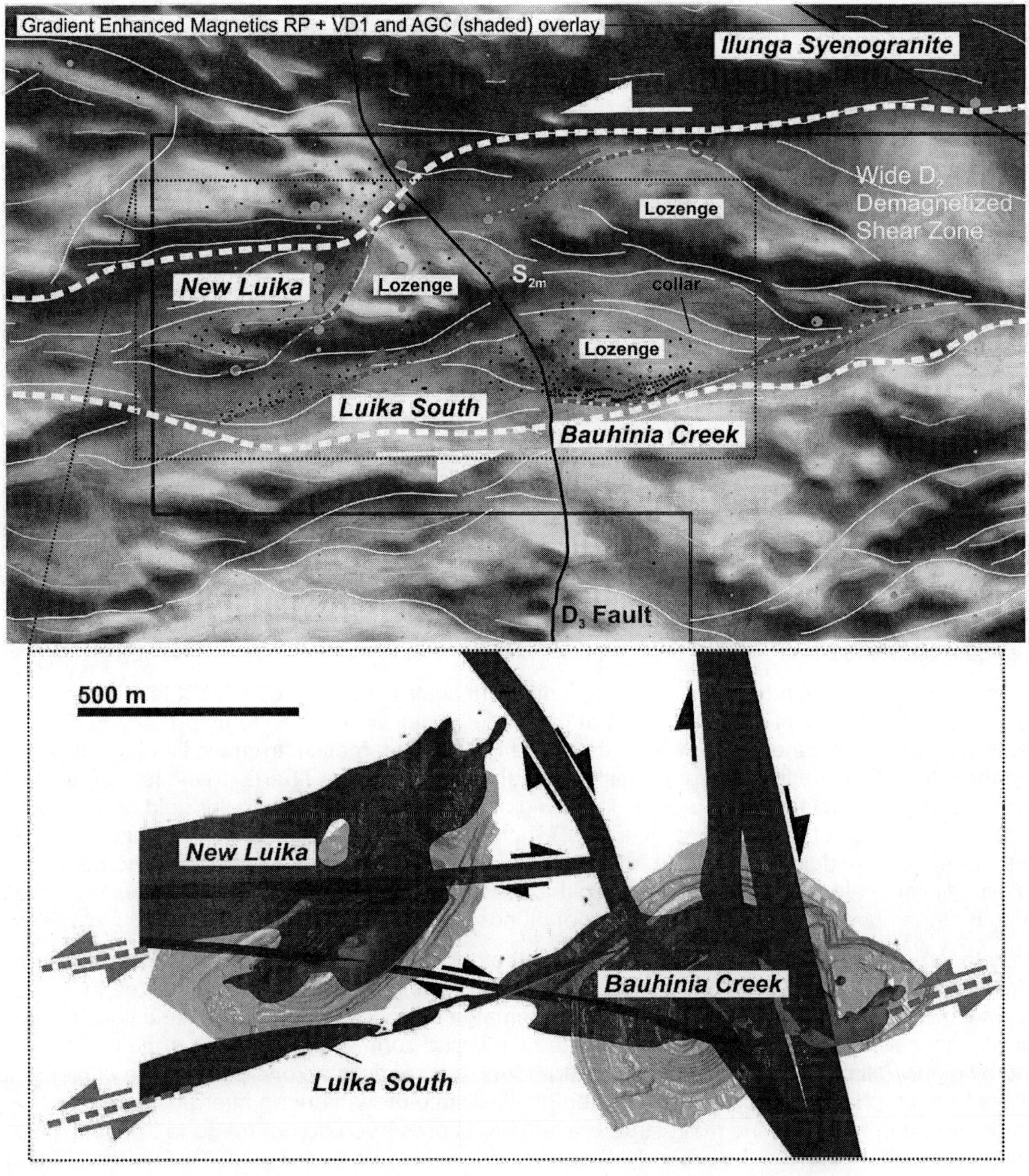

S10
Patterns of Gold Distribution, Pure Gold Mine, Red Lake Greenstone Belt, Ontario

Darcy Baker[1], Phil Smerchanski[2]
1. Equity Exploration Consultants Ltd., Vancouver, BC, Canada, 2. Pure Gold Mining Inc., Vancouver, BC, Canada

Historical data for ~13,500 drill holes for the Pure Gold Mine in Red Lake, Ontario, were well catalogued in a robust database but lacked focused interpretation, making comparison to other Archean orogenic gold deposits and determination of controls on gold deposition challenging. Past work focused on mining-scale interpretation used to follow known lodes down-plunge during nearly 40 years of underground production. Published research was preliminary and based on brief site visits during the early 2000s when underground access was limited and drill hole data not yet compiled.

Upon acquiring the project in 2014, Pure Gold Mining began systematic exploration of the large mineral system with property-wide geophysical, geochemical, and extensive geological surveys along with focused lithogemchemical and petrological studies, building a framework for interrogation of detailed deposit-scale information recorded by previous workers. A coherent structural model was required tie these large data sets together.

Anchored by the results of new drilling (~1,000 holes) by Pure Gold and relogging of all available archival core (~600 holes) the mine geology was reinterpreted in three dimensions on the recreated mine grid (~180 sections spaced at 100 feet) and on 30 level plans to ~1.5 km below surface. The historical lithology database was leveled and recoded to focus on identifying the main rock types and their distribution and relationship to gold orebodies.

We found consistent, repeating patterns meaningful at different scales. At the property scale, gold is distributed within planar zones that are continuous at the scale of kilometres despite being only ~2-50 metres wide. High-grade gold zones occur where these planes transect mafic/ultramafic lithological contacts, and the intersection defines the predominant oreshoot geometry. A quartz-phyric felsic unit conspicuously occupies the gold-bearing planes, including many areas with spectacular gold enrichment. We interpret these planes as structures that controlled emplacement of the felsic porphyries and also acted as the hydrothermal fluid conduits responsible for extensive alteration and later silicification and gold deposition. At a smaller scale (10s of metres), the gold lodes are modified into lensoidal shapes that locally define minor fold hinges and internal steeper ore-shoot plunges.

Careful interpretation of key lithological contracts through the dense network of drilling and extrapolation into lesser drilled areas results in a coherent geometrical pattern with gold demonstrably related to lithological contacts. The planar structures provided the primary control on hydrothermal fluid flow; however, they are typically not conspicuous within drill core. Instead, a broad zone within the mine trend is characterized by highly foliated and altered rock. In our model, this fabric is an overprint on the earlier gold system. Relict alteration phases (carbonate) promoted phyllosilicate growth during later postmineral compressive deformation such that the main structural elements preserved do not relate to the gold deposition event but rather modified the orebody shapes and locally remobilized gold. This understanding was critical to developing a new resource base for the mine currently under construction and expected to be in commercial production by early 2021.

S11
Application of Alteration Facies and Infrared Spectroscopy (IRS) in Structural Geology Analyses

Anna Fonseca
Principal Consultant (Geology), SRK Consulting, Almaty, Kazakhstan

Alteration features can provide important clues to decipher the sense and amount of displacement in a fault during ore emplacement and optimize exploration strategies to target structurally favourable zones. This contribution emphasizes the use of infrared spectroscopy (IRS) and specifically point analyses of outcrops and mine exposures to incorporate invisible alteration information in the analysis of structural controls on epithermal mineralization.

Hydrothermal alteration forms two groups:

1. Visible alteration features currently require expert identification of diagnostic facies in outcrop, drill core, or mine exposures. In the epithermal environment they include features such as paleosurface indicators, paleowater table, and steam-heated silicification facies. The recognition of contrasting visibly recognizable paleodepth alteration indicators on opposite sides of a fault can provide an estimate of fault displacement magnitudes.
2. Invisible alteration features include a wide variety of data obtained through ground or remote sensing analytical methods such as IRS, geochemical analyses, and various geophysical techniques. These alteration data can be more efficiently digested by a data scientist than by a field geoscientist, if sound geologic concepts and models are incorporated into the data sampling program.

Neither of the two groups provides a flawless methodology. Instead, the most robust input into structural geology analyses rests in a combination of visible and invisible alteration features.

Whereas the most widespread use of point IRS analyses consists of the identification of mineral species, the greatest value to structural geologists lies in batch extractions of semiquantitative and qualitative spectral parameters that are independent of mineral identifications. Specifically, high sericite and to a lesser degree high kaolinite crystallinity indices highlight faults that were active during hydrothermal events, albeit these faults may only be mineralized along segments that constitute structural or chemical traps. The presence of ammonium in spectra, regardless of its host mineral, is known to represent barren paleosurface alteration; yet where the ammonium is structurally trapped, it highlights high-permeability zones. In certain cases, the juxtaposition of incompatible spectral features on opposite sides of a fault can provide information on postmineralization displacement sense and magnitude.

Despite yielding semiquantitative data, field-based point IRS data collection has important practical advantages over quantitative methods such as portable XRF. Namely, IRS instruments do not pose health hazards and therefore lack transport and use restrictions. Relative to remote sensed ASTER or Worldview-2/3 data, point IRS has the advantages of providing higher spectral resolution data not impacted by the atmosphere. Additionally, the method has low sensitivity to cross-sample contamination and requires no sample preparation.

One of the main limitations of IRS lies on the arduous labour required to distinguish hydrothermal from metamorphic alteration features, rendering IRS application more arduous in older mineralized belts, where postmineral orogenesis may have imbricated phyllosilicates of different origins.

To illustrate, I contrast successful and unsuccessful structural geology applications of ground-based visible alteration and point data IRS in epithermal deposits that underwent significantly different tectonic evolutions. Cenozoic deposits of the Mexican backbone emplaced after the Laramide/Sevier compression are contrasted to Permian deposits of the Angren District of Uzbekistan.

S12

Structural Modeling in 2D and 3D and with Time (4D) for Mineral Exploration and Mine Development

Armelle Kloppenburg
4DGeo | Applied Structural Geology, The Hague, Netherlands

Structures play a significant role in mineral systems by contributing to the plumbing system, the underlying connected network of fluid pathways. They contribute to both the system charge, i.e., the transport and focussing of mineral-bearing fluids from a (upper mantle) source, up through the crust, as well as contributing to the trap by creating space as well as impermeability with favourable trapping geometry. Less favoured are postmineralization structures that displace part of a mineralized volume. Hence, for optimal exploration and mine development, it is essential to have both a valid interpretation of the present-day structural framework as well as an understanding of how that framework changed shape through time with respect to the timing of mineralization.

Structural interpretation benefits from identifying the appropriate structural style(s). Comparing a limited data set with known, theoretical structural models and kinematic concepts helps to recognise patterns, structural hierarchy, and identifying options (and alternatives) for the dominant structural style(s) in areas without direct observations (yet). Where geologic histories are long and complex, this is not always trivial. In some cases it helps to restore the effects of one deformation phase to see through to a geometry at an earlier moment in time.

Complications at a range of scales may hinder interpretation. At deposit scale there is the mechanical stratigraphy—the relative strength of the various (stratigraphic) units, which dictates the amount and level(s) of detachment. Detachments have a direct effect on structural style, both in extension and shortening. At regional scale, complications are caused by older, inherited, weaknesses in underlying rocks that—when reactivated (particularly within a different regional stress field)—may lead to complex structures in overlying younger rocks, causing lateral breaks in structural trends. Of course Sod's law has it that these locations are often favourable for mineralization. Complexities at both scales obstruct predictability and interpretation in between and away from individual data points.

Interpretation uncertainty can be mitigated by both 2D, but especially 3D, seismic reflection data. Seismic can help constrain the structural framework geometry at the level of the deposit and deeper, beyond drill hole control. Acquisition and processing of seismic data in hard-rock terrain is not trivial (steep dips, high and heterogeneous velocities) and must be carefully designed and tailored to optimise results, particularly for the imaging of steep structures, but is done successfully.

The structural model can be further improved by digital kinematic analysis. This starts by restoring the effects of deformation to see if things fit back together to test for internal consistency (balancing), which then allows quantifying palaeogeometry by taking models back in geologic time step-by-step, restoring the effects of folding and faulting, (igneous) intrusion, sedimentation, and erosion. Individual structures and their role in the mineral system through time can be tested, quantified, and better understood. This helps reduce operational risks, both for minerals (brownfield) exploration and geotechnical aspects.

A number of examples will illustrate the usefulness of 2D, 3D, and 4D structural models.

The structural modelling challenge: Meaningfully 'connect the dots' with limited data

S13

Automated Core Imaging: Opportunities for Integrating Hyperspectral Mineralogy and Structural Orientation Data for More Robust Structural Interpretations

Cassady L. Harraden, Laurel A. Stothers, Cari Deyell-Wurst
Corescan, Vancouver, BC, Canada

Automated core imaging technologies are transforming how geoscientists log core. Multisensor platforms collect visual, geochemical, and/or mineralogical data, making informed geologic and engineering decisions in near real time. New developments also allow for measurements of core morphology and structural features (fractures, faults, and veins) that can provide vital information for the successful exploration and development of ore deposits.

Traditional practice requires geoscientists to collect structural data manually, from macro- or microscale inspection. This is labor intensive, time consuming, and can lead to incomplete or inconsistent results. The ability to automatically measure the orientation of structures in drill core using digital data sets including photographs or digital surface models is a significant advancement. These automated methods increase not only the speed and efficiency of data collection but also the density of structural measurements that can be collected downhole. While orientations of these features are important, the ability to integrate key structural and hyperspectral mineralogical data from logging systems provides an opportunity to maximize value from recovered core and make better, informed geologic and engineering decisions.

Here, we present a new approach to structural data collection by integrating mineralogical data from a hyperspectral imaging system to automatically identify structures in drill core. Structural identification and orientation measurements are completed using previously established contouring, morphological filtering, and adaptive feature thresholding methods. Data are derived from coregistered, high-resolution, 3D profiler and core imagery from an automated hyperspectral imaging system. Because the hyperspectral mineralogy data is coregistered, the mineralogy of structures can be integrated with their orientation values. This method determines the spatial extents of structural features in the image, then applies a series of multiple-distance buffers to a set of regions proximal to the defined structural feature. A query of the coregistered, hyperspectral mineralogical data can then be performed within these buffered distances. The outputs are integrated structural orientation measurements with hyperspectral mineralogical data that can be used to delineate fracture sets or vein types even in deposits with seemingly uniform structural orientations (Fig. 1).

The accurate characterization of structural features is key to predicting the geotechnical behavior of an orebody. These data are fundamental for mine planning decisions and affect the economics of ore recovery. Additionally, these data are used by exploration teams to map vein distribution and density and understand fluid pathways, controls on mineralization, and deposit geometry. Combining structural and mineralogical data to assist in mineral exploration is common; however, innovation in the automated data acquisition of these key data sets provides a novel, fit-for-purpose solution. The application of automated core imaging technology to structural measurements provides the opportunity for increased and targeted data points to be acquired over shorter time frames, improving sampling statistics to provide robust data sets for informed decision-making.

Fig. 1. Example of open joint structural orientations from an orogenic gold deposit measured using an automated hyperspectral imaging system (Corescan's HCI System) showing the orientation only (A) and the orientations colored by primary hyperspectral alteration assemblage within and immediately proximal (within 5 mm) to the structure (B).

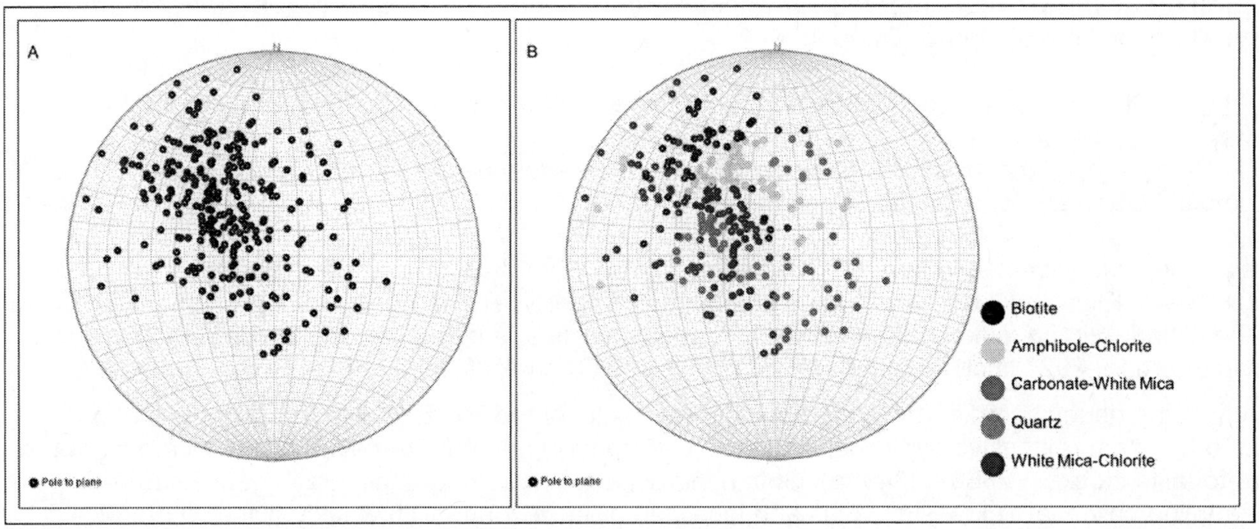

S14

3D Structural Modeling Using Oriented Drill Core: A Best-Practice Case Study from the Ormaque Discovery, Val-d'Or Mining Camp, Québec

Sean McKinley[1], Chris R. Siron[2], Jacques Simoneau[3], Pape Mactar-Dieng[3], Nancy LaFrance[3], Tim Baker[1], Peter Lewis[1]

1. Eldorado Gold Corp., Vancouver, BC, Canada, 2. Green Mountain Exploration LLC, Colchester, VT, USA, 3. Eldorado Gold Quebec, Val-d'Or, QC, Canada

The Ormaque deposit is located in the Val-d'Or mining camp of northwestern Québec and is the most recent discovery at Eldorado Gold Corporation's Lamaque property. The deposit sits approximately midway between the historic Sigma mine and the currently producing Triangle mine. The maiden inferred resource, published in February 2021, included 2.62 Mt at a grade of 9.53 g/t Au for 0.803 Moz of contained gold.

The geology of the deposit area is dominated by volcanic rocks of the ca. 2704 Ma Val-d'Or Formation, which is intruded by an extensive synvolcanic diorite (C-porphyry) and younger syn- to late-tectonic gabbro and diorite-tonalite stocks (~2686-2684 Ma). Gold mineralization at Ormaque occurs in gently south-dipping quartz-tourmaline-carbonate extension veins and localized breccia zones with weakly disseminated to massive tourmaline-pyrite replacement halos. The vein system occurs within the C-porphyry at the contact with volcaniclastic rocks to the north and is enveloped by an Fe-chlorite alteration footprint as defined by spectroscopy (SWIR analysis). Individual extension veins have widths ranging from several cm up to 2 m and inferred east-west strike lengths of up to 650 m and are vertically stacked from a depth of about 150 m to at least 600 m below surface. Moderately south-dipping shear or hybrid shear-extension veins locally interconnect some extension vein segments. Ductile to semibrittle, east-west-striking and steeply north-dipping shear zones anastomose through the C-porphyry and may represent preexisting and/or synmineral structures that controlled vein formation and higher-grade domains.

Interpretation of the geologic and structural framework of the Ormaque vein system involved collection of oriented drillcore measurements and modeling using Seequent's Leapfrog Geo software. Core orientation followed rigorous QA/QC monitoring prior to logging and data collection. Planar structural data were grouped and filtered by structural type and style. Orientation data were statistically analyzed to determine representative vein and shear zone orientations and to inform the construction of structural form interpolant surfaces, which allowed for 3D visualization of structural trends. Intervals containing extensional quartz-tourmaline-carbonate veins were modeled by compositing Au-bearing intercepts based on a cutoff grade (0.5 g/t Au) over the entire composite and a minimum width of 0.5 m. Composites were allowed to incorporate multiple thin veins and a maximum of 4.5 metres of dilution. Composites were grouped into 3D volumes guided by the trends established by the structural form interpolants. This compositing approach, in tandem with detailed geologic observations, oriented core measurements, and structural form interpolant visualization, allowed for the accurate and consistent modeling of vein geometry and established the geologic framework on which the initial resource estimate was made. The workflow also provides a testable and rapidly updatable model for continued exploration and resource drilling on the Lamaque property.

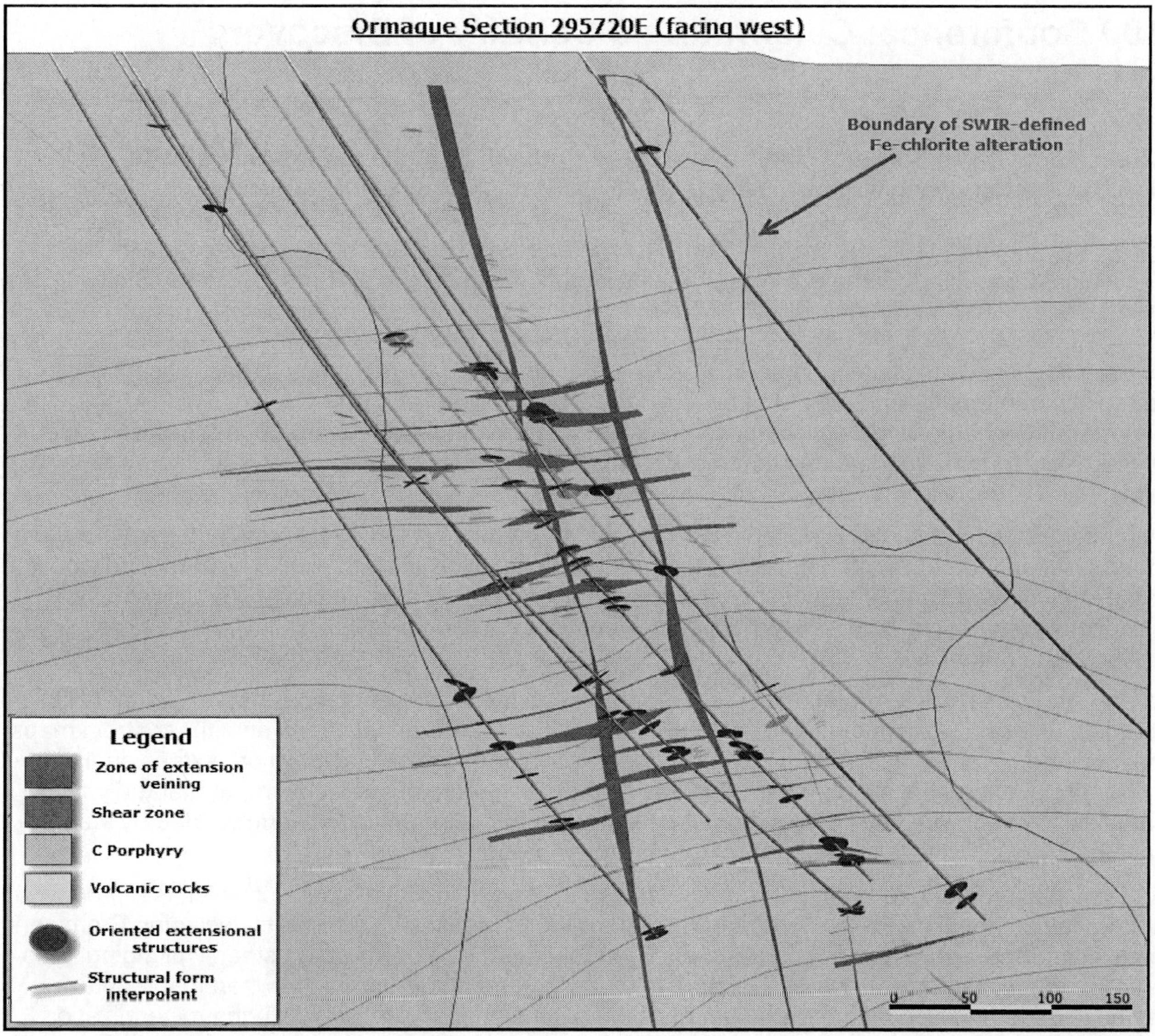
Ormaque Section 295720E (facing west)
Boundary of SWIR-defined Fe-chlorite alteration
Legend
Zone of extension veining
Shear zone
C Porphyry
Volcanic rocks
Oriented extensional structures
Structural form interpolant
0
50
100
150

S15

Stress-Strain Modeling as Part of a Mineral Systems Approach to IOCG Exploration and Target Generation: A Case Study from the Mt. Woods Inlier, Gawler Craton, South Australia

Jonathan N. Gloyn-Jones[1], Ian J. Basson[1, 2], Corné Koegelenberg[1], Ben Stoch[1], Michael-John McCall[1]
1. Tect Geological Consulting, Somerset West, South Africa, 2. Department of Earth Sciences, Stellenbosch University, Stellenbosch, South Africa

Hydrothermal IOCG deposits contain numerous phases of hydrothermal brecciation, cataclastic milling, pervasive replacement, multistage alteration assemblages, and polymetallic (Fe, Cu, Au, Ba, F, U, Ce, and La) mineralization. Orebodies typically comprise breccias or stockworks, with sharp boundaries and steep or vertical pipe-like or columnar geometries that show spatial relationships to deep-seated structures or their intersections.

Structural reinterpretation of publicly available high-resolution geophysical and drill hole data over the Mt Woods Inlier, Gawler Craton, South Australia, allows for the application of finite element analysis to simulate the relative distribution and magnitude of stress-strain conditions during a geologically brief NNW-SSE-oriented extensional event (ca. 1595-1590 Ma), coincident with IOCG-hydrothermal fluid flow and mineralization in the region.

Structural and lithological reinterpretation of the Mounts Woods Inlier area encompassing Prominent Hill was undertaken and used as input into the finite element analysis. Differential stress and shear strain maps across the modeled terrane highlight regions that were predisposed to strain localization, extensional failure, and fluid throughput during the simulated mineralization event. The role of stress risers, or markedly more competent units, is exemplified by a large, sigmoidal-shaped mafic-ultramafic body surrounded by steeply dipping, alternating host units to the N and NW of the Prominent Hill area.

Stress-strain maps are integrated with other data sets and interpretation layers, one of which is a proposed structural-geometrical relationship that is apparent in many world-class IOCG-deposits, including Prominent Hill, Olympic Dam, Sossego, Salobo, Cristalino, and Candelaria. These occur at the steeply plunging, pipe-like intersection of a conjugate-extensional system of faults, shears and/or contacts, wherein the obtuse angle may have been bisected by the maximum principal extensional axis (viz., σ_3) during mineralization. Several other layers were also used for the generation of targets, such as distance from major shear zones, angular tolerances of geometrically favourable fabrics and/or structures, favourable host lithologies, and proximity to tectonostratigraphic contacts of markedly contrasting competency. The result is an integrated targeting index and heat map for IOCG prospectively within the Mount Woods Inlier.

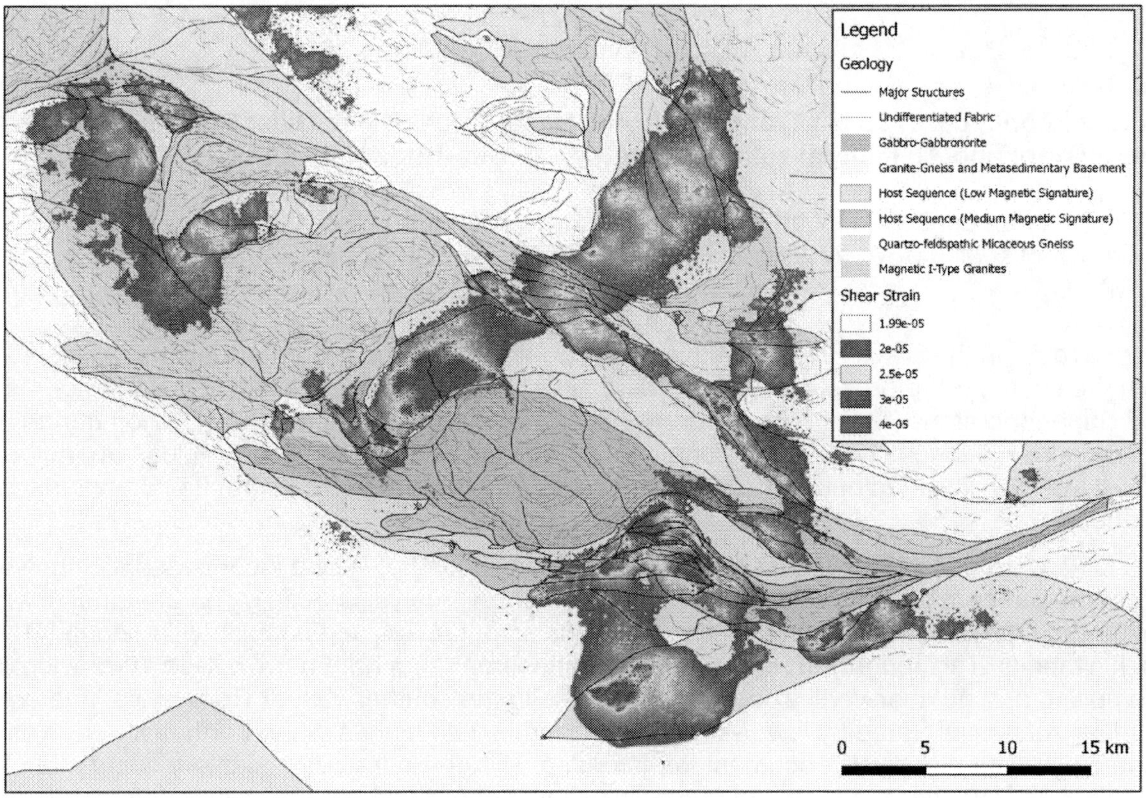

Legend
Geology
Major Structures
Undifferentiated Fabric
Gabbro-Gabbronorite
Granite-Gneiss and Metasedimentary Basement
Host Sequence (Low Magnetic Signature)
Host Sequence (Medium Magnetic Signature)
Quartzo-feldspathic Micaceous Gneiss
Magnetic I-Type Granites
Shear Strain
1.99e-05
2e-05
2.5e-05
3e-05
4e-05
0
5
10
15 km

ST.190
Dakotan Tectonic Zone: ca. 1730 Ma Transpressional Shear Zone, Black Hills, SD: Implications for Late Tectonic History and Au Mineralization in Southern Trans-Hudson Orogen

Stephen T. Allard[2], Brian Lentz[1, 3], Robert Bergmann[1, 3], Eric Nowariak[3], Chris Nicosia[3]
1. F3 Gold, Minneapolis, MN, USA, 2. Winona State University, Winona, MN, USA, 3. Big Rock Exploration, Minneapolis, MN, USA

Paleoproterozoic rocks in the Black Hills of southwestern South Dakota record deformation associated with accretion from the south by Proterozoic terranes (Yavapai Orogeny) followed closely by suturing of the Wyoming and Superior cratons (Black Hills Orogeny; ca. 1785–1715 Ma). Orthogonal collision during the latter event is responsible for ca. 1750 Ma regional metamorphism and NW-striking F_2 folds forming the regional grain of the hills. Post-tectonic intrusion of the ca. 1715 Ma Harney Peak granite punctuates the cratonic suturing.

Rare F_3 cross folds were described by past workers as local reworking of F_2 and inconsequential; however, our results from more recent detailed mapping and microstructural analysis, focusing on penetrative fabrics along the eastern margin of the uplift, recognize the F_3 folds as pervasive and coupled with strong NNW-striking mylonitic fabrics. This roughly 10-km-wide zone varies from abundant narrow 1- to 10-m-wide shear zones anastomosing around near-vertical isoclinal F_3 hinges in the southeast near Rockerville, to a series of ≥4-km-wide shear zones separated by lower-strain zones and F_3 fold hinges in the northeast near Nemo. Outcrop and microstructural analysis document left-lateral, east-side-up transpression. Ongoing mapping west of the original discovery has recognized numerous similar shear-fold coupled high-strain zones across the hills supporting an uplift-wide transpressional zone.

Furthermore, identical fabric and kinematic indicators are recognized in Archean basement rocks near Nemo previously thought to have escaped Proterozoic deformation. The lack of D_2 structures in the Archean basement signals a change from thin-skinned to deep-seated basement-involved deformation coincident with a switch from orthogonal to transpressive collision. This major D_3 structure, now named the Dakotan Tectonic Zone (DTZ), is constrained by published ages to post D_2 (1750 +/- 10 Ma) and pre-Harney Peak intrusions at 1715 +/- 3 Ma. Evidence that stresses associated with the DTZ were transferred inboard to the Wyoming Province is seen to the southwest in the Hartville Uplift and Laramie Mountains where structures with similar timing and kinematics overprint Cheyenne Belt deformation.

Metamorphism during D_3 is scarce and limited mostly to grain-size reduction and fabric development; however, static post-D_3 hydrothermal alteration leading to downgrade metamorphic assemblages along with oxide/sulfide mineralization in quartz-rich veins is ubiquitous along D_3 structures. This alteration dissipates within meters perpendicular to D_3 structures, consistent with these structures acting as hydrothermal fluid conduits. The temporal proximity to Harney Peak intrusion and the through-crustal nature of these structures allows for either the granite or a deep crustal source for the hydrothermal fluids.

The giant Homestake Gold Deposit occurs along strike and within mapped expressions of the eastern DTZ structure, with mineralization controlled by similar fold-and-shear structures. Timing for gold mineralization in Homestake is dated at 1736 +/- 8 Ma by others, within permissive ages for alteration along the DTZ. If DTZ structures are controlling mineralization, then the accepted stratabound model for mineralization at Homestake may need to be reconsidered, and the wide-spread presence of DTZ deformation across the hills opens the possibility of similar lode gold mineralization elsewhere in the uplift.

ST.191
New U-Pb Sphene Ages Confirm Synchronous Gold Mineralization and Dakotan Tectonic Zone Deformation at Homestake Mine and the Black Hills Region, South Dakota, USA

Stephen T. Allard[2], Eric Nowariak[1], Kevin R. Chamberlain[4], Robert Bergmann[3, 1], Brian Lentz[3, 1]
1. Big Rock Exploration, Minneapolis, MN, USA, 2. Winona State University, Winona, MN, USA, 3. F3 Gold, Minneapolis, MN, USA, 4. University of Wyoming, Laramie, WY, USA

The Paleoproterozoic tectonic and deformational history of the Black Hills, South Dakota, has historically been interpreted to be dominated by 1780-1750 Ma deformation (D_1 and D_2) related to the accretion of Paleoproterozoic volcano-sedimentary terranes and subsequent suturing of the Archean Wyoming and Superior provinces. Recent publications, however, document the Dakotan Tectonic Zone (DTZ), a system of pervasive north-northwest-striking transpressional shear zones expressed along the eastern margin of the Black Hills that overprint regional D_2 structures and fabrics and attributed to the final suturing of the Wyoming and Superior Provinces. Reconnaissance mapping and sampling reveal that DTZ shear zones intersect the Homestake Gold Mine and several gold deposits and occurrences that lie in the eastern and southwestern Black Hills.

This study presents new IDTIMS U/Pb metamorphic sphene dates of 1710.2 +/- 3.8 and 1712 +/- 12 Ma from DTZ deformation zones. The dated samples include isoclinally refolded quartz-garnet-amphibole-sphene-pyrrhotite calc-silicates interlayered with pelitic schists within the Paleoproterozoic Mayo Formation and mylonitic actinolite-microcline-pyroxene-biotite-sphene-carbonate calc-silicate from the Paleoproterozoic Crow Formation, respectively. In both samples, microtextural and mineralogical analysis shows the dated sphene crystals overgrew metamorphic fabrics and is associated with metasomatism, suggesting static growth syn- or postdeformation with the dates representing minimum ages for deformation.

With these new dates, the bracket for DTZ deformation and final suturing of the Wyoming and Superior Provinces can be constrained to ca. 1750-1710 Ma. DTZ deformation ages described here are coeval with published ca. 1730 Ma ages for gold mineralization at the Homestake Gold Mine, supporting a temporal and spatial relationship between DTZ deformation and shear-hosted gold mineralization in the Black Hills. Regionally, this ca. 1720 Ma tectonic event is also observed to the west in the southeastern Wyoming Province where structural analysis and geochronology have described a NW-SE compressional-transpressional deformational event along the eastern margin that postdates the accretion of island arcs along the Cheyenne Belt and suggests the Wyoming and Superior Provinces had separate secular Paleoproterozoic histories before DTZ deformation during final suturing at ca. 1720 Ma.

ST.192
A Near-Mine Discovery Revealed by 3D Seismic Data and a New Integrated Structural Model at the Darlot Gold Deposit, Yilgarn Craton, Western Australia

Ashleigh L. Ball
University of Tasmania, Mosman Park, WA, Australia

A 3D seismic survey was completed across the Darlot gold deposit of the world-class Yilgarn Craton, Western Australia, with the aim of improving the understanding of the structural setting and controls on mineralization. The survey was highly successful in delineating structure and provided significant insights into the relative timing of mineralized structures as well as their extents away from the known deposit. When integrated with existing data sets a new structural and geologic model was generated that was extrapolated into the regional exploration space. This facilitated the generation of numerous near-mine and greenfields targets, the testing of which led to a near-mine discovery.

The 3D seismic survey was completed across a 5- x 5-km area to a 5-km depth with a seismic line resolution of 30 m. The seismic lines successfully depicted structure of various orientations as well as regional folds in the stratigraphy. Known ore-bearing structures could be traced for considerable distances and depths away from the deposit in conjunction with new, previously unidentified parallel structures. A significant insight was the identification of a large EW-striking structure that hosted a large lamprophyre dike around which the deposit was centred. This structure was far more extensive than previously recognized, and the localization of mineralization as well as the lamprophyre dike and a suite of fractionated granites along the intersection of this structure with a regional shear suggested this intersection may have been fundamental to the deposit's formation.

The 3D seismic data was integrated with existing geophysical, geochemical, drilling, and mapping data sets to generate a new structural and geologic model. This model could be extrapolated into the regional space with the 3D seismic providing a higher confidence in interpretation than existing 2D data sets. The extrapolated model gave new context to known geochemical anomalies, historical workings, and mineralized drill intercepts and rejuvenated exploration. Numerous new near-mine and regional targets were generated, including along previously overlooked EW-striking structures.

One of the near-mine targets to be tested was a newly identified structure situated parallel to a known mineralized fault. Seismic reflectors formed a sigmoidal pattern that was analogous to the geometry of known vein lodes in the existing deposit. This was drill tested, and a significant new zone of mineralization was identified, resulting in a near-mine orebody discovery. Due to the depth of required holes to test the structure, the 3D seismic data was not only key in identifying the broad target area but also in the design of drill holes to ensure the target was accurately intercepted.

The 3D seismic survey across the Darlot deposit was pivotal in the creation of a new structural and geologic model and the generation of multiple new exploration targets. Although still a relatively new tool in hard-rock exploration, when combined with existing geologic data sets it proved to be highly valuable in adding to and expanding the understanding of the Darlot deposit and ultimately in exploration success.

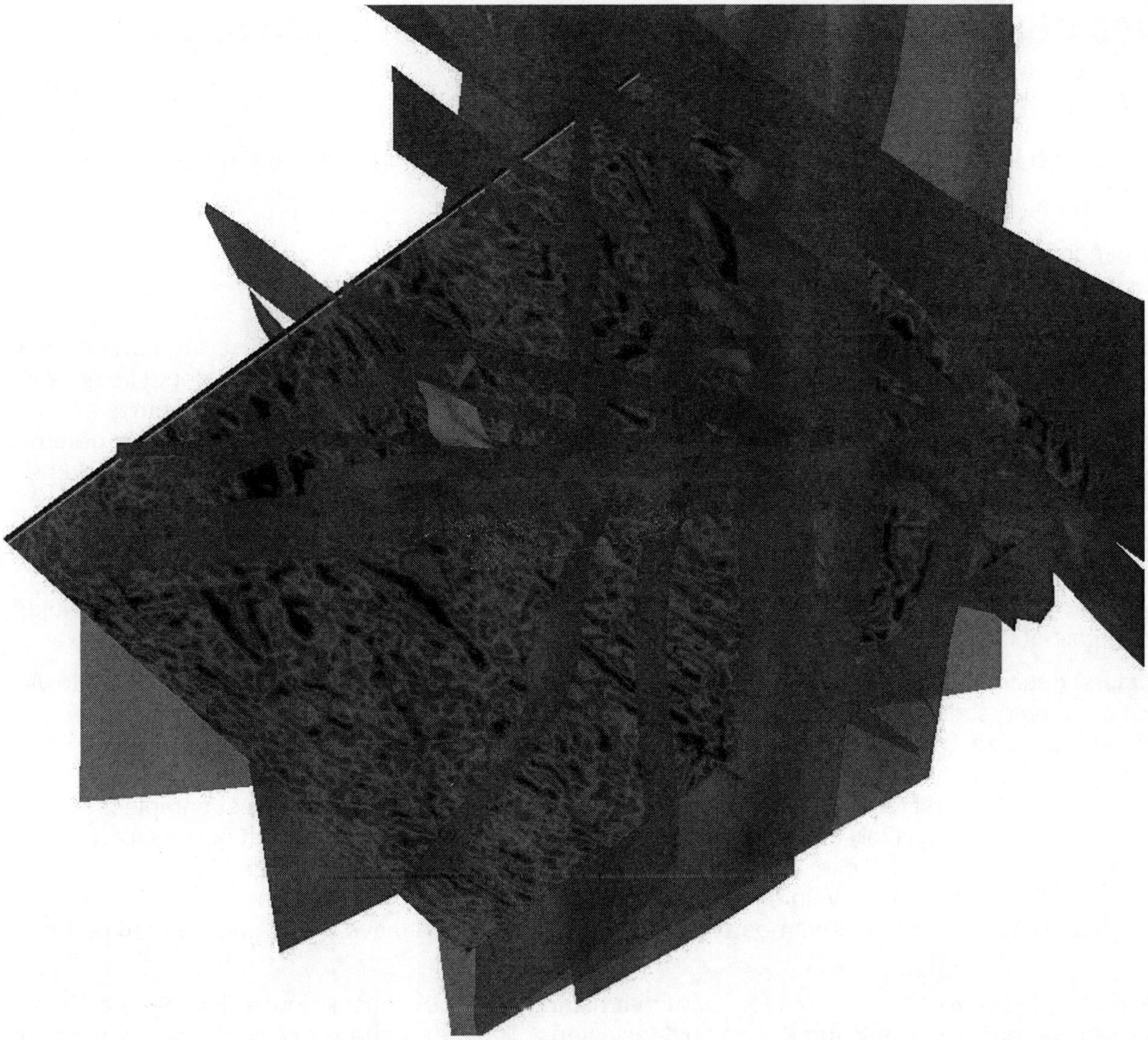

ST.193
Stress and Strain in Quartz Veins along the Neoarchean Cadillac-Larder Lake Fault Zone, Abitibi, Canada

Crystal Brochard, Michel Jébrak, Stéphane De Souza, Daniele L. Pinti
Universite du Quebec a Montreal (UQAM), Montreal, QC, Canada

Most Archean gold deposits are associated with major structural faults such as the Cadillac-Larder Lake fault zone (CLLFZ, Superior craton) and the Boulder-Lefroy Fault (Yilgarn craton). Gold deposits are related to second-order structures, which host hydrothermal quartz vein systems. Emplacement depth is an important parameter in the crustal continuum model for orogenic gold deposits but is complex to determine in deformed terranes. The source of fluids in orogenic gold deposits also lacks consensus and could be related to deep-seated metamorphic, magmatic, or supracrustal fluids.

A total of 40 quartz veins were sampled in 23 orogenic gold deposits along the CLLFZ in Abitibi, from Val-d'Or (QC) to Kirkland Lake (ON). The CLLFZ is a major crustal structure stretching onto 250 km in the Archean Abitibi greenstone belt. The aim of this study is to determine the nature of the mineralized fluids and the dynamic of the CLLFZ after original vein crystallization.

Microstructural description of quartz veins using transmitted light microscopy and EBSD analysis allows determination of recrystallization textures, dominant dynamic recrystallization (DRX) mechanism, as well as recrystallized grain size and flow stress. Samples were collected at several depths in drill cores, as well as from underground mines and outcrops. The host rock varies from intermediate intrusive to mafic volcanic and sedimentary rocks. First results show evidence of dynamic recrystallization in most samples. Samples analyzed with EBSD display a log-normal distribution of recrystallized grain sizes that is consistent with data for natural samples such as mylonite. A 3D test shows a variation of the mean recrystallized grain size in one direction (surface plane in the vein direction). A sample in a fold hinge (Dubuisson deposit) shows variation of both DRX mechanism and mean recrystallized grain size relative to the undeformed part of the vein.

Quartz vein samples were also selected for vacuum crushing and noble gas isotopes (He, Ne, Ar, Kr, Xe) and CO_2, H_2O contents were measured from fluid inclusions. The inert behavior of noble gases prevents them from chemically interacting with their surroundings, whereas noble gas isotopes can be fractionated during geologic processes such as magmatic differentiation or hydrothermal boiling. The relative abundance and isotopic composition of noble gases in fluid inclusions can be helpful to determine the origin of these fluids. Preliminary results show lack of He and Ne, explained by their high diffusivity in quartz and the age of the samples. Two groups of samples are distinguished by high CO_2 contents and relatively little radiogenic $^{40}Ar/^{36}Ar$ ratio of less than 5,000, and CO_2-poor samples with $^{40}Ar/^{36}Ar$ ratio >18,000. Elemental ratios Kr/Ar and Xe/Ar are higher than that expected in aqueous fluids and possibly indicate fractionation by boiling or contribution from a sedimentary component, as often documented in Paleozoic and Cenozoic counterparts.

Fig. 1. Postprocessed EBSD map of sample SIL1 (Silidor deposit, Rouyn-Noranda), using MTEX add-on in Matlab and modified matlab code from Cross et al. (2017). Red grains depict relict grains or porphyroclasts; blue ones show the recrystallized grains. Dominant DRX mechanism is subgrain rotation in this case.

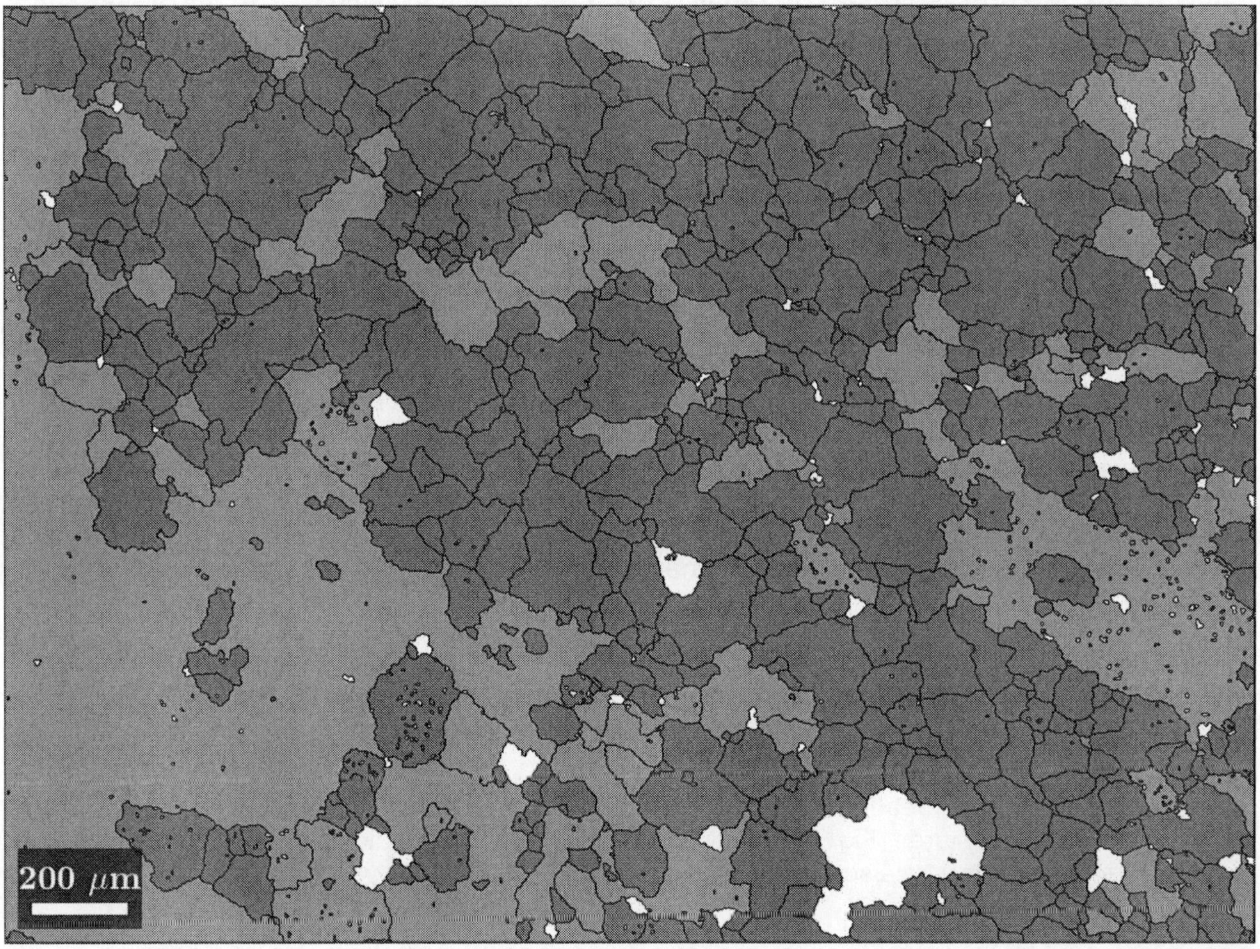

ST.195
Application of X-Ray Computed Microtomography for Quartz-Sulfide Stockwork Study: Drazhnoye Gold Deposit (Yakutia, Russia)

Viktoria Chikatueva[1], Dmitry Korost[2], Andrey Chitalin[1], Nikita Stepanov[2]
1. Institute of Geotechnology, Moscow, Russia, 2. Deco-service, Moscow, Russia

The Drazhnoe orogenic-type gold deposit is located in Yakutia, Russia. Several gold-bearing stockwork systems within the NNW left-lateral shear zone were explored. To study the relationship of the geologic elements of the stockwork, half of the drill core sample was studied using the X-ray computed tomography method (CT). The drill core sample is 20 cm long and 6 cm in diameter. The sample was scanned by a medical tomograph at two energy levels (80 and 130 keV). Postprocessing of CT data made it possible to obtain a reliable distribution of minerals in the sample. As a result of the study, we obtained a detailed stereological model (tomogram) that allowed us to isolate quartz and quartz-carbonate veins, to divide sulfide mineralization into pyrite and arsenopyrite, and also to isolate gold (Fig. 1A). As a result of the analysis of both the sample and the obtained tomogram, all the selected veins and sulfide mineralization were divided into generations. After that, every generation of each mineral was analyzed and, considering all the revealed structural relationships and knowledge about the mineralogical composition of mineralization, we distinguished the structural and mineral paragenesis into various stages and substages:

Stage 1 (pre-mineralization). Quartz veins were formed in horizontal compressional stress conditions.

Stage 2 (mineralization). The mineralization stage, at early substage, was characterized by the formation of metasomatic pyrite and arsenopyrite in terrigenous sedimentary rocks. These sulfides are controlled by cleavage and bedding and they also fill the microfractures. At this substage, veins were also formed in horizontal compressional stress conditions (Fig. 1B).

During the second substage, gold-bearing quartz-carbonate-sulfide veins were formed. These veins crosscut previously formed structures (Fig. 1B). Native gold grains are associated with large quantities of early pyrite and arsenopyrite and are also localized in quartz-carbonate-pyrite veins.

During the final, third substage, flat-dipping veins with coarse-crystalline quartz were formed. The older structures are intersected by the veins. In some sections of the deposit, such veins are gold-bearing.

As a result of the study, the stereological models for each stage and substage of mineralization were created. This made it possible to study the distribution of minerals in the core, to investigate the morphology of the selected structures, and to make a kinematic interpretation of their formation. This contributes to developing a more detailed, realistic, and reliable three-dimensional model of the deposit.

Fig. 1. A -general view of the half core sample that has been examined and its stereological model; B - stereological models of selected stages and substages; in the orange frame - detailing the location of native gold.

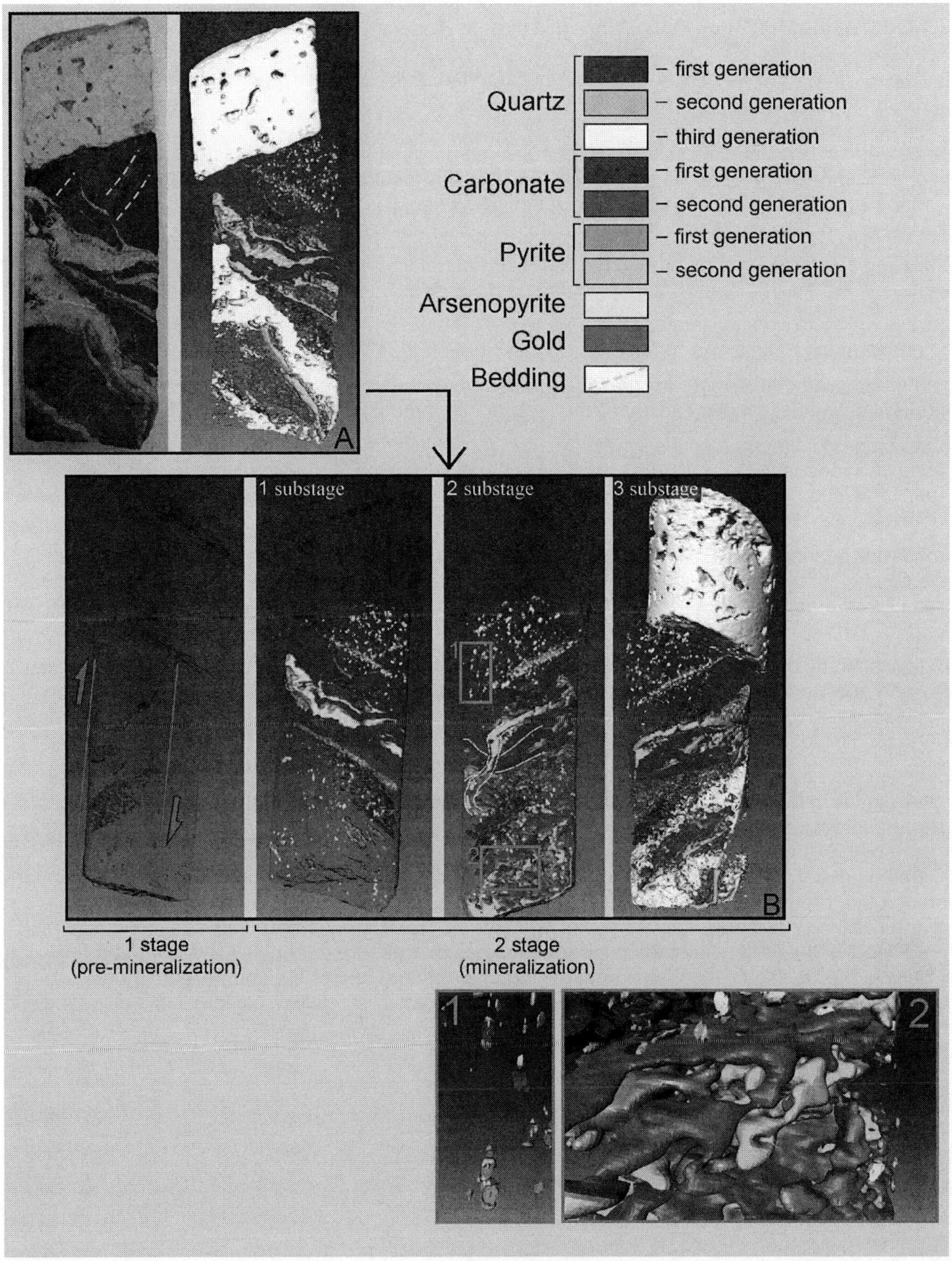

ST.196
Structural Model of the Drazhnoe Gold Deposit, Yakutia, Russia

Andrey Chitalin, Evgeny Grishin, Dmitry Sivkov, Viktor Usenko, Evgeny Fomichev, Viktoria Chikatueva
Institute of Geotechnology, Moscow, Russia

The Drazhnoe deposit is situated in Yakutia, Russia. It is an orogenic-type gold deposit with a structural-lithological control of ore mineralization (Fig. 1A). The deposit is localized in the Adicha-Taryn regional-scale fault zone, along which gold deposits are concentrated. The deposit is confined to the northeast limb and hinge of the Taryn syncline, which is composed of the Upper Triassic silty sandstones and siltstones.

Four gold-bearing stockwork systems within the NNW left-lateral shear zone were recognized and explored. This zone crosses the Adicha-Taryn Fault by angle 20° clockwise.

The contours of the stockwork systems correspond to the areas of pyritization. Linear flat horizontal and low-angled quartz stockworks are grouped in arc-like systems (Fig. 1B).

Highest gold grades were discovered in cores of some quartz stockworks in the pyrite-arsenopyrite reach areas where NW oblique-slip faults and later transverse folds and dextral strike-slip faults intersects.

There are two stages of deformation distinguished. First (pre-ore) stage includes multiordinal cleavage folds F1 and ductile cleavage faults with S-C fabrics. Faults planes dip to the SW and NE at medium to steep angles. Transverse and diagonal strike-slip faults that are part of the structural paragenesis of the first stage of deformation are also distinguished

At the second (ore-bearing) stage of deformation under the conditions of horizontal compression-vertical extension, extended inclined quartz stockworks of overthrust kinematics and the associated subhorizontal stretching stockworks were formed. Veins with rich gold mineralization are often visible in core parts of stockworks (Fig. 1C).

The mineralogical study of sulfide mineralization was carried out on samples by the methods of computer tomography and microtomography. Mismatching of structural plans of gold and arsenopyrite distribution in quartz veins was revealed. Microstructural trends affecting gold distribution were visible in different stress fields, then quartz veinlets were formed.

Axes of folds F1 of the early substage $D\text{-}1^1$ of the first stage of deformation extend northwestward bend smoothly and slightly in the plan due to shear deformation at the late substage $D1^2$ of the first stage in the Adicha-Taryn Fault destroyed zone. The fault is probably formed in the late of the first stage of deformation as a wide zone of thickening of viscous reverse-sinistral strike-slip faults R extending NW. At the same stage conjugated dextral strike-slip faults R' were formed. Points where R and R' intersect predetermined the position of future ore-bearing stockworks. At the second (ore-bearing) stage of deformation D2 the direction of the horizontal compression vector has changed to SE. The mineralized sinistral strike-slip fault zone with regular distributed of gold-bearing stockwork systems was formed. Folds F2 of northeastern extension were formed within the zone in a transtention environment. In this superimposed mineralized shear zone, the ore-bearing stockwork systems like push-up structures between sinistral strike-slip faults R were formed (Fig. 1-D).

Fig.1. A-Drazhnoe deposit location; B-Cross-section of the arc-like stockwork system. The Gold Channel indicates concentration of highest gold grades; C-structural stereogram of the poles to gold bearing veinlets, lower hemisphere; D-structural model of the Drazhnoe deposit.

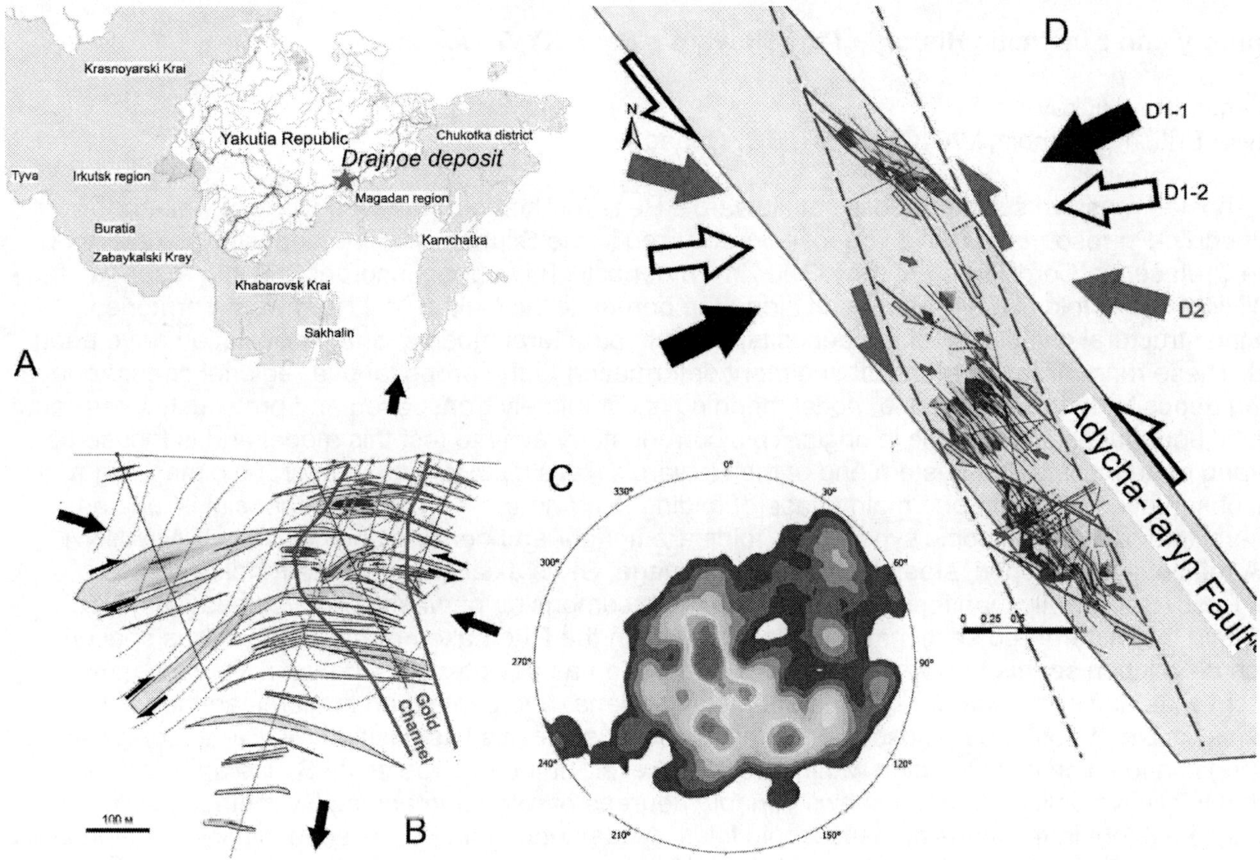

ST.198
The Geometry and Kinematic History of the Howard's Pass XY Deposits

Darius Kamal, Ken Hickey
University of British Columbia, VANCOUVER, BC, Canada

The shale-hosted massive sulfide deposits of Howard's Pass are home to the world's most well-endowed undeveloped Zn-Pb resources. Sulfide deposition occurred in the Silurian, and the deposit was deformed during the Cretaceous Cordilleran orogeny. The Zn-Pb deposits have been incorporated into a series of upright WNW-striking folds on a 40-km trend along the border of the Yukon and Northwest Territories. Owing to the structural complexity of the deposits, multiple structural models for their evolution have been proposed. These range from Silurian soft-sediment deformation to the proposal of a regional shear zone hosting the deposits. In the shear zone model, bedding is completely transposed and previously interpreted stratigraphic boundaries are tectonic in origin. This current study aims to test this model and is focused on the XY group of deposits at the eastern end of the Howard's Pass district. Lithostratigraphic mapping and structural observations indicate one main phase of folding, F_1, and the XY group of deposits is located on the southern limb of a macroscopic syncline. F_1 folds are upright and gently plunging to the WNW-NNW (Fig. 1). A regionally developed, steep NE-dipping cleavage, S_1, is axial planar to the F_1 folds across Howard's Pass. S_1 typically manifests as a slaty cleavage comprising pervasively developed dissolution seams. Within the mineralized sections at XY, located within the Duo Lake Formation, S_1 forms spaced sulfide-rich dissolution seams that were previously interpreted as dewatering structures. These seams are up to 0.5 cm wide and show coarser-grained sphalerite, galena, and pyrite than the laminae they cut across. S_1 is visible at both the mesoscopic and microscopic scale, and its position relative to compositional layering (S_0) is a function of its location within a fold; however, at the outcrop scale S_0 has an enveloping surface that is always oblique to S_1. It shows variable degrees of refraction across S_0 contacts, which is especially prevalent in the limbs of mesoscopic folds. A later kink fabric, S_k, is seen locally at a high angle to S_1 and may be genetically linked to movement on WNW- and NNE-striking faults that overprint the F_1 folds. Mapping constraints show that the faults exhibit normal dip-slip movement. Structural mapping shows that the geometry of the deposits is primarily controlled by folding and that contacts between units are stratigraphic in origin. S_1 and S_k are the only foliations mapped. No shear zone fabrics were identified and there is no evidence that bedding has been transposed.

Fig. 3. Geologic map of the 2020 XY region field area. Stereonet of all bedding data collected in mapping area is shown in the lower right (lower hemisphere projection, equal area). From Kamal and Hickey (2020).

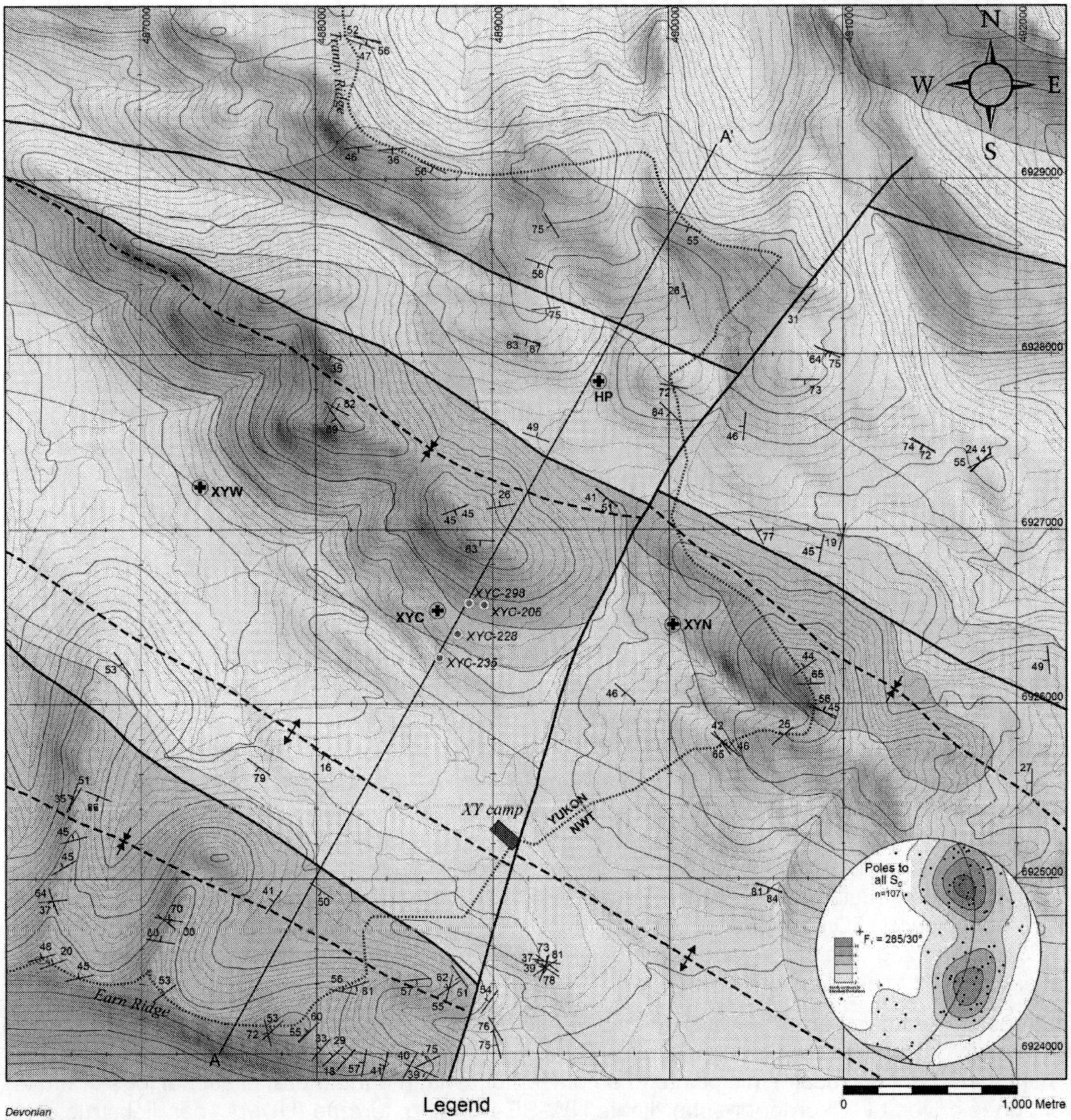

Legend

Devonian

Earn Group
Undifferentiated mudstone and sandstone. Dark to light grey finely laminated mudstone with <5 mm grey-black silty layers. Weathered outcrops show distinct maroon bands. Interbedded with cherty lenses 1-20 cm and silty limestone up to 1 m. Often forms shale when breaking along dense cleavage planes. Mudstone is variably calcareous. Sandstone is fine-grained with small, rounded chert and quartz granules. Matrix is beige-yellow with orange weathering and consists of subrounded chert clasts and lithic fragments. Seen as piles of rubble and not *in situ*. Massive sandstone blocks are cut by large quartz veins. Poorly sorted conglomerate is seen just outside of the field area.

Silurian to Early Devonian

Steel Formation
Medium to dark grey mudstone with strong orange weathering strongly expressed in outcrop and in talus. Small, <1 cm dark grey silty layers show "wispy" texture that does not weather as strongly. Associated with weathered pyrite cubes with quartz-calcite pressure shadows. Contains coarse, crystalline limestone concretions ranging from 1 cm up to 2 m. Concretions exhibit a strong, fetid odor when a fresh surface is exposed.

Early Ordovician to Silurian

Duo Lake Formation
Dark grey to black, recessively weathering, mudstone. Fine-grained, carbonaceous and locally calcareous. Interbedded with thin <1 cm dark grey silty bands, chert lenses up to 1 m and limestone beds from 1 cm to 2 m. Contains 1 cm blue bands of phosphate. Very fissile and forms pencil cleavage in outcrop. Locally graptolitic towards the top. Mineralization is hosted within the "Active Member" and sulphides (sphalerite, galena, pyrite) are seen in sedimentary beds and in spaced, sulphide rich dissolution seams that are axial planar to folds.

Cambrian to Early Ordovician

Rabbitkettle Formation
Dominantly fine-grained, crystalline light grey limestone layers up to 5 cm with a "wavy" texture interbedded in a rusty orange silty, argillaceous matrix with visible quartz and feldspar. Matrix is variably dolomitic. Different varieties show much thinner layers with laminae thickness <1 cm and no dolomite. Local layers of massive crystalline limestone with no internal textures.

Cambrian

Vampire Formation
Recessive weathering, dark grey to brown, mudstone, siltstone and interbedded sandstone. Not mapped as part of this study, but compiled from Gordey and Anderson, (1993)

XYC — Pb-Zn deposit

XYC-228 — Drillhole

84 — Bedding, S_0

Contact

Fault

Trace of S_1 axial plane cleavage (cross section)

F_1 synform

F_1 antiform

AM — Active member of Duo Lake (on cross section)

Universal Transverse Mercator Projection, Zone 9N, North American Datum 1983.

ST.199

Regional Structural Controls and Emplacement of the Miduk Cu Porphyry from High-Resolution Structural Mapping and Fully Constrained 3D Implicit Geologic Modeling

Corné Koegelenberg[1, 2], Ian J. Basson[1, 2], Michael-John McCall[1], Payam Ekhlaspour[3], Alireza Siami[3], Ali Rahnama[3]
1. Tect Geological Consulting, Somerset West, South Africa, 2. Stellenbosch University, Stellenbosch, South Africa, 3. Kavakaray Novin Madan Consulting Engineers, Kerman, Iran

The Miduk Cu Porphyry is located 85 km along strike from the world-class Sarcheshmeh Cu porphyry and forms one of a suite of major, actively mined Cu porphyries within the Urumieh-Dokhtar Magmatic Arc (UDMA). Miduk's Cu mineralization remains significantly open at depth and opencast mining, along with future underground operations, are planned over a 20-year life-of-mine cycle. Exploration of analogous satellite Cu porphyry systems in Miduk's vicinity is also underway, adding to the resolution of regional-scale endowment of a fertile Cu-mineralized porphyry system and associated crustal- to meso-scale structural emplacement mechanisms.

Since 2017, comprehensive structural mapping and implicit 3D geologic modeling of the deposit and its country rocks was undertaken. This resulted in the integration of over 12,000 structural measurements, from exposed pit faces and excavations, UAV-based photogrammetry surveys, and regional outcrops, with drill hole data and mine-wide downhole ATV surveys and macrostructural logging, which produced the first fully constrained implicit 3D geologic model of the deposit, depicting structure, country rocks, alteration, and mineralization zones.

The structural model depicts predominantly synmineralization brittle deformation, characterized by domain-bounding, steeply dipping, N-S- to NNE-SSW-trending, sinistral strike-slip faults (D_2) that crosscut earlier, conjugate NW-SE and NE-SW strike-slip faults (D_1). Overall, the geometry and kinematics of D_2 are consistent with R', and D_1 with R and P shears as part of a right-lateral Riedel shear configuration. This reflects the progressive development of right-lateral transpressional tectonics along the UDMA and associated lateral escape structures such as the Sanandaj-Sirjan Suture Zone, from the late Miocene (<12 Ma) onwards. A conspicuous 3D radial fracture/joint pattern provides further insight into the emplacement and geometry of mineralized vein stocks above the quartz-diorite cupola and the development of the associated alteration and mineralization envelopes.

Structural features elucidate preferential reactivation of brittle R' shears and associated extensional fractures as the main driving/throttling mechanism for dilation and thereby emplacement of the Miduk Cu porphyry system. On a regional scale, this points to emplacement within right-lateral dilational jogs or strike-slip step-over zones driven by 1st-order or subordinate NW-SE-trending, terrane (UDMA)-parallel structures. These findings provide invaluable insight not only for existing Cu porphyry deposits along the UDMA, their ongoing mining, and their geotechnical design, but also for regional and mine-scale exploration strategies.

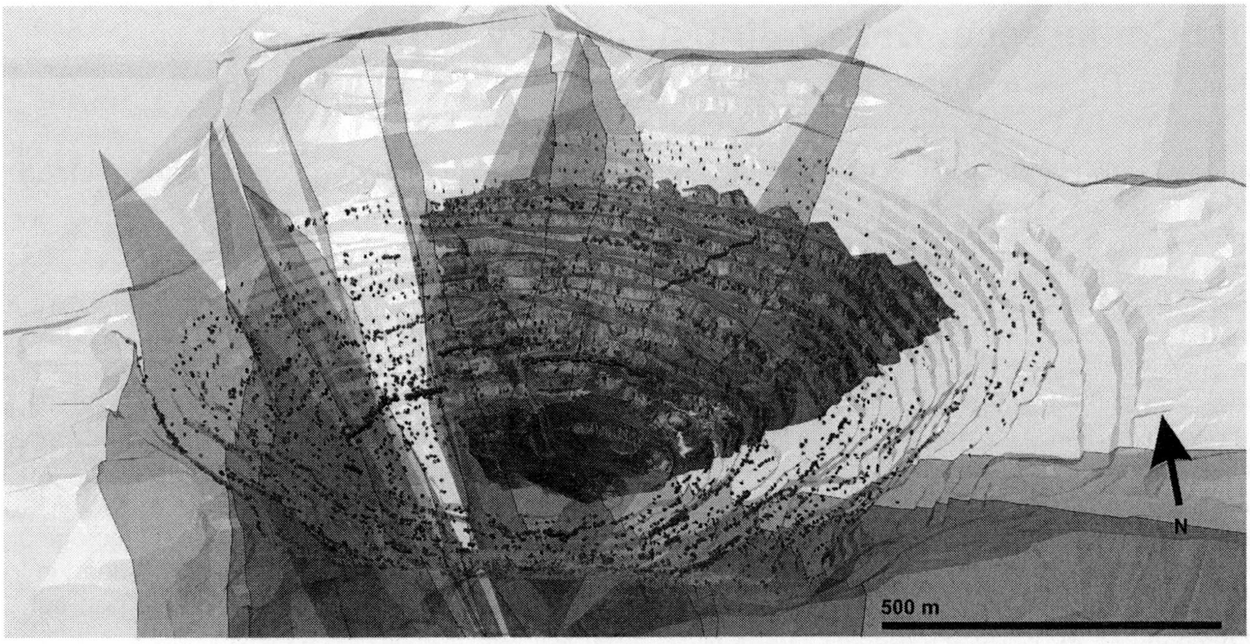
N
500 m

ST.200
Structural and Lithological Overview of a Polygenetic Mélange in an Evolving Lufilian Arc Foreland Basin, Kolwezi, DRC

Michael-John McCall[1], Ian Basson[1, 2]
1. TECT Geological Consulting, Cape Town, South Africa, 2. University of Stellenbosch, Cape Town, South Africa

The Kolwezi area of the Democratic Republic of the Congo is host to world-class stratabound Cu (±Co, Ni, U, Pb, Zn) mineralization. This mineralization is hosted in extremely large fragments of coherent lower Roan Group units. Long-debated genetic models for the development of this deposit include 1) tectonic mélange (friction breccias), 2) halokinesis or salt tectonics, and 3) sedimentary mélange. Historical data at K.O.V. Mine, spanning 70 years and comprising diamond and reverse-circulation drilling, structural and lithological face mapping, downhole televiewer data, and macro-structural logging, has been reanalysed and used in the construction of a fully constrained 3D implicit model of structures and lithologies, which sheds new light on the genesis of this world-class deposit.

Several features, at a variety of scales, provide evidence of its genesis: 1) structures inherited from residence time within the fold-and-thrust belt of the hinterland; 2) block-in-matrix fabrics; 3) soft-sediment deformation structures; 4) matrix-supported, polymictic conglomerates; 5) discordant and unconformable sedimentary relationships; 6) occurrence of multiple subtypes of R.A.T; 7) different styles of shearing at and away from the base of fragments; 8) large-scale juxtaposition of fragments, possibly complicated by dewatering of the pile; 9) variable geometries, sizes, and attitudes of fragments; and 10) KOV's unique structural setting in the foreland of the main arc.

This study concludes that gravity-driven mass-transport processes, concomitant with the development of a polygenetic sedimentary mélange, occurred within a foreland basin, a portion of which is preserved at Kolwezi, during Pan African orogenesis (ca. 550 Ma). Variable degrees of postemplacement modification of the primary sedimentary mélange resulted in predominantly brittle folding, low-grade shearing of fragments during continued approximately N-S-directed orogenesis, tightening of the "Kolwezi Klippe" and impingement of fragments against one another accompanied by dewatering of the pile and remobilization of metals.

Figure – Rounded, slumped fragment at the base of the Ore Zone within R.A.T, Lilas, indicating unconsolidated nature of the underlying host R.A.T. sequence at the time of incursion or arrival of the overlying fragments, resulting in soft-sediment deformation.

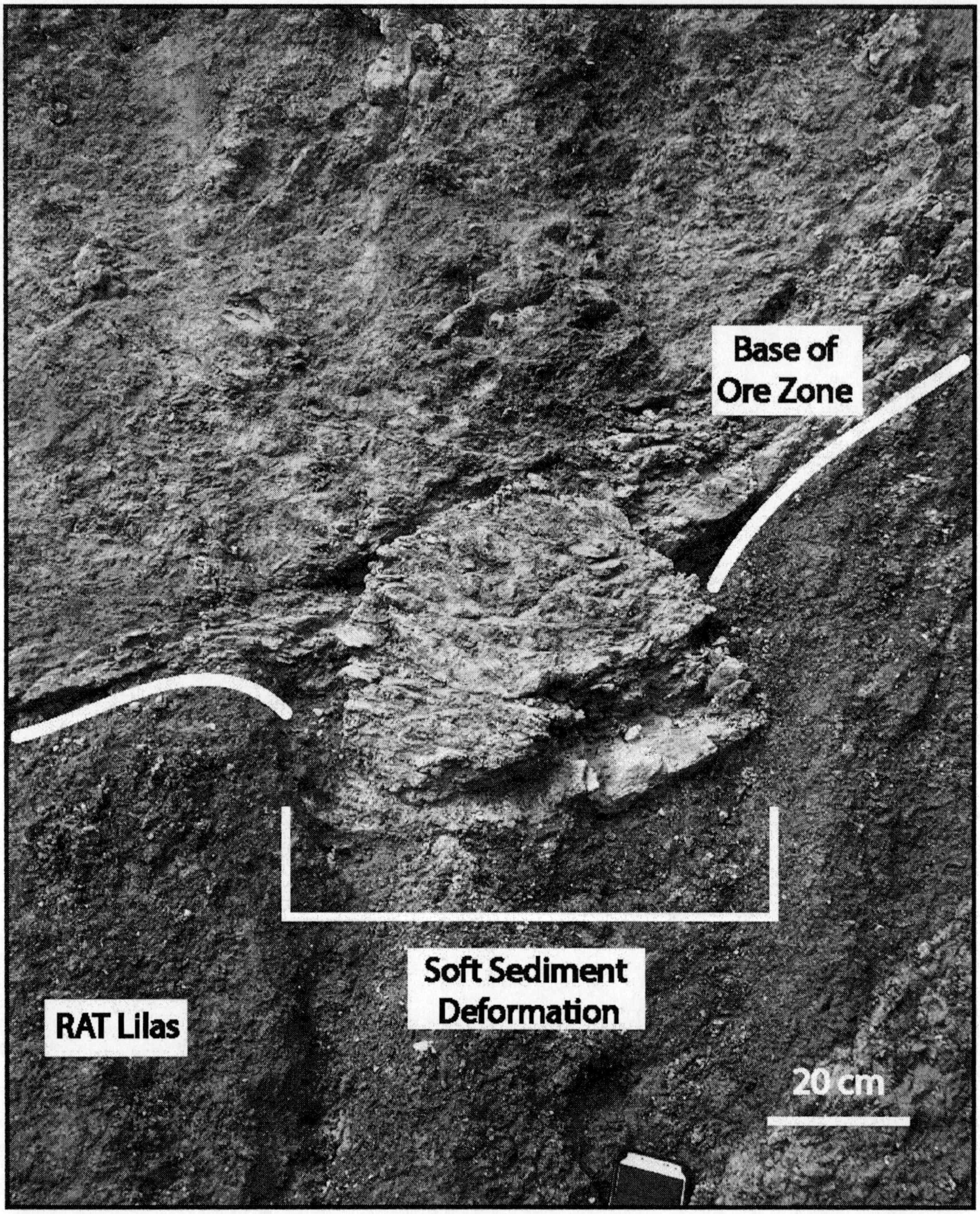
Base of
Ore Zone
Soft Sediment
Deformation
RAT Lilas
20 cm

This page intentionally left blank.

ST.202

Geochemical and Mineralogical Variation of Gold Mineralization Across the Dakotan Tectonic Zone, Black Hills, South Dakota – A Crustal Continuum of Orogenic Gold Deposits

Eric Nowariak[2], Robert Bergmann[1, 2], Brian Lentz[1, 2], Chris Nicosia[2]
1. F3 Gold, Minneapolis, MN, USA, 2. Big Rock Exploration, Minneapolis, MN, USA

The Black Hills of South Dakota hosts the world-class 60 Moz Homestake Mine (the type locality for iron formation-hosted gold deposits). Recent recognition of a brittle-ductile shear zone system across the entire Black Hills, the Dakotan Tectonic Zone (DTZ), has provided an updated framework for exploring shear-hosted orogenic gold mineralization in the region. Orogenic gold deposits in this Paleoproterozoic volcano-sedimentary basin resulted from complex polyphase metamorphism, deformation, hydrothermal fluid flow, and granitic magmatism during arc accretion and subsequent suturing of the Superior and Wyoming Provinces.

Gold occurrences across the greater Black Hills are found proximal to mapped DTZ structures in varied geodynamic settings, within varied host lithologies subject to metamorphic conditions ranging from upper amphibolite to subgreenschist facies, each with distinct geochemical associations. The occurrences and deposits in the eastern and central Black Hills are characterized by host rocks including siliceous metagraywackes, carbonaceous phyllites, and iron formations with sericite-graphite-tourmaline+/-chlorite alteration and an intimate Au-As(+/-Ag, Sb, Pb, Zn) association, with a 0.56 log-scale correlation coefficient for samples within a 500-km^2 area. Using the As-in-arsenopyrite geothermometer, mineralization temperature estimates from the Homelode Trend, Homestake Mine, and the Keystone Trend range from <290°-470°C. Though all ore depositional textures indicate a postpeak metamorphic timing, relative differences in estimated mineralization temperatures correlate well to differences in peak metamorphic and textural characteristics observable in the field and thin sections. In contrast, in the southwestern Black Hills, gold mineralization is hosted in shear zones within upper amphibolite facies alkalic mafic rocks and pelitic schists, with strong correlations between Au-Bi-Te-V and pervasive tourmaline(schorl-dravite)-graphite-sillimanite alteration. Sillimanite alteration, along with mineralogical textures within mineralized quartz veins, may be attributable to immiscible Bi melts and/or carbonic fluids, suggesting mineralization temperatures above 500°C. The postmetamorphic timing of host shear zones, tourmaline alteration, and high-temperature estimates for mineralization are consistent with mineralization driven by the intrusion of Harney Peak Granite, which is also responsible for nearby tourmaline-rich pegmatite dikes, satellite intrusions, and metasomatic tourmalinites. Together, these gold deposits and occurrences represent orogenic gold subtypes including epizonal, mezozonal, and hypozonal deposits and provide an example of multiple gold mineralization styles emplaced at distinct crustal levels in diverse host rocks within a single district—a departure from the iron formation dominated model previously ascribed to the area.

ST.204

Progressive Shearing and Its Control on the Cavanacaw Gold Vein Deposit, Northern Ireland: Interpreting a Complex Vein System and Implications for Regional Metallogeny

James I. Shaw[1], Taija Torvela[1], Mark R. Cooper[2], Graham A. Leslie[3], Robert J. Chapman[1]
1. University of Leeds, Leeds, United Kingdom, 2. Geological Survey of Northern Ireland, Belfast, United Kingdom, 3. British Geological Survey, Edinburgh, United Kingdom

Gold-quartz vein systems that propagate within progressive shear zones can exhibit complex geometric and kinematic frameworks. This can easily lead to the development of overcomplicated structural models and/or misinterpretations regarding when these vein deposits formed in respect to the tectonic and metallogenic evolution of an area. In this presentation, we present a case study where this has been the case: the Cavanacaw gold-silver-lead vein deposit (~0.5 Moz Au) in Northern Ireland (Fig. 1) (Galantas Gold Corp, 2014). Past studies have interpreted Cavanacaw to have had a long-lived geologic history: the broadly N-S-striking sinistral vein system is inferred to have initiated during Grampian compression (ca. 470-465 Ma) and to have mineralised though the injection of several fluid pulses during the Palaeozoic (Siluro-Devonian and Carboniferous) (Cliff and Wolfenden, 1989; Earls et al., 1996; Parnell et al., 2000). These studies further document broad similarities in paragenesis and fluid chemistry to infer a temporal link between Cavanacaw and the nearby Curraghinalt deposit (>6 Moz Au) that is located ca. 27 km to the NE (Dalradian Resources Inc., 2018). Here, we present a new genetic model for the Cavanacaw deposit based on geologic observations and the detailed structural analysis of historic trench maps. This model is notably simpler than what has previously been proposed.

We demonstrate that the subvertical vein system at Cavanacaw strikes broadly N to NW and propagated within an overall ENE-WSW-striking sinistral shear zone system. The veins initiated as NNE-SSW-striking dilational veinlets and were rotated anticlockwise to their current N to NW strike orientations through progressive shearing. We show that as the veins were rotated through the normal of the shear zone they buckled to have the opposite sense of vergence to the overall shear direction. In areas where the veins were rotated enough to strike perpendicular to the overarching maximum principal stress, the veins clearly accommodated late internal thrusting. As orogen-parallel, ENE-WNW-striking sinistral shearing is widely documented to have occurred across NW Ireland during the late-Scandian event of the Caledonian orogeny (430-400 Ma), we interpret the Cavanacaw deposit to be Siluro-Devonian in age (Dewey and Stratchan, 2003).

Despite the interpretations of previous studies (e.g., Earls et al., 1996; Parnell et al., 2000), the structural model presented for Cavanacaw demonstrates that the deposit is younger than the nearby Curraghinalt deposit, which is now respectively constrained to a postorogenic "collapse" episode following the Grampian event (ca. 462.7-452.8 Ma) (Rice et al., 2016; Shaw et al., in prep). Instead, we highlight that the vein system at Cavanacaw propagated under a common palaeostress field to the Cononish, Croagh Patrick, and Clontibret gold vein deposits that are located further afield in the Laurentian Caledonides of Ireland and Scotland.

Fig. 1. Deposit-scale geologic map of the Cavanacaw deposit.

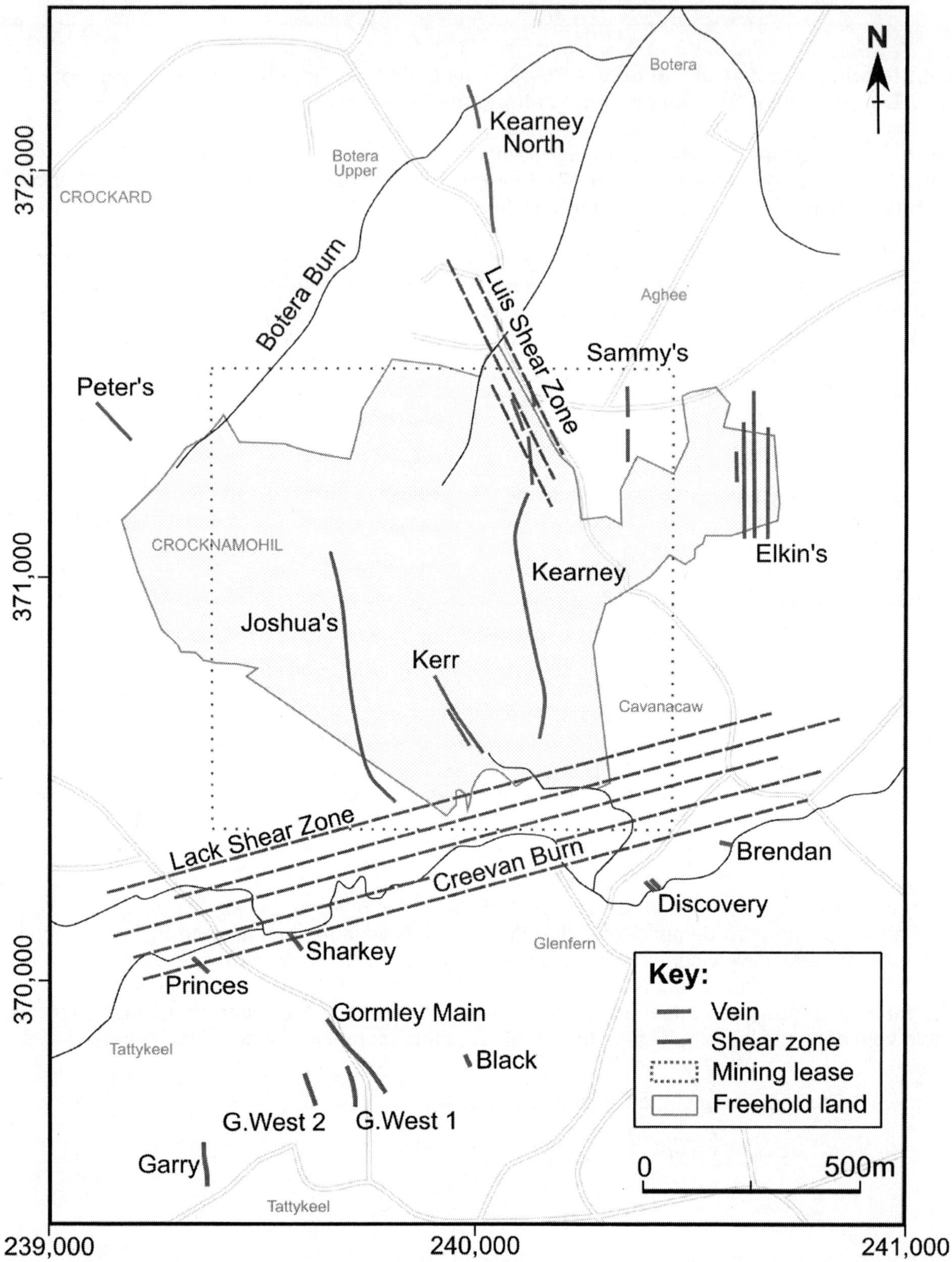
N
Botera
Kearney North
Botera Upper
CROCKARD
Botera Burn
Luis Shear Zone
Aghee
Sammy's
Peter's
CROCKNAMOHIL
Elkin's
Kearney
Joshua's
Kerr
Cavanacaw
Lack Shear Zone
Creevan Burn
Brendan
Discovery
Sharkey
Glenfern
Princes
Gormley Main
Black
Tattykeel
G.West 2
G.West 1
Garry
Tattykeel
Key:
Vein
Shear zone
Mining lease
Freehold land
0
500m
372,000
371,000
370,000
239,000
240,000
241,000

ST.205
Regional Structural Controls on the Evolution of Low-Sulfidation Epithermal Au-Ag Mineralization in Sindirgi, Western Turkey

Ahmet K. Sener[1], Berkin Uğurlu[2], Izak van Coller[1], Erdin Bozkurt[3]
1. Ariana Resources plc, London, United Kingdom, 2. Galata Madencilik San. ve Tic. Ltd., Ankara, Turkey, 3. Middle East Technical University, Ankara, Turkey

Western Turkey hosts several low-sulphidation epithermal gold-silver deposits of Miocene age, several having been mined since the early 2000s. In the Sindirgi District, Balikesir Province, a series of deposits occur along a NE-oriented trend referred to as the Sindirgi Gold Corridor (SGC). This corridor terminates against the Simav Fault to the SW and contains four principle epithermal quartz vein fields constituting a total vein length of 51.2 km, two of which are examined here. The veins show various orientations and represent hydrothermal deposition at a variety of different crustal depths.

These systems are interpreted to have been generated under a N-S extensional regime approximately coeval with, or just postdating, volcanic activity. The deposits are primarily hosted by rocks of the Sindirgi Volcanic Complex, which comprises a thick sequence of dacitic and rhyodacitic-rhyolitic welded pyroclastics and felsic intrusions, which occur as domes and dikes. These volcanic rocks overlie sedimentary and ophiolitic rocks of dominantly Upper Cretaceous age.

Two large-scale regional structures, the Bigadic and Karabahce faults, trend WNW and appear to dissect the SGC. These structures probably represent fragmented splays of the Simav Fault, a major fault zone that defines the half-graben to the north. Elsewhere within the corridor, other NNE- and NE-trending structures occur as prominent faults. The orientation of veins varies from NW-SE, N-S, NE-SW, to E-W, in order of decreasing economic importance.

The mineralization shows various morphological and textural features suggesting deposition through several phases of hydrothermal activity, controlled by the development of several key structures and the relative position of the palaeowater table. The >0.4 Moz (Au equivalent) Kiziltepe Mine is located at the SW end of the SGC (lat.39°16'45"N, long.28°15'00"E) and contains four dominantly NW-trending vein zones. These zones are characterized by massive and multiple subparallel quartz veins, including stockworks and breccias in places. Weakly banded crustiform-colloform (with ginguro) vein textures are associated with the highest gold grades.

At the opposite end of the SGC, 23 km NE of Kiziltepe, lies the <0.1 Moz Kizilcukur deposit (lat.39°23'05"N, long.28°27'51"E). Like Kiziltepe, the veins trend largely NW, but are hosted by an Upper Cretaceous basalt unit, exposed to the north of the regional WNW-trending Karabahce Fault. Gold mineralisation is associated with manganese carbonates and base metal concentrations an order of magnitude higher than those at Kiziltepe. The veins display distinct banding and are typically more crystalline, with well-developed dog-tooth textures, suggesting a deeper level of deposition than at Kiziltepe.

Three decades of mineral exploration data comprising geologic, geochemical, and geophysical information are utilized to improve interpretation of the regional and local structural controls on the distribution of mineralization. Particular focus is applied to the kinematic relationships between regional and local structures, the definition of pre- and synmineralization structures, and the confirmation of structures significant to the control of hydrothermal fluid flow. This will establish links between the orientation, kinematics, and relative age of key structures and the evolution of the epithermal deposits in the region and, consequently, will aid the development of predictive models for future mineral exploration.

ST.206

Geologic Setting and Timing of Mineralization at the World-Class Banded Iron Formation-Hosted Amaruq Gold Deposit, Churchill Province, Nunavut, Canada

Manon A. Valette[1], Stéphane De Souza[1], Patrick Mercier-Langevin[2], Olivier Côté-Mantha[3], Marjorie Simard[4], Natasha Wodicka[5], Vicki McNicoll[5], Patrice Barbe[3]
1. Université du Québec à Montréal, Montréal, QC, Canada, 2. Geological Survey of Canada, Quebec, QC, Canada, 3. Agnico Eagle MInes Ltd., Val-d'Or, QC, Canada, 4. Agnico Eagle Mines Ltd., Rouyn-Noranda, QC, Canada, 5. Geological Survey of Canada, Ottawa, ON, Canada

Precambrian gold deposits associated with Algoma-type banded iron formations (BIF) are complex systems in which lithological and structural traps are the principal controls on the location and geometry of ore zones. One such deposit is the 6.1 Moz Amaruq gold deposit, which is part of the polydeformed and metamorphosed Neoarchean (ca. 2.73–2.63 Ga) Woodburn Lake greenstone belt, within the Rae craton of the western Churchill province, Canada. The host rocks at Amaruq consist of a moderately to steeply dipping, 250-m-thick tectonostratigraphic sequence comprising BIF interlayered with greywacke, siltstone, black shale, and mafic-ultramafic magmatic rocks, which have been affected by Archean and Paleoproterozoic orogens. The ore zones at Amaruq are predominantly controlled by pre-ore tight and isoclinal F_1 folds and D_1 shear zones that are overprinted by syn-ore upright and isoclinal northeast-trending and plunging F_2 folds and associated D_2 shear zones. Progressive strain accommodation during D_2 led to the development of northwest-verging thrusts and recumbent F_2 folds. Subsequent deformation consists of open to chevron-style, northeast- and southwest-plunging F_3 folds, and north-south and east-west D_4 and D_5 normal faults and shear zones, respectively. The Amaruq gold deposit comprises the Whale Tail and IVR zones, which are characterized by contrasting styles of mineralization and associated metamorphosed calcic and potassic hydrothermal alteration halos. The IVR and Whale Tail zones are located in the hinge zone and along the limb of a deposit-scale F_2 fold, respectively. The mineralized zones all contain pyrrhotite and calcite and consist of 1) stratabound (Ca-Fe)-amphibole-biotite-epidote-garnet±arsenopyrite-gersdorffite replacement-style zones in chert-poor BIF layers associated with ultramafic rocks; 2) (Ca-Fe)-amphibole-stilpnomelane-biotite-chlorite-arsenopyrite±löllingite-bearing zones of diffuse veining ("silica-flooding") in chert-rich BIF units interlayered with black shale and siltstone beds; and 3) shear and extensional quartz±calcite veins with (Ca-Fe)-amphibole and/or sericite/biotite in high-strain zones close to sheared and/or folded contacts between volcanic and sedimentary rocks. The first and second styles of mineralization are more common in the Whale Tail zone, whereas the third style is better developed in the IVR zone. The metamorphic paragenesis of silicate (e.g., amphiboles and micas) and sulphide minerals (e.g., arsenopyrite and löllingite) provide evidence for pre- to early-D_2 mineralization and syn-D_2 peak metamorphism. Rheological contrasts between BIF, sedimentary, and mafic-ultramafic rocks contributed to strain partitioning, which channeled ore-forming fluid flow in structurally and chemically favorable lithologies such as BIF and chert. Previous research in the western Churchill province indicated that gold was introduced during the Trans-Hudson orogen (ca. 1.91–1.81 Ga). However, a series of Re-Os model ages as old as 2.3 Ga were obtained at the Meliadine deposit, and as old as 2.6 Ga at Amaruq (this study), thus suggesting a multiphased hydrothermal history. The timing of the gold mineralization and the possibility of an Archean gold heritage represent key issues that are being addressed through this study to help in the development of improved geologic and exploration models for orogenic gold deposits in the Churchill province and other Archean terranes.

ST.209
Predicting Clusters of Gold Mineralization Outside of Resource Model Domains Using Assay Data of Varying Density and Machine Learning Techniques

Farzi Yusufali[1], Ilia Sucholutsky[1, 2]
1. Stratum, Toronto, ON, Canada, 2. University of Toronto, Toronto, ON, Canada

Predicting clusters of mineralization and optimally placed drilling targets for in-mine exploration is a valuable strategy to expand an asset's reserve for reporting and planning purposes. In this paper, we propose a machine learning (ML)-based protocol that accepts drill hole data and chip data to produce extrapolated grade predictions for blocks outside the boundaries of resource model domains. We train a neural network using the assayed data from the mine site to predict clusters of gold mineralization outside of the data domain (i.e., make grade predictions for blocks with no assayed samples and outside the variography boundaries). We demonstrate the proposed method's ability to find clusters of mineralization while reducing the noise from the nugget effect from live asset gold-grade data. While chip data bias and drill hole density introduces error to ML's prediction, the results of the temporal benchmarking of data and live test with short-range drill targets to confirm the prediction clearly show the ML's robustness against varying data densities and generalizability given the lack of overfitting (Fig. 1). It is notable that large clusters of mineralization were verified using optimally placed drill holes constrained by the mine's underground infrastructure and existing stopes.

Fig. 1. Clusters of Au mineralization accessible to existing infrastructure where clusters in planned development are corroborated and unique clusters are verified using drill targets.

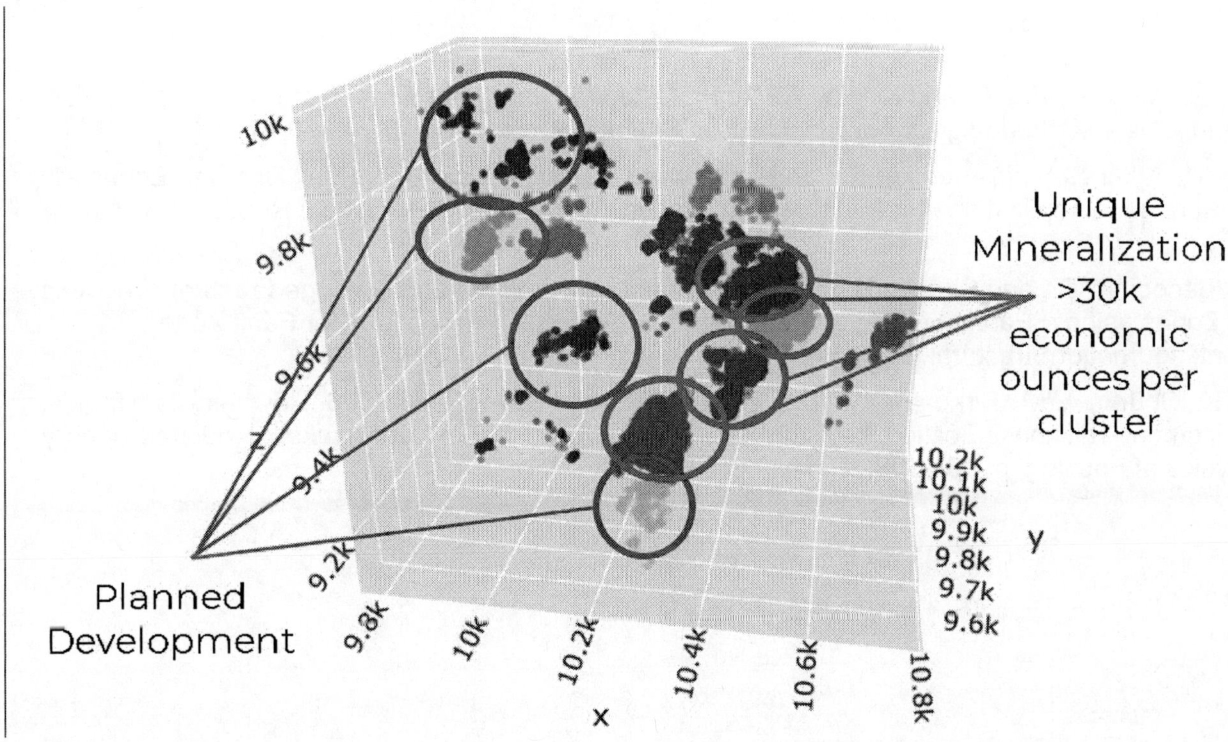

ST.210
Fracture Density and Damage Zone Thickness Associated with Faults at the Resolution Copper Porphyry Deposit

Zacharie A. Zens, Jennifer Evans
Resolution Copper, Superior, AZ, USA

Faults are heterogeneous and anisotropic structures that generally consist of A) a fault core where most displacement is accommodated; B) one or more principal slip surface(s) within the fault core; and C) a surrounding fracture zone, which develops during displacement. Fracture density and damage zone thickness is variable and is at least partially controlled by host-rock lithology and depth (Hara et al., 2017). Fracture density is anomalously high in fault and damage zones and decays as a function of distance from the fault core to background levels (Faulkner et al., 2011) (Fig. 1A, B).

Fracture density and damage zones associated with faulting in unaltered sedimentary rocks is fairly well understood (e.g., Choi et al., 2015; Shipton et al., 2006), but is poorly studied in faults associated with metallic ore deposits. Here, we characterize over 600 fault and damage zones from drill core at the Resolution Copper Porphyry Deposit to better understand the primary controls on fracture density and damage zone thickness.

Fracture zones were identified by enhanced fracture density relative to background fracturing and may correspond to acoustic differences or mineralogical changes (Fig. 1C, D). Fracture zone boundaries were defined as the intersection point between two different gradients of cumulative fracture counts (Fig. 1B). The true stratigraphic thickness of damage zones was calculated from the average orientation of fault breccias within the damage zone.

References:

Choi, J. H., Edwards, P., Ko, K., and Kim, Y. S., 2015, Definition and Classification of Fault Damage Zones: A Review and a New Methodological Approach: Earth Science Reviews, v. 152, p. 70-87.

Faulkner, D. R., Mitchell, T. M., Jensen, E., and Cembrano, J., 2011, Scaling of Fault Damage Zones with Displacement and the Implications for Fault Growth Process: Journal of Geophysical Research, v. 116, p. 1-11.

Hara, A. P., Jacobi, R. D., and Sheets, H. D., 2017, Predicting the Width and Average Fracture Frequency of Damage Zones using a Partial Least Squares Statistical Analysis: Implications for Fault Zone Development: Journal of Structural Geology, v. 98, p. 38-52.

Shipton, Z. K., Soden, A. M., Kirkpatrick, J. D., Bright, A. M., and Lunn, R. J., 2006, How thick is a fault? Fault Displacement-Thickness Scaling Revisited. In Abercrombie, R. (Ed) Earthquakes: Radiated Energy and the Physics of Faulting, p. 193-198.

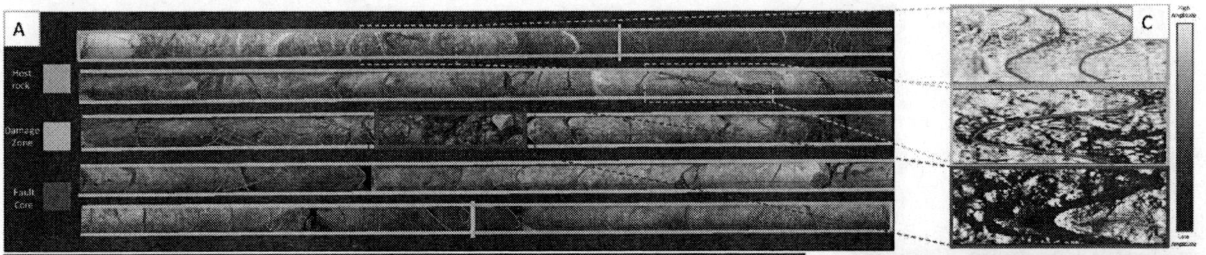

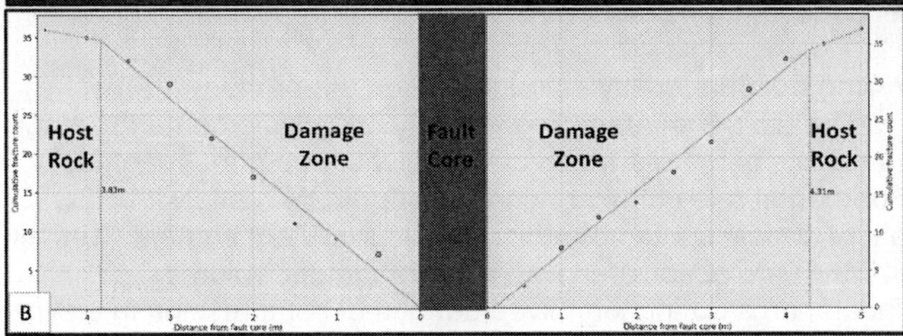

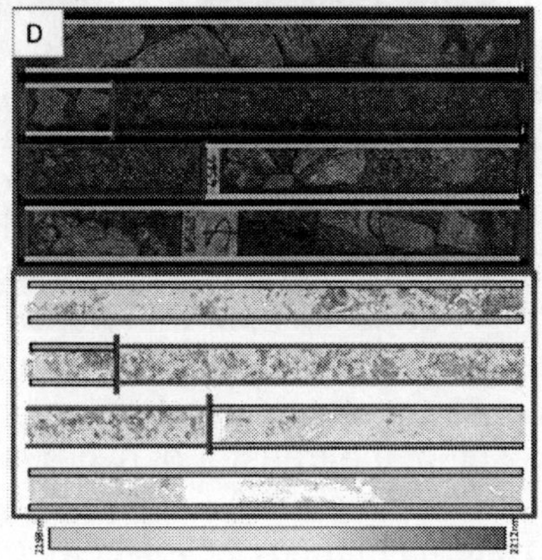

Figure 1. An example of the methodology and datasets used in this study to characterize fault and damage zones. A) Fault zone consisting of highly fractured damage zone and a brecciated fault core with 1-5cm diabase clasts in a fault gouge matrix. The host rock is locally foliated diabase with potassic (biotite and rutile after pyroxene and amphibole) and phyllic (white mica and kaolinite after plagioclase; quartz-sericite-pyrite±chlorite veins) alteration. Fractures are outlined in red. B) Bivariate plot of cumulative fracture counts against distance from the fault core. The damage zone boundary is defined by the intersection point between the two slope gradients. C) Acoustic borehole image of the undamaged host rock, damage zone, and the fault core. D) Half core from the fault in A with hyperspectral imagery from CoreScan showing the wavelength position of the 2200nm absorption feature. Higher wavelengths in the fault core suggest more Fe^{+2} and Mg^{+2} substituting in the octahedral site within micas.

R1
Magmatic controls on metal endowments of porphyry Cu-Au deposits

Massimo Chiaradia
Department of Earth Sciences - University of Geneva, Geneva, Switzerland

In this review I illustrate the processes controlling magma fertility in the formation of porphyry Cu-Au deposits. By magma fertility I mean here all magmatic parameters (e.g., metal and volatile contents, magma and fluid volumes) that may result in higher amounts of metals which are exsolvable from the magma. Mantle source processes seem to play a fundamental role in the enrichment of primary melts with H2O, S and Cl, all essential ingredients to form porphyry deposits, but no particular role in Cu enrichment.Cu-rich porphyry Cu-Au deposits (i.e., with Au/Cu ~4*10-6) are associated with large magmatic volumes accumulated in the lower thick crust of continental arcs during long-lived periods of compression in a syn-subduction environment. Mineralization occurs after such accumulations have reached significant volumes and is the result of the transfer of hydrous magmas from deep to shallower crustal levels, probably favored by tectonic stress changes.Au-rich porphyry Cu-Au deposits (i.e., with Au/Cu ~80*10-6) are associated with magmatic systems that have evolved at overall shallower crustal levels and for this reason can be found in geodynamic settings characterized by thinner crust (e.g., island arcs with intermediate crust thickness) and/or in variably extensional settings occurring above a slab-metasomatized mantle (post-subduction setting, extensional syn-subduction setting). The six largest Au-rich porphyry Cu-Au deposits (>~1300 tons Au) are associated with variably alkaline magmas, that are typical of post-subduction and/or extensional settings, suggesting a petrogenetic control on the Au-rich nature of these deposits.

R2

Mesozoic to Cenozoic Metallogenic and Magmatic Evolution of the Lesser Caucasus and the Eastern Pontides

Robert Moritz[1], Marc Hässig[1], Marion Grosjean[1], Samvel Hovakimyan[1], Hervé Rezeau[3], François Turlin[1], Şafak U. Sönmez[1], Okan Delibas[2]
1. University of Geneva, Geneva, Switzerland, 2. Hacettepe University, Ankara, Turkey, 3. Massachusetts Institute of Technology, Cambridge, MA, USA

The Eastern Pontides and the Lesser Caucasus are two adjacent mountain ranges that belong to a continuous metallogenic belt (Fig. 1) and that are particularly well endowed in copper, gold, and molybdenum. The metallogenic and magmatic evolution occurred in successive and distinct steps, as the geodynamic setting evolved from subduction to postcollision.

During the Jurassic and Early Cretaceous, the Eastern Pontides and the Somkheto-Karabagh belt were the active margins along the southern Eurasian plate during subduction of the northern Neotethys (Fig. 1). A calc-alkaline magmatic arc and porphyry Cu and intermediate- to high-sulfidation epithermal deposits were mainly emplaced in the Somkheto-Karabagh belt, in Azerbaijan and Armenia, during arc maturation and thickening, but they are also recognized in the Eastern Pontides. This is evidence that during the Early Cretaceous, both mountain ranges were part of the same continuous metallogenic and magmatic belt.

The Jurassic to Early Cretaceous evolution was followed by a metallogenic and magmatic lull during ~40 m.y., between ~130 and ~90 Ma. This lull coincides with the opening of the eastern Black Sea and Transcaucasian basins and was possibly related to submarine plateau accretion along the southern Eurasian margin, which blocked the subduction of the northern Neotethys.

Subduction of the northern Neotethys resumed during the Late Cretaceous resulting in abundant Turonian to early Campanian mafic and silicic magmatism along different segments of the Turkish Eastern Pontides and in the Georgian Bolnisi district, northern Lesser Caucasus (Fig. 1). This magmatic evolution during arc rifting was accompanied by the formation of volcanogenic massive sulfide and epithermal ore deposits. During the waning stages of the northern Neotethys subduction, porphyry deposits were emplaced at ~77-76 Ma in the Eastern Pontides, coeval with high-K calc-alkaline to shoshonitic magmatism.

The northern Neotethys was closed during the Late Cretaceous to Paleocene, as a consequence of the accretion of the Gondwana-derived Tauride-Anatolide platform and South Armenian block with the Eurasian margin (Fig. 1). During postcollisional evolution, early Eocene adakite-like magmatism affected the Eastern Pontides and the Georgian Bolnisi district, which was accompanied by postcollisional porphyry mineralization in the Eastern Pontides.

After closure of the northern Neotethys, the active subduction jumped to the south of the Tauride-Anatolide platform and the South Armenian block. Middle Eocene calc-alkaline magmatism related to the subduction of the southern Neotethys resulted in porphyry Cu-Mo and epithermal deposits mainly in the South Armenian block, where it was controlled by dextral strike-slip tectonics. The middle Eocene magmatic and ore-formation stage, related to subduction of the southern Neotethys, is also recorded in the Eastern Pontides.

After collision of the Arabian plate with the Eurasian margin, the final evolution resulted in late Eocene to Oligocene high-K calc-alkaline to shoshonitic magmatism and late Oligocene to Miocene adakite-like magmatism, accompanied by postcollisional porphyry Cu-Mo and epithermal deposits within the South Armenian block (Fig. 1).

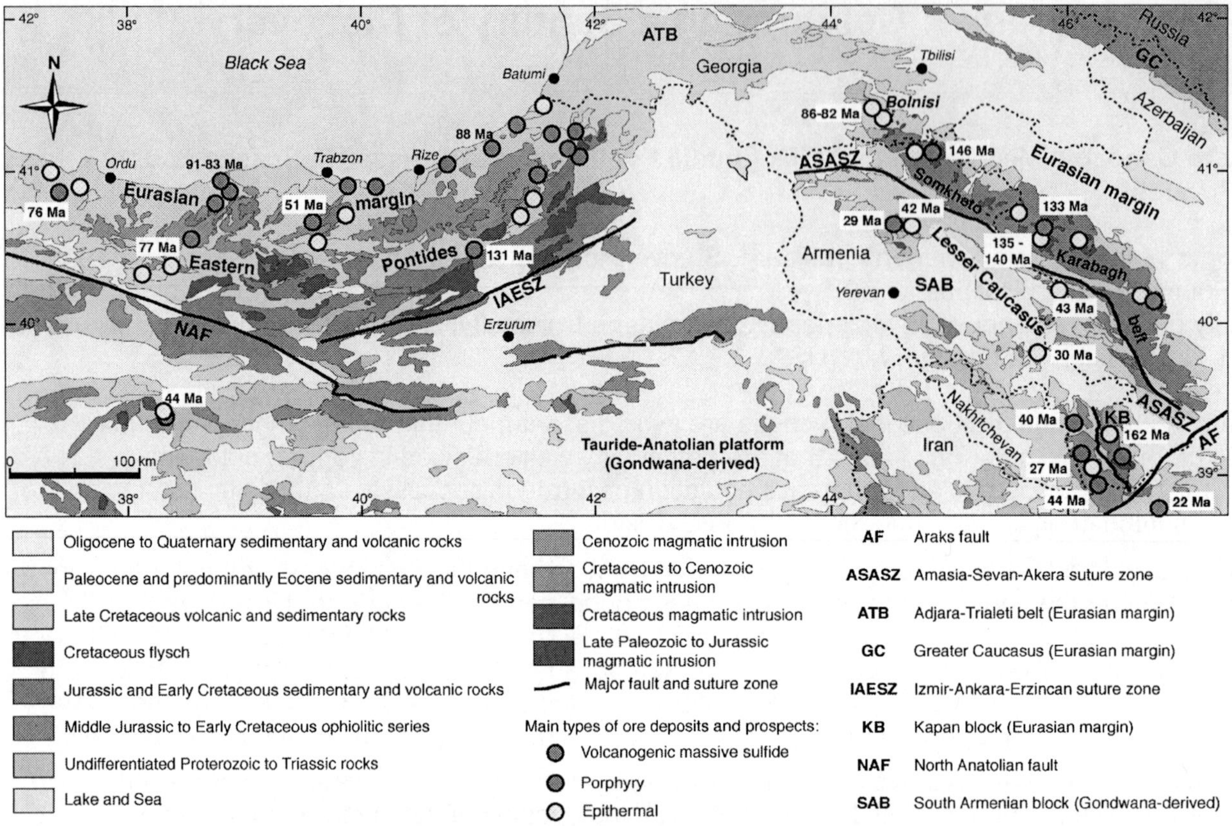

Figure 1 – Geology of the Eastern Pontides, the Lesser Caucasus, and adjacent tectonic zones, with ages and locations of major ore deposits and prospects.

R3

Regional Magmatic Evolution and Metal Fertility of Igneous Rocks from the Kerman Porphyry Belt, SE Iran: Insights from Whole-Rock and Zircon Geochemistry and Geochronology

Ali Sholeh[1], Jeremy P. Richards[2], David R. Cooke[3], William Powell[4]
1. Department of Geology, Faculty of Basic Sciences, Tarbiat Modares University, Tehran, Iran, 2. Mineral Exploration Research Centre, Harquail School of Earth Sciences, Laurentian University, Sudbury, ON, Canada, 3. ARC Research Hub for Transforming the Mining Value Chain and CODES, Centre for Ore Deposit and Earth Sciences, University of Tasmania, Hobart, TAS, Australia, 4. Rio Tinto, Bundoora , Melbourne, VIC, Australia

The center of the Mesozoic-Cenozoic Neotethyan orogenic belt is located in Iran, where the porphyry-mineralized Urumieh Dokhtar Magmatic Arc (UDMA) extends from Urumieh in the NW to Dokhtar in the SE. The Kerman Porphyry Belt (KPB) is a well-explored and studied part of the Neotethyan orogeny and is located to the southwest of the UDMA. In the KPB, the main period of porphyry mineralization was early Middle Miocene, synchronous with the terminal collision of the Afro-Arabian and Eurasian plates. The KPB hosts several world-class porphyry Cu deposits, mostly associated with postcollisional calc-alkaline magmas including Sar Cheshmeh (13.6 ± 0.1 Ma; 1.7 Gt at 0.65% Cu, 0.03% Mo, 0.06 g/t Au) and Meiduk (12.5 ± 0.1 Ma; 500 Mt at 0.86% Cu, 0.01% Mo, 0.08 g/t Au), together with several smaller/less explored deposits such as Takht Gonbad (~26 Ma; ~60 Mt at 0.57% Cu).

In this study, we focused on the whole-rock and zircon geochemistry and geochronology of fertile and infertile igneous rocks that crop out in the region between Meiduk, Sar Cheshmeh, and Takht Gonabd in the KPB. Forty-six samples of fresh and altered volcanic and intrusive rocks were collected from surface outcrops and from diamond drill core at Takht Gonbad and Meiduk for whole-rock and zircon geochemistry analysis and geochronology. The samples were classified into nine different groups including Takht Gonabd porphyry (group 1), Takht Gonbad regional (group 2), Kuh-e-Panj (group 3), Sar Cheshmeh porphyry (group 4), Sar Cheshmeh regional (group 5), Meiduk porphyry (group 6), Meiduk regional (group 7), Rafsanjan 100k geologic map igneous rocks (group 8), and Neogene igneous suites (group 9).

The ages of zircons range from 42.6 ± 0.33 Ma (Dacite Lava) to 1.53 ± 0.17 Ma (andesite flow) from the Rafsanjan 100k geologic map area. Whole-rock assays have been combined with zircon chemistry to determine variations through time and differences between fertile and infertile magmatic suites by means of zircon and whole-rock REE trends, redox-sensitive Eu, and Ce enrichment/depletion. These data can be used to predict variations in magmatic water contents and magmatic oxygen fugacity as a proxy to porphyry fertility of the magmatic suites. The results showed maturation in arc magmatism from ~20 Ma in the KPB, which led to more evolved magmatism at later stages of arc history. The Pleistocene (1.6 ± 0.11 Ma) quartz-biotite-plagioclase-hornblende dacite porphyry flow dome near Sar Cheshmeh has the highest Sr/Y and La/Yb ratios, indicating early hornblende fractionation and higher water contents. These are the signatures of fertile magmas capable of producing porphyry mineralization.

Fig. 1. Total alkali silica classification diagram (Le Maitre, 1989) showing nine igneous suites from the Kerman Porphyry Belt.

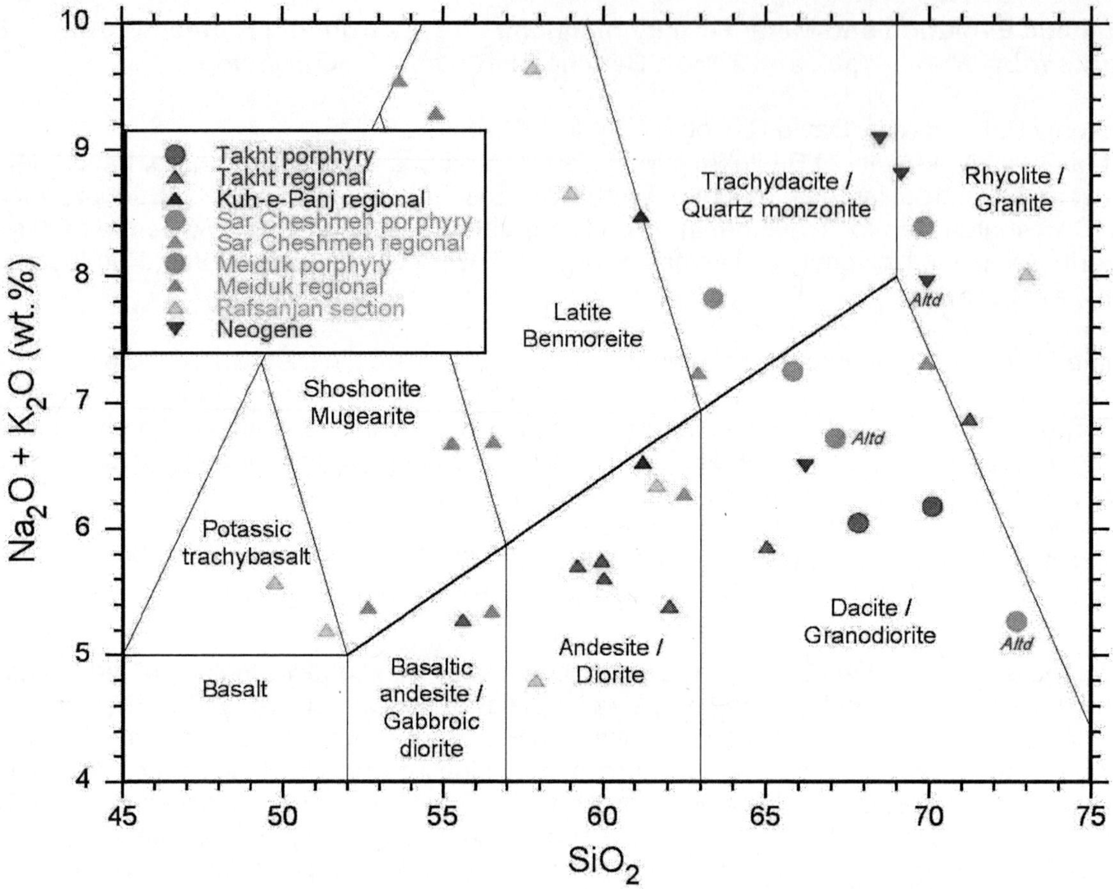

R4
Using Trace Elements in Zircon to Recognize Variability in Porphyry Magma Fertility Signatures Within and Between Arcs and Arc Settings in British Columbia, Arizona, and Chile

Taylor J. Ledoux, Craig J. Hart, Robert G. Lee
MDRU - Mineral Deposit Research Unit, Earth, Ocean and Atmospheric Sciences, University of British Columbia, Vancouver, BC, Canada

Porphyry copper deposit formation depends on magmatic characteristics related to its oxidation state, temperature, metal, water, chlorine, and sulphur content. Trace elements in zircon (TEZ) signatures and zircon grain morphology are robust indicators of many of these attributes. Zircon composition can provide porphyry-fertile signatures and identify magmatic processes that favor economic porphyry mineralization, allowing for focused exploration targeting. However, the scale of variability of fertility indicators within an evolving arc, along an arc segment, or across a migrating arc has not been well established. Similarly, fundamental questions such as, "What is the variability of TEZ fertility characteristics between arcs or arc segments with different basement types or crustal thicknesses?" are uncertain. What are their links to the petrogenetic, magmatic, and tectonic processes that control the magma formation?

Identifying fertile magmatic signatures and interpreting favourable magmatic processes requires that they can be resolved and differentiated from background arc signatures. Furthermore, the effective application of TEZ signatures as an exploration tool requires the evaluation of large areas with few samples. We utilize fluvial sand and glacial till samples to evaluate entire drainage basins or up-ice regions, compared to bedrock samples from economic terrain. Such materials have sampled numerous rock units and can provide an array of zircon populations that can provide a range of TEZ signatures needed to evaluate a target region.

Zircons have been extracted from rock, stream, and glacial till samples from three prolific porphyry copper belts, then analyzed by LA-ICP-MS. Our results identify the variability of TEZ within arc environments as indicators of porphyry fertility for numerous magmatic bodies within each of the three regions: south-central British Columbia, east-central Arizona, and central Chile.

The Late Triassic-Early Jurassic Quesnellia island-arc terrane in southern BC has three subparallel magmatic axes that young to the east. Older (229-206 Ma), large, composite and zoned, calc-alkaline granodiorite-tonalite suite batholiths emplaced 3-7 km deep with Cu-Mo deposits transitioned (204-200 Ma) to smaller, composite, shallowly emplaced (~1 km) alkaline syenite-diorite suite plutons and intrusive complexes with Au-rich Cu-Au deposits. Subsequent more inboard Early Jurassic (202-193 Ma) large, composite and zoned, calc-alkaline to high-K granodiorite-quartz diorite suite batholiths emplaced at 3- to 5-km depths host Cu-Mo and Cu-Au deposits are the youngest arc component.

East-central Arizona hosts Laramide-aged continental-margin magmatism (~75-61 Ma) with associated porphyry copper deposits within granodioritic to granitic plutons. The magmas likely have variable sources and evolution but with a large crustal component indicated by abundant inherited zircons within the magmas.

Central Chile's late Miocene-early Pliocene Cu belt hosts giant porphyry copper deposit clusters. Cenozoic magmatism progressively evolved from an Oligo-Miocene tholeiitic arc to a more calc-alkaline affinity during the Mio-Pliocene. Porphyry copper deposits preferentially formed during intrusive and volcanic stages during the late constructive period in the waning stages of the arc.

The ability to recognize fertile arc magmas and their variability within and between different arc settings with TEZ signatures will enable explorers to distinguish magmas that have the potential to form economic porphyry Cu deposits from those that are barren.

R5
Crustal Evolution, Tectonics, and Magma Fertility of the Paleoproterozoic Alta Floresta Mineral Province, Amazonian Craton

Veronica Trevisan[1], Roberto Xavier[1], Steffen Hagemann[2, 3], Robert Loucks[2], João Motta[1], Andreas Petersson[3, 4], Anthony Kemp[2, 3], Gonzalo Henríquez[2], Luis Parra-Avila[3], Jian-Feng Gao[5], Rafael Assis[6]
1. Institute of Geosciences, University of Campinas, Campinas, SP, Brazil, 2. Centre for Exploration Targeting, The University of Western Australia, Crawley, WA, Australia, 3. School of Earth Sciences, The University of Western Australia, Crawley, WA, Australia, 4. Swedish Museum of Natural History, Stockholm, Sweden, 5. State Key Laboratory of Ore Deposit Geochemistry, Chinese Academy of Sciences, Guiyang, China, 6. Institute of Geosciences, University of São Paulo, São Paulo, SP, Brazil

The Alta Floresta Mineral Province (AFMP) constitutes a belt approximately 500 km long by 100 km wide in the southern portion of the Amazonian Craton. The province comprises numerous compositionally heterogeneous acidic intrusions, acidic to intermediate volcanic rocks, and subordinate volcaniclastic rocks and granitic gneisses. These rocks are hosts to several disseminated and vein-type intrusion-hosted Au-Cu-Mo, Cu-Mo-Au, and Au-Zn-Pb-Cu primary deposits and prospects. These mineral occurrences have been interpreted as Paleoproterozoic porphyry and epithermal-style mineral systems. Despite abundant information in the literature about the hydrothermal alteration, ore mineralogy, fluid regime, mechanisms of metals precipitation, and timing relationships between mineralization and magmatism, relatively little is known about the source and evolution of the magmatism using an integrated approach across the eastern portion of the province and at a regional scale. This situation impedes understanding of the magmatic history, establishing links to regional-scale tectonic events, and identifying potentially Cu- and Au-fertile geologic units. The main objective of this study is to constrain the crustal evolution and magma fertility potential of the eastern portion of the AFMP by reconstruction of the chronochemical evolution of the magmatism, reinterpretation of the stratigraphy and new characterization of magmatic framework and geodynamic evolution, examination of whole-rock and zircon chemical compositions to characterize Cu- and Au-ore-related magmas, and identification of geologic units potentially prospective for Cu and Au mineralization. The obtained results revealed that i) magmatism in the eastern portion of the AFMP occurred from 2037 to 1757 Ma; ii) the region evolved within an accretionary orogenic system with tectonic switching that alternates between low- and normal-angle subduction; iii) precious- and base-metal porphyry- and epithermal-style mineral systems formed from 1.83-1.75 Ga during crustal extension or relaxation; iv) the eastern portion of the province correlates with the eastward Tapajós Mineral Province and records magmatism of the Western Amazonia Igneous Belt; v) the evolution of this region and other Proterozoic terranes worldwide suggests that parts of the Columbia supercontinent operated in similar fashion to Phanerozoic Andean-type accretionary orogens; vi) the whole-rock geochemistry represents an exploration tool to assess Cu- and Au prospectivity with some limitations because of the strong hydrothermal alteration that affects the rocks across the province; vii) zircon chemical analyses can be used as an efficient exploration tool to identify geologic units potentially fertile for Cu and Au; viii) the integration of whole-rock and zircon geochemistry with geochronology information is particularly useful to track the tectonic stress regime (e.g., compressive versus noncompressive) over time in the province; ix) the zircon geochemistry suggests that the Colíder group and, to a lesser extent, the Matupá and Teles Pires suites are the geologic units prospective for Cu and Au in the eastern portion of the province due to their high water content and high oxidation state; x) the identification of potentially prospective hydrous and oxidized magmas for Cu and Au mineralization across the eastern AFMP is encouraging and can help guide exploration for Cu and Au (e.g., porphyry- and epithermal-style deposits) in the southern Amazonian Craton.

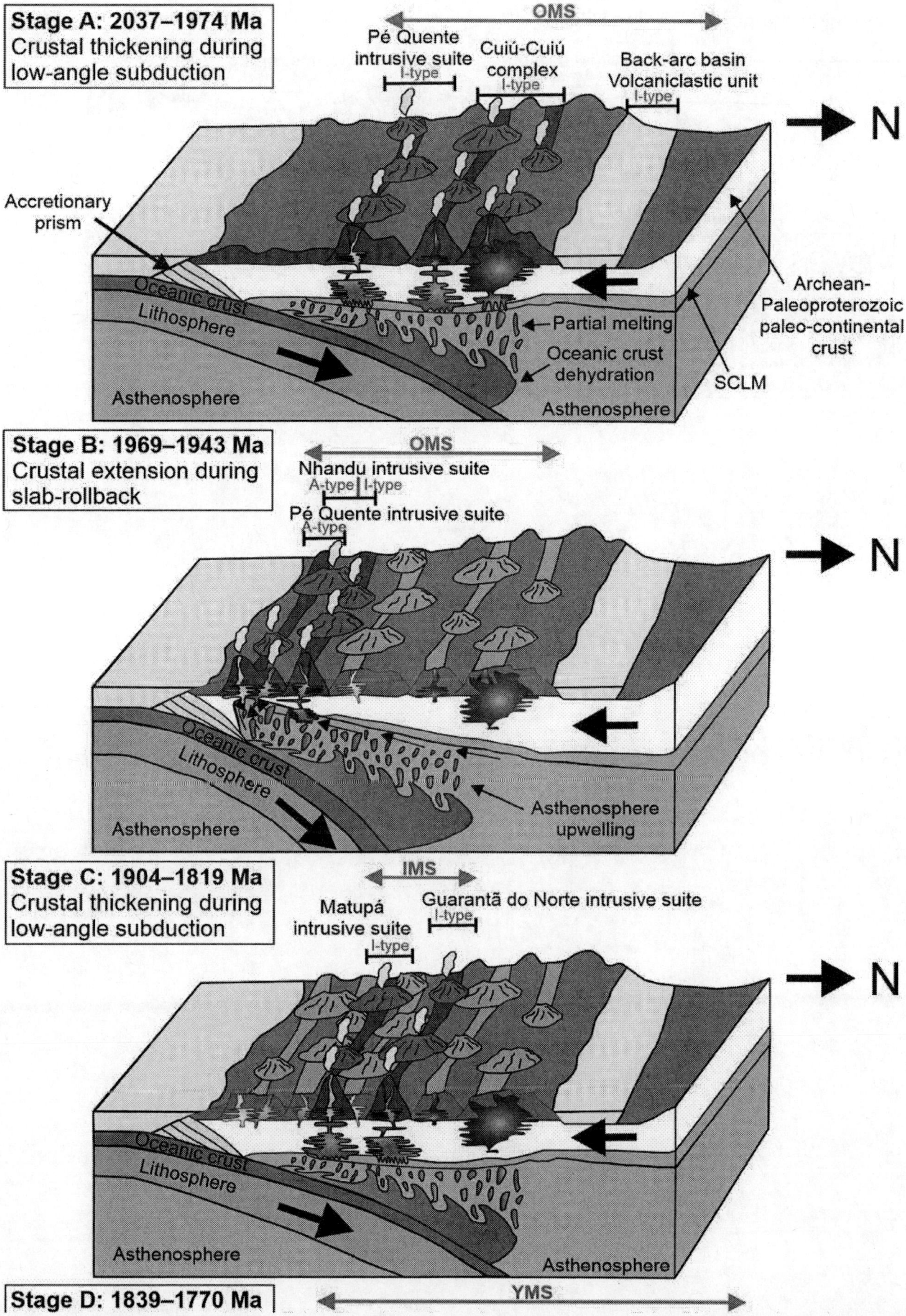

Stage A: 2037–1974 Ma
Crustal thickening during low-angle subduction
OMS
Pé Quente intrusive suite
I-type
Cuiú-Cuiú complex
I-type
Back-arc basin Volcaniclastic unit
I-type
N
Accretionary prism
Oceanic crust
Lithosphere
Partial melting
Oceanic crust dehydration
Asthenosphere
Asthenosphere
Archean-Paleoproterozoic paleo-continental crust
SCLM
Stage B: 1969–1943 Ma
Crustal extension during slab-rollback
OMS
Nhandu intrusive suite
A-type
I-type
Pé Quente intrusive suite
A-type
N
Oceanic crust
Lithosphere
Asthenosphere upwelling
Asthenosphere
Stage C: 1904–1819 Ma
Crustal thickening during low-angle subduction
IMS
Matupá intrusive suite
I-type
Guarantã do Norte intrusive suite
I-type
N
Oceanic crust
Lithosphere
Asthenosphere
Asthenosphere
Stage D: 1839–1770 Ma
YMS

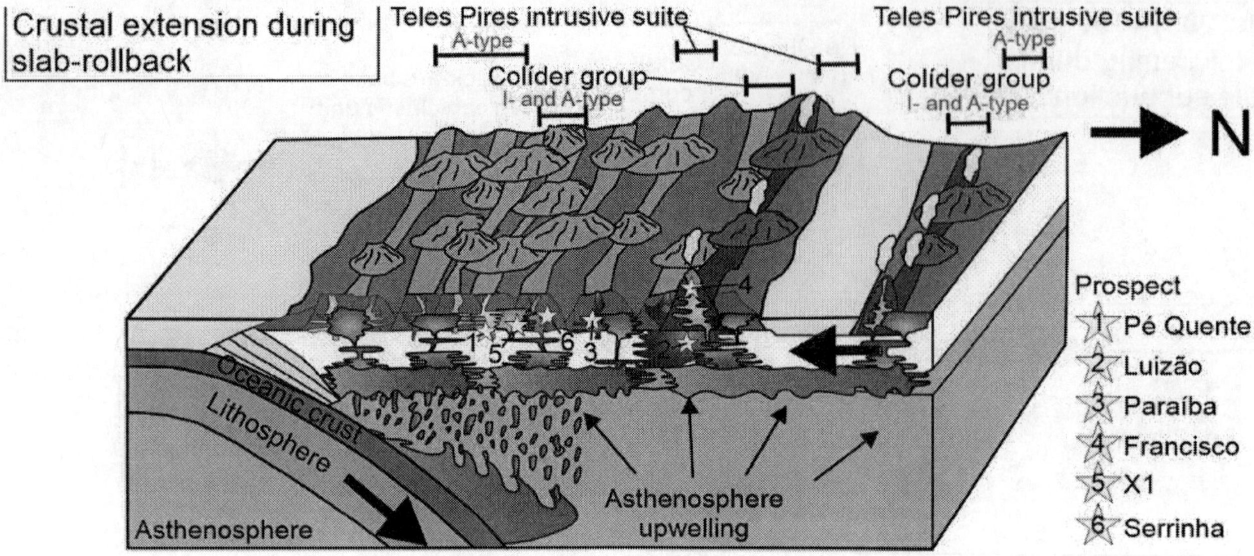
Crustal extension during slab-rollback
Teles Pires intrusive suite
A-type
Colíder group
I- and A-type
Teles Pires intrusive suite
A-type
Colíder group
I- and A-type
N
Prospect
1 Pé Quente
2 Luizão
3 Paraíba
4 Francisco
5 X1
6 Serrinha
Oceanic crust
Lithosphere
Asthenosphere
Asthenosphere upwelling

R6

Evolution of Metal Endowment in the Porphyry Au-Cu-Mo Deposits of the Long-Lived Sulphurets District, Canada: Implications for Fertile Magma Sources

Michelle E. Campbell[1, 2], John H. Dilles[1], Robert Creaser[3]
1. Oregon State University, Corvallis, OR, USA, 2. Seabridge Gold Ltd., Toronto, ON, Canada, 3. University of Alberta, Edmonton, AB, Canada

The Sulphurets district porphyry Au-Cu-Mo deposits, located in northwestern British Columbia, were formed in the Early Jurassic, during a prolific period of arc magmatism in the Stikine terrane. The four known deposits within the district are spatially and genetically associated with high-K calc-alkaline to alkaline Texas Creek suite porphyry intrusions. Owned by Seabridge Gold Inc., the porphyry ores are unusually gold rich and contain one of the largest global porphyry gold endowments, with >1,200 t Au (proven and probable reserves). New Re-Os molybdenite ages elucidate the age of mineralization at the Kerr (201.4 ± 0.9 Ma, 203.1 ± 0.9 Ma), Sulphurets (194.8 ± 0.9 Ma, 194.9 ± 0.9 Ma, 195.1 ± 0.9 Ma), and Iron Cap (194.1 ± 0.9 Ma, 194.3 ± 0.8 Ma) deposits, complementing existing geochronology data for mineralization at the Mitchell-Snowfield deposit (190.3 ± 0.8 to 192.0 ± 1.0 Ma; Febbo et al., 2019). The Re-Os molybdenite geochronology data thus reveal that the Sulphurets district was exceptionally long-lived, featuring one of the longest known lifespans among worldwide porphyry copper districts (11 ± 3.5 m.y.).

The determination of the chronology of deposit formation within the district reveals several compelling temporal trends. Notably, the Au/Cu and the Mo/Cu metal ratios of the Sulphurets district deposits increase progressively from the oldest Kerr deposit (Au/Cu $\approx 5.4 \cdot 10^{-5}$; Mo/Cu $\approx 1.2 \cdot 10^{-3}$; based on measured and indicated resource models) to the youngest Mitchell-Snowfield deposit (Au/Cu $\approx 4.3 \cdot 10^{-4}$; Mo/Cu $\approx$ 0.05). Furthermore, the volumetrically important synmineral plagioclase-hornblende diorite to monzodiorite Texas Creek suite porphyry intrusions become incrementally more evolved in the district deposits over time. The evolution of magmatic compositions is noted in systematic variations of whole-rock K_2O values and incompatible trace element ratios, such as Ce/V. The temporal covariation between increasing Au/Cu and Mo/Cu ratios with increasingly evolved and K-rich synmineral intrusions among the district deposits suggests an intimate association between magmatic composition and the potential for Au and Mo enrichment in porphyry Cu-(Au-Mo) deposits. Based upon the expansive ~11 m.y. history of the district, which surpasses the possible lifespans for upper crustal magmatic systems, the observed evolution of magmatic compositions must be linked to processes occurring in the mid- to deep-crust.

Febbo, G. E., Kennedy, L. A., Nelson, J. A. L., Savell, M. J., Campbell, M. E., Creaser, R. A., Friedman, R. M., Van Straaten, B. I., and Stein, H. J., 2019a, The evolution and structural modification of the supergiant Mitchell Au-Cu porphyry, Northwestern British Columbia: Economic Geology, v. 114, no. 2, p. 303–324.

R7

CA-TIMS and LA-ICP-MS (zircon) Characterization of Host Volcanic Rock Sequences and Porphyry Intrusions in the Eastern Timok Porphyry Cu-Au District

Alan J. Wainwright[1], Robert G. Lee[1], Craig J. Hart[1], Richard M. Friedman[2]
1. Mineral Deposit Research Unit - UBC, Vancouver, BC, Canada, 2. Department of Earth, Ocean and Atmospheric Sciences, The University of British Columbia, Vancouver, BC, Canada

The eastern Timok area of Serbia is underlain by Upper Cretaceous intrusive and volcanic rocks that compose the Timok Magmatic Complex. These rocks host large copper-gold porphyry and high-sulfidation epithermal deposits, including Cerovo, Veliki Krivelj, Borska Reka, Tilva Ros, and Cukaru Peki. We have characterized igneous rock suites from the eastern Timok and have assigned them to chronometrically constrained igneous rock sequences. We divide the rock units into 1) an intermediate composition volcaniclastic host-rock sequence, 2) a hornblende megacrystic porphyritic diorite phase, and 3) a biotite diorite porphyry phase. The Hb diorite and Bt diorite porphyry units form an axis of porphyry intrusions spatially associated with 18 km of mineralization centers of the eastern Timok (Cukaru Peki-Cerovo trend).

Chemical abrasion TIMS (zircon) data indicate that mineralized andesite volcaniclastic host rocks for the Borska Reka/Tilva Ros deposits are 85.0 ± 0.1 to 85.6 ± 0.1 Ma. The two porphyry phases are each 82 ± 1 Ma (LA-ICP-MS; zircon), consistent with crosscutting relationships. In the Bor Cu-Au camp, the Bt diorite porphyry predates hydrothermal alteration as it is cut by potassium-feldspar veins. The Hb diorite porphyry is the host rock for the Cerovo Cu-Au deposit and is therefore also premineralization.

The Ti-in-zircon thermometer indicates that most analytical points are between 600° and 800°C. The porphyry intrusions lie within a more restricted range (650°-750°C), whereas most of the outlier data (hotter and cooler than this range) are from the volcaniclastic host-rock zircons. Most of the Yb/Gd ratios in these rocks fall between 30 and 75, and the Th/U ratios mostly lie in the range of 0.15 to 0.70. The three rock units each display distinct zircon trace element characteristics and trends.

Zircon trace element data such as Ce/Ce_N^* values >100 and Eu/Eu_N^* values >0.4 (associated with high magmatic oxidation states) have been observed in porphyry copper deposits globally. Trace element in-zircon data from the eastern Timok indicate that Eu/Eu_N^* values for the magmatic phases are typically 0.4 to 1 and that cerium anomalies lie mostly between 0 and 400.

The eastern Timok rocks are characterized by high-oxidation-state magma chemistry. Eruption and emplacement of supracrustal host rocks occurred at ~85 Ma at Bor, which was followed by intrusion of porphyry phases within 2 to 4 million years afterwards. In the Bor and Cerovo areas, alteration and Cu-Au mineralization are younger than 82 ± 1 Ma. Discrete magmatic lineages are identified in the eastern Timok (based on zircon trace element composition), despite short time frames for emplacement and overlapping spatial distribution within the Cukaru Peki-Cerovo trend. The trace element data distinguish between the host volcaniclastic rocks and the porphyry intrusive phases that postdate them. Low-viscosity, water- and volatile-rich dynamic magma chambers may yield igneous rock complexes such as the Timok, with mixed zircon populations from multiple chemical domains, emplaced in short time periods.

R8

The Onto Cu-Au Discovery, Eastern Sumbawa, Indonesia: A World-Class Middle Pleistocene Lithocap-Hosted Covellite-Pyrite Deposit

David Burrows[1], PT Sumbawa Timur Mining Exploration Team[2]
1. Vale Global Exploration, Sudbury, ON, Canada, 2. Sumbawa Timur Mining, Jakarta, Indonesia

In 2013, a PT Sumbawa Timur Mining (80% Vale, 20% PT Antam) diamond drill program testing an extensive advanced argillic alteration lithocap within the Hu'u project on eastern Sumbawa Island discovered a giant hybrid high-sulphidation epithermal/porphyry copper-gold deposit under ~500 m of relatively unaltered, barren andesitic cover. Copper occurs almost exclusively as covellite with disseminated pyrite and in pyrite-covellite veins and veinlets, and gold occurs in the both the covellite and associated pyrite.

Cu-Au mineralization occurs in a tabular block measuring at least 1.5 x 1 km, with a vertical thickness of ≥1 km largely within the top of a series of coalesced porphyry stocks (60% of resource) and in a surrounding premineral polymictic diatreme breccia. A December 2019 resource estimate for Onto reported an indicated resource of 0.76 billion tonnes at 0.93% Cu, 0.56 g/t Au, and 350 ppm As and an inferred resource of 0.96 billion tonnes at 0.87% Cu, 0.44 g/t Au, and 350 ppm As for 15.3 Mt of contained copper and 27 million oz of gold in both resource categories. In addition to these resources, an exploration target of 0.6 to 1.7 billion tonnes at 0.4 to 0.7% Cu and 0.2 to 0.3 g/t Au has been delineated in a surrounding envelope based upon 64 drill holes (http://sumbawatimurmining.co.id/mineral-resource-estimate-statement/).

The porphyry-style mineralization, related to an intense AB quartz vein stockwork, and high-sulfidation-style mineralization overlap spatially and temporally, and more than 90% of the resource is hosted in either quartz-alunite or quartz-pyrophyllite ± diaspore alteration with only minimal potassic and chlorite ± sericite alteration on the margins and very deep within the system. It is exceptionally young and formed rapidly in the middle Pleistocene, and much of the geological system and alteration is intact without significant erosion.

We will discuss the tectonomagmatic setting and the unique insights provided by this deposit into porphyry-epithermal transitions.

R9
Porphyry Deposits May Not Require a Unique Melt

Ijaz Ahmad, Pedro J. Jugo, Jeremy P. Richards
Laurentian University, Sudbury, ON, Canada

Porphyry deposits are a major source of Cu, Mo, and Au and are associated with felsic to intermediate intrusions emplaced in the upper crust (2–5 km) during arc magmatism. Sulfide saturation in the lower crust is thought to play a vital role in the fertility of such magmas by scavenging chalcophile elements (Cu, Au, PGEs) due to their higher partitioning into sulfide melt, resulting in depletion of these elements in the derivative melts that ascends to the upper crust. We test this hypothesis in the Kohistan island arc, northern Pakistan, which exposes two distinct lower crustal sections in the south (Jijal and Chilas) and one mid to upper crustal section in the north (Kohistan batholith). Whole-rock geochemistry data from the lower crustal sections show depletion in Os and Ir compared to the mantle, suggesting that these elements were retained by residual monosulfide solid solution in the mantle source. However, Pt, Pd, Au, and Cu are undepleted relative to the mantle, suggesting extraction from the mantle source. In contrast, the mid to upper crustal section is diminished in Pt, Pd, Au, whereas Cu maintains mantle-like concentration in these rocks. We interpret that this is caused by the segregation of sulfides in the lower crust, as in situ LA-ICP-MS analysis of sulfides shows ppm level concentration of these elements. Modeling the effect of sulfide on the metal budget of the Kohistan magmas indicates that segregation of small amounts (< 0.17 wt %) of sulfides would not significantly affect the Cu budget of the melt that could rise to the upper crust. Such melts, although depeleted in PGEs, would contain sufficient amounts of Cu to form porphyry Cu deposits. Thus, systems like the Kohistan batholith could potentially generate a deposit of 4–5 Mt Cu with an extraction efficiency of 70 to 100%, confirming that these deposits may not require a unique (metal-rich) melt.

R10
Epithermal Alteration in the Yellowstone Hydrothermal System, Wyoming, USA

Peter B. Larson[1], Jerry Fairley[2], Jarred Zimmerman[1]
1. Washington State University, Pullman, WA, USA, 2. University of Idaho, Moscow, ID, USA

The Yellowstone Caldera hydrothermal system, Yellowstone Park, Wyoming, USA, is found over an area greater than 6,000 km^2, includes more than 10,000 thermal features, and was active since at least 400,000 years ago. There are no reported ages of hydrothermal minerals in the interval between this age and caldera eruption at 640,000 years ago. Hydrothermal fluid flow is strongly controlled by permeable zones such as contacts between postcollapse flows and along faults and fractures. Drilling into the hydrothermal system in the past has shown that the modern hydrothermal fluids rise along the hydrostatic boiling curve. The Grand Canyon of the Yellowstone River incises into the older altered rocks in the caldera and exposes them over a cross section of more than 300 meters. Protoliths in the canyon are the postcollapse Tuff of Sulfur Creek and the Canyon Flow rhyolites that erupted about 470 ka. Glaciation has removed the upper part of this exposure. The canyon walls provide a unique opportunity to study the shallower part of an older phase of the Yellowstone hydrothermal system. Two alteration assemblages are exposed in the walls. A shallower assemblage has kaolinite as the dominant clay and includes silica either as quartz or opal and locally alunite, dickite, and barite. The deeper assemblage has illite as the dominant clay with quartz and in places hydrothermal feldspars (hyalophane, buddingtonite, and adularia). Coarse vuggy quartz crystals and meter-wide quartz veins are found in one area in the lower alteration assemblage. Pyrite and marcasite are widespread as disseminations or with opal or quartz in veins. Very minor fine-grained sphalerite was noted in only a few places. Significant K was lost from and Ba was added to the rocks during kaolinite alteration. Illite alteration produced losses in divalent cations such as Sr, Ba, and Pb. One alunite from the upper assemblage yielded a $^{40}Ar/^{39}Ar$ age of 154,000 years. This suggests the hydrothermal activity was ongoing during the Bull Lake-Pinedale interglaciation. The vertical variation in alteration mineralogy suggests that a thermal and geochemical gradient was present during alteration and is most likely related to temperature variations in a rising and boiling hydrothermal fluid, similar to modern hot springs at Yellowstone. Preliminary fluid inclusion homogenization temperatures in quartz range from 190° to 280°C, and mean temperatures show a vertical gradient over 250 meters that parallels the hydrostatic boiling temperatures found in the older drill holes. These data also suggest that several hundred meters of the shallowest altered rocks have been removed by Pinedale glaciation. Hot springs are now active at the base of the canyon, and the hydrothermal system may have been continuously active over the past several hundred thousand years and during canyon incision.

R11
Crystal Mush Dikes as Conduits for Porphyry Copper Deposit-Forming Fluids

Lawrence C. Carter[1], Ben J. Williamson[1], Simon R. Tapster[2], Catia Costa[3], Geoffrey W. Grime[3], Gavyn K. Rollinson[1]
1. Camborne School Of Mines, University of Exeter, Cornwall, United Kingdom, 2. British Geological Survey, Nottingham, United Kingdom, 3. University of Surrey Ion Beam Centre, Guildford, United Kingdom

Porphyry deposit formation is associated with large, long-lived magmatic systems (e.g., Cline and Bodnar, 1991; Tapster et al., 2016) where mineralising fluids are mainly derived from melt or mush reservoirs at depths of ca. 5 to 15 km (Richards, 2005; Sillitoe, 2010). How these fluids are transported upwards and focused into narrow, relatively shallow (ca. 2-5 km deep) zones of economic mineralisation is poorly understood (Dilles, 1987; Sillitoe, 2010). Although field and laboratory-based observations can offer useful insights (Pistone et al., 2020), our ability to understand whole systems is limited as in nearly all cases only small fragments of their 3D architecture are exposed (Seedorff et al., 2008).

An exception to this is the archetypal Yerington porphyry system, Nevada, which has been tilted to reveal a unique vertical section from paleosurface to a depth of ca. 8 km, through at least 4 porphyry deposits (Dilles, 1987). This provides a rare opportunity to study temporal relationships in the deeper parts of the magmatic system, important in porphyry formation, and the magmatic-hydrothermal transition (Carter et al., 2021). We show evidence for hypogene fluid exsolution and mineralisation within aplite dikes, including the presence of unidirectional solidification textures (Kirwin, 2005) and miarolitic cavities (e.g., Candela, 1997), which have not been discussed previously in the context of mineralisation (e.g., Dilles, 1987). From our detailed scanning electron microscopy and cathodoluminescence imaging, and electron microprobe and total ion beam analyses to determine trace elements in quartz, we show that the groundmass of these aplite dikes contains a wormy, interconnected generation of quartz that directly feeds mineralised miarolitic cavities. We interpret this wormy texture as marking paleopathways between earlier crystallised magmatic minerals through which mineralising fluids flowed, within what we referred to as "crystal mush dikes" (Carter et al., 2021). These acted as conduits for the upward flow of porphyry deposit-forming fluids from more evolved/water-rich internal regions of the underlying Luhr Hill granite. The textures we describe provide the first substantive petrographic evidence for the transport of fluids through crystal mush dikes, which should be considered in any new model for the formation of porphyry- and similar-type deposits and may be important in the degassing of volcanic systems.

References

Candela, P.A., 1997, J. Petrol., 38, 1619-1633.

Carter, L.C. et al., 2021, Commun. Earth Environ., 2, 59.

Cline, J.S. & Bodnar, R.J., 1991, J. Geophys. R., 96, 8113-8126.

Dilles, J.H., 1987, Econ. Geol., 82, 1750-1789.

Kirwin, D., 2005, IAGOD Guidebook, 11, 63-84.

Pistone, M. et al., 2020, Front. Earth Sci., 8, 175.

Seedorff, E. et al., 2008, Econ. Geol. 103, 939-956.

Sillitoe, R.H., 2010, Econ. Geol., 105, 3-41.

Tapster, S. et al., 2016, Earth Planet. Sci. Lett. 442, 206-217.

R12
Multiple Formation Environments of Advanced Argillic Alteration and Exploration Implications

Antonio Arribas[1], Jeffrey Hedenquist[2]
1. University of Texas - El Paso, El Paso, TX, USA, 2. University of Ottawa, Ottawa, ON, Canada

Advanced argillic alteration, defined by Hemley as advanced hydrolytic, includes alunite, aluminosilicates (kaolinite, halloysite, dickite, pyrophyllite, andalusite), diaspore, anhydrite, and zunyite and topaz. One or more of these minerals form in five distinctly different acidic (pH ≤1 to 4-5) geologic environments, most at depths <500 m (Fig. 1). 1) Vapors that discharge from volcanic vents contain magmatic SO_2 and HCl; where condensed into local groundwater, hypogene H_2SO_4 and HCl-rich solutions form with a pH of <1-1.5 at <250°C. Isochemical dissolution of permeable host rock occurs due to the high solubility of Al and Fe hydroxides at pH <2, aside from a siliceous residue. This residual quartz, commonly vuggy in texture, is barren of metals due to low metal contents in high-temperature, low-pressure volcanic vapor. Alunite, kaolinite, and dickite (±pyrophyllite ±diaspore at higher temperatures) form a halo to the residual quartz. This barren lithocap horizon follows permeable lithologic units near feeder structures and may become mineralized if metal-bearing white mica-stable liquid subsequently ascends to this level. 2) Boiling of geothermal solutions generates vapor with CO_2 and H_2S, and where condensed above the water table in the vadose zone, atmospheric O_2 oxidizes H_2S to sulfuric acid, forming a steam-heated solution of pH 2-3. Kaolinite and alunite horizons form above the water table at ~100°C, and amorphous silica derived within the vadose zone replaces the aquifer with opal as the condensate follows the hydraulic gradient. 3) By contrast, where this vapor condenses below the water table, the CO_2 in solution forms carbonic acid (H_2CO_3) with pH of 4-5. This marginal carapace of condensate forms intermediate argillic alteration of clays and Fe carbonates: smectite at shallow depths and interstratified clays to illite ± chlorite at higher temperature. At sufficiently high H_2CO_3 concentration, kaolinite becomes stable at several 100s m depth, forming illite-kaolinite halos to quartz veins, potentially mistaken as having formed in an advanced argillic environment (e.g., steam-heated sulfate or supergene overprint). 4) The deepest environment, ~1 km, is associated with the top of porphyry deposits, where the metal-bearing fluid that forms white-mica alteration continues to ascend and cool, forming pyrophyllite (±diaspore ±dickite) near the base of the residual quartz and advanced argillic lithocap. 5) The final setting is supergene, due to posthydrothermal weathering and oxidation of pyrite, locally creating pH <1 liquid from high concentrations of H_2SO_4 within the vadose zone, where kaolinite and alunite plus Fe oxyhydroxides form.

Interpretation of the formational environment of alteration, using mineralogy, textures, and morphology, is critical to the effective exploration for epithermal, porphyry, and other hydrothermal ore deposits. For example, hypogene quartz-alunite-kaolinite forms halos to residual quartz lithocaps, which may be mineralized or barren, whereas pyrophyllite-white mica underlies lithocaps but indicates proximity to porphyry deposits. Kaolinite-alunite steam-heated blankets, typically barren, may overlie ore zones. Too often the terms advanced argillic, argillic, etc., are used on maps and sections, thus losing the significance of specific alteration minerals and morpholgies, with two or even three different origins of advanced argillic alteration possible in one deposit.

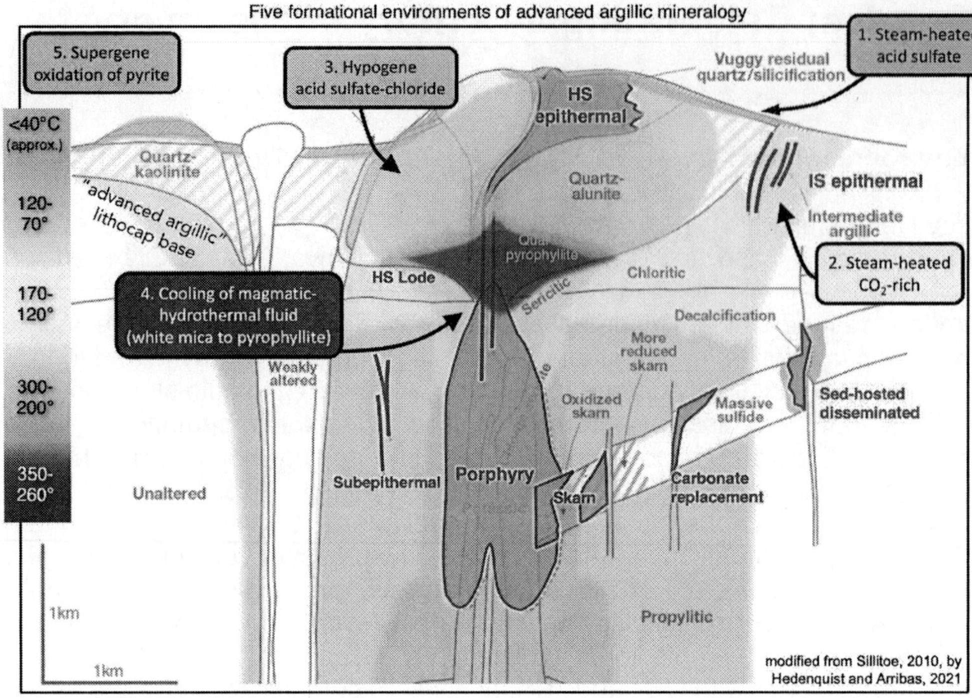

Five formational environments of advanced argillic mineralogy
5. Supergene oxidation of pyrite
3. Hypogene acid sulfate-chloride
1. Steam-heated acid sulfate
Vuggy residual quartz/silicification
HS epithermal
Quartz-alunite
IS epithermal
Intermediate argillic
2. Steam-heated CO_2-rich
<40°C (approx.)
120-70°
170-120°
300-200°
350-260°
Quartz-kaolinite
"advanced argillic" lithocap base
HS Lode
Chloritic
Sericitic
4. Cooling of magmatic-hydrothermal fluid (white mica to pyrophyllite)
Decalcification
More reduced skarn
Weakly altered
Oxidized skarn
Massive sulfide
Sed-hosted disseminated
Subepithermal
Porphyry
Skarn
Carbonate replacement
Unaltered
1km
1km
Propylitic
modified from Sillitoe, 2010, by Hedenquist and Arribas, 2021

R13
Evolution of a Laramide Porphyry Cu Cluster near Tucson, Arizona

Roy Greig, Mark Barton
University of Arizona, Tucson, AZ, USA

A new structural reconstruction of the Tucson, Arizona, area, combined with synthesis of new and available geochronological, petrological, and geologic data, demonstrates that at least seven porphyry centers in the area share many characteristics and represent a super cluster that formed as part of a longer-term igneous history. Exposures from 0- to ~15-km paleodepth, in four variably tilted mountain ranges, display late Cretaceous to Eocene-age (Laramide) igneous rocks with a >30 Ma history, which can be split into three distinct phases, which have comparable total volumes but different compositions, styles, and related hydrothermal features.

An early phase of magmatism (Group 1), between 76 and 69 Ma, consisted of compositionally diverse (basalt-rhyolite), widespread volcanism, including large felsic calderas and smaller volumes of lavas, and emplacement of compositionally zoned diorite-granite plutons. Related porphyry-type hydrothermal features (potassic and sericitic alteration, proximal Cu and distal Zn-Pb mineralization) are in general weakly developed and formed no major ore deposits.

Group 2 igneous centers (65-57 Ma) are compositionally homogeneous, felsic (granodiorite-granite; rhyodacite), and characterized by relatively voluminous plutonism and minor volcanism (lavas and small ash-flow tuffs). Porphyry-type hydrothermal features are well developed at shallow levels (potassic and sericitic alteration) and variably developed at deeper levels (coarse muscovite), in all of these centers, but only a few centers formed significant porphyry Cu deposits (e.g., Sierrita, Mission-Pima-Twin Buttes, Rosemont). Though the relative development of alteration types varies, the footprint of the porphyry systems is roughly proportional to contained metals.

Group 3 magmatism (57-42? Ma) is characterized by large sheets of peraluminous (biotite ± muscovite ± garnet) granite emplaced at depths of 4-15 km. These contain numerous simple pegmatites but lack significant hydrothermal systems or other mineralization.

The igneous history and geochemistry suggests that the variation in igneous compositions between age groups was the result of the proliferation of crustal melts, which increasingly filtered out mafic inputs to the upper crust. This may have prevented mafic recharge, which has been linked elsewhere to caldera-forming eruptions, thus controlling the contrasting fate of large magma chambers and their contained metals and volatiles: several of those in Group 1 erupted, forming calderas, while those in Group 2 formed large porphyry Cu deposits. The contrasting fates of these chambers may also relate to their apparently differing geometry, as can be inferred from tilted caldera and pluton exposures.

The similarities between the large and small Group 2 porphyry systems suggest that the volatile and metal contents of Group 2 magmas were roughly uniform and that accumulation of sufficient magma volumes was the key driver of productivity. This may have been facilitated by focusing of magma inputs in a manner analogous to that seen in other maturing arc volcanic centers.

This history of Tucson area parallels regional patterns and illustrates how productive porphyry Cu clusters emerge from a spectrum of tectonic and igneous processes operating on a range of temporal and spatial scales.

R14
The Late Miocene Middle Cauca Au-Cu Porphyry/Epithermal Belt, NW Colombia: Tectonomagmatic History and Controls on Mineralization

Hildebrando Leal-Mejia[1], Thomas Bissig[2], Craig Hart[1], Robet P. Shaw[3], Richard Friedman[4], Robert Creaser[5], Janet Gabites[4]
1. Mineral Deposit Research Unit - The University of British Columbia, Vancouver, BC, Canada, 2. Bissig Geoscience Consulting, Vancouver, BC, Canada, 3. Independent Consultant, Kelowna, BC, Canada, 4. Pacific Centre for Isotopic and Geochemical Research – The University of British Columbia, Vancouver, BC, Canada, 5. Canadian Centre for Isotopic Microanalysis, University of Alberta, Vancouver, BC, Canada

The late Miocene Middle Cauca Cu-Au porphyry/epithermal Belt (MCB) and adjacent areas are recognized as a highly prospective region within the Colombian Andes, being the focus of significant exploration since 2003. Although recent publications address the characterization of individual districts/deposits within the region, understanding the regional spatial and temporal tectonomagmatic history and the controls on the mineralization is a key aspect to support mineral exploration.

The ca. 120-km-long N-S-trending MCB comprises historically known and recently discovered deposits along the middle Cauca River valley, with porphyry-associated mineralization styles including porphyry Au-Cu (e.g., El Poma, Quinchía, Caramanta, La Mina, and Titiribí), porphyry Cu-Au (e.g., Nuevo Chaquiro), and LS-IS epithermal Au-Ag-Zn (Pb-Cu) (e.g., from S to N, Miraflores and Chuscal near Quinchía, Marmato, Yarumalito, El Zancudo near Titiribí) deposits. Adjacent areas also include important late Miocene deposits such as the La Colosa porphyry Au and the Buriticá vein-type Ag-Au-Zn (Pb, Cu).

Porphyry and epithermal mineralizations are structurally controlled and spatially associated with weakly to moderately porphyritic hornblende-biotite plagioclase diorite-granodiorite-quartz monzonite stocks intruding the Romeral melange basement, as well as the overlying the early-middle Miocene Amagá Formation and the middle-late Miocene Combia Formation.

U-Pb CA-TIMS and LA-ICP-MS zircon ages obtained for intrusions across the MCB constrain the magmatism between ca. 12 and 4 Ma, with early magmatism (12-10 Ma) starting in the south at El Poma and Quinchía, followed by northward expansion along the MCB to reach the Titiribí district by 7.5 Ma and subsequent eastward migration to Caramanta-Yarumalito and Marmato (ca. 7.2-6.0 Ma).

Re-Os molybdenite and ^{40}Ar-^{39}Ar sericite analyses confirm the close temporal relationship of mineralization with the late Miocene magmatic event with multiple mineralizing episodes, most prolific between ca. 9.3 and 7.3 Ma.

Lithogeochemistry of igneous rocks associated with porphyry districts and age-equivalent porphyry intrusions unrelated to known mineralization confirm the mantle-derived, metaluminous, calc-alkaline affinity of the magmatism, associated with subduction of the Nazca Plate beneath the Pacific Margin of northwestern South America. However, subtle but important geochemical variations in both N-S and W-E directions are evident, including increasing SiO_2 content from N to S and characteristic “adakite-like” and “porphyry-fertile” geochemical signatures. Igneous rocks from the Titiribí District have a subtle but distinctly more alkaline character when compared to other districts to the south. Such a subtle alkalic character is also observed in igneous rocks from the Buriticá epithermal deposit.

The high Au/Cu ratios observed in the Miocene porphyry-related ore deposits are attributed to shallow emplacement of porphyry systems, as inferred from the temporal relationship of water-rich oxidized porphyry stocks with only slightly older host-volcanic rocks and from the proximity and similar age of epithermal and porphyry mineralization. At the Titiribí district, the Au-rich nature of mineralization can potentially be attributed to the mildly alkalic nature of magmatism. The Cu-poor but Au-rich nature of the giant La Colosa porphyry deposit is best explained by the relatively reduced nature of the causative intrusive rocks hosted in strongly reducing Paleozoic carbonaceous schists of the Cajamarca-Valdivia terrane.

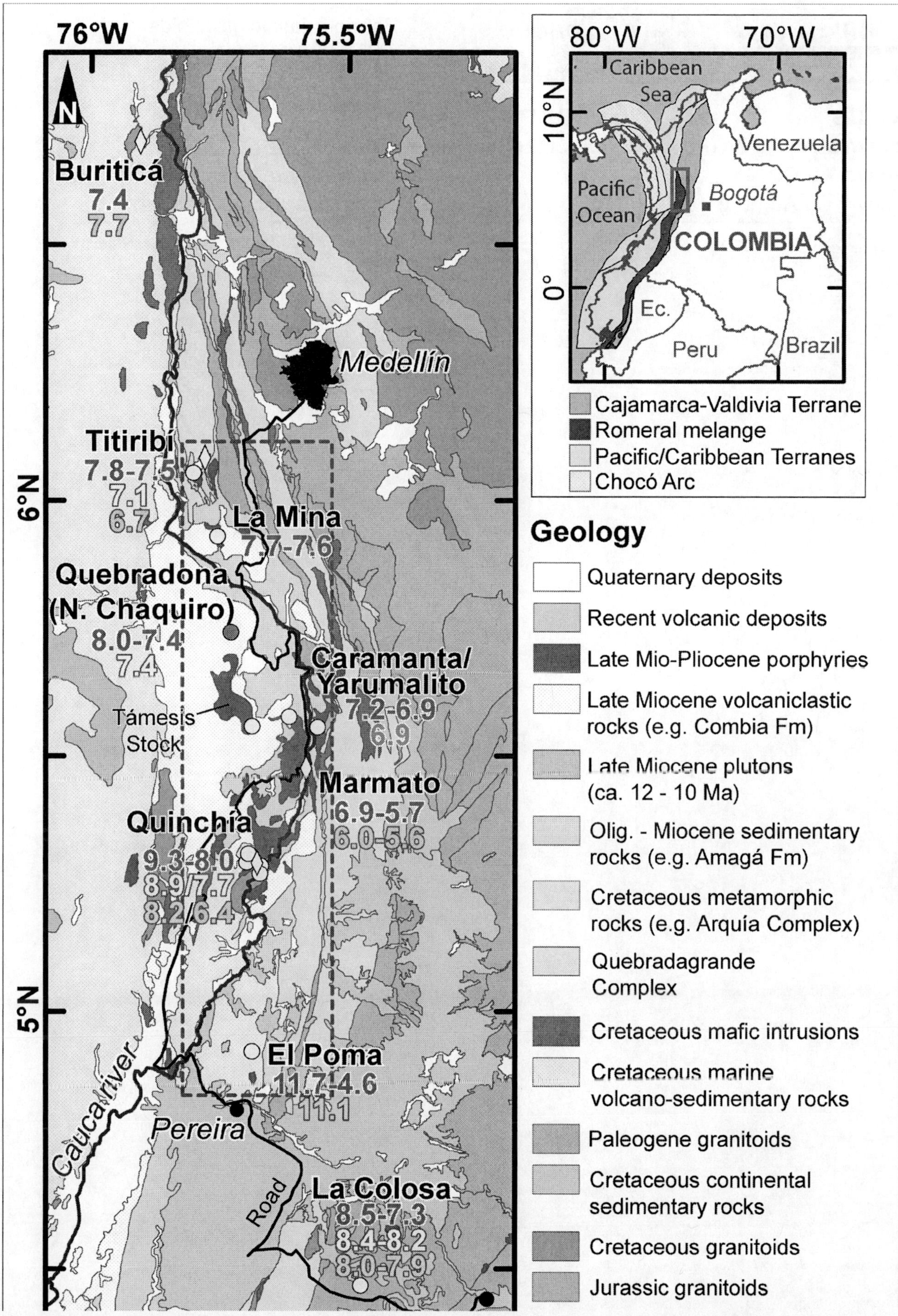
76°W
75.5°W
N
Buriticá
7.4
7.7
Medellín
Titiribí
7.8-7.5
7.1
6.7
La Mina
7.7-7.6
6°N
Quebradona
(N. Chaquiro)
8.0-7.4
7.4
Caramanta/
Yarumalito
7.2-6.9
6.9
Támesis
Stock
Marmato
6.9-5.7
6.0-5.6
Quinchía
9.3-8.0
8.9/7.7
8.2/6.4
5°N
El Poma
11.7-4.6
11.1
Cauca river
Pereira
Road
La Colosa
8.5-7.3
8.4-8.2
8.0-7.9
80°W
70°W
10°N
0°
Caribbean
Sea
Pa.
Venezuela
Pacific
Ocean
Bogotá
COLOMBIA
Ec.
Peru
Brazil
Cajamarca-Valdivia Terrane
Romeral melange
Pacific/Caribbean Terranes
Chocó Arc
Geology
Quaternary deposits
Recent volcanic deposits
Late Mio-Pliocene porphyries
Late Miocene volcaniclastic rocks (e.g. Combia Fm)
Late Miocene plutons (ca. 12 - 10 Ma)
Olig. - Miocene sedimentary rocks (e.g. Amagá Fm)
Cretaceous metamorphic rocks (e.g. Arquía Complex)
Quebradagrande Complex
Cretaceous mafic intrusions
Cretaceous marine volcano-sedimentary rocks
Paleogene granitoids
Cretaceous continental sedimentary rocks
Cretaceous granitoids
Jurassic granitoids

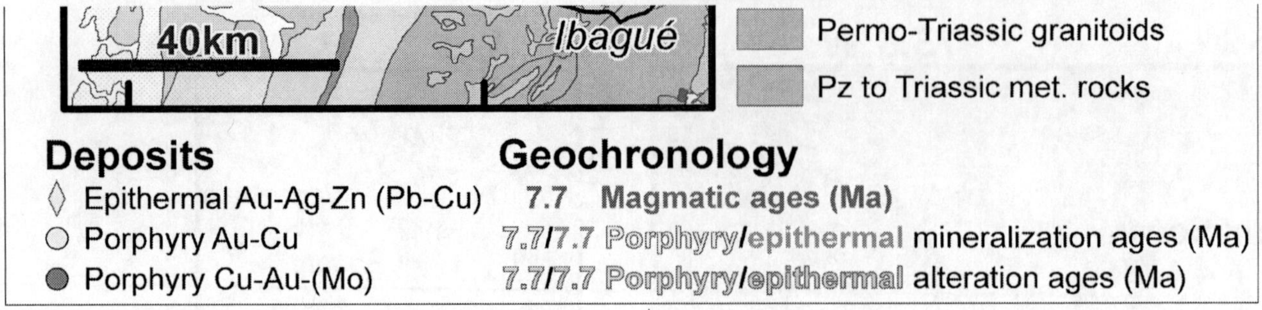
40km
Ibagué
Permo-Triassic granitoids
Pz to Triassic met. rocks
Deposits
Epithermal Au-Ag-Zn (Pb-Cu)
Porphyry Au-Cu
Porphyry Cu-Au-(Mo)
Geochronology
7.7 Magmatic ages (Ma)
7.7/7.7 Porphyry/epithermal mineralization ages (Ma)
7.7/7.7 Porphyry/epithermal alteration ages (Ma)

R15
PGE Chemistry and Magma Fertility of El Teniente Porphyry Copper Deposit, Central Chile

Yamila Cajal, Ian H. Campbell
Australian National University, Acton, ACT, Australia

The El Teniente Porphyry Copper Deposit, located in Central Chile, is one of the biggest concentrations of copper in the world with ~100 Mt of metal. Although there are many factors that might have contributed to the high metal content of this deposit (e.g., tectonic setting, length of magmatic activity, wall-rock composition), this study evaluates the role the timing of sulphide saturation, relative to volatile exsolution, plays in El Teniente's magma fertility.

Previous studies have determined the onset of sulphide saturation using platinum group elements (PGEs) because they behave similarly to chalcophile elements (e.g., Cu and Au) during magmatic differentiation. However, the PGEs better preserve the original geochemistry of intrusions in magmatic-hydrothermal systems because they are less mobile in hydrothermal fluids. These studies also suggest that PGE geochemistry can be used to distinguish between mineralized and barren systems.

The magmatism at El Teniente includes a large premineralization intrusive complex containing older mafic facies and a younger and smaller core of hydrous intrusions, as well as five younger and smaller pulses of felsic intrusions linked to mineralization. Cu-mineralized magmatic-hydrothermal breccia pipes and postmineralization mafic lamprophyre dikes are also present in the district. The wide range of magmatic compositions make El Teniente an ideal case to study the magmatic evolution of a porphyry deposit.

We have studied rock samples from the different intrusions from El Teniente District, located within ~12 km of the El Teniente mine, which range in composition from gabbro (5.3 wt % MgO) to dacite porphyries (0.6 wt % MgO). Whole-rock major and trace elements concentrations have been determined by XRF and LA-ICP-MS, respectively. The PGE concentrations are at ultratrace level and they have been determined using the nickel-sulphide fire assay, isotope dilution method.

Preliminary results show that the concentrations of PGEs decrease with the decrease of MgO, which suggests that sulphide saturation in the El Teniente magmatic system occurred prior to the MgO content of the magma falling below 5.3 wt % (MgO content of the most primitive sample analysed). This early sulphide saturation seems not to be affecting the ability of the magmatic system to form an ore deposit, which may be because the amount of sulphide that separated from the melt was too small to have a significant effect on the copper content of the magma. However, the early sulphide saturation decreased significantly the Au content of the magma, due to its higher partition coefficient in sulphide melts, leading to the formation of a Cu-dominated porphyry deposit, rather than a Cu-Au deposit.

The PGE data also confirms that the Pd/MgO against Pd/Pt and Pd/MgO versus Y diagrams proposed by Park et al. (2019) and Hao et al. (2019), respectively, can be used to discriminate between barren and fertile suites in this region. We will also report new geochronological and geochemical data from zircons from regional intrusions in the El Teniente District, which will be compared with the results from ore-related intrusions associated with this and other porphyry Cu and Cu-Au deposits.

R16
Lithospheric Architecture of the Central Andes and the Localization of Giant Porphyry Copper Deposits

Alex Farrar[1], Matthew Cracknell[1], David R. Cooke[1], Jon Hronsky[2, 3], José Piquer[4]
1. Centre for Ore Deposit and Earth Sciences (CODES), University of Tasmania, Hobart, TAS, Australia, 2. Centre for Exploration Targeting (CET), University of Western Australia, Perth, WA, Australia, 3. Western Mining Services, Perth, WA, Australia, 4. Universidad Austral de Chile, Valdivia, Chile

The central Andes (14°S-35°S) is the most important copper province on the planet, accounting for 40% of the world's annual copper supply. The decades leading up to the year 2000 were a period of abundant greenfields porphyry copper discovery in the central Andes. However, since 2000, reports of significant new greenfields porphyry copper deposits (PCD) in the central Andes have drastically declined. Giant PCDs tend to group in discrete multideposit geographic clusters of similar age, indicating that exceptional transient geologic processes have affected localised regions of the crust prior to and during the age of mineralisation and that the formation of giant porphyry deposits is nonrandom. Key predictive geologic features of giant porphyry Cu deposits are the structural pathways (basement faults) that focus fluid flow from the lower to upper crust. Nevertheless, these translithospheric faults (TLFs) are notoriously difficult to identify in the field due to their subtle surficial characteristics and continental scale. As a result, the notion of TLFs has been treated with scepticism by some members the geologic community. If TLFs are a fundamental control on porphyry localisation, then knowledge of the geographic areas where favourable permeability and geodynamic preconditioning established effective structural architectures for giant PCD formation would be of great benefit to mineral explorers, especially in regions with large amounts of postmineral cover, such as the central Andes.

This research focuses on identifying and mapping the continental-scale translithospheric structural architecture of the central Andes by integrating multiple geoscience data sets supported by field observations. Data sets used in the analysis include geophysical inputs such as airborne magnetics, regional gravity, and seismic epicentres, as well as geologic reconstructions through time from the Proterozoic to present, which map out inherited basement architecture as well as regions of rapid crustal thickening or thinning. Our field work demonstrates that the surface expression of a TLF is a consistent zone of structural features that are traceable for >100 km in strike and 10-40 km in width and is spatially consistent with the trace of lineaments in regional geophysical and geologic data sets. This is interpreted to reflect the upper crustal propagation of the underlying zone of basement weakness through younger sequences in the geologically active convergent plate margin. A striking characteristic of the results of this study are the nested scales of structural features that define a TLF and the regular orientation and spacing to TLF distribution, both displaying characteristics of self-organizing critical systems. A prominent relationship exists with the location of known giant porphyry deposit camps occurring where two or more TLFs meet. Such regions are inferred to have been zones of deep dilation, triggered during transient changes to the regional stress field. Regions adjacent to the intersection of two or more TLFs that overlap with the magmatic arc during metallogenic epochs are deemed to represent valid exploration targets in this model.

R17

Sulfide Saturation in Thick Arc Crust Plays No Role for Porphyry Cu Deposit Formation

Rui Wang[1], Jeremy P. Richards[2], Jingbo Zhang[1]
1. China University of Geosciences Beijing, Beijing, China, 2. Laurentian University, Sudbury, ON, Canada

Cu, as the moderately incompatible element, is poorly endowed in the bulk continental crust that is ultimately derived from primary basaltic arc. The paradox exists that the bulk continental crust with low $Mg^{\#}$ has much lower Cu than the basaltic high-$Mg^{\#}$ mantle melts. Recent studies suggest the missing Cu has been remained as Cu-bearing sulfides at the base of the arc, especially a thick arc. However, a thin island arc with relatively high Cu contents is thought to have late or no sulfide saturation during magmatic evolution. If this was the case, then porphyry Cu deposits should not preferentially form from calc-alkaline magmas in mature arcs with thick crust, but they do. The early sulfide saturation appears to play no role for upper crustal porphyry Cu formation during the arc period. A large porphyry Cu deposit can readily be formed from normal arc magmas containing only 50 ppm Cu. Cu contents in intermediate calc-alkaline magmas may indeed be lower than those in equivalent tholeiitic magmas, but the former are not devoid of Cu.

We suggest the shallow-crustal exsolution of a volatile phase from evolved, hydrous arc magmas, which efficiently scavenges Cu from the melt (or early crystallized sulfides) and transports it into the hydrothermal environment, where a separate set of variables control the precipitation of Cu sulfides to form porphyry ore deposits. Conversely, in immature arcs, less hydrous tholeiitic magmas will be less likely to form voluminous subvolcanic plutons or to exsolve a volatile phase prior to eruption and will therefore be less likely to form subvolcanic porphyry Cu deposits, despite their relatively high Cu contents.

R18
Distinctive Chemical Characteristics and Petrogenesis of Gold-Ore-Forming Arc Magmas

Robert R. Loucks, Marco L. Fiorentini
Centre for Exploration Targeting, University of Western Australia, Crawley, WA, Australia

Arc magmas that exsolve gold-ore-forming hydrothermal fluids are distinguished from infertile ordinary arc magmas by higher contents of lithophile elements that partition strongly into melt relative to residues during partial melting of mantle peridotite. Figure 1A shows published whole-rock analyses representing mafic-to-felsic differentiation series in Phanerozoic gold-ore-productive igneous complexes compared with differentiation series in large arc segments apparently barren of significant hydrothermal gold mineralisation. Due to relative enrichment in mantle-incompatible lithophile elements, gold-fertile magmas have steeper average slopes on mantle-normalised spidergrams (Fig. 1B). The slope can be represented by the ratio of a highly incompatible element on the left (e.g., Ba or Nb) to a less incompatible element farther to the right (e.g., Zr or Y), as illustrated in Figure 1A. Such plots can identify other gold-fertile igneous complexes and arc segments as exploration targets. Spidergram trends of metasomatically "refertilised" average subcontinental lithospheric mantle worldwide diverge leftward from the average asthenospheric mantle (NMORB source) in the same sense as the divergence between gold-fertile and gold-infertile mafic arc magmas in Figure 1B. This is evidence that melting of old metasomatic veins in lithospheric mantle is important in the petrogenesis of gold-ore-forming arc magmas. Published experimental data show that the wet solidus of peridotite is tens of degrees below the solidus temperatures of Fe-Ni(-Cu) monosulfide solid solution (mss) at upper mantle pressures. The experiments show that Au partitions preferentially into hydrous mafic silicate melts relative to mss-bearing restite during low-percentage melting of hydrated mantle peridotite at temperatures *below* the solidus of mantle mss. The partition coefficient of Au between crystalline mss and hydrous mafic silicate melt is ~60, so if the mass fraction of mss in peridotite is <1/60 (as is nearly always the case), then the Au content in hydrous mafic silicate melt is greater than in mss-bearing mantle restite. At temperatures *above* the sulphide solidus, Au partitions into sulphide melt relative to hydrous mafic silicate melt by an enrichment factor of about 10,000, so higher-temperature mafic silicate melts have very low Au content as they segregate from molten-sulphide-bearing mantle restite. Relatively Au-, Ba-, Nb-, and K-rich, near-solidus melts of asthenospheric mantle tend to freeze in transit through the cooler lithospheric mantle, producing a long-term accumulation of metasomatic veins and dikes in typical subcontinental lithospheric mantle. Metasomatic veins in lithospheric mantle have lower melting temperatures than their refractory peridotite matrix, so they can be selectively remelted in appropriate tectonic situations, producing magmas that are unusually fertile for magmatic-hydrothermal gold metallogeny.

Fig. 1(A) For unaltered or little-altered whole-rock analyses of samples from arc igneous complexes, ratios of a highly mantle-incompatible lithophile element to a moderately incompatible element are effective in discriminating magmas in barren arc segments (grey) from igneous complexes that produced major hydrothermal gold-rich deposits of varied types—porphyry, high-sulphidation to low-sulphidation epithermal, Tintina- and Carlin-types. (B) Relative abundances of incompatible lithophile elements in primitive arc magmas (6-12 wt % MgO) that are spatially and temporally closely associated with magmatic-hydrothermal gold mineralisation are compared to primitive magmas in unmineralised arc segments.

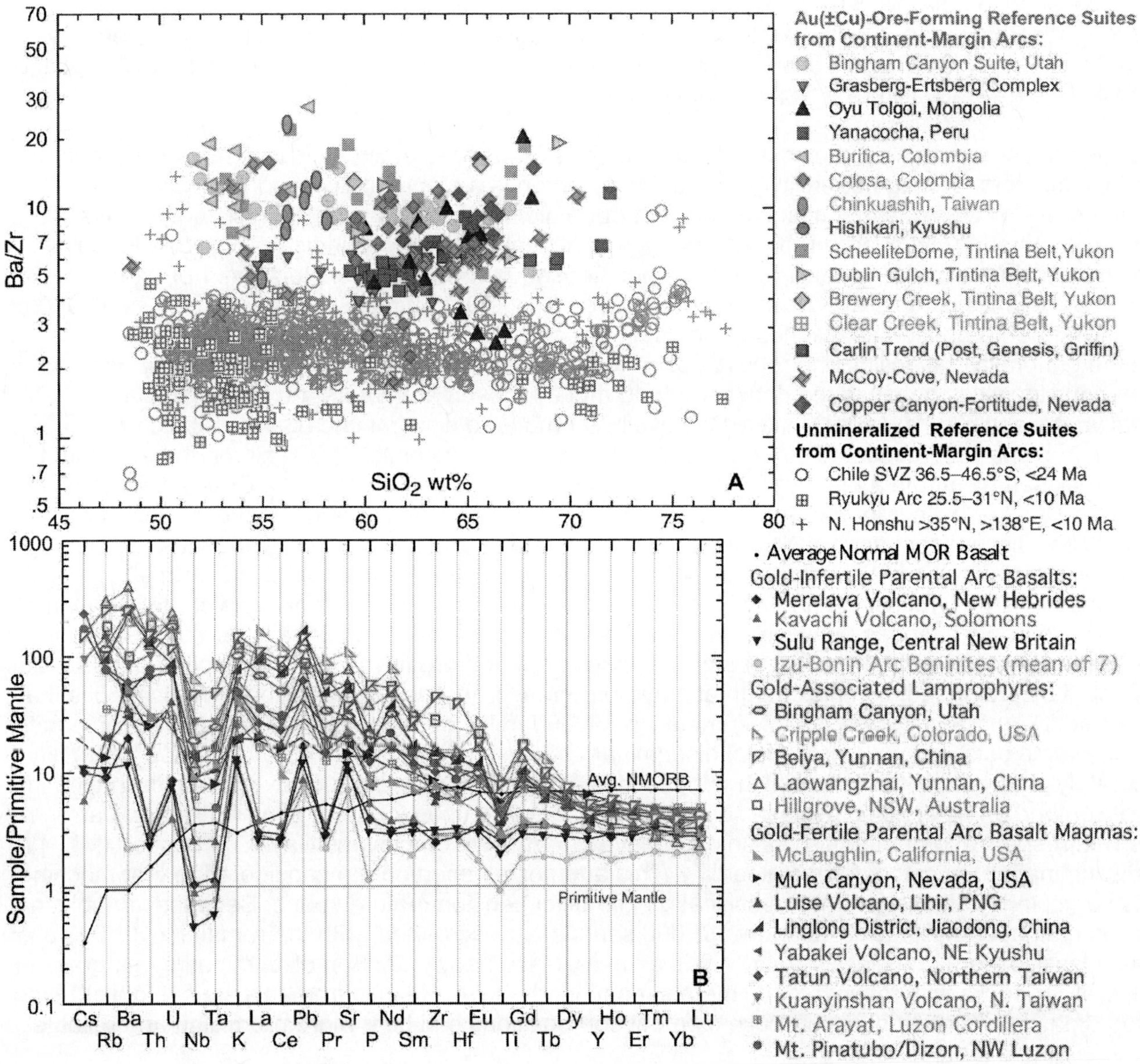

R19
Promising Fertility Indicators of Magmatic Systems: An Integrated Study of Zircon Geochemistry and Multilayer Perception

Si-Yuan Li[1], Si-Shen Li[3], Jian-Wei Li[1, 2]
1. School of Earth Resources, China University of Geosciences, Wuhan, China, 2. State Key Laboratory of Geological Processes and Mineral Resources, China University of Geosciences, Wuhan, China, 3. College of Automobile Engineering, Jilin University, Changchun, China

Recent studies have increasingly shown that some minerals (e.g., apatite, zircon, rutile, magnetite, epidote) are potential indicators for magma fertility. Among those, zircon is a common accessory mineral in intermediate to felsic intrusions that hosts a variety of trace elements and is typically resistant to physical and chemical weathering. As such, zircon may serve as an ideal petrogenetic indicator, oxygen fugacity proxy, and mineral indicator for source rock fertility and consequently mineral exploration. For example, Ce anomalies, Eu anomalies, and several trace element ratios (e.g., 10,000*(Eu/Eu*)/Y, (Ce/Nd)/Y, and Dy/Yb) of zircons have been proposed as diagnostic indicator for porphyry Cu (±Mo ±Au) productive magmas. However, the generality, reliability, and robustness of those parameters remain undetermined or need to be tested in various localities because the data used to formulate those parameters are not sufficiently representative. Meanwhile, few similar attempts have been made to discriminate unmineralized from mineralized intrusions with respect to other metals. In this study, we compile 18,000 zircon trace element analyses of rock samples from a large variety of deposits (Au, Cu series covering Cu, Cu-Au, Cu-Mo, Cu-Mo-Au, Cu-Mo-W, Cu-Pb-Zn, Mo, Fe, Sn, W, W-Sn, W-Mo, W-Cu-Mo) and unmineralized intrusions in various localities. Discriminant analysis in various violin plot shows that zircons from Cu-series deposits statistically have high values of Ce anomalies, Ce^{4+}/Ce^{3+} ratios, Eu anomalies, 10,000*(Eu/Eu*)/Y, (Ce/Nd)/Y, and low Dy/Yb ratios, confirming previous studies. However, all those parameters fall in a large range with considerable overlaps between unmineralized and mineralized intrusions in terms of Cu series. This misclassification may reflect the fact that metal enrichment and ore formation is a function of numerous factors affected by prolonged, complex magmatic-hydrothermal processes. An additional implication is that when a data set is sufficiently large, the discrimination for Cu fertility of magmatic rocks may be problematic. As such, large data sets and new methods of discriminant analysis are needed to evaluate magma fertility more accurately. The results further show that those indicators, even they are effective when discriminating Cu-mineralized and unmineralized rocks in some cases, cannot be used for separating fertile and barren intrusions with respect to other metals. As an alternative attempt, we employ multilayer perception (MLP) to develop discriminative indicators for metal fertility. This attempt is based on the consideration that machine learning is particularly suitable for data classification in multidimension feature space. Because variable in MLP function cannot be null, 6,000 out of the 18,000 samples are screened, with the remaining 12,000 used for the data classification. The classification yields an overall accuracy of 82% in discriminating mineralized and unmineralized rocks with respect to the metals mentioned above. Results from this study highlight that MLP-based data classification of zircon trace element compositions can be a more promising and reliable approach to evaluate magma fertility.

ST.211
Geology, Geochronology, and Alteration of the High-Sulfidation Epithermal Au-Ag Deposit and Porphyry Cu-Mo Occurrence at the Kirazlı District, Biga Peninsula, Turkey

Ali Aluç[1, 2], Robert Moritz[1], Ilkay Kuşcu[2], Alexey Ulianov[5], Mehtap Karcı[4], Dave Selby[3]
1. University of Geneva, Geneva, Switzerland, 2. Mugla Sitki Kocman University, Mugla, Turkey, 3. University of Durham, Durham, United Kingdom, 4. Dogu Biga Madencilik Subsidiary of Alamos Gold INC., Ankara, Turkey, 5. University of Lausanne, Lausanne, Switzerland

The Kirazlı district is located at the center of the Biga peninsula, which is a major metallogenic province at the northwestern tip of the Anatolian plate. The district contains a high-sulfidation (HS) epithermal Au-Ag deposit (0.33 Mt at 0.71 Au) and a porphyry Cu-Mo occurrence hosted by late Eocene-Oligocene magmatic rocks. There are five main target zones, including Kirazlı main, Çatalkaya, Rock Pile, Kale, and easternmost Iri. A NNW-trending HS orebody hosted by an Oligocene volcanic-volcaniclastic sequence covers large areas at the Kirazlı main, Çatalkaya, and Rock Pile. The sequence consists mainly of basaltic andesite lava flow and lithic/crystal tuff. A middle Eocene undifferentiated intrusive rock has been intersected by drilling at Kale, which hosts the porphyry Cu-Mo occurrence and is highly deformed by a conjugate fault system. LA-ICP-MS U-Pb zircon dating of the host units has yielded ages of 31.9 ± 0.5 to 32.7 ± 0.3 Ma and 41.5 ± 0.5 to 40.5 ± 0.3 Ma for the epithermal and porphyry systems, respectively.

The host lithological units in the Kirazlı district have been affected by pervasive hydrothermal alteration. A mineral-based alteration map was obtained by the portable infrared mineral analyzer (PIMA) studies using surface and drill core samples. Kirazlı Main has a proximal to distal zoned alteration pattern, including silicification, alunite-, dickite-, and more regional kaolinite-rich alteration assemblages, and subsidiary montmorillonite., Pyrophyllite- and sericite-rich alteration zones are more prominent towards the Çatalkaya and Kale zones, at topographically lower elevations. They document a transition from epithermal to porphyry environments. The Au-Ag mineralization occurs as disseminations, small veins, and veinlets within argillaceous and silicified (mainly vuggy and massive) host rocks and hydrothermal breccia in the HS environment. The Cu-Mo mineralization occurs as disseminations in a sericite-rich alteration zone and porphyry-type veins in the potassic alteration zone overprinted by sericite-chlorite and younger argillic alteration assemblages in the Kale zone. Re-Os geochronology on two molybdenite samples from quartz-pyrite-molybdenite veins at the porphyry mineralization yielded a 33.6 ± 0.2 Ma age.

ST.212
Geologic, Geochemical, and Geophysical Characteristics of the Dalli Porphyry Copper-Gold Deposit in Central Iran and Implications for Exploration

Hooshang Asadi Haroni[1, 2], Yongjun Lu[3, 1], Maryam Veiskarami[2]
1. Centre for Exploration Targeting and Australian Research Council Centre of Excellence for Core to Crust Fluid Systems (CCFS), School of Earth Sciences, The University of Western Australia,, Crawley, WA, Australia, 2. Department of Mining Engineering, Isfahan University of Technology, Isfahan, Iran, 3. Geological Survey of Western Australia, 100 Plain Street, East Perth, WA, Australia

The recently discovered Dalli porphyry copper-gold deposit (17 Mt at 0.5% Cu and 0.65 g/t Au) is located in the central segment of the Urumieh-Dokhtar Magmatic Arc (UDMA) in Iran. The UDMA hosts several major porphyry Cu (Mo-Au) systems and is part of the Tethyan metallogenic belt. This research aims characterize the deposit in order to guide further drill targets and define exploration criteria for similar mineralization along the UDMA. Dalli was initially identified by Rio Tinto, then an Iranian junior company discovered two major mineralized porphyry centers by drilling. The mineralization is hosted by several NE-trending quartz-diorite porphyry stocks intruded into a wall-rock andesite porphyry. The mineralized porphyries show surface expressions of 190 x 225 m in a conical hill, and 250 x 400 m in a twin hill (Fig. 1).

Lithogeochemical analyses of the least altered Dalli intrusions and volcanic rocks indicate high Sr/Y (49-61) and Eu/Eu* (0.89-0.92), features typical of Cu porphyries. LA-ICP-MS U-Pb dating of zircons of the mineralized quartz diorites and andesite porphyry yielded magmatic crystallization ages of 16.0-15.4 Ma. The zircon trace element concentrations of Dalli are characterized by high Eu/Eu* (0.3-0.8), (Ce/Nd)/Y (0.01-0.3) and 10,000*(Eu/Eu*)/Y (2-15) ratios (Fig. 1), similar to fertile porphyry suites such as the giant Sar-Cheshmeh and Qulong porphyry Cu deposits along the Tethyan metalogenic belt. This suggests that the Dalli intrusions are fertile and require extensive deep drillings to define their real potential.

Disseminated and stockwork Cu-Au mineralization at Dalli is mostly related to potassic alteration, comprising magnetite and well-developed late K-feldspar and biotite, and quartz-sericite-specularite overprint, surrounded by extensive barren argillic and propyllitic alterations. In the peripheries of the porphyry centers, there are some N-trending vuggy quartz veins, which host epithermal Au-Ag-As-Sb-Bi-Pb-Zn-Cu mineralization. Geochemical analyses of drill core samples showed that the core of the porphyry stocks is low grade, whereas the high-grade disseminated and stockwork mineralization (~1% Cu and ~1.2 g/t Au) occurred within 40 m on both sides of the contact between the porphyry stocks and wall-rock andesite porphyry. These contact zones are favorable drill targets for high-grade copper and gold mineralization.

There is a strong correlation between Cu and Au, and both show a second-order correlation with Fe, As, Pb, and Zn at the Dalli deposit. A ground magnetic survey revealed two significant magnetic anomalies, which are associated with intensive mineralized potassic alterations in quartz-diorite intrusions and their contacts with andesite porphyries. A relatively weaker E-trending magnetic anomaly, showing no surface porphyry expressions, was identified in the reduced-to-the-pole magnetic map of the area. This magnetic anomaly is associated with a lithocap, consisting of argillic alteration and surface expressions of epithermal mineralization, and is indicative of a possible hidden porphyry center. The discovery and characteristics of the Dalli deposit highlight the importance of the UDMA as a potentially significant, economically important, but relatively underexplored porphyry province.

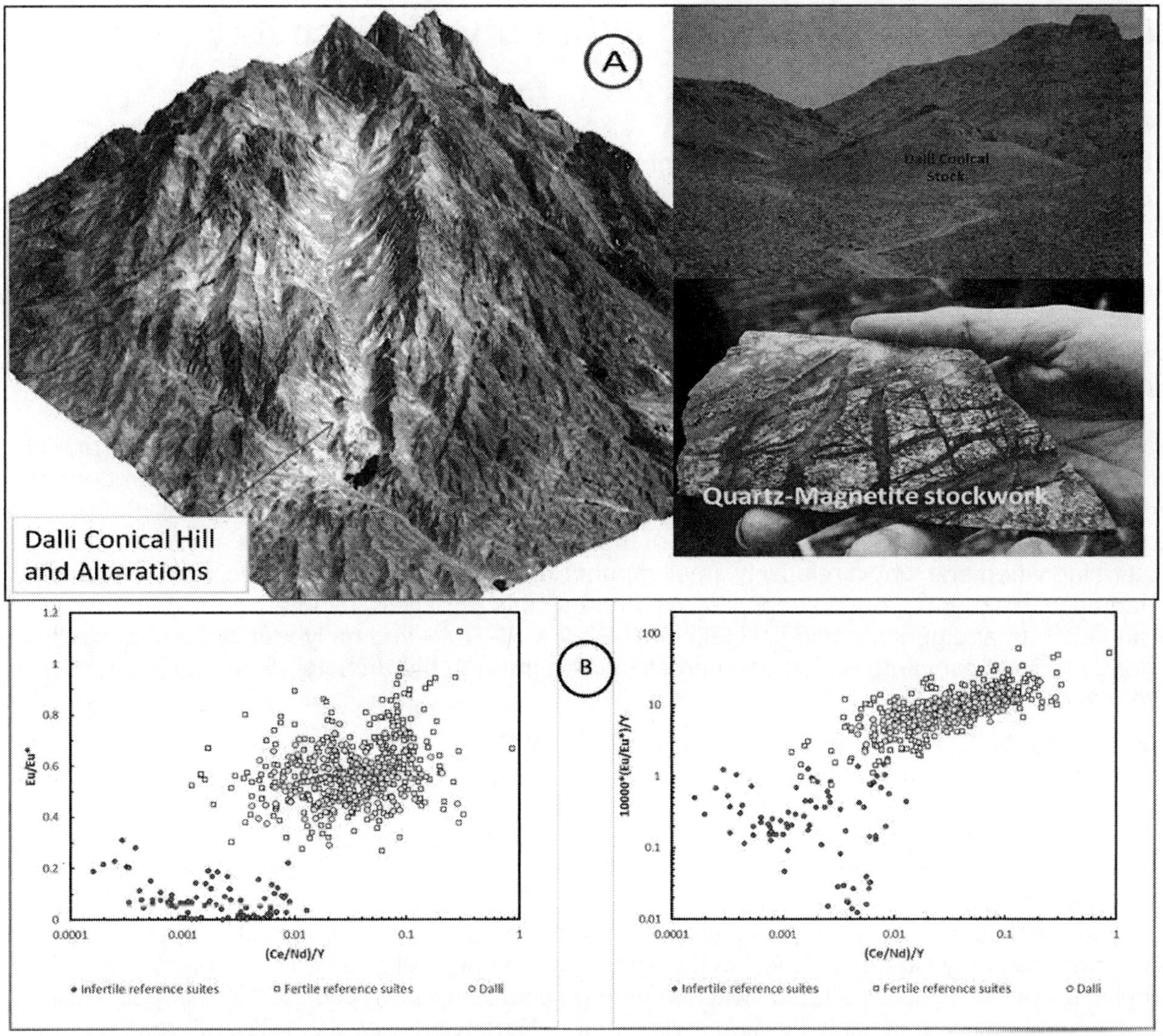

Fig. 1 A -Satellite Image and field overviews of the Dalli deposit, B- Zircon trace element composition of the Dalli porphyries in comparison with the infertile and giant fertile reference porphyry suites.

ST.213
Abrupt Switch in Magmatic Plumbing to Tap Porphyry Copper-Fertile Magmas

Lawrence C. Carter[1], Simon R. Tapster[2], Ben J. Williamson[1, 3], Yannick Buret[3], David Selby[4, 5], Daniel Parvaz[6, 7]

1. Camborne School Of Mines, University of Exeter, Cornwall, United Kingdom, 2. British Geological Survey, Nottingham, United Kingdom, 3. Natural History Museum, London, United Kingdom, 4. Durham University, Durham, United Kingdom, 5. State Key Laboratory of Geological Processes and Mineral Resources, Wuhan, China, 6. Selfrag AG, Kerzers, Switzerland, 7. Lightning Machines, Manchester, United Kingdom

Porphyry-type Cu deposits are amongst the most intensively studied ore systems (Richards, 2018) and will play a major role in the shift towards a green economy. They form from large, long-lived, multicomponent magmatic complexes, mostly in subduction-related settings (e.g.,Seedorff et al., 2005; Sillitoe, 2010; Wilkinson, 2013). Due to limited or patchy exposure of their 3D architecture (Seedorff et al., 2008), it is often difficult to determine when and why a relatively small number of systems, and more specifically certain batches of magma within them, produce deposits. We address this from field, mineralogical, geochemical (whole-rock and zircon), and geochronological (zircon U-Pb CA-ID-TIMS and molybdenite Re-Os) studies of the unique, comparatively complete ~8-km-deep cross section through the archetypal Yerington porphyry system, Nevada (Dilles, 1987).

We show that long-lived (>1.5 m.y.), relatively shallow (~1-8 km; Dilles, 1987) emplacement of the premineralisation McLeod Hill quartz monzodiorite and Bear quartz monzonite plutons, from a mid-crustal source (~15-20 km depth), rapidly switched (over ~100 k.y.) to slightly deeper intrusion (~4-8 km; Dilles, 1987) of more evolved magmas from a lower crustal source (~20-40 km depth) to form the Luhr Hill granite. Inward crystallisation of the latter produced Cu- and volatile-rich magmas, which were emplaced upwards as multiple generations of aplite dikes. These acted as crystal mush conduits (Carter et al., 2021) for exsolving magmatic-hydrothermal fluids from large volumes of underlying magma, over a period of up to ~1.1 m.y., to form the four known porphyry Cu deposits within the apices of the Luhr Hill granite. This framework, where porphyry deposit formation relied on a rapid switch in magma source and evolution to more copper-fertile compositions and the focussing of fluids through crystal mush dikes (Carter et al., 2021), offers new constraints for exploration models.

References

Carter, L.C. et al., 2021, Commun. Earth Environ., 2, 59.

Dilles, J.H., 1987, Econ. Geol., 82, 1750-1789.

Richards, J.P., 2018, Econ. Geol., 113, 1225-1233.

Seedorff, E. et al., 2005, Econ. Geol., 100, 251-298.

Seedorff, E. et al., 2008, Econ. Geol., 103, 939-956.

Sillitoe, R.H., 2010, Econ. Geol., 105, 3-41.

Wilkinson, J.J., 2013, Econ. Geol., 6, 917-925.

ST.214

Rare Earth Element and Sulfur Partitioning Between Apatite and Hydrothermal Fluids as a Function of Temperature, Fluid Composition, and Oxygen Fugacity

Justin Casaus[1], Daniel Harlov[2], Adam Simon[1], Brian Konecke[3]
1. University of Michigan, Ann Arbor, MI, USA, 2. GeoForschungsZentrum, Potsdam, Germany, 3. Astromaterials Research and Exploration Science (ARES), NASA Johnson Space Center, Houston, TX, USA

The mineral apatite—$Ca_{10}(PO_4)_6(F,Cl,OH)_2$—is nearly ubiquitous in terrestrial and lunar geologic systems and can incorporate nearly one-third of the periodic table of elements, including polyvalent sulfur (S) and as much as 10 wt % rare earth elements (REEs). Polyvalent S (e.g., S^{6+}, S^{2-}) is critical for the transport and enrichment of base and precious metals in igneous and hydrothermal mineral deposits. Because of their magnetic and luminescent properties, REEs are important for the manufacturing and development of renewable energy and consumer technologies. Work by Sadove et al. (2019) on metasomatized apatite from the magnetite-sulfide Philips Mine in New York revealed regions with secondary REE-rich phases and elevated REE and S concentrations associated with dominantly S^{6+} (as determined by S-XANES). They propose that the elevated concentrations of S and REEs are the result of apatite metasomatism by an oxidized hydrothermal fluid. To our knowledge, no experimental studies have explored the effects of temperature, fluid composition, and oxygen fugacity (fO_2) on S and REE partitioning between apatite and hydrothermal fluids at PTX conditions relevant for upper crustal systems.

Here, we will discuss results from experiments performed at 100 MPa, 500°–800°C and variable fluid compositions (mHCl, mH_2SO_4, $Na_2Si_2O_5$) buffered at the Ni-NiO mineral equilibrium (where S is present as both reduced and oxidized sulfur) and new data from experiments buffered at M-H (where the S present is dominantly oxidized) and Co-CoO (where the S present is dominantly reduced). The experiments are designed to constrain the partitioning of S and REEs between fluid and fluorapatite [$Ca_{10}(PO_4)_6(F)_2$]. Run product apatite grains were analyzed by EPMA, SEM EDX mapping, S-XANES maps of S oxidation states, and CL images. The data from experiments buffered at NNO revealed altered regions of apatite that exhibit depleted Na, Si, and S concentrations and leaching of REEs via dissolution-reprecipitation of the starting apatite, which resulted in the growth of secondary REE phases, e.g., monazite. The apatite was more reactive with HCl-bearing fluids than H_2SO_4-bearing fluids, and the leaching of Na, Si, S, and the REEs was buffered by the presence of Na and Si in the fluid. The data from experiments buffered at NNO also reveal that apatite-fluid reactivity decreases with increasing temperature for mHCl-bearing fluids, which can be explained by the increased association of HCl at elevated temperatures. The mineral apatite is not currently sought after as a source of REEs; however, as the need for these energy-critical metals is projected to increase, an understanding of the PTX conditions that result in REE-rich apatite is of interest.

ST.216
Geologic Setting and Mineralization of Gicik Low-Sulfidation Epithermal Au Deposit (North-Central Anatolia)

Akın Çil, Ali İmer
Department of Geological Engineering, Middle East Technical University, Ankara 06800, Turkey

The Turkish part of the Mesozoic-Cenozoic Western Tethyan Metallogenic Belt contains a large number of epithermal systems mainly clustered in western and northeastern Anatolia. The Gicik gold deposit is located within a rather unexplored section of this belt in north central Anatolia, about 15 km north of Ankara city center. Gold mineralization is hosted in middle Eocene agglomerates and massive to flow-banded andesite to dacite lavas overlying the Variscan basement of the southern margin of the Laurasian Sakarya Zone. Volcanic rocks are medium-K, calc-alkaline in character and show relative depletions in high field strength elements and relative enrichments in large-ion lithophile and light rare earth elements. These characteristics indicate that middle Eocene magmatism at Gicik was developed in a subduction-related setting.

Gold mineralization occurs in a series of NNE-trending, vertical to subvertical siliceous veins (up to 10 meters thick), which have been localized along dextral reverse faults. Fault plane measurements indicated two distinct fault sets: the first oriented NE-SW and hosting the mineralized veins and the second set oriented in NW-SE to E-W, offsetting the mineralized structures. According to the paleostress analyses of fault-slip data, vein-hosted gold mineralization was interpreted to have developed under NW-directed transpression.

Widespread hydrothermal alteration is characterized by vein-proximal silicification consisting of chalcedony/fine-grained massive to colloform-banded quartz to coarser-grained gray quartz as well as late amethystic quartz, surrounded by wider zones of argillic alteration represented by quartz-illite-kaolinite assemblages. District-scale propylitic alteration, characterized by chlorite-epidote-carbonate-quartz-pyrite assemblages, envelops the aforementioned alteration domains.

Siliceous veins display typical textures of low-sulfidation epithermal systems including colloform banding, hydrothermal brecciations, stockwork veinlets, and lattice-bladed textures. Mineralized samples from the surface have been extensively oxidized into hematite and goethite. Relatively fresh exposures along steep valleys, however, contain hypogene sulfide phases including pyrite, arsenopyrite, and Ag tetrahedrite together with native gold. Supergene covellite and digenite locally overprints Ag tetrahedrite hosted in colloform banded veins. Electron microprobe analysis revealed that significant concentrations of gold and silver (up to ~700 ppm and ~8.5 wt %, respectively) are contained within sulfide/sulfosalt minerals.

K-Ar and $^{40}Ar/^{39}Ar$ dating of igneous and hydrothermal minerals suggest that volcanic activity and epithermal-style alteration occurred contemporaneously at ~44 Ma during the terminal stages of widespread Paleogene volcanism in north central Anatolia. This period of volcanism is considered to have been triggered either by collisional processes or by breakoff of the downgoing northern Neotethyan slab following collision between Sakarya Zone and the Tauride-Anatolide Block. Overall deposit characteristics of the Gicik low-sulfidation epithermal system are analogous with similarly aged gold deposits of the collisional belt located in Biga Peninsula and the eastern Pontides. Thus, we propose that middle Eocene tectonomagmatic conditions along the central Sakarya Zone were also favorable for generation of low-sulfidation (and possibly intermediate-sulfidation) epithermal mineralization.

ST.217
A New Discovery in Biga Peninsula: Geologic, Mineralogical-Petrographic, and Fluid Inclusion Characteristics of Çatalçam (Soma-Manisa) Au-Pb-Zn-Cu Mineralization

Ramazan Sarı[1], Şahset Küçükefe[1], Zehra Deveci Aral[1], Gülcan Bozkaya[2], Ömer Bozkaya[2], Fatih Bademler[1], Elif D. Bayrakçıoğlu[1], Cahit Dönmez[1], Serkan Özkümüş[1]
1. General Directorate of Mineral Research and Exploration, Ankara, Turkey, 2. Pamukkale University, Denizli, Turkey

Biga Peninsula, located in the NW Turkey and one of the main porphyry and epithermal mineralization belts, recorded multiphase deformation and magmatism events caused by successive opening and closing of the Paleo- and Neotethys oceans. Following the closure of the Neotethyan Ocean, Eocene-aged arc-related magmatism developed and continued until the Lower Miocene. The arc-related magmatism in the region is the source of porphyry and epithermal-ype polymetallic mineralizations.

In the south of Biga Peninsula, within the scope of the exploration project actively conducted by MTA, a diorite porphyry intrusion (Çatalçam Intrusion) that cuts the Lower and Middle Miocene-aged volcanic rocks was mapped for the first time, and Au-Pb-Zn-Cu mineralization, which is thought to have developed related to this intrusion, was discovered. Crosscutting and/or telescopic potassic (magnetite, biotite, undulated quartz), phyllic (sericite, quartz), and argillic (kaolinite, barite, quartz, and calcite) alteration zones are developed in porphyry body and host volcanic rocks. Stockwork vein-veinlets containing pyrite, chalcopyrite, sphalerite, and galena, as well as gold, have formed in the Çatalçam Intrusion and host rocks. While A-type quartz vein-veinlets are formed in the deep levels, crosscutting quartz, calcite, dolomite, and barite vein-veinlets are observed in the middle and upper levels. Based on the fluid inclusion data from sphalerite, quartz, and barite; the temperatures of the ore-forming solutions are collected in three groups (150°-220°C: argillic, 250°-350°C: phyllic and >400°C: potassic). Inclusions containing hematite, chalcopyrite, and salt solid phases as well as liquid and gas phases indicate that mineralization formed related to the intrusion. Fluid inclusion data displays that the temperatures and salinity of the ore-bearing solutions in the early stage (porphyry) are higher, and they are lower in the late stage (epithermal). As a result, Çatalçam mineralization developed in two different phases: porphyry Au-Cu and the overlapping epithermal Zn-Pb-Cu-Au mineralization. This intrusion and mineralization are important discoveries that will affect the future of the mineral exploration strategy in the region and expand the target time range and area.

ST.218

Processes and Mineralogy Involved in the Porphyry-Epithermal System in California Prospect, Tolima (Colombia)

Iván Mateo Espinel Pachón[1], Mónica Á. López[1], Andrés F. González Durán[2], Juan C. Molano Mendoza[2], Juan D. Alarcón Rodriguez[2]
1. University of Geneva, Geneva, Switzerland, 2. Universidad Nacional de Colombia, sede Bogotá, Bogotá, Colombia

California prospect is located 200 km west from Bogotá, Colombia, and 8 km SE from the La Colosa gold-rich porphyry. Within this area, gold was extracted from quartz veins by artisanal miners in the 1960s. In this study, the mineralization and processes were characterized using field work, petrography, metallography, X-Ray diffraction, electron probe microanalyzer (EPMA), Raman spectroscopy, and VNIR-SWIR spectroscopy.

The host rocks in the area are muscovite-quartz-graphite schists of Palaeozoic age, which were mineralized by the emplacement of at least two porphyritic dacites. Phyllic alteration was identified in the area, accompanied by disseminated sulfides (mainly pyrite, and base metals in minor amount). Quartz veins are cutting the disseminated mineralization and contain Fe-rich sphalerite, galena, chalcopyrite, pyrrhotite, marcasite, tetradymite, and pyrite.

There is an important structural control of the mineralization. Most of the veins found in the field are related to extensional structures, which cut the foliation of the schists. Tectonic breccias with clasts of mineralized dacites were also found.

Intermediate product was identified in this area. This texture was associated with the redox change of the fluid generated by the reduced host rock (graphite schists). The GGIMFis thermometer was used in the Fe-rich sphalerite, resulting in temperatures of 250°C. This temperature was corroborated with illite crystallinity analysis.

The tellurides in this area were not reported before, and they are found as micrometric (30 μ) inclusions in pyrite.

This area is an interesting target for exploration and drilling because it can be related to the last stage of the La Colosa porphyry system and could lead to the discovery of new mineralized areas.

Mrc
200 microns

ST.220

New Insights into the Genesis of Apatite of the Sorkhe-Dizaj Magnetite Deposit (Tarom, NW Iran) Constrained by Fluid Inclusion Investigation

Shahin Ghahramani[1], Alexandre Tarantola[2], Hubert Whitechurch[1]
1. Strasbourg University/School and Observatory of Earth Science (EOST) Institute of Global Physics of Strasbourg (IPGS), Strasbourg, France, 2. Université de Lorraine, CNRS, GeoRessources, Vandœuvre-lès-Nancy, France

The Sorkheh-Dizaj magnetite-apatite deposits are situated in the Alborz-Azarbaijan magmatic belt (NW Iran). Eocene calc-alkaline volcanic rocks (Karaj Formation) are intruded, shortly after their emplacement, by felsic (granodiorite, monzonite, granite) plutons, over the subduction zone of the Neo-Tethys. They are of "type I" metaluminous, amphibole calc-alkaline (ACG) to potassic calc-alkaline (KCG) granitoid, of K-rich calc-alkaline affinity.

The magnetite-apatite mineralization is of iron oxide-apatite Kiruna type and occurs in massive vein types, irregular veinlets, or stockworks. They were formed under high oxygen fugacity, over the magnetite-hematite buffer. They have a high concentration of REEs (up to 8,000 times the chondritic value) and show a strong LREE/HREE ratio with a negative Eu anomaly.

Petrography and microthermometry of ~120 apatite fluid inclusions indicate entrapment of primary and secondary generations of inclusions. The variable volumes of vapor in the primary inclusions suggest that the fluids were boiling at the time of growth of apatite.

Low first-melting temperature (-50° to -28°C) suggests mixing of H_2O-NaCl-$CaCl_2$ in the fluids. Primary inclusions have homogenization temperatures ($T_{h\ LV\text{-}L}$) with an apparent bimodal repartition (~360°-380°C and ~180°-200°C). The salt content is in the range 1-40 wt % NaCl and 0-10.4 wt % $CaCl_2$. The homogenization temperatures of secondary fluid inclusions range between 115° and 234°C. The secondary inclusions have a salinity ranging between 3 and 22 wt % NaCl equiv, and in the H_2O-NaCl-$CaCl_2$ system, between 4.5 and 8.8 wt % NaCl and between 2.6 and 14.7 wt % $CaCl_2$, slightly richer in $CaCl_2$ than the primary inclusions. The calculated pressure at entrapment conditions is less than 200 bar.

Raman spectrometry reveals that the gas phase is composed of pure H_2O without any trace of CO_2, CH_4, N_2, H_2, and H_2S. The compositions of microsolids detected in the primary inclusions reveal the presence of magnetite, apatite, chalcopyrite, xenotime, calcite, celestite, goethite, and hematite in different proportions. This suggests that the fluids circulating around the apatites were oversaturated in the elements constituting these solids.

These observations suggest that high temperatures and high-salinity magmatic fluids, observed in the primary inclusions, rich in Fe-REE-P-S and metals, exsolved from felsic magmas. Moving upward in the already solidified pluton at low depth, they encountered other fluids of lower salinity and temperature coming from the surface (meteoritic, seawater, and/or basinal fluids). These fluids do not mix completely, as pure water has been detected in some primary inclusions besides primary fluid inclusions with the highest salinity.

Over the intrusion and laterally, sulfide veins filled fractures at lower temperatures. These fluids correspond in salinity and homogenization temperature to the secondary fluids in the apatite and those filling the quartz-sulfide veinlets crosscutting the Fe apatite deposits.

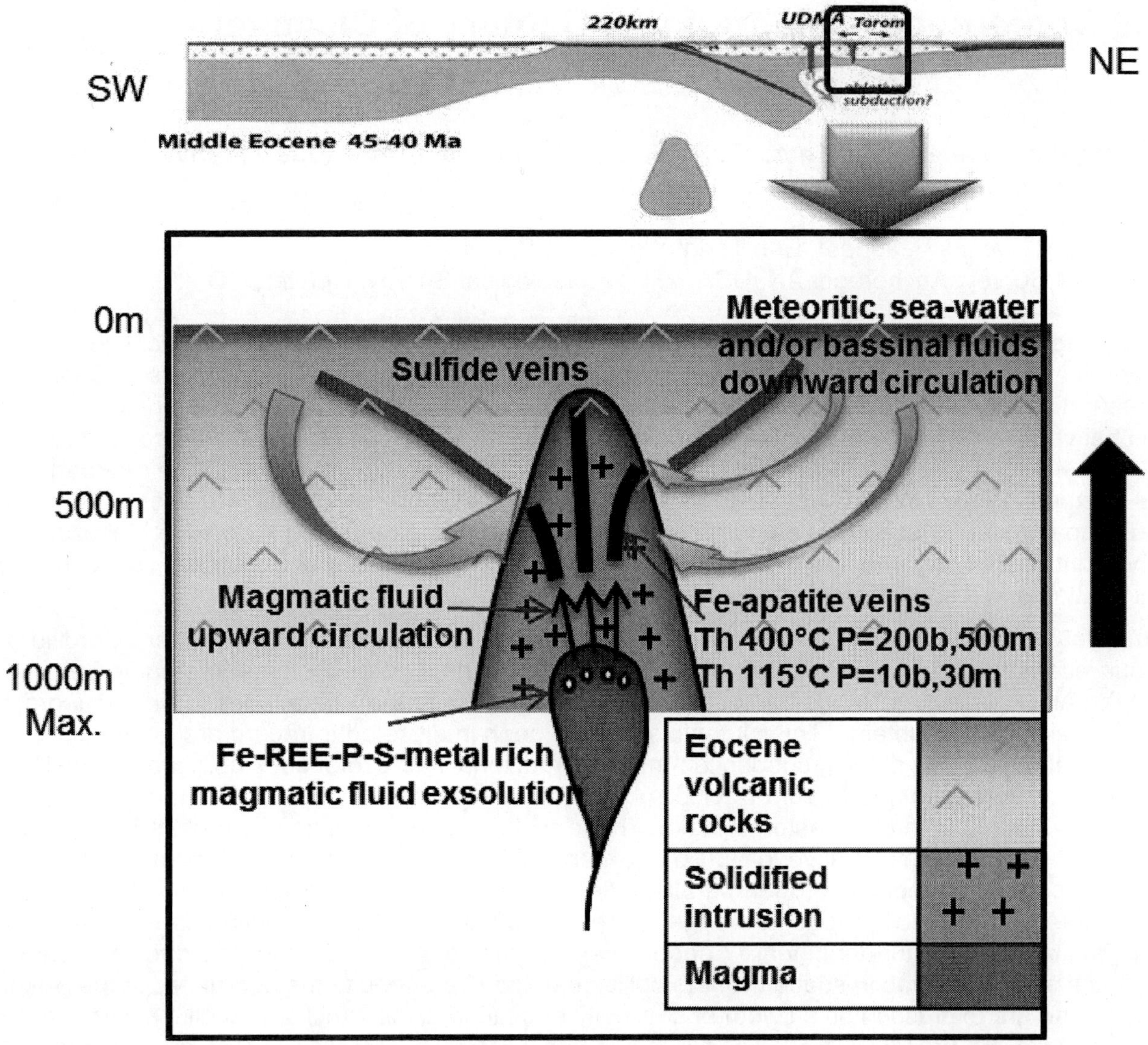
220km
UDMA
Tarom
SW
NE
subduction?
Middle Eocene 45-40 Ma
0m
500m
1000m
Max.
Sulfide veins
Meteoritic, sea-water
and/or bassinal fluids
downward circulation
Magmatic fluid
upward circulation
Fe-apatite veins
Th 400°C P=200b,500m
Th 115°C P=10b,30m
Fe-REE-P-S-metal rich
magmatic fluid exsolution
Eocene
volcanic
rocks
Solidified
intrusion
Magma

ST.223

Reconstructing the Geometry and Tectonic Setting of Mid to Late Cretaceous Porphyry Formation in Eastern Interior Alaska

Douglas C. Kreiner[1], James V. Jones[1], Erin Todd[1], Jonathan S. Caine[2]
1. U.S. Geological Survey, Anchorage, AK, USA, 2. U.S. Geological Survey, Denver, CO, USA

Understanding the original, regional geometry of Mesozoic magmatic belts in Alaska is a critical component for interpreting the tectonic setting in which porphyry deposits formed. Hundreds of kilometers of Cenozoic strike-slip separation along major faults has substantially disrupted or obscured possible tectonic connections between terranes, overlap assemblages, and regional igneous suites. Here we consider a geological reconstruction that restores ~400 km of dextral separation on the Denali fault. The restored Cretaceous geometry places the Talkeetna Mountains (presently in south-central Alaska) directly south of the Yukon-Tanana upland (presently in eastern interior Alaska), bringing regionally extensive Mesozoic magmatic belts into better alignment and providing a framework for interpreting geochemical, isotopic, and metallogenic patterns in a spatially realistic framework.

Early and middle Cretaceous magmatism (~135-90 Ma) occurred along the entire northwestern Cordillera margin. In our reconstructed geometry, rocks of this age range define a ~1,000-kilometer arcuate belt in Alaska from the Alaska-Aleutian Range batholith in southwest Alaska to the Yukon border, with a magmatic gap across the Talkeetna Mountains. This plutonic suite has been interpreted to record arc magmatism above a moderately to steeply dipping subduction margin. Additional mid-Cretaceous collisional intrusive suites are emplaced inboard of the Circum-Pacific subduction margin define a 40- to 100-km-wide belt across the Yukon-Tanana upland in eastern Alaska. These rocks are characterized by crustal Sr and Nd isotope ratios and elevated U/Yb relative to Nb/Yb in zircon. Collisional intrusive suites are interpreted to have been emplaced during regional orogen-parallel extension in the thickened back arc and record collapse of a Jurassic-Cretaceous orogenic wedge. Between ~90 and 80 Ma, subduction configuration changed, shifting magmatism further inboard (modern-day north) into the mid-Cretaceous back arc. Late Cretaceous arc magmas overlap in space with the collisional mid-Cretaceous crustal granites in the Yukon-Tanana upland and are expressed as a swath of shallowly emplaced dikes, stocks, and dike swarms and coeval volcanic rocks in a broadly NW-SE belt extending into western Yukon. Arc plutons are characterized by distinctly more juvenile Sr and Nd isotope ratios and lower U/Yb relative to Nb/Yb in zircon compared with the collisional plutons.

Porphyry systems formed throughout the duration of Cretaceous magmatism in eastern Alaska. However, the most prolific porphyry formation (e.g., Pebble, Orange Hill, Bond Creek) occurred in association with arc magmatism in districts southeast and southwest of the Yukon-Tanana upland. In contrast, porphyry Mo systems with local W-rich skarn occurrences are more prevalent in the Yukon-Tanana upland in association with collisional mid-Cretaceous granites. Late Cretaceous porphyry systems in the upland and in western Yukon are associated with the inboard calc-alkaline granodiorite to monzonitic intrusions and consist of porphyry Cu-Mo deposits with variable Au enrichments. In eastern Alaska, inferred changes in the geometry of the subduction interface are interpreted to have controlled spatial and temporal variations in the distribution and style of porphyry systems along and across the strike of the margin.

ST.224

Late Cretaceous Superimposed Porphyry Systems in the Yukon Cordillera: Insights from the Klaza Epithermal System, Dawson Range

Well-Shen Lee[1], Daniel J. Kontak[1], Jeremy P. Richards[1], James Crowley[2], Robert R. Creaser[4], Terry Spell[3], Andrew DuFrane[4]
1. Laurentian University, Sudbury, ON, Canada, 2. Boise State University, Boise, ID, USA, 3. University of Nevada Las Vegas, Las Vegas, NV, USA, 4. University of Alberta, Edmonton, AB, Canada

The Canadian Cordillera of British Columbia and contiguous Yukon contains a substantial metal endowment, including many porphyry Cu-Au ± Mo deposits (PCDs) clustering at specific intervals in the geological record. Understanding specifically one of these metalliferous periods, the Late Cretaceous event in the Yukon, was previously hindered due to the resolution of conventional geochronologic methods in addition to the lack of modern petrogenetic models and data for magmatic suites. By integrating field studies with newer high-quality age and lithogeochemical data, it is now possible to address the temporal and spatial distribution of Late Cretaceous PCDs. This work will help to refine models for PCD formation, which will contribute to brownfield and grassroots exploration for such deposits in the Yukon.

In this study, new field-based observations are integrated with geochronological, lithogeochemical, and mineral paragenetic data from the Klaza-Mt. Nansen district with data additional data collected from the giant Casino deposit and the Freegold Mountain district to update the temporal and genetic understanding of the Late Cretaceous PCDs in west-central Yukon. High-quality age data generated for zircon (LA-ICP-MS, CA-TIMS), molybdenite (Re-Os) and muscovite (Ar-Ar) display a bimodal distribution for timing of intrusions, mineralization, and alteration. The early-Late Cretaceous pulse (80-74 Ma) coincides with PCD formation (Casino, Revenue, Nucleus, Kelly) and the Casino suite intrusive activity, whereas the late-Late Cretaceous pulse (72-69 Ma) correlates with IS epithermal deposits (Au-Ag-Pb-Zn; Klaza, Mt. Nansen, Frog) and a previously unrecognized pulse of Cu-Mo porphyry-type mineralization (Cyprus), which may also overprint earlier mineralization (Kelly, Casino).

The four intrusive suites noted display weak arc-like signatures, with strong depletions in HREEs relative to LREEs, signifying delayed plagioclase fractionation. A progressive overall depletion in REEs is observed as intrusive suites evolve from the Late Triassic to the late-Late Cretaceous with each successive suite displaying a greater degree of HREE depletion. Differentiation of the intrusive suites is possible using La/Yb vs. Yb and Zr/TiO_2 vs. Nb/Y plots. Thus, the Prospector Mountain suite contains the highest La/Yb values and is slightly more alkaline relative to the Casino suite. Both suites plot in the adakitic field due to increased Sr/Y, signifying increased garnet fractionation and water content. These results suggest increasingly higher pressures of melting for each successive magmatic suite.

While the Casino suite intrusions have long been understood to be highly prospective for ca. 75 Ma porphyry Cu-Au potential, the younger, less voluminous Prospector Mt. suite intrusions have been significantly overlooked. We propose (based on current field and geochronologic evidence) that several localities in the Dawson Range display evidence of superimposed Casino- and Prospector Mt. suite magmatism and hydrothermal acitvity. These superimposed porphyry systems contain heightened potential for enriched hypogene ore shells at depth, thus representing significant untapped resources in the Yukon Cordillera.

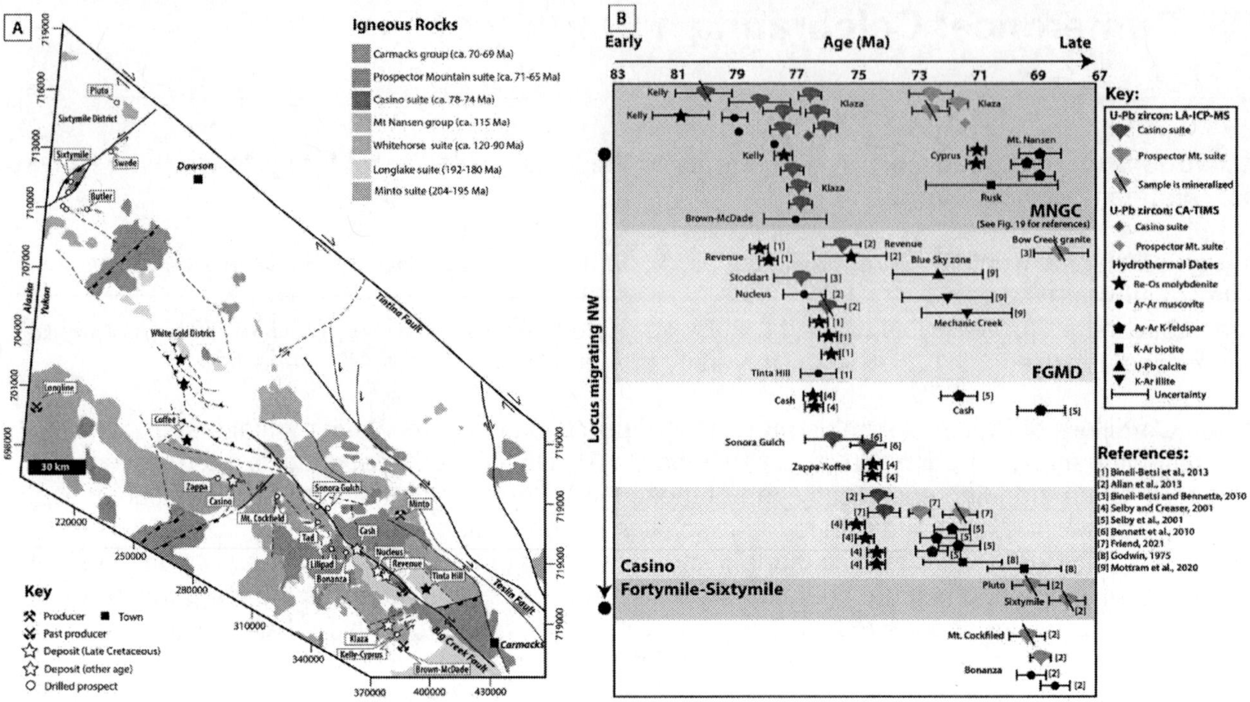
A
Igneous Rocks
Carmacks group (ca. 70-69 Ma)
Prospector Mountain suite (ca. 71-65 Ma)
Casino suite (ca. 78-74 Ma)
Mt Nansen group (ca. 115 Ma)
Whitehorse suite (ca. 120-90 Ma)
Longlake suite (192-180 Ma)
Minto suite (204-195 Ma)
Pluto
Sixtymile District
Sixtymile
Swede
Butler
Dawson
Alaska
Yukon
Tintina Fault
White Gold District
Longline
Coffee
30 km
Zappa
Casino
Sonora Gulch
Mt. Cockfield
Minto
Tad
Cash
Lilipad
Nucleus
Revenue
Bonanza
Tinta Hill
Teslin Fault
Big Creek Fault
Klaza
Kelly-Cyprus
Brown-McDade
Carmacks
Key
Producer
Town
Past producer
Deposit (Late Cretaceous)
Deposit (other age)
Drilled prospect
B
Early
Age (Ma)
Late
83
81
79
77
75
73
71
69
67
Locus migrating NW
Kelly
Klaza
Cyprus
Mt. Nansen
Rusk
Brown-McDade
MNGC
(See Fig. 19 for references)
Revenue
Blue Sky zone
Bow Creek granite
Stoddart
Nucleus
Mechanic Creek
Tinta Hill
FGMD
Cash
Sonora Gulch
Zappa-Koffee
Casino
Fortymile-Sixtymile
Pluto
Sixtymile
Mt. Cockfiled
Bonanza
Key:
U-Pb zircon: LA-ICP-MS
Casino suite
Prospector Mt. suite
Sample is mineralized
U-Pb zircon: CA-TIMS
Casino suite
Prospector Mt. suite
Hydrothermal Dates
Re-Os molybdenite
Ar-Ar muscovite
Ar-Ar K-feldspar
K-Ar biotite
U-Pb calcite
K-Ar illite
Uncertainty
References:
[1] Bineli-Betsi et al., 2013
[2] Allan et al., 2013
[3] Bineli-Betsi and Bennette, 2010
[4] Selby and Creaser, 2001
[5] Selby et al., 2001
[6] Bennett et al., 2010
[7] Friend, 2021
[8] Godwin, 1975
[9] Mottram et al., 2020

ST.227
Alteration Zonation and Genesis of the Houkeng Lithocap, East China

Qiuping Liu[1], Lejun Zhang[1], Juxing Tang[2], JingJing Dai[2]
1. ARC Research Hub for Transforming the Mining Value Chain (TMVC),University of Tasmania, Hobart, TAS, Australia, 2. Institute of Mineral Resources, Chinese Academy of Geological Sciences, Beijing, China

Lithocaps are wildly distributed in southeastern China, but no research has been done related to the lithocaps before. So we are trying to explore for potential porphyry-epithermal deposits among these lithocaps.

We conducted SWIR mapping at Houkeng lithocap using Terrespec 3 aiming to understand vertical and horizontal hydrothermal alteration zonation. The results indicate the alteration grades vertically from corundum-sericite at depth, through quartz-sericite-pyrophyllite, to alunite-kaolinite assemblages at shallow levels. Laterally, vuggy quartz-rutile (Fig. 1a) develops in the core, grading outward to advanced argillic (alunite-pyrophyllite-kaolinite-diaspore-aluminum phosphate and sulphate minerals) and argillic (kaolinite-illite-montmorillonite) alteration. Moreover, The SWIR 1,480-nm absorption peak of alunite at Houkeng ranges from 1,478 to 1,493 nm. The absorbtion peak shifts to higher wavelengths when Na/(Na + K) are higher, which is consistent with the research done in Lenpato lithocap by Chang (2011). Backscattered electron images, CL, and EMPA analyses of alunites show that alunites are typically zoned, with the core high in Ca, Na, and P contents, whereas K and Al are enhanced in the rim. Locally there are small APS minerals included in alunite.

Pyrophyllite, whose SWIR spectrum absorption peak at 2,167 nm, shows two origins: sericite retrograde alteration (Fig. 1d) and kaolinite metamorphism (Fig. 1e). Corundum (Fig. 1h) is altered to sericite by hydrothermal alteration, and later dickite (Fig. 1g) crosscutting advanced argillic alteration indicates a retrograde process.

Alunite from Houkeng shows average $\delta^{18}O$ = 8.29‰ and δD= -52.83‰ that suggest a dominant magmatic water origin. Alunite yields $\delta^{34}S$ values from 16.1 to 19.4‰ (average 17.81‰), and the associated pyrite $\delta^{34}S$ values vary from -6.6 to -10.5‰ (average -8.55‰). Average $\Delta^{34}S_{Alun-Py}$ between alunite and pyrite is 26.36‰, further indicating their magmatic origin.

Secondary copper mineral veins (Fig. 1i) developed near the Houkeng lithocap indicate a potential for porphyry and high sulfidation in and below the lithocap.

Fig.1. Hand specimen photo, microphotographs, and SEM photos of alteration assemblages in Houkeng lithocap: a. vuggy quartz with coarse-grained pyrite; b. grey quartz and white alunite and fine grained pyrite; c. cross-polarized photograph of alunite and kaolinite; d. SEM photo of muscovite transforming into pyrophyllite; e. SEM photo of kaolinite transforming into pyrophyllite; f. cross-polarized photograph of diaspore, dickite; g. late dickite vein crosscutting advanced argillic alteration; h. SEM photo of corundum transformed into muscovite; i. secondary Cu minerals in Houkeng lithocap. Abbreviations: Alu = alunite, Cor = corundum, Dic = dickite, Dis = diaspore, Kao = kaolinite, Py = pyrite, Pyr = pyrophyllite, Q = quartz.

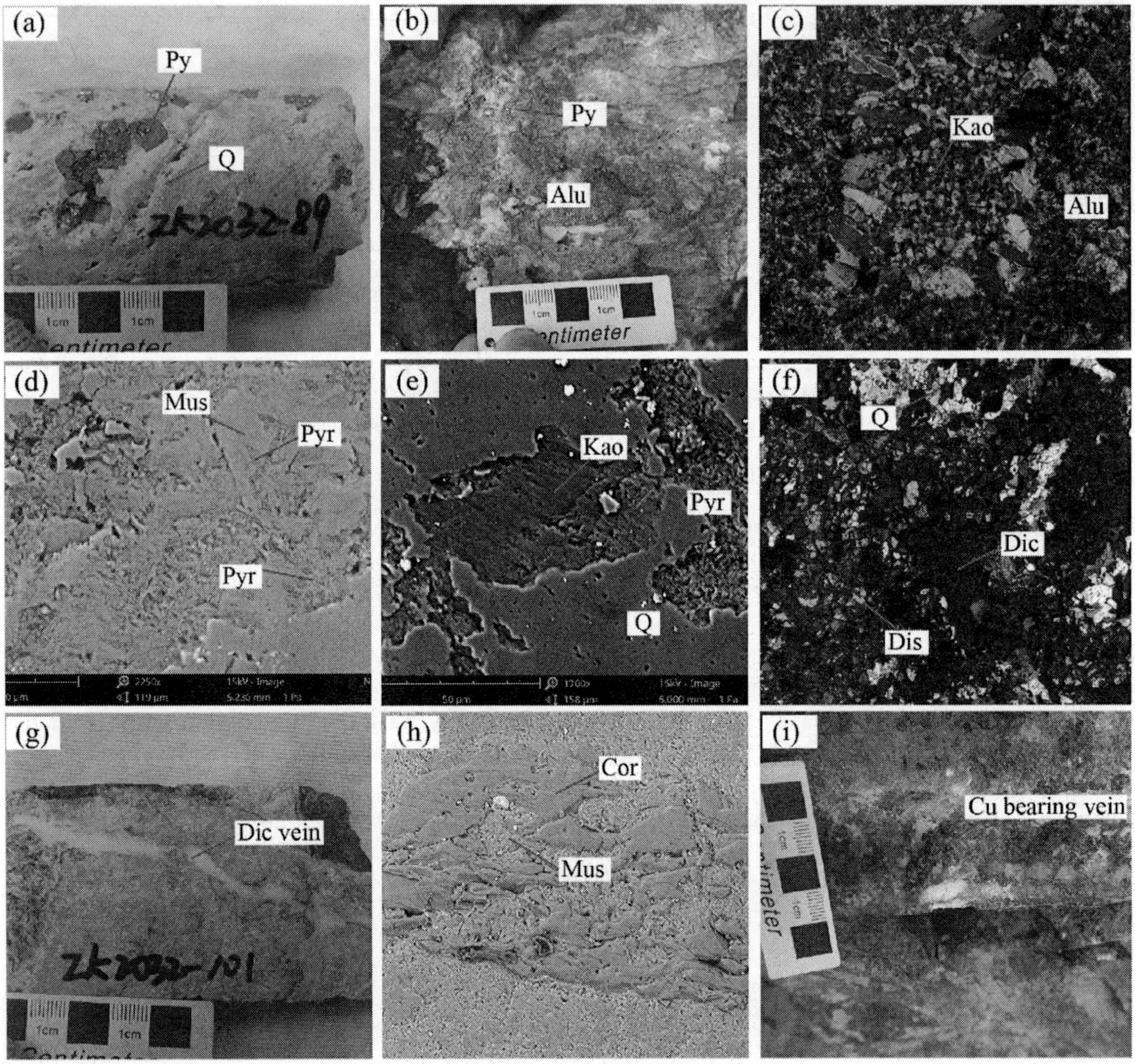

ST.229
Metal Earth in Chibougamau: Neoarchean Magmatism and Its Importance for Mineralizing Processes

Lucie Mathieu
Université du Québec à Chicoutimi - UQAC, Chicoutimi, QC, Canada

Magmatic fluids are an important carrier of precious and base metals in many mineralized systems (porphyry deposits, intrusion-related gold systems – IRGS, volcanogenic massive sulfides – VMS). The importance of fluids exsolved from magmas remains a source of much debate for some contexts, such as IRGS and orogenic gold deposits. Also poorly quantified is the importance of intermediate to felsic magmatism in the metal budget of Neoarchean greenstone belts. These issues are addressed using the Chibougamau area, NE corner of the Abitibi greenstone belt, as an example, which is characterised (1) by an abundance of Cu-Au and Au magmatic-hydrothermal mineralizing systems (porphyries and IRGS). The Chibougamau area also (2) displays several TTD (tonalite-trondhjemite-diorite) suites, while TTG (TT-granodiorite) are more abundant in the rest of the Abitibi greenstone belt, as shown by a recent interpretation of geochemical data compiled at the scale of the Abitibi belt (Mathieu et al., 2020; doi. 10.3390/min10030242). Understanding the importance of magmatism in mineralising processes requires a comprehension of magmatic systems and notably of the physicochemical parameters that favor metal uptake by magma. This contribution will highlight some of the first measurements of oxygen fugacity and volatile content of Neoarchean magmas located in the Abitibi greenstone belt, using zircon and apatite chemistry. Through the measurement of these parameters, comparison between TTG and TTD suites, as well as sanukitoïds and alkaline magmatism, becomes possible, and a discussion of the potential amount of metals that these ancient magmas can carried is proposed.

Fig. 1. Schematic geologic model showing the distribution of mineralizing systems of the synvolcanic period – volcanogenic massive sulfide (VMS), porphyry and epithermal systems – and of the syntectonic period – intrusion-related gold system (IRGS) and orogenic gold system (OGS). Three main stages of the evolution of the Abitibi greenstone belt are shown: (a) synvolcanic; (b) early syntectonic; and (c) late syntectonic periods. Figure from Mathieu (2021) - doi. 10.3390/min11030261

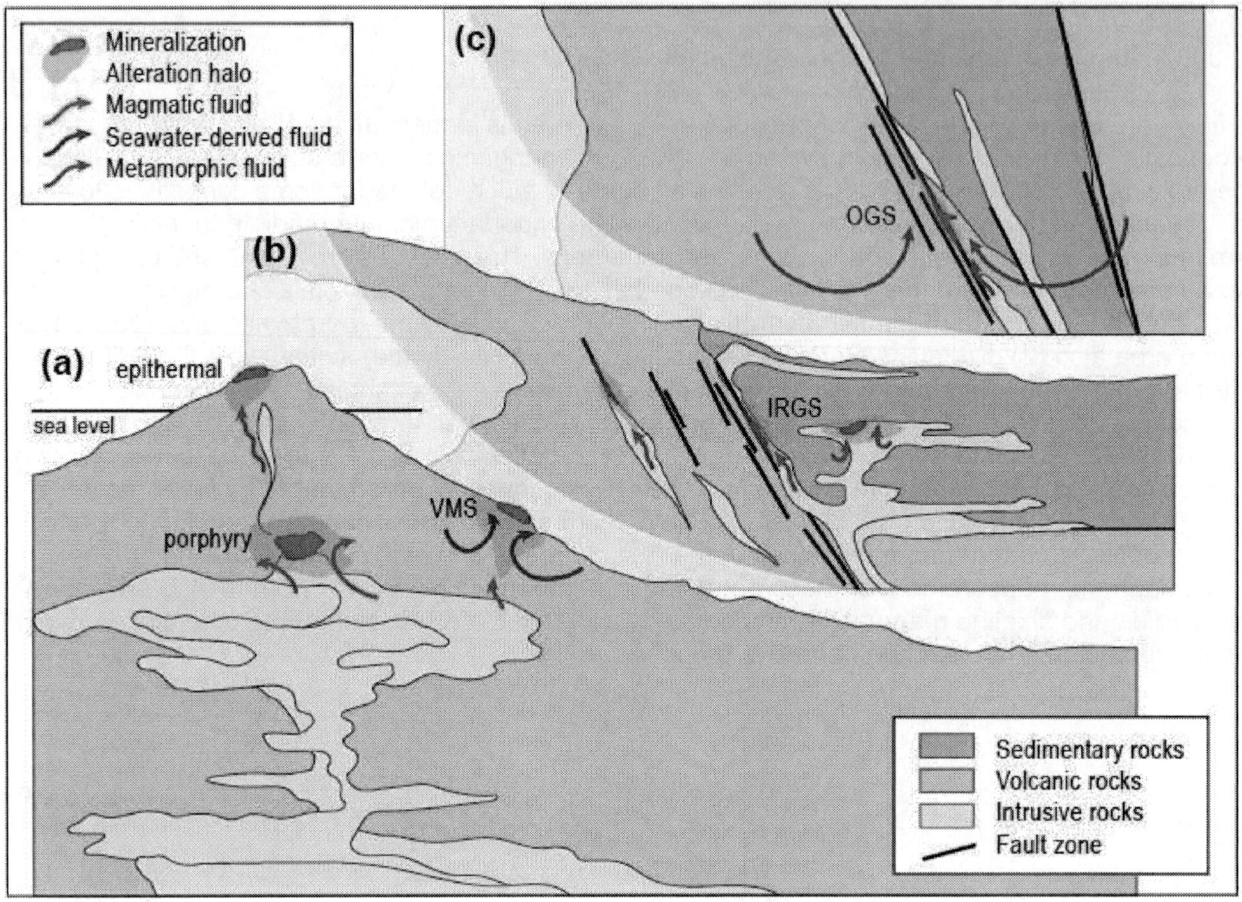

ST.230
Geologic, Geochronological, Mineralogical, and Geochemical Evidence from the Cretaceous Los Negritos Porphyry Copper Mineralized System, Chile

Miguel Montes[1, 2], Gema R. Olivo[2], Claire M. Chamberlain[5], Juliana C. Marques[3], Robert A. Creaser[4]
1. Teck Resources Chile Limitada, Santiago, Chile, 2. Department of Geological Sciences and Geological Engineering, Queen's University, Kingston, ON, Canada, 3. Departamento de Geologia, Instituto de Geociências, Universidade Federal do Rio Grande do Sul, Porto Alegre, RS, Brazil, 4. Department of Earth and Atmospheric Sciences, Edmonton, AB, Canada, 5. Anglo American, Vancouver, BC, Canada

The Los Negritos porphyry copper mineralized system is located ~4 km to the northeast of the Carmen de Andacollo Mine, the largest porphyry deposit in the Chilean Cretaceous metallogenic belt. This study included field work, mineralogical, geochemical, and geochronological investigations to determine the spatial distribution, signature, and overall evolution of the Los Negritos mineralized system. Mineralization at Los Negritos is hosted by andesite of the Quebrada Marquesa Formation and a series of at least four early to intramineral porphyritic intrusive rock types (named P1, P2, P3, P4), all dated by SHRIMP U-Pb in zircon: plagioclase quartz biotite porphyry (P1a: ca. 110 Ma and P1b: ca. 108 Ma); plagioclase biotite porphyry (P2: ca. 106 Ma); and quartz plagioclase biotite porphyry (P3: ca.106 Ma). These units are cut by late- to postmineral plagioclase-hornblende porphyritic rocks (P4a: ca. 106 Ma and P4b: ca. 106 Ma).

The earliest intrusive units (P1) were affected by an initial stage of K-feldspar-biotite alteration and associated chalcopyrite, molybdenite (Re-Os age ca. 109 Ma), and minor gold. Surrounding volcanic host rocks were overprinted by chlorite-epidote-dominated alteration. Subsequent to P2 and P3 intrusion, an albitic alteration followed by a second stage of potassic alteration affected intrusive units and volcanic wall rocks. These early stages of alteration were overprinted by copper-molybdenum-bearing chlorite-sericite alteration at ca. 107 Ma (Re-Os age in molybdenite) and by quartz-sericite-pyrite veins, respectively, in the southwest and northeast areas. Weak albite-calcite alteration, spatially associated with sulfosalts and distributed along the margins of P3, overprinted the quartz-sericite-pyrite facies.

The intrusive rock units at the Los Negritos and Carmen de Andacollo deposits are geochemically classified as diorite to granodiorite with a calc-alkaline affinity and formed in a volcanic arc setting from partial melting of a metasomatized mantle wedge. They are interpreted to be cogenetic and related to a common long-lived magma chamber that was emplaced during a period of tectonic inversion known as the Subhercynian, Peruvian, or Pacific event. This was synchronous with the formation of world-class IOCG deposits in Chile and Peru and iron oxide-apatite (IOA) deposits, representing a major metallogenic event. The integration of these results with the evolution of the Andean metallogenic belts will assist in understanding the relationship between porphyry and IOCG systems and implications for exploration models.

ST.231

Unveiling the Long-Lived Fertility Signature of Cu-Au Porphyry Systems: Insights from Apatite and Zircon at Tampakan, Philippines

Luis A. Parra-Avila[1, 2], Johannes Hammerli[1, 2, 3], Anthony I. Kemp[2], Bruce Rohrlach[4], Robert Loucks[2], Yongjun Lu[5], Ian S. Williams[6], Laure Martin[7], Marco L. Fiorentini[1, 2]
1. Centre for Exploration Targeting, School of Earth Sciences, The University of Western Australia, Perth, WA, Australia, 2. Australian Research Council Centre of Excellence for Core to Crust Fluid Systems, School of Earth Sciences, The University of Western Australia, Perth, WA, Australia, 3. Institute of Geological Sciences, University of Bern, Bern, Switzerland, 4. Sunstone Metals Ltd., Milton, QLD, Australia, 5. Geological Survey of Western Australia, Perth, WA, Australia, 6. Research School of Earth Sciences, The Australian National University, Canberra, ACT, Australia, 7. Centre for Microscopy, Characterisation and Analysis, The University of Western Australia, Perth, WA, Australia

The present study focuses on the use of accessory minerals such as zircon and apatite to investigate ore-forming processes relevant to porphyry systems. These are an essential source of a variety of base and precious metals such as Cu, Au, and Mo. Despite the importance of these mineral systems and the large number of studies available in the scientific literature, the metallogenic events that lead to their development remain elusive[1,2]. The spatiotemporal information of the ore-forming pulses and magmatic events that induce fertility remains enigmatic and difficult to fingerprint. Zircon and apatite from the world-class Tampakan porphyry and high-sulfidation system (2,940 million tons grading 0.52% Cu, 0.19 g/t Au, 64 ppm Mo) were characterized using a combined analytical approach that included whole-rock geochemistry, radiogenic isotopic systems (Hf in Zr/ Nd in Ap), and O isotopes, as well as trace element signatures in zircon. The data validate previously identified fertility indicators[3,4] and highlight the presence of elevated Cl and S concentrations in apatite associated with Cu porphyry mineralization. In addition, the combined Hf, O, and Nd isotope analyses in zircon and apatite indicate that the melt source associated with the Tampakan porphyry and high-sulfidation system remain essentially unchanged for a large period of time, showing a fertile signature before and after the ore-forming event, for a time window of approximately 7 m.y. Furthermore, the mantle source is juvenile and does not show any evidence of contamination with an older crustal component. Overall, our study indicates that the combination of zircon and apatite in situ geochemical characterization provides valuable information to refine targets to explore porphyry systems on a regional and district scale.

References

[1]Richards JP (2003) Tectono-magmatic precursors for porphyry Cu-(Mo-Au) deposit formation. Economic Geology 98(8):1515-1533

[2]Cooke D, Wilkinson J (2014) 13.14 Geochemistry of Porphyry Deposits. Treatise on Geochemistry, Second Editionth edn Elsevier, Oxford:357-381

[3]Lu YJ, Loucks RR, Fiorentini ML, McCuaig TC, Evans NJ, Yang ZM, Hou ZQ, Kirkland CL, Parra-Avila LA, Kobussen A (2016) Zircon compositions as a pathfinder for porphyry Cu ± Mo ± Au deposits, . Society of Economic Geologists Special Publication No. 19:329-347

[4]Mao M, Rukhlov AS, Rowins SM, Spence J, Coogan LA (2016) Apatite trace element compositions: a robust new tool for mineral exploration. Economic Geology 111(5):1187-1222

ST.232
Stable Isotopic Study of the Drake Goldfield, Northeastern NSW, Australia

Hongyan Quan[1], Ian Graham[1], Rohan Worland[2], Lewis Adler[3], Christian Dietz[4]
1. School of Biological, Earth and Environmental Sciences, UNSW Sydney, Kensington, NSW, Australia, 2. White Rock Minerals Ltd., Ballarat, VIC, Australia, 3. Bioanalytical Mass Spectrometry Facility, UNSW Sydney, Kensington, NSW, Australia, 4. Central Science Laboratory, University of Tasmania, Hobart, TAS, Australia

The Drake Goldfield is located in the southern New England Fold Belt of northeastern New South Wales, Australia. It contains a number of low-sulfidation epithermal precious metal deposits including those at Mt. Carrington, White Rock, and Red Rock, with a current mineral resource containing 23.2 Moz Ag and 341 koz Au. These deposits occur exclusively within the Drake Volcanics, a 60- × 20-km NW-SE-trending sequence of Late Permian shallow volcanics and related epiclastics, which drilling suggests are at least 600 meters thick. The Drake Volcanics are centered on a circular magnetic feature, the Drake Quiet Zone, which mapping confirmed to be a collapsed volcanic caldera structure. Although known and mined for over 100 years, there have been few detailed studies on deposits associated with the Drake Volcanics, the only comprehensive study being that on the Red Rock Field, one of the smallest within the larger Drake Goldfield.

A total of 92 fresh carbonate vein samples were microdrilled from diamond drillcores from across the field and at various depths, and pXRD analysis of these identified five species: ankerite, calcite, dolomite, magnesite, and siderite. Overall, the $\delta^{13}C_{PDB}$ varies little in comparison to the $\delta^{18}O_{smow}$, which is highly variable. Carbon and oxygen isotope analysis show that the range of C and O compositions are I) ankerite: $\delta^{13}C_{PDB}$ from -9.33 to -5.92‰ and $\delta^{18}O_{smow}$ from +5.81 to +7.99 ‰; II) calcite: $\delta^{13}C_{PDB}$ from -21.32 to -1.42‰ and $\delta^{18}O_{smow}$ from -0.92 to +17.94 ‰; III) dolomite: $\delta^{13}C_{PDB}$ from -13.02 to -5.15‰ and $\delta^{18}O_{smow}$ from +3.40 to +10.59‰; IV) magnesite: $\delta^{13}C_{PDB}$ from –8.94 to -5.37‰ and $\delta^{18}O_{smow}$ from +8.15 to +15.84‰; and V) siderite: $\delta^{13}C_{PDB}$ from -13.71 to -10.44‰ and $\delta^{18}O_{smow}$ from +14.12 to +17.11‰. In addition, a total of 36 sulfide samples (primarily sphalerite and pyrite) from diamond drillcores from across the Drake Goldfield were microdrilled for S isotope analysis. Overall, these have a wide range in $\delta^{34}S_{CDT}$ values from -16.54 to 2.10‰.

The carbon and oxygen isotope results indicate that the carbon is mainly of magmatic origin and little affected by the widespread and pervasive low-temperature alteration and atmospheric precipitation. Calcite shows a widespread near-horizontal distribution in oxygen isotope values, suggesting derivation of oxygen from a large range of sources including meteoric, low-temperature hydrothermal, and possibly magmatic. For magnesite, the spread in oxygen isotope values once again forms a linear array towards more positive $\delta^{18}O$ values, suggesting derivation from low-temperature alteration fluids. Dolomite and ankerite have similar trends, both opposite to that for magnesite, trending towards more negative oxygen isotope values and with much less variation. Sulfur isotope ratios of sulfide minerals indicate that the sulfur was most likely derived from at least two different sources—sedimentary and magmatic—the relative importance of each varying from one deposit to another.

ST.235

Alteration, Mineralization, and Host-Rock Geochemistry of the Yanıklı Epithermal Prospect, Eastern Pontides, NE Turkey

Şafak U. Sönmez[1], Robert Moritz[1], Ümit Aydın[2], Serdar Keskin[2], François Turlin[1]
1. University of Geneva, Department of Earth Science, Geneva, Switzerland, 2. General Directorate of Mineral Research and Exploration of Turkey, Ankara, Turkey

The western Tethyan Metallogenic Belt (wTMB) consists of different segments characterized by voluminous magmatism from the Jurassic-Cretaceous to the Miocene. The Eastern Pontides in northeastern Turkey is one of the most remarkable segments of the wTMB and hosts different types of magmatic-hydrothermal deposits, including porphyry Cu, epithermal Au-base metal, skarn, and volcanogenic-hosted massive sulfide (VHMS). The Yanıklı prospect is one of the major target areas in the Artvin province within the Eastern Pontides. The prospect has been explored by the MTA (General Directorate of Mineral Research and Exploration) since 2018 by surface sampling and a total of 18 drill holes have been completed in the prospect so far.

The present study aims at constraining the alteration pattern, the spatial distribution and timing between the host-rock units, and the different metal associations. An approximately 20-km^2 area was mapped at a scale of 1:5,000 and includes the geologic units and alteration zones. The samples were collected from the main volcanic/volcaniclastic and intrusive rocks, together with the different hydrothermal alteration types. The samples have been analyzed for both geochemical purposes and petrographic examinations. The composition and temporal relationships between alteration minerals were determined via the Terraspec HALO mineral spectrometer.

The different rock types at the Yanıklı prospect include 1) predominantly rocks of intermediate and subsidiary mafic geochemical composition, with porphyritic-textured volcanic rocks, 2) felsic porphyritic-textured and flow-banded rocks, and 3) intermediate volcaniclastic rocks. In addition to these units, different types of dikes were mapped and identified as mafic to intermediate in composition.

The interpretation of the Terraspec measurements enabled us to identify three main mineral alteration zones, including argillic, propylitic, and phyllic alteration zones. Moreover, these alteration zones can be subdivided into different mineral assemblages and alteration intensities. The argillic, prophylitic, and phyllic alteration assemblages dominantly consist of (i) kaolinite and smectite group minerals, (ii) a chlorite-epidote-zeolite mineral assemblage, and (iii) white-mica group minerals, including illite, muscovite, paragonite, etc., respectively.

The metal association was investigated through 16 drill holes. Our preliminary results show that there are two main different metal phases. The first metal phase consists of a Cu-rich metal zone hosted by a predominantly propylitic alteration zone, which also includes a Pb and Zn mineralization. In contrast, the second phase is enriched in Zn and Pb related to the sericitic alteration zone. Gold mineralization is mainly associated with the second mineralization phase, which also contains weak, subsidiary Cu assemblages.

ST.236

The Continuum Between Porphyry and Carbonate-Replacement-Type Mineralization Exemplified by the Salinbas Au-Ag and Ardala Cu-Au (Mo-Re) Deposits, NE Turkey

Ahmet K. Sener[1], Izak van Coller[2], Berkin Uğurlu[3], Fabien Rabayrol[2], Robert Lee[2]
1. Ariana Resources plc, London, United Kingdom, 2. Mineral Deposit Research Unit, University of British Columbia, Vancouver, BC, Canada, 3. Galata Madencilik San. ve Tic. Ltd., Ankara, Turkey

The >1 Moz Salinbas Project Area (SPA) is located at lat.41°07'30"N, long.41°56'30"E, within the Pontide Metallogenic Province, 20 km east of Artvin in northeastern Turkey. The SPA is located midway along a ca. 35-km-long NNE trend of mineralization, marked at its southern end by the Hot Maden Au-Cu deposit (~4 Moz Au equivalent). The SPA is characterized by a transition in mineralization styles from porphyry to epithermal, including skarnoid replacement in the vicinity of the intrusions (Fig. 1). The SPA comprises the Salinbas deposit (10 Mt at 2.03 g/t Au, 10.2 g/t Ag for a total of 0.65 Moz of gold) and the Ardala porphyry (16 Mt at 0.6 g/t Au, 0.22% Cu, and 0.014% Mo).

The Salinbas deposit is represented by a 5- to 10-m-thick 20° ENE-dipping mineralized body (1.4 x 0.5 km), largely emplaced along the unconformable contact between folded Late Cretaceous (ca.100 Ma) Ziyarettepe Formation and Late Palaeocene (ca. 56 Ma) Kizilcik Formation. The Ziyarettepe Formation comprises massive fossiliferous limestones, whereas the overlying Kizilcik Formation consists of an intercalated sequence of conglomerates, limestones, siltstones, and mudstones (including black shales). This style of mineralisation is interpreted to be a carbonate replacement type and is sulphide rich to gossanous in character, selectively occurring within an irregular polymictic horizon. A steeply plunging breccia-pipe style of mineralisation is developed beneath part of the Salinbas deposit, emanating from the Ardala porphyry.

Contiguous with but at a lower elevation to the east of Salinbas lies the Ardala Cu-Au (Mo-Re) Porphyry Complex, characterised by a nested series of Eocene quartz-diorite stocks, which intrude the Upper Cretaceous to Paleocene volcano-sedimentary sequence. Exposed parts of the porphyry measure 600 x 700 m, displaying a well-developed potassic alteration core (including phyllic overprint with overlying argillic alteration), with lateral extensions beneath cover. Sulfide mineralization formed at 50.4±0.2 Ma (Re-Os on molybdenite), relating to active magmatism during 51-48 Ma constrained by LA-ICP-MS U-Pb dating on zircon. Precious metal- and base metal-bearing skarns and related disseminated mineralization in the host rocks are also evident, particularly at the northern margins.

A magmatic-hydrothermal event, which coincided with the intrusion of both mineralized and unmineralised porphyries in the APC, is interpreted to represent the source of the sulphide-rich mineralising fluids. This event also resulted in the deposition of volcanic units of the Avcilar Formation during the Early Eocene (56-48 Ma), which lie unconformably over the Kizilcik Formation. Zircon trace element ratios from a mineralized granodiorite yield $Eu/Eu_N^* > 0.58$ and $Ce/Ce^* > 101$, suggesting a high magma oxidation state. Furthermore, to the immediate south of the main mineralized porphyry lies an intrusion distinguished by a its rhenium content with up to 2 ppm Re and >400 ppm Mo in zones up to 10 m wide. Understanding the genetic relationship between proximal and distal mineralization, particularly with respect to fluid flow and decreasing geothermal gradients, contributes to improved exploration targeting in the area.

Fig. 1. Schematic three-dimensional model of Salinbas, showing the relative position of the Salinbas Zone and the Ardala Porphyry.

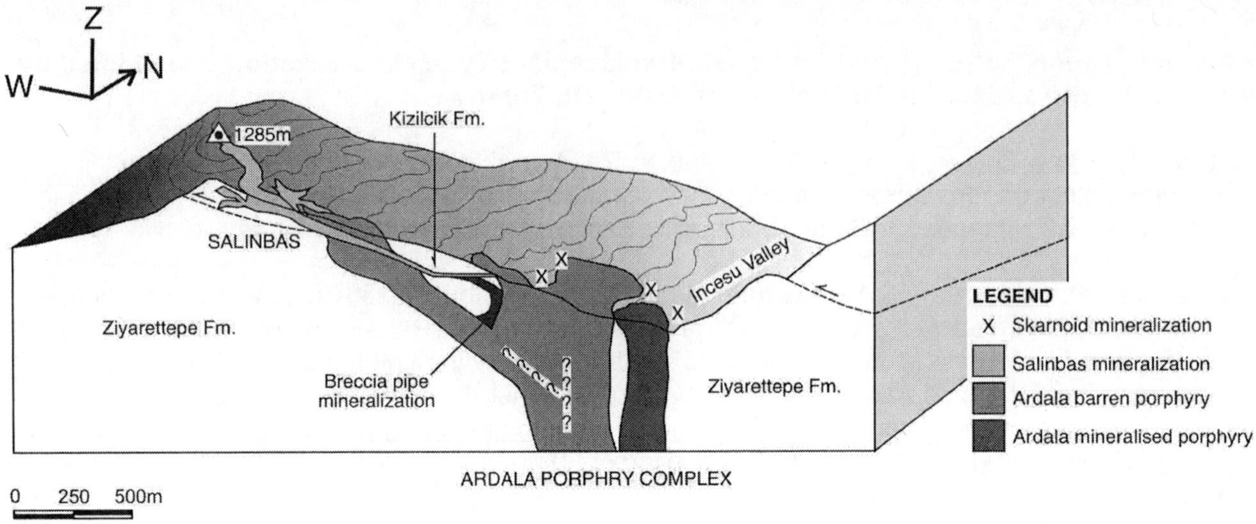

AUTHOR INDEX

Society of Economic Geologists (SEG)
7811 Shaffer Parkway
Littleton, CO 80127-3732
USA

CURRAN ASSOCIATES INC.
proceedings
.com

ISBN 978-1-7138-6509-4
9 781713 865094